Springer Monographs in Mathematics

Editors-in-Chief

Isabelle Gallagher, Paris, France
Minhyong Kim, Oxford, UK

Series Editors

Sheldon Axler, San Francisco, USA
Mark Braverman, Princeton, USA
Maria Chudnovsky, Princeton, USA
Tadahisa Funaki, Tokyo, Japan
Sinan C. Güntürk, New York, USA
Claude Le Bris, Marne la Vallée, France
Pascal Massart, Orsay, France
Alberto A. Pinto, Porto, Portugal
Gabriella Pinzari, Padova, Italy
Ken Ribet, Berkeley, USA
René Schilling, Dresden, Germany
Panagiotis Souganidis, Chicago, USA
Endre Süli, Oxford, UK
Shmuel Weinberger, Chicago, USA
Boris Zilber, Oxford, UK

This series publishes advanced monographs giving well-written presentations of the "state-of-the-art" in fields of mathematical research that have acquired the maturity needed for such a treatment. They are sufficiently self-contained to be accessible to more than just the intimate specialists of the subject, and sufficiently comprehensive to remain valuable references for many years. Besides the current state of knowledge in its field, an SMM volume should ideally describe its relevance to and interaction with neighbouring fields of mathematics, and give pointers to future directions of research.

More information about this series at http://www.springer.com/series/3733

Denis-Charles Cisinski · Frédéric Déglise

Triangulated Categories of Mixed Motives

 Springer

Denis-Charles Cisinski
Fakultät für Mathematik
Universität Regensburg
Regensburg, Bayern, Germany

Frédéric Déglise
Institut de Mathématiques de Bourgogne
CNRS, Université de Bourgogne
Dijon, France

ISSN 1439-7382   ISSN 2196-9922   (electronic)
Springer Monographs in Mathematics
ISBN 978-3-030-33244-0   ISBN 978-3-030-33242-6   (eBook)
https://doi.org/10.1007/978-3-030-33242-6

Mathematics Subject Classification (2010): 14F42, 19E15, 14C35, 14C15, 14C25, 19D55

© Springer Nature Switzerland AG 2019
This work is subject to copyright. All rights are reserved by the Publisher, whether the whole or part of the material is concerned, specifically the rights of translation, reprinting, reuse of illustrations, recitation, broadcasting, reproduction on microfilms or in any other physical way, and transmission or information storage and retrieval, electronic adaptation, computer software, or by similar or dissimilar methodology now known or hereafter developed.
The use of general descriptive names, registered names, trademarks, service marks, etc. in this publication does not imply, even in the absence of a specific statement, that such names are exempt from the relevant protective laws and regulations and therefore free for general use.
The publisher, the authors and the editors are safe to assume that the advice and information in this book are believed to be true and accurate at the date of publication. Neither the publisher nor the authors or the editors give a warranty, expressed or implied, with respect to the material contained herein or for any errors or omissions that may have been made. The publisher remains neutral with regard to jurisdictional claims in published maps and institutional affiliations.

This Springer imprint is published by the registered company Springer Nature Switzerland AG
The registered company address is: Gewerbestrasse 11, 6330 Cham, Switzerland

Contents

Introduction

- A   Historical background xi
  - A.1   The conjectural theory described by Beilinson xi
  - A.2   Voevodsky's motivic complexes xii
  - A.3   Morel and Voevodsky's homotopy theory xiii
  - A.4   Voevodsky's cross functors and Ayoub's thesis xiv
  - A.5   The Grothendieck six functors formalism xv
- B   Voevodsky's motivic complexes xviii
- C   Beilinson motives xx
  - C.1   Definition and fundamental properties xx
  - C.2   Constructible Beilinson motives xxii
  - C.3   Comparison theorems xxiii
  - C.4   Realizations xxvi
- D   Detailed organization xxviii
  - D.1   The Grothendieck six functors formalism (Part I) xxviii
  - D.2   The constructive part (Part II) xxxii
  - D.3   Motivic complexes (Part III) xxxiii
  - D.4   Beilinson motives (Part IV) xxxvi
- E   Developments since the first arXiv version xxxvii
  - E.1   Nisnevich motives with integral coefficients xxxvii
  - E.2   Étale motives with integral coefficients and ℓ-adic realization xxxviii
  - E.3   Motivic stable homotopy theory with rational coefficients ... xxxix
  - E.4   Duality, weights and traces xl
  - E.5   Enriched realizations xl
- Thanks xli
- Notations and conventions xli

Part I Fibred categories and the six functors formalism

1   General definitions and axioms 3

		1.1	$\mathscr{P}$-fibred categories	3
		1.1.a	Definitions	3
		1.1.b	Monoidal structures	10
		1.1.c	Geometric sections	13
		1.1.d	Twists	15
	1.2		Morphisms of $\mathscr{P}$-fibred categories	17
		1.2.a	The general case	17
		1.2.b	The monoidal case	18
	1.3		Structures on $\mathscr{P}$-fibred categories	20
		1.3.a	Abstract definition	20
		1.3.b	The abelian case	22
		1.3.c	The triangulated case	23
		1.3.d	The model category case	26
	1.4		Premotivic categories	28
2			Triangulated $\mathscr{P}$-fibred categories in algebraic geometry	31
	2.1		Elementary properties	32
	2.2		Exceptional functors, following Deligne	36
		2.2.a	The support axiom	36
		2.2.b	Exceptional direct image	38
		2.2.c	Further properties	43
	2.3		The localization property	46
		2.3.a	Definition	46
		2.3.b	First consequences of localization	48
		2.3.c	Localization and exchange properties	50
		2.3.d	Localization and monoidal structure	53
	2.4		Purity and the theorem of Voevodsky–Röndigs–Ayoub	57
		2.4.a	The stability property	57
		2.4.b	The purity property	62
		2.4.c	Duality, purity and orientation	68
		2.4.d	Motivic categories	78
3			Descent in $\mathscr{P}$-fibred model categories	82
	3.1		Extension of $\mathscr{P}$-fibred categories to diagrams	82
		3.1.a	The general case	82
		3.1.b	The model category case	85
	3.2		Hypercovers, descent, and derived global sections	98
	3.3		Descent over schemes	109
		3.3.a	Localization and Nisnevich descent	110
		3.3.b	Proper base change isomorphism and descent by blow-ups	112
		3.3.c	Proper descent with rational coefficients I: Galois excision	114
		3.3.d	Proper descent with rational coefficients II: separation	125
4			Constructible motives	130
	4.1		Resolution of singularities	130

		4.2	Finiteness theorems 133
		4.3	Continuity 145
		4.4	Duality 153

Part II Construction of fibred categories

5	Fibred derived categories 169
	5.1 From abelian premotives to triangulated premotives 169
	5.1.a Abelian premotives: recall and examples 169
	5.1.b The t-descent model category structure 171
	5.1.c Constructible premotivic complexes 181
	5.2 The $\mathbf{A}^1$-derived premotivic category 186
	5.2.a Localization of triangulated premotivic categories . 186
	5.2.b The homotopy relation 192
	5.2.c Explicit $\mathbf{A}^1$-resolution 197
	5.2.d Constructible $\mathbf{A}^1$-local premotives 201
	5.3 The stable $\mathbf{A}^1$-derived premotivic category 203
	5.3.a Modules 203
	5.3.b Symmetric sequences 205
	5.3.c Symmetric Tate spectra 207
	5.3.d Symmetric Tate Ω-spectra 210
	5.3.e Constructible premotivic spectra 216
6	Localization and the universal derived example 219
	6.1 Generalized derived premotivic categories 219
	6.2 The fundamental example 223
	6.3 Nearly Nisnevich sheaves 224
	6.3.a Support property (effective case) 225
	6.3.b Support property (stable case) 228
	6.3.c Localization for smooth schemes 229
7	Basic homotopy commutative algebra 230
	7.1 Rings 230
	7.2 Modules 236

Part III Motivic complexes and relative cycles

8	Relative cycles 247
	8.1 Definitions 247
	8.1.a The category of cycles 247
	8.1.b Hilbert cycles 249
	8.1.c Specialization 252
	8.1.d Pullback 256
	8.2 Intersection-theoretic properties 264
	8.2.a Commutativity 264
	8.2.b Associativity 265
	8.2.c Projection formulas 267

		8.3	Geometric properties 268
			8.3.a Constructibility 268
			8.3.b Samuel multiplicities 272
9	Finite correspondences 279
	9.1	Definition and composition 279
	9.2	Monoidal structure 285
	9.3	Functoriality 286
		9.3.a Base change 286
		9.3.b Restriction 287
		9.3.c A finiteness property 288
	9.4	The fibred category of correspondences 289
10	Sheaves with transfers 290
	10.1	Presheaves with transfers 291
	10.2	Sheaves with transfers 292
	10.3	Associated sheaf with transfers 294
	10.4	Examples 302
	10.5	Comparison results 304
		10.5.a Change of coefficients 304
		10.5.b Representable qfh-sheaves 305
		10.5.c qfh-sheaves and transfers 305
11	Motivic complexes 308
	11.1	Definition and basic properties 308
		11.1.a Premotivic categories 308
		11.1.b Constructible and geometric motives 310
		11.1.c Enlargement, descent and continuity 312
	11.2	Motivic cohomology 315
		11.2.a Definition and functoriality 315
		11.2.b Effective motivic cohomology in weight 0 and 1 .. 317
		11.2.c The motivic cohomology ring spectrum 322
	11.3	Orientation and purity 324
	11.4	The six functors 328

Part IV Beilinson motives and algebraic K-theory

12	Stable homotopy theory of schemes 335
	12.1	Ring spectra 335
	12.2	Orientation 335
	12.3	Rational category 338
13	Algebraic K-theory 338
	13.1	The K-theory spectrum 338
	13.2	Periodicity 339
	13.3	Modules over algebraic K-theory 340
	13.4	K-theory with support 341
	13.5	The fundamental class 343
	13.6	Absolute purity for K-theory 344

		13.7	Trace maps 346
14	Beilinson motives 350
	14.1	The γ-filtration 350
	14.2	Definition 352
	14.3	Motivic proper descent 357
	14.4	Motivic absolute purity 359
15	Constructible Beilinson motives 360
	15.1	Definition and basic properties 360
	15.2	The Grothendieck six functors formalism and duality 362
16	Comparison theorems 363
	16.1	Comparison with Voevodsky motives 363
	16.2	Comparison with Morel motives 367
17	Realizations 375
	17.1	Tilting 375
	17.2	Mixed Weil cohomologies 380

References 391
Index 398
Notation 405
Index of properties of $\mathscr{P}$-fibred triangulated categories 406

Introduction

A Historical background

A.1 The conjectural theory described by Beilinson

In a landmark paper, [Beĭ87], A. Beilinson states a series of conjectures which offers a complete renewal of the traditional theory of pure motives invented by A. Grothendieck. Namely, he proposes to extend the notion of pure motives to that of mixed motives with two models in mind: mixed Hodge structures defined by P. Deligne on the one hand [Del71, Del74] and perverse sheaves on the other hand, defined in [BBD82]. One of Beilinson's main innovations, motivated by the second model, is to consider a triangulated version of mixed motives in which one could hope to find the more involved theory of abelian mixed motives through the concept of t-structures. This hoped for theory was conjecturally described by Beilinson in [Beĭ87, 5.10] under the name of *motivic complexes*.

It was modeled (see *loc. cit.*, Section A) on the theory of étale ℓ-adic sheaves and their derived category as introduced fifty years ago by Grothendieck and M. Artin. The major achievement of Grothendieck and his collaborators in [AGV73] was to define a theory of coefficient systems relative to any scheme with a collection of operations, $f_*, f^*, f_!, f^!, \otimes, Hom$, satisfying a set of formulas now called the *Grothendieck six functors formalism* (see Section A.5 in this introduction for more details).[1] This formalism, formulated in the language of triangulated categories, ultimately encodes a very general duality theory. Note however that the complete duality theory for ℓ-adic sheaves was completed only recently by the work of Gabber [ILO14].

The theory was also conjectured to be deeply linked with Quillen's algebraic K-theory (see [Beĭ87, 5.10, §B]). In fact, up to torsion and for a regular scheme S,

---

[1] The full derived formalism of ℓ-adic complexes was fully established much later after [AGV73] though, by Ekedahl in [Eke90].

the ext-groups between two Tate motives over S should coincide with Adams graded parts of Quillen's algebraic K-theory.[2]

The ideas of Beilinson were very fecund and not long after the publication of [Beĭ87], three candidates for a triangulated category of mixed motives were proposed, respectively by M. Hanamura [Han95, Han04, Han99], M. Levine [Lev98], and V. Voevodsky [Voe96, Voe98, VSF00]. As a matter of fact, Voevodsky introduced two variants: using the h-topology (obtained by allowing proper surjective maps as coverings together with Zariski coverings), he defined a candidate for a theory of étale motivic sheaves [Voe96]. Inspired by his knowledge of these and by his work on rigidity results with Suslin [SV96], he introduced a more Zariski local version [VSF00] which is the one fitting in his approach to the proof of the Milnor conjecture and of the Bloch–Kato conjecture. In this book, we will focus on Voevodsky's theories.

A.2 Voevodsky's motivic complexes

As briefly alluded to above, Voevodsky's first attempt at defining the category of motivic complexes, in his 1992 Harvard thesis, introduces the fundamental process of $\mathbf{A}^1$-localization, which amounts to making the affine line contractible in the category of mixed motives, by analogy with the topological case. It also involves the use of the h-topology which was to become fundamental in the area of motives and cohomology. These two ingredients given, Voevodsky defined the triangulated category of (effective) h-motives over any base in [Voe96].

However, Voevodsky was aware that his definition will give the correct answer to Beilinson's conjectural construction only with rational coefficients (he was aware that the torsion part of the theory of h-motives would be closely related to Grothendieck's étale cohomology).[3] In [VSF00, chap. 5], he introduces another definition of motivic complexes over a perfect field with integral coefficients, still using the $\mathbf{A}^1$-localization process but, this time, introducing the notion of Nisnevich sheaves with transfers and their derived category (see [MVW06] for a detailed exposition). At this stage, all the properties foreseen by Beilinson are established for this integral category over a perfect field, except for the construction of the motivic t-structure.[4] It remained to extend this definition to arbitrary bases and to establish the Grothendieck six functors formalism.

The path in this direction was laid down by Voevodsky in [Voe10a], where he uses the theory of relative cycles introduced by Suslin and Voevodsky in [VSF00] to extend the definition of a sheaf with transfers. This definition was also exploited by

---

[2] See page xxiii below for the precise statement.

[3] This is made precise by Suslin and Voevodsky [SV96] over a field, and we developed this idea in full generality in [CD16], using the main results and constructions of the present book.

[4] This hoped for t-structure is described in [Voe92, Hyp. 0.0.21]. Moreover, Voevodsky proved in [VSF00] that such a t-structure does not exist with integral coefficients; however, it should exist with rational coefficients, and, more generally, for h-motives with integral coefficients.

A Historical background xiii

Ivorra in [Ivo07] to extend Voevodsky's construction of geometric motivic complexes over any base, avoiding the use of sheaves with transfers. Nevertheless, constructing the Grothendieck six functors formalism for this definition remained untouched at this point.

A.3 Morel and Voevodsky's homotopy theory

Soon after the introduction of Voevodsky's motivic complexes, F. Morel and Voevodsky introduced the more general theory of $\mathbf{A}^1$-homotopy of schemes [MV99] whose design is to extend the framework of algebraic topology to algebraic geometry and is built around the $\mathbf{A}^1$-localization tool. It is within this theory that another important tool in motivic homotopy theory was introduced: the $\mathbf{P}^1$-stabilization process.[5] From the purely motivic point of view, this amounts to inverting the Tate motive $\mathbf{Z}(1)$ for the tensor product. From the homotopical point of view, this operation is much more involved and reveals the theory of spectra, objects which incarnate cohomology theories in algebraic topology. These two processes, of $\mathbf{A}^1$-localization and $\mathbf{P}^1$-stabilization, applied to the category of simplicial Nisnevich sheaves, led to the stable $\mathbf{A}^1$-homotopy category of schemes (see [Jar00], or the last chapter of [Jar15], for instance, or [Rob15, Hoy17] for more modern approaches), a triangulated category with integral coefficients, defined over any base, which generalizes the category of motivic complexes.[6]

Over a perfect field, and with rational coefficients, the relation between $\mathbf{A}^1$-homotopy invariant sheaves and motives was clarified in an unpublished paper of Morel [Mor06] (with precise statements but without proofs): the rational stable $\mathbf{A}^1$-homotopy category contains the stable (*i.e.* $\mathbf{P}^1$-stable) version of the category of motivic complexes as an explicit direct factor, called the +-*part* of the stable homotopy category (that is, the part where the algebraic Hopf fibration acts as in oriented cohomology theories).[7] Then Morel introduces this +-part as a good candidate for the rational version of the triangulated category of motives ([Mor06, paragraph at the end of p.2]). We will dub the objects of this category *Morel motives*.

In the language of motivic stable homotopy theory, as initiated by Spitzweck in [Spi01], a natural candidate for the category of motivic sheaves is the homotopy category of modules over the motivic ring spectrum which represents motivic cohomology. With integral coefficients, O. Röndigs and P.A. Østvær showed that, over a field of characteristic 0, this category of modules is equivalent to the $\mathbf{P}^1$-stable

---

[5] At that time, with the impulse of Voevodsky's theory, the general process of ⊗-inverting an object such as a topological circle of $\mathbf{P}^1$ in a homotopy-theoretic way was quickly fully documented; see [Hov01].

[6] Heuristically, the essential difference between stable $\mathbf{A}^1$-homotopy and motivic complexes is the presence of transfers in the latter case.

[7] One of the goals of this book is to provide a proof of the generalization to arbitrary base schemes of Morel's claim; see Theorem 11 in this introduction and its corollary.

category of motivic complexes (see [RØ08a]).[8] This ring spectrum was introduced by Voevodsky [Voe98, §6.1], using the theory of relative cycles. It is defined over any base and one is led to consider the category of modules over this ring spectrum as a possible definition of the integral triangulated category of motives.

A.4 Voevodsky's cross functors and Ayoub's thesis

The definitive step towards the six functors formalism in motivic homotopy theory was taken up by Voevodsky in a series of lectures where he laid down the theory of *cross functors*. The main theorem of this theory consists in giving a criterion on a system of triangulated categories indexed by schemes, equipped with a basic functoriality, to be able to construct exceptional functors ($f_!, f^!$) satisfying the properties required by the Grothendieck six functors formalism. In particular, the system of triangulated categories must satisfy three notable properties: the $\mathbf{A}^1$-*localization property*, the $\mathbf{P}^1$-*stability property* and the *localization property*. Unfortunately, only an introductory part on this theory was released (see [Del01]) in which the basic setup is established but which does not contain the proof of the main result.

The writing of this theory was accomplished by J. Ayoub in the first half of his thesis (see [Ayo07a]). Ayoub uses the axioms laid down by Voevodsky: he calls a system of triangulated categories satisfying the properties alluded to above a *homotopy stable functor*. Moreover, he goes beyond the original result of Voevodsky: apart from the complete theory of cross functors (concerned with $f_!, f^!$), he also studied monoidal structures, *constructibility* properties and their stability under the six operations, *homotopy t-structures* and *specialization functors* such as the vanishing cycle functor. The main example of a stable homotopy functor is the stable $\mathbf{A}^1$-homotopy category. This was established independently by Röndigs [Rön05] and Ayoub [Ayo07a], who both also derived two fundamental properties: the one of relative purity and the proper base change isomorphism. One readily deduces that the category of Morel motives is also a homotopy stable functor. Furthermore, Ayoub's axiomatic approach allows a uniform treatment which also applies to many natural variations of the stable $\mathbf{A}^1$-homotopy category (as recalled in [Ayo07b], we may vary the topology as well as the coefficients in which sheaves take their values). However, despite its already great level of generality, Ayoub's work does not allow us to reach all the constructions of interest. For instance, it only provides the construction of the functors $f_!$ and $f^!$ when f is quasi-projective, and the finiteness and duality theorems only apply under hypotheses (such as absolute purity) which are far from being obvious in practice (Ayoub only discusses this issue for schemes of finite type over a perfect field, in which case this follows from the property of relative purity). Moreover, the techniques recalled in the third chapter of Ayoub's

---

[8] See also Theorem 8 in this introduction for an extension of their result to an arbitrary base, at the price of working with rational coefficients. For fields of characteristic $p > 0$, this has been extended to $\mathbf{Z}[1/p]$-linear coefficients by Hoyois, Kelly and Østvær in [HKØ17]. Finally, using the results of the present book, this has been extended to regular schemes of equal characteristic in [CD15].

thesis do not explain how to construct examples out of sheaves equipped with extra structures, such as transfers, which are fundamental tools to understand how algebraic cycles play a role in $\mathbf{A}^1$-homotopy theory. The extra technicalities related to this problem (such as having a derived tensor product as well as derived pull-back functors for suitable notions of $\mathbf{P}^1$-stable sheaves with transfers) were addressed via two approaches: the first one, by Röndigs and Østvær [RØ08a], uses homotopy theory together with enriched category theory, while the second one, due to the authors [CD09], uses abstract methods of homotopical algebras applied to cochain complexes in Grothendieck abelian categories. A second kind of problem which is not addressed in the early work of Ayoub is representability for K-theory, or for Chow groups, according to Beilinson's vision.

This is why, in order to discuss the original approach of Voevodsky to motivic sheaves alluded to above (using h-sheaves or Nisnevich sheaves with transfers), as well as to prove the absolute purity theorem, we had to take over from scratch many of the basic constructions. This also led us to reach a greater level of generality (avoiding unnecessary quasi-projectivity hypotheses) as well as more precise computations. To be fair, we should mention that, after a first version of the present text has been made public in 2009, Ayoub [Ayo14] reproved some of the representability results as well as the absolute purity theorem of this book in the particular case of the étale version of the motivic stable homotopy theory with rational coefficients. We should also mention right away that the integral version of Voevodsky's h-motives is only fully understood in a sequel to the present book [CD16].

Finally, we would like to end this section by recalling that the problem of constructing triangulated categories of motives related to Chow groups with integral coefficients and which define a homotopy stable functor is still an open problem. For instance, it is by no means obvious that the category of modules over the motivic homotopy ring spectrum meets the requirements of a homotopy stable functor. In fact, this latter property can be seen to be equivalent to Conjecture 17 of Voevodsky in [Voe02b], which states the stability of the motivic homotopy ring spectrum by pull-backs; this is made precise in this book in Proposition 11.4.7, as an application of our main constructions.

A.5 The Grothendieck six functors formalism

A.5.1 We now give the precise formulation of the *Grothendieck six functors formalism* (although we do not describe all the coherences yet). As presented here, it is extracted from the properties of the derived category of ℓ-adic sheaves.[9]

A triangulated category $\mathscr{T}$, fibred over the category of schemes, satisfies the *Grothendieck six functors formalism* if the following conditions hold:

1. There exists three pairs of adjoint functors as follows:

---

[9] It also coincides with formulas gathered by Deligne in an unpublished note which he graciously shared with us.

$f^* : \mathscr{T}(X) \rightleftarrows \mathscr{T}(Y) : f_*$, f any morphism,

$f_! : \mathscr{T}(Y) \rightleftarrows \mathscr{T}(X) : f^!$, f any separated morphism of finite type,

$(\otimes, Hom)$, symmetric closed monoidal structure on $\mathscr{T}(X)$.

The functors of type f^* are monoidal.

2. There exists a structure of a covariant (resp. contravariant) 2-functors on $f \mapsto f_*$, $f \mapsto f_!$ (resp. $f \mapsto f^*$, $f \mapsto f^!$).
3. There exists a natural transformation

$$\alpha_f : f_! \longrightarrow f_*$$

which is an isomorphism when f is proper. Moreover, α is a morphism of 2-functors.

4. For any smooth separated morphism $f : X \longrightarrow S$ in $\mathscr{S}$ of relative dimension d, there exists a canonical natural isomorphism

$$\mathfrak{p}'_f : f^* \xrightarrow{\sim} f^!(-d)[-2d]$$

where $?(-d)$ denotes the inverse of the Tate twist iterated d-times. Moreover $\mathfrak{p}'$ is an isomorphism of 2-functors.

5. For any cartesian square in $\mathscr{S}$:

$$\begin{array}{ccc} Y' & \xrightarrow{f'} & X' \\ g' \downarrow & \Delta & \downarrow g \\ Y & \xrightarrow{f} & X, \end{array}$$

such that f is separated of finite type, there exist natural isomorphisms

$$g^* f_! \xrightarrow{\sim} f'_! g'^*,$$
$$g'_* f'^! \xrightarrow{\sim} f^! g_*.$$

6. For any separated morphism of finite type $f : Y \longrightarrow X$, there exist natural isomorphisms

$$(f_! K) \otimes_X L \xrightarrow{\sim} f_!(K \otimes_Y f^* L),$$
$$Hom_X(f_!(L), K) \xrightarrow{\sim} f_* Hom_Y(L, f^!(K)),$$
$$f^! Hom_X(L, M) \xrightarrow{\sim} Hom_Y(f^*(L), f^!(M)).$$

(Loc) For any closed immersion $i : Z \longrightarrow S$ with complementary open immersion j, there exists a distinguished triangle of natural transformations as follows:

$$j_! j^! \xrightarrow{\alpha'_j} 1 \xrightarrow{\alpha_i} i_* i^* \xrightarrow{\partial_i} j_! j^![1],$$

A Historical background

where $\alpha'_?$ (resp. $\alpha_?$) denotes the co-unit (resp. unit) of the relevant adjunction.

A.5.2 The next part of the Grothendieck six functors formalism is concerned with duality. Historically, this is the initial motivation behind this formalism, as it appears in its first account, Hartshorne's notes on Grothendieck's 1963/64 seminar, [Har66]. It is considered more axiomatically, in the case of étale sheaves, in [Gro77, Exp. I].[10] In *loc. cit.*, Grothendieck states the fundamental property of absolute purity and indicates its fundamental link with duality. We state these properties as natural extensions of the properties given in the preceding section; assume $\mathscr{T}$ satisfies the properties listed above:

(7) *Absolute purity.*– For any closed immersion $i : Z \longrightarrow S$ of regular schemes of (constant) codimension c, there exists a canonical isomorphism:
$$\mathbb{1}_Z(-c)[-2c] \xrightarrow{\sim} i^!(\mathbb{1}_S),$$
where $\mathbb{1}$ denotes the unit object for the tensor product.

(8) *Duality.*– Let S be a regular scheme and K_S be any invertible object of $\mathscr{T}(S)$. For any separated morphism $f : X \longrightarrow S$ of finite type, put $K_X = f^!(K_S)$. For any object M of $\mathscr{T}(X)$, put $D_X(M) = Hom(M, K_X)$.

  a. For any X/S as above, K_X is a dualizing object of $\mathscr{T}(X)$: the canonical map
  $$M \longrightarrow D_X(D_X(M))$$
  is an isomorphism.

  b. For any X/S as above, and any objects M, N of $\mathscr{T}(X)$, we have a canonical isomorphism
  $$D_X(M \otimes D_X(N)) \simeq Hom_X(M, N).$$

  c. For any morphism between separated S-schemes of finite type $f : Y \longrightarrow X$, we have natural isomorphisms
  $$D_Y(f^*(M)) \simeq f^!(D_X(M)),$$
  $$f^*(D_X(M)) \simeq D_Y(f^!(M)),$$
  $$D_X(f_!(N)) \simeq f_*(D_Y(N)),$$
  $$f_!(D_Y(N)) \simeq D_X(f_*(N)).$$

A.5.3 The last property we want to exhibit as a natural extension of the Grothendieck six functors formalism is the compatibility with projective limits of schemes. The basis for the next statement is [AGV73, Exp. VI], though it does not appear explicitly. As in the case of the duality property, it should involve some finiteness assumption (constructibility) on the objects of $\mathscr{T}$. Note the formulation below is valid for an arbitrary triangulated monoidal category $\mathscr{T}$ fibred over schemes.

---

[10] The duality properties are stated in unpublished notes of Deligne, as part of the complete formalism.

(9) *Continuity.–* Let $(S_\alpha)_{\alpha \in A}$ be an essentially affine projective system of schemes. Put $S = \varprojlim_{\alpha \in A} S_\alpha$.
Then the canonical functor

$$2\text{-}\varinjlim_\alpha \mathscr{T}(S_\alpha) \longrightarrow \mathscr{T}(S)$$

is an equivalence of monoidal triangulated categories.

The purpose of this book is to discuss such a formalism in various contexts of motivic sheaves.

B Voevodsky's motivic complexes

The primary goal of this treatise is to develop the theory of Voevodsky motives, integrally over any base scheme,[11] within the framework of sheaves with transfers. Actually, we can define Voevodsky's motives with coefficients in an arbitrary ring Λ and prove all the results stated below in that case, but we restrict this presentation to integral coefficients for simplicity.

After refining and completing the Suslin–Voevodsky theory of relative cycles, we introduce the category $\mathscr{S}m_{Z,S}^{cor}$ of integral finite correspondences over smooth S-schemes and the related notion of (Nisnevich) sheaves with transfers over a base scheme S (Definition 10.4.2) as in the usual case of a perfect base field. Following the idea of stable homotopy, we define the triangulated category $\mathrm{DM}(X)$ of *stable motivic complexes* (see Definition 11.1.1) as the $\mathbf{P}^1$-stabilization of the $\mathbf{A}^1$-localization of the derived category of the (Grothendieck) abelian category of sheaves with transfers over S.

One easily gets that the fibred category DM is equipped with the basic functoriality needed by the cross-functor formalism. The main difficulty is the localization property, labelled (Loc) in Section A.5.1. Unfortunately, though all the functors involved in the formulation of (Loc) are well defined for DM, we can only prove this property when S and Z are smooth over some base scheme (see Proposition 11.4.2). In particular, the formalism of stable homotopy functors does not apply. However, we are able to construct the six operations for DM using the method of Deligne, used in [AGV73, XVII], and partially get the Grothendieck six functors formalism:

Theorem 1 (see Theorem 11.4.5) *The triangulated category* DM, *fibred over the category of schemes, satisfies the following of the properties stated in Section A.5.1:*

- *properties (1), (2), (3) (i.e. the construction of $f_!$ and $f^!$ in DM_Λ for any separated morphism of finite type f),*
- *property (4) when f is an open immersion or f is projective and smooth,*
- *property (5) when g is smooth or f is projective and smooth,*

---

[11] In this introduction, all schemes will be assumed to be noetherian of finite dimension.

B Voevodsky's motivic complexes

- *property (6) when f is projective and smooth,*
- *property (Loc) when S and Z are smooth over some common base scheme.*

One of the applications of this theory is that we get a well-defined integral motivic cohomology theory for any scheme X:

$$H^{n,m}_{\mathcal{M}}(X, \mathbf{Z}) = \mathrm{Hom}_{\mathrm{DM}(X)}\left(\mathbb{1}_X, \mathbb{1}_X(m)[n]\right),$$

which enjoys the following properties (see Section 11.2):

- it admits a ring structure, the pullback maps associated with any morphism of schemes compatible with this ring structure,
- it admits push-forward maps with respect to projective morphisms between schemes which are smooth over some common base, or with respect to some finite morphisms (for example finite flat; see Section 11.2.4),
- it coincides with Voevodsky's motivic cohomology groups when X is smooth over a perfect field (see Example 11.2.3); in particular one gets the following identification with higher Chow groups:

$$H^{n,m}_{\mathcal{M}}(X, \mathbf{Z}) = CH^m(X, 2m - n),$$

- it admits Chern classes and satisfies the projective bundle formula,
- it admits a localization long exact sequence associated with a closed immersion of schemes which are smooth over some common base.

As in the classical case, any smooth S-scheme X admits a motive $M_S(X)$ in $\mathrm{DM}(S)$. Moreover, one defines the Tate motive $\mathbb{1}_S(1)$ as the reduced motive of $\mathbf{P}^1_S$. We define the category of *constructible motives* $\mathrm{DM}_c(S)$ as the thick triangulated subcategory of DM generated by the objects of the form $M_S(X)(n)$ for a smooth S-scheme X and an integer $n \in \mathbf{Z}$, where $?(n)$ refers to the n-th Tate twist. One gets the following generalization of the classical result obtained by Voevodsky over a perfect field:

Theorem 2 (see Theorem 11.1.13) *A motive M in $\mathrm{DM}(S)$ is constructible if and only if it is compact.*[12]

The category $\mathrm{DM}_c(S)$ is equivalent to the category obtained from the bounded homotopy category of the additive category $\mathcal{S}m^{cor}_{\mathbf{Z},S}$ in the following way:

- take the Verdier quotient modulo the thick triangulated subcategory generated by:
  - for any Nisnevich distinguished square
    $$\begin{array}{ccc} W & \xrightarrow{k} & V \\ {}^{g}\downarrow & & \downarrow{f} \\ U & \xrightarrow{j} & X \end{array}$$
    of smooth S-schemes:
    $$[W] \xrightarrow{g_* - k_*} [U] \oplus [V] \xrightarrow{j^* + f^*} [X]$$

---
[12] Recall that M is compact if the functor $\mathrm{Hom}(M, -)$ commutes with arbitrary direct sums.

– for any smooth S-scheme X, $p : \mathbf{A}^1_X \longrightarrow X$ the canonical projection:

$$[\mathbf{A}^1_X] \xrightarrow{p_*} [X],$$

- *invert the Tate twist,*
- *take the pseudo-abelian envelope.*

The triangulated category $\mathrm{DM}_c(X)$ is stable under the operations f^* for all f, and f_*, when f is smooth projective, as well as $\otimes$, but we cannot prove the stability for the other operations of DM and *a fortiori* do not get the duality properties (7) and (8) of the Grothendieck six functors formalism.

However, we are able to prove the continuity property (9) for the category DM_c:

$$2\text{-}\varinjlim_{\alpha} \mathrm{DM}_c(S_\alpha) \simeq \mathrm{DM}_c(S),$$

with the restriction that the transition morphisms of (X_α) are affine and dominant (see Theorem 11.1.24). Note this result allows us to extend the comparison of motivic cohomology with higher Chow groups to arbitrary regular schemes of equal characteristics.

Finally, although we could not prove all the expected properties of the six operations in DM_Λ, we prove that the six operations behave as expected in DM_Λ if and only if Conjecture 17 of Voevodsky in [Voe02b] is true; see Proposition 11.4.7.

C Beilinson motives

C.1 Definition and fundamental properties

As anticipated by Morel, the theory of mixed motives with rational coefficients is much simpler and we succeed in establishing a complete formalism for them. However, there are many candidates for **Q**-linear motivic sheaves over a scheme X: there are Voevodsky's h-motives $\mathrm{DM}_{h,\mathbf{Q}}(X)$, Voevodsky's motivic sheaves constructed out **Q**-linear sheaves with transfers $\mathrm{DM}_\Lambda(X, \mathbf{Q})$, Morel motives $\mathrm{SH}_\mathbf{Q}(X)_+$, **Q**-linear étale motives $\mathrm{D}_{\mathbf{A}^1,\acute{e}t}(X, \mathbf{Q}) \simeq \mathrm{SH}_{\acute{e}t,\mathbf{Q}}(X)$ (also introduced by Morel, and used at length by Ayoub). Our goal is not only to prove that the six operations act on each of these candidates, but also to compare all these versions of motivic sheaves with one another. In fact, our strategy consists in producing yet another candidate for **Q**-linear motivic sheaves, namely that of *Beilinson motives* $\mathrm{DM}_\mathrm{B}(S)$, for which we can prove all the features we want for it, and use them to compare Beilinson motives with all the other versions of **Q**-linear motivic sheaves mentioned above.

More precisely, we construct, out of the rational motivic stable homotopy category and the ring spectrum associated with rational Quillen K-theory, a **Q**-linear triangulated category $\mathrm{DM}_\mathrm{B}(X)$, which we call the *triangulated category of Beilinson*

C Beilinson motives   xxi

motives (see Definition 14.2.1). Essentially by construction, in the case where X is regular, we have a natural identification

$$\mathrm{Hom}_{\mathrm{DM}_\mathrm{B}(X)}(\mathbf{Q}_X, \mathbf{Q}_X(p)[q]) \simeq Gr^p_\gamma K_{2p-q}(X)_\mathbf{Q},$$

where the right-hand side is the graded part of the algebraic K-theory of X with respect to the γ-filtration.

These groups were first regarded by Beilinson as the rational motivic cohomology groups. We call them the *Beilinson motivic cohomology groups*.

Part of the interest of our definition is that the localization property (Loc) can be easily deduced from its validity for the stable homotopy category. Therefore, the cross-functor formalism and more generally, our generalization of the results of Ayoub can be applied to DM_B.

Theorem 3 (see Corollary 14.2.11 and Theorem 2.4.50) *The standard Grothendieck six functors formalism (see Section A.5.1) is verified by the fibred triangulated category* DM_B.

Concerning duality for Beilinson motives, we first deduce from Quillen's localization theorem in algebraic K-theory the absolute purity theorem:

Theorem 4 (see Theorem 14.4.1) *The absolute purity property (see Section A.5.2(7)) holds for* DM_B.

As said before, this result is not enough to establish duality for Beilinson motives. We first have to use descent theory and resolution of singularities (as first explained by Grothendieck in [Gro77, I.3]). Using the existence of trace maps in algebraic K-theory, we prove the following result:

Theorem 5 (h**-descent, see Theorem 14.3.3 and Theorem 4.4.1**) *Consider a finite group G and a pullback square of schemes*

$$\begin{array}{ccc} T & \xrightarrow{h} & Y \\ g \downarrow & & \downarrow f \\ Z & \xrightarrow{i} & X \end{array}$$

in which Y is endowed with an action of G over X. Put $U = X - Z$ and assume the following three conditions are satisfied:

(a) The morphism f is proper and surjective.
(b) The induced morphism $f^{-1}(U) \longrightarrow U$ is finite.
(c) The morphism $f^{-1}(U)/G \longrightarrow U$ is generically radicial.

Put $a = f \circ h = i \circ g$. Then, for any object M of $\mathrm{DM}_\mathrm{B}(X)$, we get a canonical distinguished triangle in $\mathrm{DM}_\mathrm{B}(X)$:

$$M \longrightarrow i_* i^*(M) \oplus f_* f^*(M)^G \longrightarrow a_* a^*(M)^G \longrightarrow M[1],$$

where $?^G$ means the invariants under the action of G, and the first (resp. second) map of the triangle is induced by the difference (resp. sum) of the obvious adjunction morphisms.

In fact, we show that this apparently simple result implies a much stronger descent property for the fibred triangulated category DM_B: descent for the h-topology, thus, in particular, étale descent flat descent, as well as proper descent. The general fact that, in the presence of the six operations, the property of **Q**-linear h-descent is essentially equivalent to the presence of a suitable theory trace maps is a key feature of this text; this is developed systematically in Chapter 3 of this book. This will be at the heart of our main comparison results explained below.

C.2 Constructible Beilinson motives

The next step towards duality for Beilinson motives is the definition of a suitable finiteness condition. As in the case of Voevodsky motives, we define the category of *Beilinson constructible motives*, denoted by $DM_{B,c}(X)$, as the thick subcategory of $DM_B(X)$ generated by the motives of the form $M_X(Y)(p) := f_! f^!(\mathbf{Q}_X)(p)$ for $f : Y \longrightarrow X$ separated smooth of finite type, and $p \in \mathbf{Z}$. This category coincides with the full subcategory of compact objects in $DM_B(X)$.[13]

The usefulness of this definition comes from the following result, which is the analog of Gabber's finiteness theorem in the ℓ-adic setting. Analogously, its proof relies on absolute purity, (a weak form of) proper descent as well as Gabber's weak uniformization theorem.[14]

Theorem 6 (finiteness, see Theorem 15.2.1) *The subcategory* $DM_{B,c}$ *is stable under the six operations of Grothendieck when restricted to excellent schemes.*

The final statement concerning the Grothendieck six functors formalism in the setting of Beilinson motives is that, when one restricts to constructible Beilinson motives and separated B-schemes of finite type for an excellent scheme B of dimension less than 2, the complete formalism is available:[15]

Theorem 7 (see Theorem 15.2.4 and Proposition 15.1.6) *The fibred category* $DM_{B,c}$ *over the category of schemes described above satisfies the complete Grothendieck six functors formalism described in Section A.5, in particular the duality property A.5.2(8) and the continuity property A.5.3(9).*

---

[13] Note the striking analogy with perfect complexes.

[14] i.e. that, locally for the h-topology, any excellent scheme is regular, and any closed immersion between excellent schemes is the embedding of a strict normal crossing divisor into a regular scheme.

[15] There is a way to avoid this extra hypothesis to get duality theorems (i.e. to work with quasi-excellent schemes over a regular base in full generality). However, this comes at the price of higher coherence results (i.e. of promoting the construction $f \mapsto f^!$ to ∞-categories). See [Cis18].

C.3 Comparison theorems

In the historical part of this introduction, we saw many approaches to the triangulated category of (rational) motives. We succeed in comparing them all with our definition of Beilinson motives.

Denote by KGL_S the algebraic K-theory spectrum in Morel and Voevodsky's stable homotopy category $SH(S)$. By virtue of a result of Riou, the γ-filtration on K-theory induces a decomposition of $KGL_{S,\mathbf{Q}}$:

$$KGL_{S,\mathbf{Q}} \simeq \bigoplus_{n \in \mathbf{Z}} H_{B,S}(n)[2n].$$

The ring spectrum $H_{B,S}$ represents Beilinson motivic cohomology. Almost by construction, the category $\mathrm{DM}_B(S)$ is the full subcategory of $SH_\mathbf{Q}(S)$ which consists of objects E such that the unit map $E \longrightarrow H_{B,S} \otimes E$ is an isomorphism. In fact, our first comparison result relates the theory of Beilinson motives with the approach of Spitzweck, Röndigs and Østvær through modules over a ring spectrum:

Theorem 8 (see Theorem 14.2.9) *For any scheme S, there is a canonical equivalence of categories*

$$\mathrm{DM}_B(S) \simeq \mathrm{Ho}(H_{B,S}\text{-mod})$$

where the right-hand side denotes the homotopy category of modules over the ring spectrum $H_{B,S}$.

The next comparison involves the h-topology: we recall that this is the Grothendieck topology on the category of schemes, generated by étale surjective morphisms and proper surjective morphisms. The first published work of Voevodsky on triangulated categories of mixed motives [Voe96] introduces the $\mathbf{A}^1$-homotopy category of the derived category of h-sheaves. We consider a $\mathbf{Q}$-linear and $\mathbf{P}^1$-stable version of it, which we denote by $\underline{\mathrm{DM}}_{h,\mathbf{Q}}(S)$. By construction, for any S-scheme of finite type X, there is an h-motive $M_S(X)$ in $\underline{\mathrm{DM}}_{h,\mathbf{Q}}(S)$. We define $\mathrm{DM}_{h,\mathbf{Q}}(S)$ as the smallest triangulated full subcategory of $\underline{\mathrm{DM}}_{h,\mathbf{Q}}(S)$ which is stable under (infinite) direct sums, and which contains the objects $M_S(X)(p)$, for X/S smooth of finite type, and $p \in \mathbf{Z}$. Using h-descent in DM_B, we get the following comparison result.

Theorem 9 (see Theorem 16.1.2) *If S is excellent, then we have a canonical equivalences of categories*

$$\mathrm{DM}_B(S) \simeq \mathrm{DM}_{h,\mathbf{Q}}(S).$$

In fact, we first prove this result for the variant of $\mathrm{DM}_{h,\mathbf{Q}}(S)$ obtained by replacing everywhere the h-topology by the qfh-topology – in the latter, also introduced by Voevodsky, coverings are generated by étale covers and finite surjective morphisms. In particular, we get an equivalence of categories: $\mathrm{DM}_{h,\mathbf{Q}}(S) \simeq \mathrm{DM}_{qfh,\mathbf{Q}}(S)$. This result allows us to link Beilinson motives with Voevodsky's motivic complexes. Let us denote by $\mathrm{DM}_\mathbf{Q}$ the rationalization of the fibred category of stable motivic complexes alluded to in Section B. Using the preceding result in the case of the qfh-topology, we prove:

Theorem 10 (see Theorem 16.1.4) *If S is excellent and geometrically unibranch, then there is a canonical equivalence of categories*

$$DM_B(S) \simeq DM_Q(S).$$

In particular, given such a scheme S, we get a description of $DM_{B,c}(S)$ as in Theorem 2 cited above. Voevodsky's integral (resp. rational) motivic cohomology is represented in $SH(S)$ by a ring spectrum $H_{\mathcal{M},S}$ (resp. $H^Q_{\mathcal{M},S}$). The preceding theorem immediately gives an isomorphism of ring spectra:[16]

$$H_{B,S} \simeq H^Q_{\mathcal{M},S}.$$

As Beilinson motivic cohomology ring spectra over different bases are compatible with pullbacks, we easily deduce the following corollary, which solves affirmatively conjecture 17 of [Voe02b] in some cases, and up to torsion:

Corollary *For any morphism $f : T \longrightarrow S$ of excellent geometrically unibranch schemes, the canonical map*

$$f^* H^Q_{\mathcal{M},S} \longrightarrow H^Q_{\mathcal{M},T}$$

is an isomorphism of ring spectra.

The next comparison statement is concerned with the approach of Morel, according to whom the category $SH_Q(S)$ can be decomposed into two factors, one of them being $SH_Q(S)_+$, that is, the part of $SH_Q(S)$ on which the map $\epsilon : S^0_Q \longrightarrow S^0_Q$, induced by the permutation of the factors in $\mathbf{G}_m \wedge \mathbf{G}_m$, acts as -1. Let S^0_{Q+} be the unit object of $SH_Q(S)_+$.

Using the presentation of Beilinson motives in terms of H_B-modules (Theorem 8 cited above) as well as Morel's computation of the motivic sphere spectrum in terms of Milnor–Witt K-theory, we obtain a proof of a statement which, in the case where S is the spectrum of a field, was claimed by Morel in [Mor06]:

Theorem 11 (see Theorem 16.2.13) *For any scheme S, the canonical map $S^0_{Q+} \longrightarrow H_{B,S}$ is an isomorphism.*

In fact, we even get the following corollary:

Corollary *For any scheme S, there is a canonical equivalence of categories*

$$SH_Q(S)_+ \simeq DM_B(S).$$

---

[16] Note in particular that, when S is regular, we get an isomorphism:

$$H^{p,q}_{\mathcal{M}}(S, \mathbf{Z}) \otimes \mathbf{Q} \simeq Gr^p_\gamma K_{2p-q}(S)_\mathbf{Q}$$

which extends the known isomorphism when S has equal characteristics. It is natural with respect to pullbacks, Gysin morphisms, and is compatible with products and Chern classes.

C Beilinson motives

Recall from Morel theory that, when -1 is a sum of squares in all the residue fields of S, ϵ is equal to $-\mathrm{Id}$ on the whole of $\mathrm{SH}_\mathbf{Q}(S)$. Thus in that particular case (e.g. S is a scheme over an algebraically closed field), the category of Beilinson motives coincides with the rational stable homotopy category. In general, we can introduce according to Morel the étale variant of $\mathrm{SH}_\mathbf{Q}(S)$, denoted by $\mathrm{D}_{\mathbf{A}^1,ét}(S,\mathbf{Q})$.[17] As locally for the étale topology, -1 is always a square, and because DM_B satisfies étale descent, we get the following final illuminating comparison statement.[18]

Corollary *For any scheme S, there is a canonical equivalence of categories*

$$\mathrm{D}_{\mathbf{A}^1,ét}(S,\mathbf{Q}) \simeq \mathrm{DM}_\mathrm{B}(S).$$

Let us draw a conclusive picture which summarizes most of the comparison results we have obtained:

Corollary *Given any scheme S, the category $\mathrm{DM}_\mathrm{B}(S)$ is a full subcategory of the rational stable homotopy category $\mathrm{SH}_\mathbf{Q}(S)$. Given a rational spectrum E over S, the following conditions are equivalent:*

(i) E is a Beilinson motive,
(ii) E is an $H_{\mathrm{B},S}$-module,
(iii) E satisfies étale descent,
(iii') (S excellent) E satisfies qfh-descent,
(iii'') (S excellent) E satisfies h-descent,
(iv) (S excellent geometrically unibranch) E admits transfers,
(v) the endomorphism $\epsilon \in \mathrm{End}(S_\mathbf{Q}^0)$ acts by $-\mathrm{Id}$ on E, i.e. $\epsilon \otimes 1_E = -1_E$.

Remark (see Corollary 14.2.16) Points (iv) and (v) are related to the orientation theory for spectra (not only ring spectra). In fact, $H_{\mathrm{B},S}$ is the universal orientable rational ring spectrum over S.

Let $\mathbf{Q}.Sm_S$ be the $\mathbf{Q}$-linear envelope of the category Sm_S. One obtains (see Example 5.3.43 in conjunction with Section 5.3.35) that the full subcategory of compact objects of $\mathrm{SH}_\mathbf{Q}(S)$ is equivalent to the category obtained from the homotopy category $\mathrm{K}^b(\mathbf{Q}.Sm_S)$ by performing the following operations:

- take the Verdier quotient modulo the thick triangulated subcategory generated by:

---

[17] In brief, this is the $\mathbf{P}^1$-stabilization of the $\mathbf{A}^1$-localization of the derived category of sheaves of $\mathbf{Q}$-vector spaces over the *lisse-étale site* of S.

[18] In particular, the finiteness theorem as well as the duality property also hold for $\mathrm{D}_{\mathbf{A}^1,ét}(-,\mathbf{Q})$. The finiteness theorem and the duality theorem may be deduced from [Ayo07a] (Scholie 2.2.34 and Theorem 2.3.73, respectively) when one restricts to quasi-projective schemes over a field or over a discrete valuation ring. Nevertheless, even if one is eager to accept such restrictions, over a discrete valuation ring, the proof relies in an essential way on the absolute purity property (Theorem 4 stated above) which is proved in the present text.

- for any Nisnevich distinguished square $\begin{smallmatrix} W & \xrightarrow{k} & V \\ g\downarrow & & \downarrow f \\ U & \xrightarrow{j} & X \end{smallmatrix}$ of smooth S-schemes:

$$\mathbf{Q}_S(W) \xrightarrow{g_*-k_*} \mathbf{Q}_S(U) \oplus \mathbf{Q}_S(V) \xrightarrow{j^*+f^*} \mathbf{Q}_S(X)$$

- for any smooth S-scheme X, $p : \mathbf{A}_X^1 \longrightarrow X$ the canonical projection:

$$\mathbf{Q}_S(\mathbf{A}_X^1) \xrightarrow{p_*} \mathbf{Q}_S(X),$$

- invert the Tate twist,
- take the pseudo-abelian envelope.

Let us denote this category by $\mathrm{D}_{\mathbf{A}^1,c}(S, \mathbf{Q})$. We finally obtain the following concrete description of Beilinson constructible motives:

Corollary *Given any scheme S, the category $\mathrm{DM}_{\mathrm{B},c}(S)$ is equivalent to the full subcategory of $\mathrm{D}_{\mathbf{A}^1,c}(S, \mathbf{Q})$ spanned by the objects E which satisfy one the following equivalent conditions:*

(i) (Galois descent) given any smooth S-scheme X and any Galois S-cover $f : Y \longrightarrow X$ of group G, the canonical map $E \otimes \mathbf{Q}_S(Y)/G \longrightarrow E \otimes \mathbf{Q}_S(X)$ is an isomorphism,
(ii) (Orientability) ϵ acts by $-\mathrm{Id}$ on E,

Recall again the following remarks:

1. When (-1) is a sum of squares in every residue field of S, conditions (i), (ii) are true for any rational spectrum E over S.
2. When S is excellent and geometrically unibranch, the category $\mathrm{DM}_{\mathrm{B},c}(S)$ is equivalent to the category of rational geometric Voevodsky motives (same definition as in Theorem 2 but replacing $\mathbf{Z}$ by $\mathbf{Q}$).

C.4 Realizations

The last feature of Beilinson motives is that they are easily realizable in various cohomology theories. To show this, we use the setting of modules over a strict ring spectrum.[19] Given such a ring spectrum $\mathscr{E}$ in $\mathrm{DM}_{\mathrm{B}}(S)$, one can define, for any S-scheme X, the triangulated category

$$\mathrm{D}(X, \mathscr{E}) = \mathrm{Ho}(\mathscr{E}_X\text{-mod}),$$

where $\mathscr{E}_X = f^*\mathscr{E}$, for $f : X \longrightarrow S$ the structural map.

---

[19] i.e. we say a ring spectrum is *strict* if it is a commutative monoid in the underlying model category.

C Beilinson motives

We then have realization functors

$$\mathrm{DM}_{\mathrm{B}}(X) \longrightarrow \mathrm{D}(X, \mathscr{E}), \quad M \mapsto \mathscr{E}_X \otimes_X M$$

which commute with the six operations of Grothendieck. Using Ayoub's description of the Betti realization, we obtain:

Theorem 12 *If $S = \mathrm{Spec}(k)$ with k a subfield of $\mathbf{C}$, and if $\mathscr{E}_{Betti}$ represents Betti cohomology in $\mathrm{DM}_{\mathrm{B}}(S)$, then, for any k-scheme of finite type, the full subcategory of compact objects of $\mathrm{D}(X, \mathscr{E}_{Betti})$ is canonically equivalent to the derived category of constructible sheaves of geometric origin $\mathrm{D}_c^b(X(\mathbf{C}), \mathbf{Q})$.*

More generally, if S is the spectrum of some field k, given a mixed Weil cohomology $\mathscr{E}$, with coefficient field (of characteristic zero) $\mathbf{K}$, we get realization functors

$$\mathrm{DM}_{\mathrm{B},c}(X) \longrightarrow \mathrm{D}_c(X, \mathscr{E}), \quad M \mapsto \mathscr{E}_X \otimes_X M$$

(where $\mathrm{D}_c(X, \mathscr{E})$ stands for the category of compact objects of $\mathrm{D}(X, \mathscr{E})$), which commute with the six operations of Grothendieck (which preserve compact objects on both sides). Moreover, the category $\mathrm{D}_c(S, \mathscr{E})$ is then canonically equivalent to the bounded derived category of the abelian category of finite-dimensional $\mathbf{K}$-vector spaces. As a by-product, we get the following concrete finiteness result: for any k-scheme of finite type X, and for any objects M and N in $\mathrm{D}_c(X, \mathscr{E})$, the $\mathbf{K}$-vector space $\mathrm{Hom}_{\mathrm{D}_c(X,\mathscr{E})}(M, N[n])$ is finite-dimensional, and it is trivial for all but a finite number of values of n.

If the field k is of characteristic zero, this abstract construction gives essentially the usual categories of coefficients (as seen above in the case of Betti cohomology), and in a sequel of this work, we shall prove that one recovers in this way the derived categories of constructible ℓ-adic sheaves (of geometric origin) in any characteristic. But something new happens in positive characteristic:

Theorem 13 *Let V be a complete discrete valuation ring of mixed characteristic, with field of functions K, and residue field k. Then rigid cohomology is a K-linear mixed Weil cohomology, and thus defines a ring spectrum $\mathscr{E}_{rig}$ in $\mathrm{DM}_{\mathrm{B}}(k)$. We obtain a system of closed symmetric monoidal triangulated categories $\mathrm{D}_{rig}(X) = \mathrm{D}_c(X, \mathscr{E}_{rig})$, for any k-scheme of finite type X, such that*

$$\mathrm{Hom}_{\mathrm{D}_{rig}(X)}(\mathbb{1}_X, \mathbb{1}_X(p)[q]) \simeq H_{rig}^q(X)(p),$$

as well as realization functors

$$R_{rig} : \mathrm{DM}_{\mathrm{B},c}(X) \longrightarrow \mathrm{D}_{rig}(X)$$

which preserve the six operations of Grothendieck.

D Detailed organization

The book is organized into four parts that we now review in more detail.

D.1 The Grothendieck six functors formalism (Part I)

The first part is concerned with the formalism described in Section A.5 above. It is the foundational part of this work.

We use the language of fibred categories (introduced in [Gro03, VI]), complemented by that of 2-functors (or pseudo-functors), in order to describe the six functors formalism. We first describe axioms which allow us to derive the core formalism – i.e. the part described in Section A.5.1 – from simpler axioms. We do not claim originality in this task: our main contribution is to give a synthesis of the approach of Deligne described in [AGV73, XVII] (see also [Har66, Appendix]) with that of Voevodsky developed by Ayoub in [Ayo07a].

Recall that a (cleaved) fibred category $\mathcal{M}$ over $\mathcal{S}$ can be seen as a family of categories $\mathcal{M}(S)$ for every object S of $\mathcal{S}$ together with a pullback functor $f^*: \mathcal{M}(S) \longrightarrow \mathcal{M}(T)$ for any morphism $f: T \longrightarrow S$ of $\mathcal{S}$.[20] Given a suitable class $\mathcal{P}$ of morphisms in $\mathcal{S}$, we set up a systematic study of a particular class of fibred categories, called $\mathcal{P}$-fibred categories (Definition 1.1.10): these are categories where for any f in $\mathcal{P}$, the pullback functor f^* admits a *left* adjoint, generically denoted by $f_\sharp$. The functor $f_\sharp$ has to be thought of as a variant of the *exceptional direct image functor*.[21]

In Section 1, we study basic properties of $\mathcal{P}$-fibred categories which will be the core of the six functors formalism, such as *base change formulas* and *projection formulas* when an additional monoidal structure is involved. These formulas are special cases of a compatibility relation between different types of functors expressed through a canonical comparison morphism. This kind of comparison morphism is generically called an *exchange morphism*. These are very versatile and appear everywhere in the theory (see Sections 1.1.6, 1.1.15, 1.1.24, 1.1.31, 1.1.33, 1.2.5). In fact, they appear fundamentally in the Grothendieck six functors formalism: in the list of properties A.5.1, they are the isomorphisms of (5), (6) and even (4). In the direction of the full Grothendieck functoriality, we introduce core axioms for $\mathcal{P}$-fibred categories that we consider minimal: the categories satisfying these axioms are called $\mathcal{P}$-*premotivic* (Section 1.4). $\mathcal{P}$-premotivic categories will form the basic setting in all this work. They will appear in three different flavors, depending on which particular kind of additional structure we consider on categories: abelian, triangulated and model categories.

---

[20] These pullback functors are subject to the usual cocycle condition; see Section 1.

[21] This kind of situation frequently happens: the analytic case (open immersions), sheaves on the small étale site (étale morphisms), Nisnevich sheaves on the smooth site (smooth morphisms).

D Detailed organization

In Section 2, we restrict our attention to the triangulated and geometric case, meaning that we consider triangulated $\mathscr{P}$-fibred categories over a suitable category of schemes $\mathscr{S}$. The aim of this section is to develop, and extend, the Grothendieck six functors formalism in this basic setting. We exhibit many properties of such fibred categories which are indexed in the appendix. Let us concentrate in this introduction on the two main properties which will correspond respectively to Deligne and Voevodsky's approach to the six functors formalism.

The first property, called the *support property* and abbreviated by (Supp), asserts that the adjoint functors of the kind f_*, for f proper, and $j_\sharp$, for j an open immersion, satisfy a gluing property that allows us to use Deligne's argument to construct the exceptional direct image functor $f_!$.[22] Several properties are derived from (Supp) and the basic axioms of $\mathscr{P}$-fibred categories. Eventually, it leads to a partial version of the six functors formalism (see Theorem 2.2.14).

The second property, most fundamental in the motivic context, is the *localization property*, abbreviated by (Loc), which is in fact part of the six functors formalism (see Section A.5.1). It has many interesting consequences and reformulations that are derived in Section 2.3.1. Note that (Loc) is also known in the literature as the "gluing formalism". Some properties that we prove in *loc.cit.* are already classical (see [BBD82]).

The most interesting consequence of (Loc) was discovered by Voevodsky: together with the usual $\mathbf{A}^1$-localization and $\mathbf{P}^1$-stabilization properties of the motivic context, it implies the complete basic six functors formalism as stated in Section A.5.1. This was proved by Ayoub in [Ayo07a]. In Section 2.4, we revisit Ayoub's proof and give some improvement of his theorems (see Theorem 2.4.50 for the precise statement):

- we remove the quasi-projectivity assumption for the existence of $f_!$, replacing it with the assumption that f is separated of finite type;
- we introduce the *orientation property* which allows us to get a simpler, more usual, form of the purity isomorphism (the one actually stated in point (4) of Section A.5.1);
- we give another proof of the main theorem in the oriented case by showing that relative purity is equivalent to some (strong) duality property in the smooth projective case (see Theorem 2.4.42);
- we directly incorporate the monoidal structure whereas Ayoub gives a separate discussion for this.

Apart from these differences, the material of Section 2.4 is very similar to that of [Ayo07a]. Moreover, in the non-oriented case, it should be clear that we rely on Ayoub's original argument for the proof of Theorem 2.4.42.

Concerning terminology, we have called a *motivic triangulated category* (Definition 2.4.45) what Ayoub calls a "monoidal stable homotopy functor" (except that Ayoub only considers operations induced by quasi-projective morphisms).

---

[22] In the context of torsion étale sheaves of [AGV73, XVII], property (Supp) is a consequence of the proper base change theorem.

The remainder of Part I is concerned with extensions of the Grothendieck six functors formalism.

In Section 3, we show how to use the setting of $\mathscr{P}$-fibred model categories as a framework to formulate Deligne's cohomological descent theory.

Except in trivial cases, objects of a derived category are not local.[23] To formulate descent theory in derived categories, Deligne's main idea was to extend the derived category of a scheme by one relative to a simplicial scheme, usually a hypercover with respect to a Grothendieck topology (see [AGV73, Vbis]). The construction consists in first extending the theory of sheaves to the case where the base is a simplicial scheme and then considering the associated derived category.

We generalize this construction to the case of an arbitrary $\mathscr{P}$-fibred category equipped with a suitable model category structure. In fact, we show in Section 3.1 how to extend a $\mathscr{P}$-fibred category over a category of schemes to the corresponding category of simplicial schemes and even of arbitrary diagrams of schemes. Most importantly, we show how to extend the fibred model structure to the case of diagrams of schemes (see Proposition 3.1.11).[24] Concretely, this means that we define a derived functor of the kind $\mathbf{L}\varphi^*$ (resp. $\mathbf{R}\varphi_*$) for an arbitrary morphism φ of diagrams of schemes. Let us underline that these derived functors mingle two different kinds of functoriality: the usual pullback f^* (resp. direct image f_*) for a morphism of schemes f together with homotopy colimits (resp. limits) of arbitrary diagrams — see the discussion in Section 3.1.12 through to Proposition 3.1.16. With that extension in hand, we can easily formulate (cohomological) descent theory for arbitrary Grothendieck topologies on the category of schemes for the homotopy category of a $\mathscr{P}$-fibred model category: see Definition 3.2.5.

The end of Section 3 is devoted to concrete examples of descent in $\mathscr{P}$-fibred model categories, and their relation with properties of the associated homotopy category, assuming it is triangulated, as introduced in Section 2. The first and most simple example corresponds to the case of a Grothendieck topology associated with a cd-structure in the sense of Voevodsky (as the Nisnevich and the cdh-topology – see [Voe10b] or Section 2.1.10). In that case, descent can be characterized as the existence of certain distinguished triangles (Mayer–Vietoris for the Zariski topology, Brown–Gersten for the Nisnevich topology): this is Theorem 3.3.2, which is in fact a reformulation of the results of Voevodsky.

We then proceed to the most fundamental case of descent in algebraic geometry, that for proper surjective maps, which allows in principle the use of resolution of singularities. In fact, the main result of the whole of Section 3 is a characterization of h-descent which allows us to reduce it, for $\mathscr{P}$-fibred homotopy triangulated categories which are rational and motivic, to a simple property easily checked in

---

[23] The first example of this fact is the circle: any non-trivial connected open subset of S^1 is contractible whereas S^1 itself is not.

[24] By restricting the morphisms of diagrams of schemes to a certain class denoted by $\mathscr{P}_{cart}$, we also show how to get a $\mathscr{P}_{cart}$-fibred model category over diagrams of schemes (Remark 3.1.21), but this is not really needed in the descent theory.

practice:[25] this is Theorem 3.3.37. Along the way, we also proved the following results, interesting on their own:

- several characterizations of étale descent (Theorems 3.3.23 and 3.3.32);
- a characterization of qfh-descent (Theorem 3.3.25) as if it was defined by a cd-structure.[26]

In fact, the last point is the heart of the proof of the main result of this section (Theorem 3.3.37). Whereas the extension of fibred homotopy categories to diagrams of schemes is not unprecedented (see [Ayo07b]), our study of proper and h-descent seems to be completely new. In our opinion, it is one of the most important technical innovations of this book.

In Section 4, we study the extension of the Grothendieck six functors formalism in *rational* motivic categories, mainly duality and continuity. As already mentioned, the general principle is not new and mainly follows the path laid by Grothendieck in [Gro77].

In the case of an abstract motivic triangulated category — which is for the purposes of descent theory the homotopy category of an underlying fibred model category — the first task is to introduce a correct property of finiteness inherent to any duality theorem. This is done following Voevodsky, as in the work of Ayoub, by introducing the notion of *constructibility* in Definition 4.2.1. The name is inspired by the étale case, but the notion of constructibility which we consider here is defined by a generation property which really corresponds to what Voevodsky called *geometric motives*: constructible motives in our sense are generated by twists of motives of smooth schemes and are stable under cones, direct factors and finite sums. Let us mention that in good cases, the property of being constructible coincides with that of being compact in a triangulated category, resounding with the theory of perfect complexes (in the context of ℓ-adic sheaves, this corresponds to "constructible of geometric origin").

The main point on constructible motives is the study of their stability under the six operations that we get from the axioms of a triangulated motivic category. This is done in Section 4.2. As in the étale case, the crucial point is the stability with respect to the operation f_*, when f is a morphism of finite type between excellent schemes. In Theorem 4.2.24, we give conditions on a motivic triangulated category so that the stability for f_* is guaranteed (then the stability under the other operations follows easily, see Theorem 4.2.29). Our proof essentially follows an argument of Gabber. The general principle, going back to [AGV73, XIX, 5.1], is to use resolution of singularities to reduce to an absolute purity statement which is among our assumptions.[27]

---

[25] This is the *separation* property defined in Definition 2.1.7. Let us mention here that it is a consequence of the existence of well-behaved trace maps (see the proof of Theorem 14.3.3).

[26] It is at the origin of the formulation of descent that we gave for DM_B in Theorem 5(b) above. A systematic approach to such generalized cd-structures is developed by Park in [Par19].

[27] Absolute purity will be proved later for Beilinson motives.

In Section 4.3, we introduce an important property of motivic triangulated categories, called *continuity*, which allows reasoning that involves projective limits of schemes. In fact, it is shown in Proposition 4.3.4 that this property implies the property (9) of the (extended) Grothendieck six functors formalism (see Section A.5.3 above). We also give a criterion for continuity (Proposition 4.3.6) which will be applied later in concrete cases and draw some interesting consequences.

Finally, Section 4.4 deals with duality itself for constructible motives, that is, property (8) of Section A.5.2. The main theorem 4.4.21 asserts that, under the same condition as Theorem 4.2.24, and if one restricts to schemes that are separated of finite type over an excellent base scheme B of dimension less than or equal to 2, then the full duality property holds (see also Corollary 4.4.24). The proof follows the same lines as the analogous Theorem 2.3.73 of [Ayo07a]. In particular, the main point is the fact that constructible motives are generated by some nice motives adapted to the use of resolution of singularities: see Corollary 4.4.3. The main difference with *op. cit.* is that we implement De Jong's equivariant resolution of singularities [dJ97], so that our assumptions are much weaker.[28]

D.2 The constructive part (Part II)

The purpose of this part is to give a method of construction of triangulated categories that satisfies the formalism described in Part I. We have chosen to mainly use the setting of derived categories. Also, we use our notion of $\mathscr{P}$-fibred categories ($\mathscr{P}$-premotivic with a good monoidal structure). Recall that this means the pullback functor f^* admits a left adjoint $f_\sharp$ when $f \in \mathscr{P}$. Essentially, $\mathscr{P}$ will be either the class of smooth morphisms of finite type or the class of all morphisms of finite type (eventually separated).

In Section 5.1, starting from a $\mathscr{P}$-premotivic abelian category $\mathscr{A}$, we first show how to prove that the associated derived category $D(\mathscr{A})$ is also a $\mathscr{P}$-premotivic category. This consists in deriving the structural functors of a $\mathscr{P}$-premotivic category, which is done by building a suitable underlying $\mathscr{P}$-fibred model category in Proposition 5.1.12. Actually, the proof of the axioms of a model category has already appeared in our previous work [CD09]. Let us mention the flavor of this model structure: we can explicitly describe cofibrations as well as fibrations, by the use of an appropriate Grothendieck topology t. This model structure is linked with cohomological t-descent (as shown later in Proposition 5.2.10). The advantage of our framework is to easily obtain the functoriality of this construction (Section 5.1.23), as well as other homotopical constructions (dg-structure: Remark 5.1.19, extension to diagrams of schemes: Section 5.1.20). In Section 5.1.c, we also describe in suitable cases the constructible objects of the derived category by a presentation similar to that of Voevodsky's geometric motives over a perfect field.

---

[28] See also footnote [15] on page xxiv, which applies to this more general setting as well.

In Section 5.2 (resp. Section 5.3) we show how to describe the $\mathbf{A}^1$-localization (resp. $\mathbf{P}^1$-stabilization) process in $\mathscr{P}$-premotivic derived categories: to any $\mathscr{P}$-premotivic abelian category $\mathscr{A}$ is associated an $\mathbf{A}^1$-derived category $D^{\mathit{eff}}_{\mathbf{A}^1}(\mathscr{A})$ (resp. $\mathbf{P}^1$-stable and $\mathbf{A}^1$-derived category $D_{\mathbf{A}^1}(\mathscr{A})$) in Definition 5.2.16 (resp. Definition 5.3.22). From the model category obtained in Section 5.1, the construction uses the classical tools of motivic homotopy theory as introduced by Morel and Voevodsky. Again, our framework allows us to get the same homotopical constructions as in the simple derived case as well as some nice universal properties. We also get a description of constructible objects under suitable assumptions: Section 5.2.d (resp. Section 5.3.e). These sections are filled with concrete examples.

In Section 6, we focus on the main (in fact universal) example of motivic derived categories, the $\mathbf{A}^1$-derived category of Morel, obtained by the process described above from the abelian premotivic category of abelian sheaves over the smooth Nisnevich site. The main point here is that one gets the localization property for this category by a theorem of Morel and Voevodsky. We give two new contributions on this topic. First we show in Section 6.1 that the $\mathbf{A}^1$-derived category can be embedded in a larger category which naturally contains objects that we can call motives of singular schemes. This is useful to state descent properties and will be essential to study h-motives. Second, we show in Section 6.3 how one can use the $\mathbf{A}^1$-derived category to obtain good properties of another premotivic derived category satisfying suitable assumptions. This will be applied to motivic complexes.

In Section 7, we go back to the case of an arbitrary monoidal $\mathscr{P}$-fibred model category $\mathscr{M}$ and explain how to use the setting of ring spectra and modules over ring spectra in the premotivic context. The main construction associates to a suitable collection of (commutative) ring spectra R in $\mathscr{M}$ a $\mathscr{P}$-fibred monoidal category denoted by $\mathrm{Ho}(R\text{-mod})$: Proposition 7.2.13. This construction will be used several times:

- in the study of algebraic K-theory (Section 13): the category of modules over K-theory is the fundamental technical tool to get motivic proper descent as well as motivic absolute purity;
- in the study of Beilinson motives when we will relate them to modules over motivic cohomology (Theorem 14.2.9);
- in the study of realizations associated with a mixed Weil cohomology (Section 17).

D.3 Motivic complexes (Part III)

This part is concerned with the constructions described above, in Section B. Our aim is to extend the definition of Voevodsky's integral motivic complexes to any base, then study their functoriality and introduce their non-effective, or rather $\mathbf{P}^1$-stable, counterpart.

Our first task, in Section 8, is to revisit the Suslin–Voevodsky theory of *relative cycles* exposed in [SVoob]. Indeed, they will be at the heart of the general construction. Our presentation is made to prepare the theory of *finite correspondences*, a particular case of relative cycles. We especially want to give a meaning to the following picture representing the composition of finite correspondences α from X to Y and β from Y to Z:

$$\begin{array}{ccc} \beta \otimes_Y \alpha & \to \beta \to Z \\ \downarrow & \downarrow \\ \alpha & \longrightarrow Y \\ \downarrow \\ X \end{array}$$

(see also (9.1.4.1)). More precisely, we want to interpret this as a diagram of cycles. Thus we are led to consider cycles (with their support) as objects of a category. Concretely, a cycle is considered as a multi-pointed scheme, each point being endowed with some multiplicity (an integer or rational number).

This conceptual shift has the advantage of allowing a treatment of cycles analogous to that of algebraic varieties, or rather schemes, promoted by Grothendieck via his study of morphisms. Thus, we replace the various groups of relative cycles introduced by Suslin and Voevodsky in *op. cit.* by properties of morphisms of cycles. Here is a list of the principal properties:

- pseudo-dominant (Definition 8.1.2), equidimensional (Section 8.1.3 and Definition 8.3.18),
- pre-special (Definition 8.1.20),
- special (Definition 8.1.28),
- Λ-universal (Definition 8.1.49).

The most intriguing property, being *pre-special*, has no counterpart in *op. cit.* Its idea comes from a mistake (fortunately insignificant) in the convention of Suslin and Voevodsky. Indeed, Lemma 3.2.4 of *op. cit.* is false whenever the base S is non-reduced and irreducible: then any fat point (x_0, x_1) and any flat S-scheme give a counter-example.[29] The explanation is that the operation of specialization along a fat point does not take into account the geometric multiplicities of the base. On the contrary, when X is flat over an irreducible scheme S, the geometric multiplicity of any irreducible component of X is a multiple of the geometric multiplicity of S. This leads us to the definition of a pre-special morphism of cycles β/α, where a divisibility condition appears in the multiplicities of β with respect to that of α.[30]

The main achievement of Suslin and Voevodsky's theory is the construction of a pullback operation for relative cycles. In our language, it corresponds to a kind

---

[29] Explicitly, take $S = Z = \mathrm{Spec}\,(k[t]/(t^2)) = \{\eta\}$, $R = (k[t])_{(t)}$. The left-hand side of the equality of 3.2.4 is $2.\eta$ while the right-hand side is η.

[30] To anticipate the rest of the construction, given a non-reduced scheme S, this will allow for the operation of pull-back along the immersion $S_{red} \longrightarrow S$ associated with the reduction of S: it simply corresponds to dividing by the geometric multiplicities of S, as the base change to S_{red} does for flat S-schemes.

of tensor product, more precisely a product of cycles relative to a common base cycle (such as, for example, the cycle $\beta \otimes_Y \alpha$ of the preceding picture). Despite our different presentation, the method to define this operation closely follows Suslin and Voevodsky's original idea: use the *flatification theorem* of Gruson and Raynaud to reduce to the case of flat base change of cycles. Recall that the key point is to find the correct condition on cycles – or rather morphisms of cycles in our language – so that one obtains a uniquely defined operation independent of the chosen flatification. This is measured by a specialization procedure (Definition 8.1.25) associated with *fat points* (Definition 8.1.22) and leads to the central notion of *special morphisms* of cycles (Definition 8.1.28). An innovation that we introduce in the theory is to give, as soon as possible, local definitions at a point in the style of EGA. This is in particular the case for the property of being special.

Once this notion is in place, one defines for a base cycle α, a special α-cycle β and any morphism $\phi : \alpha' \longrightarrow \alpha$ the relative product, denoted by $\beta \otimes_\alpha \alpha'$. Equivalently, it corresponds to the base change of β/α along ϕ (Definition 8.1.40). This notion is close to the correspondence morphisms of Section 3.2 of *op. cit.* In particular, it usually involves denominators. The last important notion, being Λ-*universal*, corresponds to cycles β/α with coefficients in a ring $\Lambda \subset \mathbf{Q}$, which keeps their coefficients in Λ after any base change.

One sees that our language is especially convenient when it is time to consider the stability of certain properties of morphisms of cycles under composition (Corollary 8.2.6) or base change (Corollary 8.1.46). Then the usual statements of intersection theory are proven in Section 8.2, still following or extending Suslin and Voevodsky: commutativity, associativity, projection formulas. This makes our relative product a good extension of the classical notion of the exterior product of cycles (over a field).

The focal point of intersection theory is the study of multiplicities. Thus we introduce *Suslin–Voevodsky multiplicities*, as those appearing as a corollary of the existence of the relative cycle $\beta \otimes_\alpha \alpha'$ (Definition 8.1.43). A very important result in the theory, already enlightened by Suslin and Voevodsky, is the fact these multiplicities can be expressed in terms of *Samuel multiplicities*.[31] In fact, independently of Suslin and Voevodsky, we prove a new criterion for the property of being special at a point involving Samuel multiplicities at the branches of the point: see Corollary 8.3.25. Roughly speaking, the multiplicities arising from Samuel's definition at each branches of the point must coincide: then this common value is simply the Suslin–Voevodsky multiplicity.

Finally, still following Grothendieck's treatment of algebraic geometry, we add to the theory of Suslin and Voevodsky the study of constructibility properties for morphisms of cycles (special and Λ-universal). Here, our categorical point of view is plainly justified. Explicitly, we prove that given a relative cycle β/α, when α is the cycle associated with a scheme S, the locus where β is special (resp. Λ-universal) is an ind-constructible subset of S (Lemma 8.3.4). This allows us to prove the good behavior of these notions with respect to projective limits of schemes (see in

---

[31] When a correct regularity assumption is added, one reduces to the usual Serre's Tor-intersection formula: see Theorem 8.3.31 and Remark 8.3.32).

particular Proposition 8.3.9). This will be the *key point* when proving the continuity property — (9) of Section A.5.3 — of the fibred category DM.

The rest of Part III consists in extending the theory of sheaves with transfers introduced by Voevodsky, originally over a perfect field, to the case of an arbitrary base and applying to it the general procedures studied in Part II to get the fibred category DM.

In Section 9, we work out the theory of finite correspondences using the formalism of relative cycles. The construction is summarized in Corollary 9.4.1: given a class of morphisms $\mathscr{P}$ contained in the class of separated morphisms of finite type and a ring of coefficients Λ, we produce a monoidal $\mathscr{P}$-fibred category, denoted by $\mathscr{P}_\Lambda^{cor}$, whose fiber over a noetherian scheme S (eventually singular) is the category of $\mathscr{P}$-schemes over S with morphisms the finite correspondences.

In Section 10, we develop the theory of sheaves with transfers along the very same line as the original treatment of Voevodsky. This time, the outcome can be summarized by Corollaries 10.3.11 and 10.3.15: given a class $\mathscr{P}$ of morphisms as above and a suitable Grothendieck topology t, we construct an abelian premotivic category $\mathrm{Sh}_t(\mathscr{P}, \Lambda)$ which is compatible with the topology t (cf. Part II); its fiber over a scheme S is given by t-sheaves of Λ-modules with transfers (in particular presheaves on $\mathscr{P}_{\Lambda,S}^{cor}$).[32] The section closes with an important comparison result, essentially due to Voevodsky, between Nisnevich sheaves with transfers and sheaves for the qfh-topology (with rational coefficients over geometrically unibranch bases): see Theorem 10.5.15.

Finally, Section 11 is devoted to gathering the work done previously and defining the stable derived category of motivic complexes DM_Λ, given an arbitrary ring of coefficients Λ. The outcome has already been described in Section B above.

D.4 Beilinson motives (Part IV)

This part contains the construction of Beilinson motives as well as the proof of all the properties stated before. It is based on the first and second parts but independent of the third one — except in the comparison statements of Section 16.1.

Section 12 contains a short review of the stable homotopy category and the notion of oriented ring spectra.

Section 13 is the heart of our construction. It contains a detailed study of the K-theory ring spectrum KGL and the associated notion of KGL-modules in the homotopical sense (based on the formalism introduced in Section 7). Using the works of several authors (most notably: Riou, Naumann, Spitzweck, Østvær), we show how the central results of Quillen on algebraic K-theory give important properties of KGL-modules: absolute purity (Theorem 13.6.3) and trace maps (Definition 13.7.4).

---

[32] The most notable topologies t that fit into this result are the Nisnevich and the cdh topologies. See Section 10.4.

In Section 14, we finally introduce the definition of Beilinson motives. Let us describe it in detail now. It is based on the process of Bousfield localization of the stable homotopy category with respect to a cohomology. This operation is fundamental in modern algebraic topology. We apply it in algebraic geometry to the rational stable homotopy category (or, what amount to the same, to the rational stable $\mathbf{A}^1$-derived category of Morel, Section 6) and to the rational K-theory spectrum $KGL_\mathbf{Q}$: the Bousfield localization of $D_{\mathbf{A}^1}(S,\mathbf{Q})$ with respect to $KGL_{\mathbf{Q},S}$ is the category of Beilinson motives $DM_B(S)$ over S (Definition 14.2.1). Using the preceding study of $KGL_\mathbf{Q}$ together with the decomposition of Riou recalled in the beginning of Section C.3, we get the main properties of the premotivic category DM_B: the h-descent theorem (14.3.4) and the absolute purity theorem (14.4.1).

Then the theoretical background laid down in Part I is applied to DM_B, given in particular the complete Grothendieck six functors formalism for constructible Beilinson motives (Section 15). Our work closes with the two main subjects described above on Beilinson motives: the comparison statements (Section 16) and the study of motivic realizations (Section 17).

E Developments since the first arXiv version

The first version of this work appeared on arXiv in December 2009.[33] For almost ten years, until its publication by Springer, it has been used in several works, and completed by several other mathematicians, who solved questions left open in the present text. For completeness, it appears to us beneficial to the reader to give an account of some of these developments which are most directly related to the present contribution. Mathematics is indeed a collective work, each part of which is destined to be used, completed, renewed or superseded.

E.1 Nisnevich motives with integral coefficients

E.1.1 *cdh-motives.–* One aim of the present work was to work out the theory of finite correspondences in the spirit of [VSF00], whose original aim was to obtain an integral theory of motivic complexes related to Chow groups. The theory of cdh-sheaves with transfers (see Proposition 10.4.8) was introduced with this motivation in mind. The theory of cdh-motives and motivic complexes was successfully developed in the equal characteristic case in [CD15], provided one inverts the residue characteristic. In this latter work, the crucial property of localization for cdh-motives is shown, as well as all the expected results: constructibility of the six operations, duality, continuity, comparison with modules over the cdh-local version of Voevodsky's motivic

---

[33] A new version was uploaded in 2012, which included the introduction, which was written in order to clear up the contributions and history on mixed motives and more specifically motivic homotopy theory.

cohomology, and the relation with higher Chow groups. To get these results, the key points are the continuity property of motivic complexes, which is proved in this book (Theorem 11.1.24), as a result of our reinforcement of the Suslin–Voevodsky theory of relative cycles (see in particular Section 8.3.a on constructibility for properties of relative cycles) together with Kelly's new motivic descent results [Kel17], which allow us to use Gabber's improvements of de Jong's alteration theorems [ILO14]. Note also that [CD15, 3.6 and 5.1] generalizes our result on Voevodsky's conjecture on base change of the motivic Eilenberg–MacLane spectrum (Corollary 16.1.7).

E.1.2 *Spitzweck's motivic cohomology spectrum.–* One of the problems with defining mixed motives as modules over Voevodsky's motivic cohomology spectrum is the compatibility of this spectrum with base change.[34] The idea of Spitzweck's paper [Spi18] is to build a spectrum which satisfies compatibility by base change; equivalently, one has to build a ring spectrum $H\mathbf{Z}$ over $S = \mathrm{Spec}\,(\mathbf{Z})$ (or more generally over a Dedekind ring) which pulls back to Voevodsky's motivic cohomology spectrum over the residue fields of S. This is what M. Spitzweck achieves with virtuosity in *loc. cit.*, therefore obtaining a convenient category of $H\mathbf{Z}$-modules which coincides with Voevodsky's original triangulated category over the residue fields of S; in fact, it also coincides with DM_{cdh} over any k-scheme, after inverting the residue characteristic of k. But the construction of Spitzweck works integrally. Moreover, by its very construction, the cohomology represented by $H\mathbf{Z}$ coincides with Bloch's higher Chow groups for smooth S-schemes. A question left open is a possible comparison with Voevodsky's motivic cohomology spectrum (which is again equivalent to Voevodsky's base change conjecture).

E.2 Étale motives with integral coefficients and ℓ-adic realization

E.2.1 *Voevodsky's motives in the étale topology and rigidity theorems.–* With rational coefficients, the comparison theorems obtained in this book (Section 16) show that varying the underlying topology is beneficial. In particular, with rational coefficients, we are not able to get the localization property for Nisnevich motivic complexes for non-geometrically unibranch base schemes, but we do get that property when replacing the Nisnevich topology by the qfh-topology, or the h-topology. This leads us to believe that the transfers will be better behaved with integral coefficients with respect to stronger topologies. In [CD16], we prove that the localization property holds for motivic complexes with torsion coefficients locally for the étale topology ([CD16, Theorem 4.3.1]). It follows that the same property holds with integral coefficients for geometrically unibranch schemes. Moreover, we prove in *loc. cit.* that, locally for the h-topology, motivic complexes with integral coefficients are perfectly well behaved and satisfy all the expected properties, as listed in Section A.5 of this intro-

---

[34] Recall again this compatibility was conjectured by Voevodsky. See Conjecture 11.2.22 for an explicit formulation. Note also that we prove the latter conjecture is actually equivalent to the localization property for (Nisnevich) motivic complexes: see Proposition 11.4.7.

E Developments since the first arXiv version

duction.[35] It is remarkable that we were able to obtain the complete Grothendieck six functors formalism for Voevodsky's original construction of étale motives, as defined in his Ph. D. thesis [Voe96], and once again show Voevodsky's visionary power. Furthermore, we also show that one recovers the theory of ℓ-adic complexes out of h-motives by the categorical process of ℓ-adic completion (see [CD16, §7.2]). This gives a new insight into ℓ-adic realization of motives.

E.2.2 *Rigidity theorems without transfers.–* Another extension of Suslin and Voevodsky's rigidity theorem to arbitrary bases is due to Ayoub, [Ayo14]. In this latter work, Ayoub studies the category introduced in the present book under the notation $D_{\mathbf{A}^1,\acute{e}t}(S,\Lambda)$ (following Morel), while he uses the notation $DA(S,\Lambda)$. The first result of *loc. cit.*, inspired by an earlier work of Röndigs and Østvær [RØ08b], is indeed a variation on the rigidity theorem, identifying the category $D_{\mathbf{A}^1,\acute{e}t}(S,\Lambda)$ for a $\Lambda = \mathbf{Z}/N\mathbf{Z}$ with N invertible on S with the derived category of the category of sheaves of Λ-modules on the small étale site of S (under suitable hypotheses on S and N). From there, one can extend the results proved in this book for $D_{\mathbf{A}^1,\acute{e}t}(S,\mathbf{Q})$ (see Theorem 16.2.18) to the case of arbitrary coefficients: absolute purity, constructibility of the six operations, and duality. Note however that, despite what is claimed in the appendix of Ayoub's article, the particular case of 2-torsion for base schemes S of mixed or positive characteristic is problematic in his approach (see [CD16, Rem. 5.5.8]).

Since then, Bachmann [Bac18b] has significantly extended the preceding rigidity theorems to torsion $\mathbf{P}^1$-stable motivic étale sheaves of *spectra*. This result also solves the aforementioned issues left open in [Ayo14].

E.3 Motivic stable homotopy theory with rational coefficients

E.3.1 *Witt sheaves.–* In this book, following Morel's insights, we have split the rational motivic stable homotopy category $SH(X)_{\mathbf{Q}} \simeq D_{\mathbf{A}^1}(X,\mathbf{Q})$ into two factors

$$(D_{\mathbf{A}^1}(X,\mathbf{Q})_+) \times (D_{\mathbf{A}^1}(X,\mathbf{Q})_-) \simeq D_{\mathbf{A}^1}(X,\mathbf{Q})$$

and we have identified the oriented part $D_{\mathbf{A}^1}(X,\mathbf{Q})_+$ with Beilinson's motives $DM_B(X)$. On the other hand, in the case where $X = \mathrm{Spec}(k)$ is the spectrum of a field, Ananyevskiy, Levine and Panin [ALP17] have identified the non-oriented part $D_{\mathbf{A}^1}(k,\mathbf{Q})_-$ with a suitable category of Witt sheaves. The conjunction of their results with ours may be seen as a motivic analog of (a trivial consequence of) a theorem of Serre that the stable homotopy groups of spheres are finite in degree > 0; see the introduction of *loc. cit.* The results of Ananyevskiy, Levine and Panin have been improved by Bachmann [Bac18a], where the comparison of $SH(k)_-$ with Witt

---

[35] Based on the results of this book, we only get Grothendieck–Verdier duality for schemes of finite type over a regular 2-dimensional excellent scheme, but this extra hypothesis has been removed by the first author in [Cis18].

sheaves is promoted to $\mathbf{Z}[1/2]$-linear coefficients. Bachmann's results follow from a nice analog of the rigidity theorem over a general base for the *real étale* topology.

E.3.2 *Rational absolute purity.–* Déglise, Fasel, Jin and Khan [DFJK19] have proved absolute purity property for the motivic sphere spectrum with rational coefficients.

E.4 Duality, weights and traces

E.4.1 *Weight structures.–* Bondarko's theory of weight structures [Bon10] has been shown to be compatible with the six operations with rational coefficients in [Héb11, Bon14]. In the setting of *cdh*-sheaves [CD15], this has been extended by Bondarko and Ivanov [BI15] to $\mathbf{Z}[1/p]$-linear coefficients in equal characteristic, where $p \geq 1$ denotes the exponent characteristic of the ground field. Such weight complexes have been used by Wildeshaus [Wil17, Wil18] in order to give unconditional constructions of motivic intersection complexes of certain Shimura varieties. They also play a role, together with the realization functors associated to mixed Weil cohomologies, in geometric representation theory, in the work of Soergel and his collaborators [SW18, SVW18].

E.4.2 *Motivic Lefschetz–Verdier trace formula.–* An obvious application of the theory of motivic sheaves and their realizations is the proof of independence of ℓ results for a wealth of trace-like constructions. A $\mathbf{Q}$-linear version of such kind of results is provided by Olsson [Ols15, Ols16], where some versions of the Motivic Lefschetz–Verdier trace formula and of characteristic classes are discussed. A slight improvement, allowing torsion, may be found in [Cis18], but a full account on integral formulas, including for characteristic classes, is settled in the recent work of Jin and Yang [JY19].

E.5 Enriched realizations

E.5.1 *Structured mixed Weil cohomologies.–* In his thesis [Dre13], Drew extends the formalism of mixed Weil cohomologies to cohomologies with values in a Tannakian category. He also defines the realization functor into algebraic $\mathscr{D}$-modules for schemes of finite type over a field of characteristic zero and proves that, for any separated smooth scheme X over a field of characteristic zero, constructible modules over de Rham cohomology in $\mathrm{SH}(X)$ embed fully faithfully into algebraic $\mathscr{D}$-modules. Drew deduces from this embedding a new purely algebraic proof of the Riemann–Hilbert correspondence, using motivic sheaves, as predicted in Example 17.2.22 in the present book. His work also provides a way to define Hodge realizations of mixed motivic sheaves; see [Dre18].

E Developments since the first arXiv version

E.5.2 *Arakelov motivic cohomology*.– Holmstrom and Scholbach [HS15] have extended the representability of algebraic de Rham cohomology to the filtered de Rham complex, and used it to define a motivic version of Arakelov cohomology. The relation with more classical versions of Arakelov cohomology and with height pairings is discussed in [Sch15].

Thanks

The authors want to heartily thank the following people for their help, motivation, corrections and suggestions during the elaboration of this text: Joseph Ayoub, Alexander Beilinson, Pierre Deligne, Brad Drew, David Hébert, Jens Hornbostel, Annette Huber-Klawitter, Bruno Kahn, Shane Kelly, Marc Levine, Georges Maltsiniotis, Fabien Morel, Paul Arne Østvær, Joël Riou, Oliver Röndigs, Valentina Sala, Markus Spitzweck, Vladimir Voevodsky, Vadim Vologodsky, Chuck Weibel, and finally, as well as particularly, Jörg Wildeshaus.

We would like to thank our editors at Springer simply for making the publication of this book possible.

D.-C. C is partially supported by the SFB 1085 "Higher Invariants" funded by the Deutsche Forschungsgemeinschaft (DFG), and F. D. by the French "Investissements d'Avenir" program, project ISITE-BFC (contract ANR-lS-IDEX-OOOB).

Notations and conventions

In every section, we will fix a category denoted by $\mathscr{S}$ which will contain our geometric objects. Most of the time, $\mathscr{S}$ will be a category of schemes which are suitable for our needs; the required hypotheses on $\mathscr{S}$ are given at the head of each section. In the text, when no details are given, any scheme will be assumed to be an object of $\mathscr{S}$.

When $\mathscr{A}$ is an additive category, we denote by $\mathscr{A}^\natural$ the pseudo-abelian envelope of $\mathscr{A}$. We denote by $C(\mathscr{A})$ the category of complexes of $\mathscr{A}$. We consider $K(\mathscr{A})$ (resp. $K^b(\mathscr{A})$) the category of complexes (resp. bounded complexes) of $\mathscr{A}$ modulo the chain homotopy equivalences and when $\mathscr{A}$ is abelian, we let $D(\mathscr{A})$ be the derived category of $\mathscr{A}$.

If $\mathscr{M}$ is a model category, $\mathrm{Ho}(\mathscr{M})$ will denote its homotopy category.

We will use the notation

$$\alpha : \mathscr{C} \rightleftarrows \mathscr{D} : \beta$$

to mean a pair of functors such that α is left adjoint to β. Similarly, when we speak of an adjoint pair of functors (α, β), α will always be the left adjoint. We will denote by

$$ad(\alpha, \beta) : 1 \longrightarrow \beta\alpha \text{ (resp. } ad'(\alpha, \beta) : \alpha\beta \longrightarrow 1\text{)}$$

the unit (resp. counit) of the adjunction (α, β). Considering a natural transformation $\eta : F \longrightarrow G$ of functors, we usually denote by the same letter η — when the context is clear — the induced natural transformation $AFB \longrightarrow AGB$ obtained when considering functors A and B composed on the left and right with F and G respectively.

In Section 8, we will assume that equidimensional morphisms have constant relative dimension.

Part I
Fibred categories and the six functors formalism

1 General definitions and axioms

1.0.1 We assume that $\mathscr{S}$ is an arbitrary category.

We shall say that a class $\mathscr{P}$ of morphisms of $\mathscr{S}$ is *admissible* if it is has the following properties.

(Pa) Any isomorphism is in $\mathscr{P}$.
(Pb) The class $\mathscr{P}$ is stable under composition.
(Pc) The class $\mathscr{P}$ is stable under pullbacks: for any morphism $f : X \longrightarrow Y$ in $\mathscr{P}$ and any morphism $Y' \longrightarrow Y$, the pullback $X' = Y' \times_Y X$ is representable in $\mathscr{S}$, and the projection $f' : X' \longrightarrow Y'$ is in $\mathscr{P}$.

The morphisms which are in $\mathscr{P}$ will be called the $\mathscr{P}$-*morphisms*.[36]

In what follows, we assume that an admissible class of morphisms $\mathscr{P}$ is fixed.

1.1 $\mathscr{P}$-fibred categories

1.1.a Definitions

Let $\mathscr{C}at$ be the 2-category of categories.

1.1.1 Let $\mathscr{M}$ be a fibred category over $\mathscr{S}$, seen as a 2-functor $\mathscr{M} : \mathscr{S}^{op} \longrightarrow \mathscr{C}at$; see [Gro03, Exp. VI].

Given a morphism $f : T \longrightarrow S$ in $\mathscr{S}$, we shall denote by

$$f^* : \mathscr{M}(S) \longrightarrow \mathscr{M}(T)$$

the corresponding pullback functor between the corresponding fibers. We shall always assume that $(1_S)^* = 1_{\mathscr{M}(S)}$, and that for any morphisms $W \xrightarrow{g} T \xrightarrow{f} S$ in $\mathscr{S}$, we have structural isomorphisms:

(1.1.1.1) $$g^* f^* \xrightarrow{\sim} (fg)^*,$$

which are subject to the usual cocycle condition with respect to composition of morphisms.

Given a morphism $f : T \longrightarrow S$ in $\mathscr{S}$, if the corresponding inverse image functor f^* has a left adjoint, we shall denote it by

$$f_\sharp : \mathscr{M}(T) \longrightarrow \mathscr{M}(S).$$

---

[36] In practice, $\mathscr{S}$ will be an adequate subcategory of the category of noetherian schemes and $\mathscr{P}$ will be the class of smooth morphisms (resp. étale morphisms, morphisms of finite type, separated or not necessarily separated) in $\mathscr{S}$.

For any morphisms $W \xrightarrow{g} T \xrightarrow{f} S$ in $\mathscr{S}$ such that f^* and g^* have a left adjoint, we have an isomorphism obtained by transposition from the isomorphism (1.1.1.1):

(1.1.1.2) $$(fg)_\sharp \xrightarrow{\sim} f_\sharp g_\sharp.$$

Definition 1.1.2 A *pre-$\mathscr{P}$-fibred category* $\mathscr{M}$ over $\mathscr{S}$ is a fibred category $\mathscr{M}$ over $\mathscr{S}$ such that, for any morphism $p : T \longrightarrow S$ in $\mathscr{P}$, the pullback functor $p^* : \mathscr{M}(S) \longrightarrow \mathscr{M}(T)$ has a left adjoint $p_\sharp : \mathscr{M}(T) \longrightarrow \mathscr{M}(S)$.

Convention 1.1.3 Usually, we will consider that (1.1.1.1) and (1.1.1.2) are identities. Similarly, we consider that for any object S of $\mathscr{S}$, $(1_S)^* = 1_{\mathscr{M}(S)}$ and $(1_S)_\sharp = 1_{\mathscr{M}(S)}$.[37]

Example 1.1.4 Let S be an object of $\mathscr{S}$. We let $\mathscr{P}/S$ be the full subcategory of the comma category $\mathscr{S}/S$ comprising objects over S whose structural morphism is in $\mathscr{P}$. We will usually call the objects of $\mathscr{P}/S$ the $\mathscr{P}$-*objects over* S.

Given a morphism $f : T \longrightarrow S$ in $\mathscr{S}$ and a $\mathscr{P}$-morphism $\pi : X \longrightarrow S$, we put $f^*(\pi) = \pi \times_S T$ using the property (Pc) of $\mathscr{P}$ (see Section 1.0.1). This defines a functor $f^* : \mathscr{P}/S \longrightarrow \mathscr{P}/T$.

Given two $\mathscr{P}$-morphisms $f : T \longrightarrow S$ and $\pi : Y \longrightarrow T$, we put $f_\sharp(\pi) = f \circ \pi$ using the property (Pb) of $\mathscr{P}$. This defines a functor $f_\sharp : \mathscr{P}/T \longrightarrow \mathscr{P}/S$. According to the property of pullbacks, $f_\sharp$ is left adjoint to f^*.

We thus get a pre-$\mathscr{P}$-fibred category $\mathscr{P}/? : S \longmapsto \mathscr{P}/S$.

Example 1.1.5 Assume $\mathscr{S}$ is the category of noetherian schemes of finite dimension, and $\mathscr{P} = Sm$. For a scheme S of $\mathscr{S}$, let $\mathscr{H}_\bullet(S)$ be the pointed homotopy category of schemes over S defined by Morel and Voevodsky in [MV99]. Then according to *op. cit.*, $\mathscr{H}_\bullet$ is a pre-Sm-fibred category over $\mathscr{S}$.

1.1.6 *Exchange structures I.–* Suppose given a pre-$\mathscr{P}$-fibred category $\mathscr{M}$.

Consider a commutative square of $\mathscr{S}$

$$\begin{array}{ccc} Y & \xrightarrow{q} & X \\ g \downarrow & \Delta & \downarrow f \\ T & \xrightarrow{p} & S \end{array}$$

such that p and hence q are $\mathscr{P}$-morphisms, using the identification of Convention 1.1.3 we get a canonical natural transformation

$$Ex(\Delta_\sharp^*) : q_\sharp g^* \xrightarrow{ad(p_\sharp, p^*)} q_\sharp g^* p^* p_\sharp = q_\sharp q^* f^* p_\sharp \xrightarrow{ad'(q_\sharp, q^*)} f^* p_\sharp$$

---

[37] We can always globally strictify the fibred category structure so that $g^* f^* = (fg)^*$ for any composable morphisms f and g, and so that $(1_S)^* = 1_{\mathscr{M}(S)}$ for any object S of $\mathscr{S}$; moreover, for a morphism h of $\mathscr{S}$ such that a left adjoint of h^* exists, and we can choose the left adjoint functor $h_\sharp$ which we feel to be the most convenient for us, depending on the situation we are dealing with. For instance, if $h = 1_S$, we can choose $h_\sharp$ to be $1_{\mathscr{M}(S)}$, and if $h = fg$, with f^* and g^* having left adjoints, we can choose $h_\sharp$ to be $f_\sharp g_\sharp$ (with the unit and counit naturally induced by composition).

1 General definitions and axioms

called the *exchange transformation* between $q_\sharp$ and g^*.

Remark 1.1.7 These exchange transformations satisfy a coherence condition with respect to the relations $(fg)^* = g^*f^*$ and $(fg)_\sharp = f_\sharp g_\sharp$. As an example, consider two commutative squares in $\mathscr{S}$:

$$\begin{array}{ccccc} Z & \xrightarrow{q'} & Y & \xrightarrow{q} & X \\ {\scriptstyle h}\downarrow & \Theta & {\scriptstyle g}\downarrow & \Delta & \downarrow{\scriptstyle f} \\ W & \xrightarrow{p'} & T & \xrightarrow{p} & S \end{array}$$

and let $\Delta \circ \Theta$ be the commutative square made by the exterior maps — it is usually called the horizontal composition of the squares. Then, the following diagram of 2-morphisms is commutative:

$$\begin{array}{ccc} (qq')_\sharp h^* & \xrightarrow{Ex(\Delta \circ \Theta)^*_\sharp} & f^*(pp')_\sharp \\ \| & & \| \\ q_\sharp q'_\sharp h^* \xrightarrow{Ex(\Theta^*_\sharp)} q_\sharp g^* p'_\sharp & \xrightarrow{Ex(\Delta^*_\sharp)} & f^* p_\sharp p'_\sharp \end{array}$$

To see this, one proceeds as follows. First, we observe that, since ad'_q is a natural transformation, for each object M of $\mathscr{M}(T)$, the square

$$\begin{array}{ccc} q'_\sharp q'^* g^*(M) & \xrightarrow{q'_\sharp q'^* g^*(ad_q(M))} & q'_\sharp q'^* g^* p^* p_\sharp(M) \\ {\scriptstyle ad'_q(g^*(M))}\downarrow & & \downarrow{\scriptstyle ad'_q(g^* p^* p_\sharp(M))} \\ g^*(M) & \xrightarrow{g^*(ad_q(M))} & g^* p^* p_\sharp(M) \end{array}$$

commutes. In other words, with a slight abuse of notation, we have the following commutative square of functors:

$$\begin{array}{ccc} q'_\sharp q'^* g^* & \xrightarrow{ad_q} & q'_\sharp q'^* g^* p^* p_\sharp \\ {\scriptstyle ad'_q}\downarrow & & \downarrow{\scriptstyle ad'_q} \\ g^*(M) & \xrightarrow{ad_q} & g^* p^* p_\sharp \end{array}$$

We then consider the diagram below, in which ad_r (resp. ad'_r) indicates the morphism obtained from the obvious unit morphism (resp. counit morphism) of the adjunction $(r_\sharp, r^*)$ by eventually adding functors on the left side or on the right side, and we can check easily that each cell below is commutative, proving our claim.

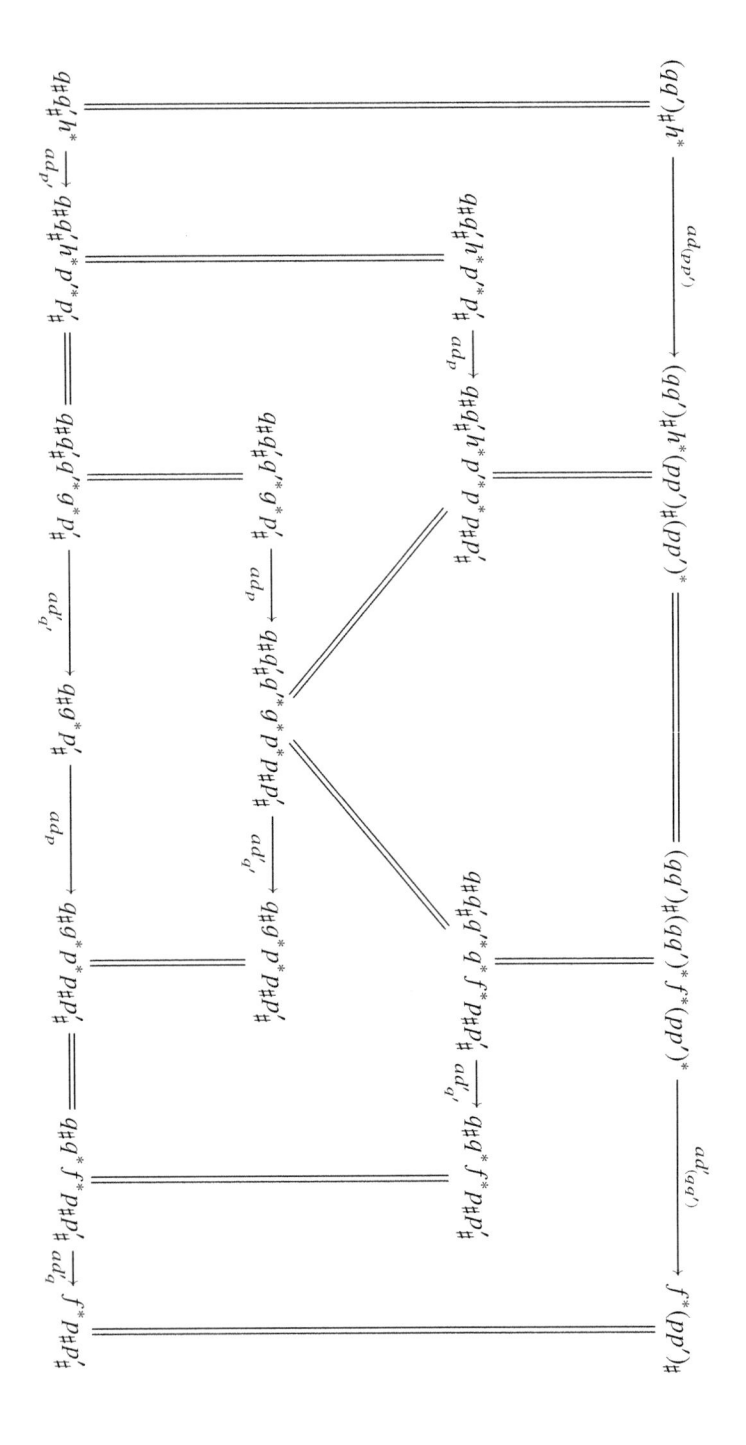

1 General definitions and axioms

Therefore, according to our abuse of notation for natural transformations, Ex behaves as a contravariant functor with respect to the horizontal composition of squares. The same is true for vertical composition of commutative squares.

Remark 1.1.8 In the sequel, we will introduce several exchange transformations between various functors. We speak of an *exchange isomorphism* when the transformation is an isomorphism. When only two kind of functors are involved, say of type a and b, we say that functors of type a and functors of type b commute when the exchange transformation is an isomorphism.

As an example (see also the next definition), when the exchange transformation $Ex(\Delta_\sharp^*)$ is an isomorphism, we simply say that f^* and $p_\sharp$ commute — or also that f^* commutes with $p_\sharp$.

1.1.9 Under the setting of Section 1.1.6, we will consider the following property:

($\mathscr{P}$-BC) *$\mathscr{P}$-base change.*– For any Cartesian square

$$\begin{array}{ccc} Y & \xrightarrow{q} & X \\ g \downarrow & \Delta & \downarrow f \\ T & \xrightarrow{p} & S \end{array}$$

such that p is a $\mathscr{P}$-morphism, the exchange transformation

$$Ex(\Delta_\sharp^*) : q_\sharp g^* \longrightarrow f^* p_\sharp$$

is an isomorphism.[38]

Definition 1.1.10 A $\mathscr{P}$-*fibred category* over $\mathscr{S}$ is a pre-$\mathscr{P}$-fibred category $\mathscr{M}$ over $\mathscr{S}$ which satisfies the property of $\mathscr{P}$-base change.

Example 1.1.11 Consider the notation of Example 1.1.4. Then the transitivity property of pullbacks of morphisms in $\mathscr{P}$ amounts to saying that the category $\mathscr{P}/?$ satisfies the $\mathscr{P}$-base change property. Thus, $\mathscr{P}/?$ is in fact a $\mathscr{P}$-fibred category, called *the canonical $\mathscr{P}$-fibred category*.

Definition 1.1.12 A $\mathscr{P}$-fibred category $\mathscr{M}$ over $\mathscr{S}$ is *complete* if, for any morphism $f : T \longrightarrow S$, the pullback functor $f^* : \mathscr{M}(S) \longrightarrow \mathscr{M}(T)$ admits a right adjoint $f_* : \mathscr{M}(S) \longrightarrow \mathscr{M}(T)$.

Remark 1.1.13 In the case where $\mathscr{P}$ is the class of isomorphisms, a $\mathscr{P}$-fibred category is what we usually call a bifibred category over $\mathscr{S}$.

Example 1.1.14 The pre-Sm-fibred category $\mathscr{H}_\bullet$ of Example 1.1.5 is a complete Sm-fibred category according to [MV99, p. 102–105, 108–110].

---
[38] In other words, f^* commutes with $p_\sharp$.

1.1.15 *Exchange structures II.*– Let $\mathscr{M}$ be a complete $\mathscr{P}$-fibred category. Consider a commutative square

$$\begin{array}{ccc} Y & \xrightarrow{q} & X \\ {\scriptstyle g}\downarrow & \Delta & \downarrow{\scriptstyle f} \\ T & \xrightarrow{p} & S. \end{array}$$

We obtain an exchange transformation:

$$Ex(\Delta_*^*) : p^* f_* \xrightarrow{ad(g^*, g_*)} g_* g^* p^* f_* = g_* q^* f^* f_* \xrightarrow{ad'(f^*, f_*)} g_* q^*.$$

Assume moreover that p and q are $\mathscr{P}$-morphisms. Then we can check that $Ex(\Delta_*^*)$ is the transpose of the exchange $Ex(\Delta_\sharp^*)$. Thus, when Δ is Cartesian and p is a $\mathscr{P}$-morphism, $Ex(\Delta_*^*)$ is an isomorphism according to ($\mathscr{P}$-BC).

We can also define an exchange transformation:

$$Ex(\Delta_{\sharp *}) : p_\sharp g_* \xrightarrow{ad(f^*, f_*)} f_* f^* p_\sharp g_* \xrightarrow{Ex(\Delta_\sharp^*)^{-1}} f_* q_\sharp g^* g_* \xrightarrow{ad'(g^*, g_*)} f_* q_\sharp.$$

Remark 1.1.16 As in Remark 1.1.7, we obtain coherence results for these exchange transformations.

First with respect to the identifications of the kind $f^* g^* = (gf)^*$, $(fg)_* = f_* g_*$, $(fg)_\sharp = f_\sharp g_\sharp$. Second when several exchange transformations of different kinds are involved. As an example, we consider the following commutative diagram in $\mathscr{S}$:

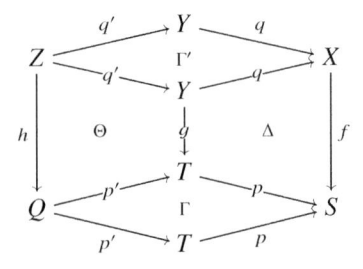

$$\begin{array}{c}\text{(diagram)}\end{array}$$

Then the following diagram of natural transformations is commutative:

$$\begin{array}{c}\text{(diagram)}\end{array}$$

1 General definitions and axioms

We leave the verification to the reader (it is analogous to that of Remark 1.1.7 except that it also involves the compatibility of the unit and counit of an adjunction).

Definition 1.1.17 Let $\mathscr{M}$ be a complete $\mathscr{P}$-fibred category. Consider a commutative square in $\mathscr{S}$

$$\begin{array}{ccc} Y & \xrightarrow{q} & X \\ {\scriptstyle g}\downarrow & \Delta & \downarrow{\scriptstyle f} \\ T & \xrightarrow{p} & S. \end{array}$$

We will say that Δ is $\mathscr{M}$-*transversal* if the exchange transformation

$$Ex(\Delta_*^*) : p^* f_* \longrightarrow g_* q^*$$

of Section 1.1.15 is an isomorphism.

Given an admissible class of morphisms Q in $\mathscr{S}$, we say that $\mathscr{M}$ has the *transversality* (resp. *cotransversality*) *property with respect to Q-morphisms*, if, for any Cartesian square Δ as above such that f is in Q (resp. p is in Q), Δ is $\mathscr{M}$-transversal.

Remark 1.1.18 Assume $\mathscr{S}$ is a sub-category of the category of schemes. When Q is the class of smooth morphisms (resp. proper morphisms), the cotransversality (resp. transversality) property with respect to Q is usually called the *smooth base change property* (resp. *proper base change property*). See also Definition 2.2.13.

According to Section 1.1.15, we derive the following consequence of our axioms:

Proposition 1.1.19 *Any complete $\mathscr{P}$-fibred category has the cotransversality property with respect to $\mathscr{P}$.*

Let us note for future reference the following corollary:

Corollary 1.1.20 *If $\mathscr{M}$ is a $\mathscr{P}$-fibred category, then, for any monomorphism $j : U \longrightarrow S$ in $\mathscr{P}$, the functor $j_\sharp$ is fully faithful. If moreover $\mathscr{M}$ is complete, then the functor j_* is fully faithful as well.*

Proof Because j is a monomorphism, we get a Cartesian square in $\mathscr{S}$:

$$\begin{array}{ccc} U & == & U \\ \| & \Delta & \downarrow{\scriptstyle j} \\ U & \xrightarrow{j} & S. \end{array}$$

Note that $Ex(\Delta_\sharp^*) : 1 \longrightarrow j^* j_\sharp$ is the unit of the adjunction $(j_\sharp, j^*)$. Thus the $\mathscr{P}$-base change property shows that $j_\sharp$ is fully faithful.

Assume $\mathscr{M}$ is complete. Similarly observe that $Ex(\Delta_*^*) : j^* j_* \longrightarrow 1$ is the counit of the adjunction (j^*, j_*). Thus, the above proposition shows readily that j_* is fully faithful. $\square$

1.1.b Monoidal structures

Let $\mathscr{C}at^{\otimes}$ be the sub-2-category of $\mathscr{C}at$ comprising symmetric monoidal categories whose 1-morphisms are (strong) symmetric monoidal functors and 2-morphisms are symmetric monoidal transformations.

Definition 1.1.21 A *monoidal pre-$\mathscr{P}$-fibred category over $\mathscr{S}$* is a 2-functor

$$\mathscr{M} : \mathscr{S} \longrightarrow \mathscr{C}at^{\otimes}$$

such that $\mathscr{M}$ is a pre-$\mathscr{P}$-fibred category.

In other words, $\mathscr{M}$ is a pre-$\mathscr{P}$-fibred category such that each of its fibers $\mathscr{M}(S)$ is endowed with the structure of a monoidal category, and any pullback morphism f^* is monoidal, with the obvious coherent structures. For an object S of $\mathscr{S}$, we will usually denote by $\otimes_S$ (resp. $\mathbb{1}_S$) the tensor product (resp. unit) of $\mathscr{M}(S)$.

In particular, we then have the following natural isomorphisms:

- for a morphism $f : T \longrightarrow S$ in $\mathscr{S}$, and objects M, N of $\mathscr{M}(S)$,

$$f^*(M) \otimes_T f^*(N) \xrightarrow{\sim} f^*(M \otimes_S N);$$

- for a morphism $f : T \longrightarrow S$ in $\mathscr{S}$,

$$f^*(\mathbb{1}_S) \xrightarrow{\sim} \mathbb{1}_T.$$

Convention 1.1.22 As in Convention 1.1.3, we will write formulas as though these structural isomorphisms are identities.

Example 1.1.23 Consider the notations of Example 1.1.4.

Using the properties (Pb) and (Pc) of $\mathscr{P}$ (see Section 1.0.1), for two S-objects X and Y in $\mathscr{P}/S$, the Cartesian product $X \times_S Y$ is an object of $\mathscr{P}/S$. This defines a symmetric monoidal structure on $\mathscr{P}/S$ with unit the trivial S-object S. Moreover, the functor f^* defined in Example 1.1.4 is monoidal. Thus, the pre-$\mathscr{P}$-fibred category $\mathscr{P}/?$ is in fact monoidal.

1.1.24 *Monoidal exchange structures I.* Let $\mathscr{M}$ be a monoidal pre-$\mathscr{P}$-fibred category over $\mathscr{S}$.

Consider a $\mathscr{P}$-morphism $f : T \longrightarrow S$, and M (resp. N) an object of $\mathscr{M}(T)$ (resp. $\mathscr{M}(S)$).

We get a morphism in $\mathscr{M}(S)$

$$Ex(f_\sharp^*, \otimes) : f_\sharp(M \otimes_T f^*(N)) \longrightarrow f_\sharp(M) \otimes_S N$$

as the composition

$$f_\sharp(M \otimes_T f^*(N)) \longrightarrow f_\sharp(f^* f_\sharp(M) \otimes_T f^*(N)) \simeq f_\sharp f^*(f_\sharp(M) \otimes_S N) \longrightarrow f_\sharp(M) \otimes_S N.$$

1 General definitions and axioms

This map is natural in M and N. It will be called the *exchange transformation* between $f_\sharp$ and $\otimes_T$.

Note also that the functor $f_\sharp$, as a left adjoint of a symmetric monoidal functor, is colax symmetric monoidal: for any objects M and N of $\mathscr{M}(T)$, there is a canonical morphism

(1.1.24.1) $$f_\sharp(M) \otimes_S f_\sharp(N) \longrightarrow f_\sharp(M \otimes_T N)$$

natural in M and N, as well as a natural map

(1.1.24.2) $$f_\sharp(\mathbb{1}_T) \longrightarrow \mathbb{1}_S.$$

Remark 1.1.25 As in Remark 1.1.7, the preceding exchange transformations satisfy a coherence condition for composable morphisms $W \xrightarrow{g} T \xrightarrow{f} S$. In fact, we get a commutative diagram:

$$\begin{array}{ccc}
(fg)_\sharp(M \otimes_S (fg)^*(N)) & \xrightarrow{Ex((fg)_\sharp^*,\otimes)} & ((fg)_\sharp(M)) \otimes_W N \\
\| & & \| \\
f_\sharp g_\sharp(M \otimes_S g^* f^*(N)) & \xrightarrow{Ex(g_\sharp^*,\otimes)} f_\sharp(g_\sharp(M) \otimes_T f^*(N)) \xrightarrow{Ex(f_\sharp^*,\otimes)} & (f_\sharp g_\sharp(M)) \otimes_W N.
\end{array}$$

As in Remark 1.1.16, there is also a coherence relation when different kinds of exchange transformations are involved. Consider a commutative square in $\mathscr{S}$

$$\begin{array}{ccc}
Y & \xrightarrow{q} & X \\
g \downarrow & \Delta & \downarrow f \\
T & \xrightarrow{p} & S
\end{array}$$

such that p and q are $\mathscr{P}$-morphisms. Then the following diagram is commutative:

$$\begin{array}{ccc}
q_\sharp g^*(M \otimes_T p^*N) & \xrightarrow{Ex(\Delta_\sharp^*)} f^* p_\sharp(M \otimes_T p^*N) \xrightarrow{Ex(p_\sharp^*,\otimes)} & f^*(p_\sharp M \otimes_S N) \\
\| & & \| \\
q_\sharp(g^*M \otimes_Y q^*f^*N) & \xrightarrow{Ex(q_\sharp^*,\otimes)} (q_\sharp g^*M) \otimes_X f^*N \xrightarrow{Ex(\Delta_\sharp^*)} & (f^* p_\sharp M) \otimes_X f^*N.
\end{array}$$

We leave the verification to the reader.

1.1.26 Under the assumptions of Section 1.1.24, we will consider the following property:

($\mathscr{P}$-PF) $\mathscr{P}$-*projection formula.*– For any $\mathscr{P}$-morphism $f : T \longrightarrow S$ the exchange transformation

$$Ex(f_\sharp, \otimes_T) : f_\sharp(M \otimes_T f^*(N)) \longrightarrow f_\sharp(M) \otimes_S N$$

is an isomorphism for all M and N.

Definition 1.1.27 A *monoidal $\mathscr{P}$-fibred category* over $\mathscr{S}$ is a monoidal pre-$\mathscr{P}$-fibred category $\mathscr{M} : \mathscr{S}^{op} \longrightarrow \mathscr{C}at^{\otimes}$ over $\mathscr{S}$ which satisfies the $\mathscr{P}$-projection formula.

Example 1.1.28 Consider the canonical monoidal weak $\mathscr{P}$-fibred category $\mathscr{P}/?$ (see Example 1.1.23). The transitivity property of pullbacks readily implies that $\mathscr{P}/?$ satisfies the property ($\mathscr{P}$-PF). Thus, $\mathscr{P}/?$ is in fact a monoidal $\mathscr{P}$-fibred category which we call *canonical*.

Definition 1.1.29 A monoidal $\mathscr{P}$-fibred category $\mathscr{M}$ over $\mathscr{S}$ is *complete* if it satisfies the following conditions:

1. $\mathscr{M}$ is complete as a $\mathscr{P}$-fibred category.
2. For any object S of $\mathscr{S}$, the monoidal category $\mathscr{M}(S)$ is closed (i.e. has an internal Hom).

In this case, we will usually denote by Hom_S the internal Hom in $\mathscr{M}(S)$, so that we have natural bijections
$$\mathrm{Hom}_{\mathscr{M}(S)}(A \otimes_S B, C) \simeq \mathrm{Hom}_{\mathscr{M}(S)}(A, Hom_S(B,C)).$$

Example 1.1.30 The $\mathscr{P}$-fibred category $\mathscr{H}_\bullet$ of Example 1.1.14 is in fact a complete monoidal $\mathscr{P}$-fibred category. The tensor product is given by the smash product (see [MV99]).

1.1.31 *Monoidal exchange structures II.–* Let $\mathscr{M}$ be a complete monoidal $\mathscr{P}$-fibred category.

Consider a morphism $f : T \longrightarrow S$ in $\mathscr{S}$. Then we obtain an exchange transformation:
$$Ex(f_*^*, \otimes_S) : (f_*M) \otimes_S N \xrightarrow{ad(f^*, f_*)} f_* f^*\big((f_*M) \otimes_S N\big)$$
$$= f_*\big((f^* f_* M) \otimes_T f^* N\big) \xrightarrow{ad'(f^*, f_*)} f_*(M \otimes_T f^* N).$$

Remark 1.1.32 As in Remark 1.1.25, these exchange transformations are compatible with the identifications $(fg)_* = f_* g_*$ and $(fg)^* = g^* f^*$. Moreover, there is a coherence relation when composing the exchange transformations of the kind $Ex(f_*^*, \otimes)$ with exchange transformations of the kind $Ex(\Delta_*^*)$ as in *loc. cit*. Finally, note that there is another class of coherence relations involving $Ex(f_*^*, \otimes)$, $Ex(\Delta_*^*)$ (resp. $Ex(f_\sharp^*, \otimes)$) and $Ex(\Delta_{\sharp*})$.

We leave the formulation of these coherence relations to the reader, based on the model of the preceding ones.

1.1.33 *Monoidal exchange structures III.–* Let $\mathscr{M}$ be a complete monoidal $\mathscr{P}$-fibred category and $f : T \longrightarrow S$ be a morphism in $\mathscr{S}$.

Because f^* is monoidal, we get by adjunction a canonical isomorphism

$$Hom_S(M, f_*N) \longrightarrow f_*Hom_T(f^*M, N).$$

Assume that f is a $\mathscr{P}$-morphism. Then from the $\mathscr{P}$-projection formula, we get by adjunction two canonical isomorphisms:

$$f^*Hom_S(M, N) \longrightarrow Hom_T(f^*M, f^*N),$$
$$Hom_S(f_\sharp M, N) \longrightarrow f_*Hom_T(M, f^*N).$$

These isomorphisms are generically called *exchange isomorphisms*.

1.1.c Geometric sections

1.1.34 Consider a complete monoidal $\mathscr{P}$-fibred category $\mathscr{M}$.

Let S be a scheme. For any $\mathscr{P}$-morphism $p : X \longrightarrow S$, we put $M_S(X) := p_\sharp(\mathbb{1}_X)$. According to our conventions, this object is identified with $p_\sharp p^*(\mathbb{1}_S)$. As the $\mathscr{P}$-fibred category $\mathscr{M}$ is complete, the functor $p_\sharp p^*$ is left adjoint to $p_* p^*$. Consider a commutative diagram of schemes in $\mathscr{S}$:

$$\begin{array}{ccc} Y & \xrightarrow{f} & X \\ & {}_q\searrow \quad \swarrow_p & \\ & S & \end{array}$$

such that p and q are in $\mathscr{P}$. In other words, f is a morphism in the category $\mathscr{P}/S$ of Example 1.1.4. Then we get a natural transformation of functors:

(1.1.34.1) $$p_* p^* \xrightarrow{ad(f^*, f_*)} p_* f_* f^* p^* = q_* q^*.$$

By adjunction, one deduces a natural transformation:

$$q_\sharp q^* \longrightarrow p_\sharp p^*,$$

which gives a morphism $M_S(Y) \xrightarrow{f_*} M_S(X)$. One can check that the relation $f_* g_* = (fg)_*$ holds — by reducing to the same assertion for the map (1.1.34.1) which follows by a standard 2-functoriality argument. Therefore, we have obtained a covariant functor $M_S : \mathscr{P}/S \longrightarrow \mathscr{M}$.

Consider a Cartesian square in $\mathscr{S}$

$$\begin{array}{ccc} Y & \xrightarrow{g} & X \\ {}_q\downarrow & \Delta & \downarrow_p \\ T & \xrightarrow{f} & S \end{array}$$

such that p is a $\mathscr{P}$-morphism. With the notations of Example 1.1.4, $Y = f^*(X)$. Then we get a natural exchange transformation

$$Ex(M_T, f^*) : M_T(f^*(X)) = q_\sharp(\mathbb{1}_Y) = q_\sharp g^*(\mathbb{1}_X) \xrightarrow{Ex(\Delta_\sharp^*)} f^* p_\sharp(\mathbb{1}_X) = f^* M_S(X).$$

In other words, M defines a lax natural transformation $\mathscr{P}/? \longrightarrow \mathscr{M}$.

Consider $\mathscr{P}$-morphisms $p : X \longrightarrow S, q : Y \longrightarrow S$. Let $Z = X \times_S Y$ be the Cartesian product and consider the Cartesian square:

$$\begin{array}{ccc} Z & \xrightarrow{p'} & Y \\ {\scriptstyle q'}\downarrow & \Theta & \downarrow{\scriptstyle q} \\ X & \xrightarrow{p} & S. \end{array}$$

Using the exchange transformations of the preceding paragraph, we get a canonical morphism

$$Ex(M_S, \otimes_S) : M_S(X \times_S Y) \longrightarrow M_S(X) \otimes_S M_S(Y)$$

as the composition

$$M_S(X \times_S Y) = p_\sharp q'_\sharp p'^*(\mathbb{1}_Y) \xrightarrow{Ex(\Theta_\sharp^*)} p_\sharp p^* q_\sharp(\mathbb{1}_Y) = p_\sharp(\mathbb{1}_X \otimes_X p^* q_\sharp(\mathbb{1}_Y))$$

$$\xrightarrow{Ex(p_\sharp, \otimes_X)} p_\sharp(\mathbb{1}_X) \otimes_S q_\sharp(\mathbb{1}_Y) = M_S(X) \otimes_S M_S(Y).$$

In other words, the functor M_S is symmetric colax monoidal.

Finally observe that for any $\mathscr{P}$-morphism $p : T \longrightarrow S$, and any $\mathscr{P}$-object Y over T, we obtain according to our convention an identification $p_\sharp M_T(Y) = M_S(Y)$.

Definition 1.1.35 Given a complete monoidal $\mathscr{P}$-fibred category $\mathscr{M}$ over $\mathscr{S}$, the lax natural transformation $M : \mathscr{P}/? \longrightarrow \mathscr{M}$ constructed above will be called the *geometric sections of $\mathscr{M}$*.

The following lemma is obvious from the definitions above:

Lemma 1.1.36 *Let $\mathscr{M}$ be a complete monoidal $\mathscr{P}$-fibred category. Let $M : \mathscr{P}/? \longrightarrow \mathscr{M}$ be the geometric sections of $\mathscr{M}$. Then:*

(i) For any morphism $f : T \longrightarrow S$ in $\mathscr{S}$, the exchange $Ex(M_T, f^*)$ defined above is an isomorphism.

(ii) For any scheme S, the exchange $Ex(M_S, \otimes_S)$ defined above is an isomorphism.

In other words, M is a Cartesian functor and M_S is a (strong) symmetric monoidal functor.

1.1.37 In the situation of the lemma we thus obtain the following isomorphisms:
- $f^* M_S(X) \simeq M_T(X \times_S T)$,
- $p_\sharp M_T(Y) \simeq M_S(Y)$,

1 General definitions and axioms

- $M_S(X \times_S Y) \simeq M_S(X) \otimes_S M_S(Y)$,

whenever it makes sense.

1.1.d Twists

1.1.38 Let $\mathscr{M}$ be a pre-$\mathscr{P}$-fibred category of $\mathscr{S}$. Recall that a Cartesian section of $\mathscr{M}$ (i.e. a Cartesian functor $A : \mathscr{S} \longrightarrow \mathscr{M}$) is the data of an object A_S of $\mathscr{M}(S)$ for each object S of $\mathscr{S}$ and of isomorphisms

$$f^*(A_S) \xrightarrow{\sim} A_T$$

for each morphism $f : T \longrightarrow S$, subject to coherence identities; see [Gro03, Exp. VI].

If $\mathscr{M}$ is monoidal, the tensor product of two Cartesian sections is defined termwise.

Definition 1.1.39 Let $\mathscr{M}$ be a monoidal pre-$\mathscr{P}$-fibred category. A set of *twists* τ for $\mathscr{M}$ is a set of Cartesian sections of $\mathscr{M}$ which is stable under tensor products (up to isomorphism), and contains the unit $\mathbb{1}$. For short, when $\mathscr{M}$ is endowed with a set of twists τ, we also say that $\mathscr{M}$ is τ-*twisted*.

1.1.40 Let $\mathscr{M}$ be a monoidal pre-$\mathscr{P}$-fibred category endowed with a set of twists τ.

The tensor product on τ induces a monoid structure that we will denote by $+$ (the unit object of τ will be written 0).

Consider an object $i \in \tau$. For any object S of $\mathscr{S}$, we thus obtain an object $t(i)_S$ in $\mathscr{M}(S)$ associated with i. Given any object M of $\mathscr{M}(S)$, we simply put:

$$M\{i\} = M \otimes_S i_S$$

and call this object the twist of M by i. We also define $M\{0\} = M$.

For any $i, j \in \tau$, and any object M of $\mathscr{M}(S)$, we define $M\{i+j\} = (M\{i\})\{j\}$. Given a morphism $f : T \longrightarrow S$, an object M of $\mathscr{M}(S)$ and a twist $i \in \tau$, we also obtain $f^*(M\{i\}) = (f^*M)\{i\}$. If f is a $\mathscr{P}$-morphism, for any object M of $\mathscr{M}(T)$, the exchange transformation $Ex(f_\sharp^*, \otimes_T)$ of Section 1.1.6 induces a canonical morphism

$$Ex(f_\sharp, \{i\}) : f_\sharp(M\{i\}) \longrightarrow (f_\sharp M)\{i\}.$$

We will say that $f_\sharp$ *commutes with* τ-*twists* (or simply twists when τ is clear) if for any $i \in \tau$, the natural transformation $Ex(f_\sharp, \{i\})$ is an isomorphism.

Definition 1.1.41 Let $\mathscr{M}$ be a complete monoidal $\mathscr{P}$-fibred category with a set of twists τ and $M : \mathscr{P}/? \longrightarrow \mathscr{M}$ be the geometric sections of $\mathscr{M}$.

We say $\mathscr{M}$ is τ-*generated* if for any object S of $\mathscr{S}$, the family of functors

$$\mathrm{Hom}_{\mathscr{M}(S)}(M_S(X)\{i\}, -) : \mathscr{M}(S) \longrightarrow \mathscr{S}et$$

indexed by a $\mathscr{P}$-object X/S and an element $i \in \tau$ is conservative.

Of course, we do not exclude the case where τ is trivial, but then, we shall simply say that $\mathscr{M}$ is *geometrically generated*.

We shall frequently use the following proposition to characterize complete monoidal $\mathscr{P}$-fibred categories over $\mathscr{S}$:

Proposition 1.1.42 *Let* $\mathscr{M} : \mathscr{S} \longrightarrow \mathscr{C}at^\otimes$ *be a 2-functor such that:*

1. *For any $\mathscr{P}$-morphism* $f : T \longrightarrow S$, *the pullback functor* $f^* : \mathscr{M}(S) \longrightarrow \mathscr{M}(T)$ *is monoidal and admits a left adjoint* $f_\sharp$ *in* $\mathscr{C}$.
2. *For any morphism* $f : T \longrightarrow S$, *the pullback functor* $f^* : \mathscr{M}(S) \longrightarrow \mathscr{M}(T)$ *admits a right adjoint* f_* *in* $\mathscr{C}$.

We consider $\mathscr{M}$ as a complete monoidal $\mathscr{P}$-fibred category and denote by $M : \mathscr{P}/? \longrightarrow \mathscr{M}$ *its associated geometric sections. Suppose given a set of twists τ such that $\mathscr{M}$ is τ-generated. Then, the following assertions are equivalent:*

(i) *$\mathscr{M}$ satisfies properties ($\mathscr{P}$-BC) and ($\mathscr{P}$-PF) (i.e. $\mathscr{M}$ is a complete monoidal $\mathscr{P}$-fibred category).*

(ii)   a. *M is a Cartesian functor.*
     b. *For any object S of $\mathscr{S}$, M_S is (strongly) monoidal.*
     c. *For any $\mathscr{P}$-morphism f, $f_\sharp$ commutes with τ-twists.*

Proof (i) $\Rightarrow$ (ii): This is obvious (see Lemma 1.1.36).

(ii) $\Rightarrow$ (i): We use the following easy lemma:

Lemma 1.1.43 *Let $\mathscr{C}_1$ and $\mathscr{C}_2$ be categories, $F, G : \mathscr{C}_1 \longrightarrow \mathscr{C}_2$ be two left adjoint functors, and $\eta : F \longrightarrow G$ be a natural transformation. Let $\mathscr{G}$ be a class of objects of $\mathscr{C}_1$ which is generating in the sense that the family of functors $\mathrm{Hom}_{\mathscr{C}_1}(X, -)$ for X in $\mathscr{G}$ is conservative.*

Then the following conditions are equivalent:

1. *η is an isomorphism.*
2. *For all X in $\mathscr{G}$, η_X is an isomorphism.*   □

Given this lemma, to prove property ($\mathscr{P}$-BC), we are reduced to checking that the exchange transformation $Ex(\Delta_\sharp^*)$ is an isomorphism when evaluated on an object $M_T(U)\{i\}$ for an object U of $\mathscr{P}/T$ and a twist $i \in \tau$. Then the property follows from (ii), Section 1.1.40 and Example 1.1.11.[39]

To prove property ($\mathscr{P}$-PF), we proceed in two steps, first proving the case $M = M_T(U)\{i\}$ and N any object of $\mathscr{M}(S)$ using the same argument as above with the help of Example 1.1.28. Then, we can prove the general case by another application of the same argument.   □

---

[39] The cautious reader will use Remark 1.1.7 to check that the corresponding map

$$M_X(U \times_T Y)\{i\} \longrightarrow M_X(U \times_T Y)\{i\}$$

is the identity.

1 General definitions and axioms

Suppose given a complete monoidal $\mathscr{P}$-fibred category $\mathscr{M}$ with a set of twists τ. Let $f : T \longrightarrow S$ be a morphism of $\mathscr{S}$. Then the exchange transformation 1.1.31 induces for any $i \in \tau$ an exchange transformation

$$Ex(f_*, \{i\}) : (f_*M)\{i\} \longrightarrow f_*(M\{i\}).$$

Definition 1.1.44 In the situation above, we say that f_* *commutes with τ-twists* (or simply with twists when τ is clear) if, for any $i \in \tau$, the exchange transformation $Ex(f_*, \{i\})$ is an isomorphism.

It will frequently happen that twists are $\otimes$-invertible. Then f_* commutes with twists as its right adjoint does.

1.2 Morphisms of $\mathscr{P}$-fibred categories

1.2.a The general case

1.2.1 Consider two $\mathscr{P}$-fibred categories $\mathscr{M}$ and $\mathscr{M}'$ over $\mathscr{S}$, as well as a Cartesian functor $\varphi^* : \mathscr{M} \longrightarrow \mathscr{M}'$ between the underlying fibred categories: for any object S of $\mathscr{S}$, we have a functor

$$\varphi_S^* : \mathscr{M}(S) \longrightarrow \mathscr{M}'(S),$$

and for any map $f : T \longrightarrow S$ in $\mathscr{S}$, we have an isomorphism of functors c_f

(1.2.1.1)
$$\begin{array}{ccc} \mathscr{M}(S) & \xrightarrow{\varphi_S^*} & \mathscr{M}'(S) \\ f^* \downarrow & \Swarrow c_f & \downarrow f^* \\ \mathscr{M}(T) & \xrightarrow{\varphi_T^*} & \mathscr{M}'(T) \end{array} \qquad c_f : f^* \varphi_S^* \xrightarrow{\sim} \varphi_T^* f^*$$

satisfying some cocycle condition with respect to composition in $\mathscr{S}$.

For any $\mathscr{P}$-morphism $p : T \longrightarrow S$, we construct an exchange morphism

$$Ex(p_\sharp, \varphi^*) : p_\sharp \varphi_T^* \longrightarrow \varphi_S^* p_\sharp$$

as the composition

$$p_\sharp \varphi_T^* \xrightarrow{ad(p_\sharp, p^*)} p_\sharp \varphi_T^* p^* p_\sharp \xrightarrow{c_p^{-1}} p_\sharp p^* \varphi_S^* p_\sharp \xrightarrow{ad'(p_\sharp, p^*)} \varphi_S^* p_\sharp.$$

Definition 1.2.2 Consider the situation above. We say that the Cartesian functor

$$\varphi^* : \mathscr{M} \longrightarrow \mathscr{M}'$$

is a *morphism of $\mathscr{P}$-fibred categories* if, for any $\mathscr{P}$-morphism p, the exchange transformation $Ex(p_\sharp, \varphi^*)$ is an isomorphism.

Example 1.2.3 If $\mathscr{M}$ is a monoidal $\mathscr{P}$-fibred category, then the geometric sections $M : \mathscr{P}/? \longrightarrow \mathscr{M}$ is a morphism of $\mathscr{P}$-fibred categories (Lemma 1.1.36).

Definition 1.2.4 Let $\mathscr{M}$ and $\mathscr{M}'$ be two complete $\mathscr{P}$-fibred categories. A *morphism of complete $\mathscr{P}$-fibred categories* is a morphism of $\mathscr{P}$-fibred categories

$$\varphi^* : \mathscr{M} \longrightarrow \mathscr{M}'$$

such that, for any object S of $\mathscr{S}$, the functor $\varphi_S^* : \mathscr{M}(S) \longrightarrow \mathscr{M}'(S)$ has a right adjoint

$$\varphi_{*,S} : \mathscr{M}'(S) \longrightarrow \mathscr{M}(S).$$

When we want to indicate a notation for the right adjoint of a morphism as above, we write

$$\varphi^* : \mathscr{M} \rightleftarrows \mathscr{N} : \varphi_*$$

the left adjoint being on the left-hand side.

1.2.5 *Exchange structures III.* Consider a morphism $\varphi^* : \mathscr{M} \longrightarrow \mathscr{M}'$ of complete $\mathscr{P}$-fibred categories.

Then for any morphism $f : T \longrightarrow S$ in $\mathscr{S}$, we define exchange transformations

(1.2.5.1) $$Ex(\varphi^*, f_*) : \varphi_S^* f_* \longrightarrow f_* \varphi_T^*,$$

(1.2.5.2) $$Ex(f^*, \varphi_*) : f^* \varphi_{*,S} \longrightarrow \varphi_{*,T} f^*,$$

as the respective compositions

$$\varphi_S^* f_* \xrightarrow{ad(f^*,f_*)} f_* f^* \varphi_S^* f_* \simeq f_* \varphi_T^* f^* f_* \xrightarrow{ad'(f^*,f_*)} f_* \varphi_T^*,$$

$$f^* \varphi_{*,S} \xrightarrow{ad(f^*,f_*)} f^* \varphi_{*,S} f_* f^* \simeq f^* f_* \varphi_{*,T} f^* \xrightarrow{ad'(f^*,f_*)} \varphi_{*,T} f^*.$$

Remark 1.2.6 We warn the reader that $\varphi_* : \mathscr{M}' \longrightarrow \mathscr{M}$ is not a Cartesian functor in general, meaning that the exchange transformation $Ex(f^*, \varphi_*)$ is not necessarily an isomorphism, even when f is a $\mathscr{P}$-morphism.

1.2.b The monoidal case

Definition 1.2.7 Let $\mathscr{M}$ and $\mathscr{M}'$ be monoidal $\mathscr{P}$-fibred categories.

A *morphism of monoidal $\mathscr{P}$-fibred categories* is a morphism $\varphi^* : \mathscr{M} \longrightarrow \mathscr{M}'$ of $\mathscr{P}$-fibred categories such that for any object S of $\mathscr{S}$, the functor $\varphi_S^* :

$\mathcal{M}(X) \longrightarrow \mathcal{N}(S)$ has the structure of a (strong) symmetric monoidal functor, and such that the structural isomorphisms (1.2.1.1) are isomorphisms of symmetric monoidal functors.

In the case where $\mathcal{M}$ and $\mathcal{M}'$ are complete monoidal $\mathcal{P}$-fibred categories, we shall say that such a morphism φ^* is a *morphism of complete monoidal $\mathcal{P}$-fibred categories* if φ^* is also a morphism of complete $\mathcal{P}$-fibred categories.

Remark 1.2.8 If we denote by $M(-, \mathcal{M})$ and $M(-, \mathcal{M}')$ the geometric sections of $\mathcal{M}$ and $\mathcal{M}'$ respectively, we have a natural identification:
$$\varphi_S^*(M_S(X, \mathcal{M})) \simeq M_S(X, \mathcal{M}').$$

1.2.9 *Monoidal exchange structures IV.* Consider a morphism $\varphi^* : \mathcal{M} \longrightarrow \mathcal{M}'$ of complete monoidal $\mathcal{P}$-fibred categories. For objects M (resp. N) of $\mathcal{M}(S)$ (resp. $\mathcal{M}'(S)$), we define an exchange transformation
$$Ex(\varphi_*, \otimes, \varphi^*) : (\varphi_{*,S}M) \otimes_S N \longrightarrow \varphi_{*,S}(M \otimes_T \varphi_S^* N),$$
natural in M and N, as the following composite
$$(\varphi_{*,S}M) \otimes_S N \xrightarrow{ad(\varphi^*, \varphi_*)} \varphi_{*,S}\varphi_S^*((\varphi_{*,S}M) \otimes_S N)$$
$$= \varphi_{*,S}((\varphi_S^*\varphi_{*,S}M) \otimes_T \varphi_S^* N) \xrightarrow{ad'(\varphi^*, \varphi_*)} \varphi_{*,S}(M \otimes_T \varphi_S^* N).$$

As in Remark 1.1.32, we get coherence relations between the various exchange transformations associated with a morphism of monoidal $\mathcal{P}$-fibred categories. We leave the formulation to the reader.

Note also that, because φ^* is monoidal, we get by adjunction a canonical isomorphism:
$$Hom_{\mathcal{M}(S)}(M, \varphi_{*,S}M') \xrightarrow{\sim} \varphi_{*,S}Hom_{\mathcal{M}'(S)}(\varphi_S^*M, M').$$

1.2.10 Consider two monoidal $\mathcal{P}$-fibred categories $\mathcal{M}$, $\mathcal{M}'$ and a Cartesian functor $\varphi^* : \mathcal{M} \longrightarrow \mathcal{M}'$ such that, for any scheme S, $\varphi_S^* : \mathcal{M}(S) \longrightarrow \mathcal{M}'(S)$ is monoidal.

Given a Cartesian section $K = (K_S)_{S \in \mathcal{S}}$ of $\mathcal{M}$, we obtain for any morphism $f : T \longrightarrow S$ in $\mathcal{S}$ a canonical map
$$f^* \varphi_S^*(K_S) = \varphi_T^*(f^*(K_S)) \longrightarrow \varphi_T^*(K_T)$$
which defines a Cartesian section of $\mathcal{M}'$, which we denote by $\varphi^*(K)$.

Definition 1.2.11 Let $(\mathcal{M}, \tau)$ and $(\mathcal{M}', \tau')$ be twisted monoidal $\mathcal{P}$-fibred categories. Let $\varphi^* : \mathcal{M} \longrightarrow \mathcal{M}'$ be a Cartesian functor as above (resp. a morphism of monoidal $\mathcal{P}$-fibred categories).

We say that $\varphi^* : (\mathcal{M}, \tau) \longrightarrow (\mathcal{M}', \tau')$ is *compatible with twists* if for any $i \in \tau$, the Cartesian section $\varphi^*(i)$ is in τ' (up to isomorphism in $\mathcal{M}'$).

Remark 1.2.12 In particular, φ^* induces a map $\tau \longrightarrow \tau'$ (if we consider the isomorphism classes of objects). Moreover, for any object K of $\mathscr{M}(S)$ and any twist $i \in \tau$, we get an identification:

$$\varphi_S^*(K\{i\}) \simeq (\varphi_S^* K)\{\varphi^*(i)\}.$$

Moreover, the exchange transformation $Ex(\varphi_*, \otimes)$ induces an exchange:

$$Ex(\varphi_*, \{i\}) : \varphi_{*,S}(K)\{i\} \longrightarrow \varphi_{*,S}(K\{\varphi^*(i)\}).$$

When this transformation is an isomorphism for any twist $i \in \tau$, we say that φ_* *commutes with twists*.

Note finally that Lemma 1.1.43 allows us to prove, as for Proposition 1.1.42, the following useful lemma:

Lemma 1.2.13 *Consider two complete monoidal $\mathscr{P}$-fibred categories $\mathscr{M}, \mathscr{M}'$ and denote by $M(-, \mathscr{M})$ and $M(-, \mathscr{M}')$ their respective geometric sections. Let $\varphi^* : \mathscr{M} \longrightarrow \mathscr{M}'$ be a Cartesian functor such that*

1. *For any scheme S, $\varphi_S^* : \mathscr{M}(S) \longrightarrow \mathscr{M}'(S)$ is monoidal.*
2. *For any scheme S, φ_S^* admits a right adjoint $\varphi_{*,S}$.*

Assume $\mathscr{M}$ (resp. $\mathscr{M}'$) is τ-generated (resp. τ'-twisted) and that φ^* induces a surjective map from the set of isomorphism classes of τ-twists to the set of isomorphism classes of τ'-twists. Then the following conditions are equivalent:

(i) φ^* is a morphism of complete monoidal $\mathscr{P}$-fibred categories.
(ii) For any object X of $\mathscr{P}/S$, the exchange transformation (cf. Section 1.2.1)

$$\varphi^* M_S(X, \mathscr{M}) \longrightarrow M_S(X, \mathscr{M}')$$

is an isomorphism.

1.3 Structures on $\mathscr{P}$-fibred categories

1.3.a Abstract definition

1.3.1 We fix a sub-2-category $\mathscr{C}$ of $\mathscr{C}at$ with the following properties:[40]

(1) The 2-functor

$$\mathscr{C}at \longrightarrow \mathscr{C}at', \quad A \mapsto A^{op}$$

sends $\mathscr{C}$ to $\mathscr{C}'$, where $\mathscr{C}'$ denotes the 2-category whose objects and maps are those of $\mathscr{C}$ and whose 2-morphisms are the 2-morphisms of $\mathscr{C}$, put in the reverse direction.

---
[40] See the following sections for examples.

(2) $\mathscr{C}$ is closed under adjunction: for any functor $u : A \longrightarrow B$ in $\mathscr{C}$, if a functor $v : B \longrightarrow A$ is a right adjoint or a left adjoint to u, then v is in $\mathscr{C}$.
(3) The 2-morphisms of $\mathscr{C}$ are closed under transposition: if

$$u : A \rightleftarrows B : v \text{ and } u' : A \rightleftarrows B : v'$$

are two adjunctions in $\mathscr{C}$ (with the left adjoints on the left-hand side), a natural transformation $u \longrightarrow u'$ is in $\mathscr{C}$ if and only if the corresponding natural transformation $v' \longrightarrow v$ is in $\mathscr{C}$.

We can then define and manipulate $\mathscr{C}$-structured $\mathscr{P}$-fibred categories as follows.

Definition 1.3.2 A $\mathscr{C}$-*structured $\mathscr{P}$-fibred category* (resp. $\mathscr{C}$-*structured complete $\mathscr{P}$-fibred category*) $\mathscr{M}$ over $\mathscr{S}$ is simply a $\mathscr{P}$-fibred category (resp. a complete $\mathscr{P}$-fibred category) whose underlying 2-functor $\mathscr{M} : \mathscr{S}^{op} \longrightarrow \mathscr{C}at$ factors through $\mathscr{C}$.

If $\mathscr{M}$ and $\mathscr{M}'$ are $\mathscr{C}$-structured fibred categories over $\mathscr{S}$, a Cartesian functor $\mathscr{M} \longrightarrow \mathscr{M}'$ is $\mathscr{C}$-*structured* if the functors $\mathscr{M}(S) \longrightarrow \mathscr{M}'(S)$ are in $\mathscr{C}$ for any object S of $\mathscr{S}$, and if all the structural 2-morphisms (1.2.1.1) are in $\mathscr{C}$ as well.

Definition 1.3.3 A morphism of $\mathscr{C}$-structured $\mathscr{P}$-fibred categories (resp. $\mathscr{C}$-structured complete $\mathscr{P}$-fibred categories) is a morphism of $\mathscr{P}$-fibred categories (resp. of complete $\mathscr{P}$-fibred categories) which is $\mathscr{C}$-structured as a Cartesian functor.

1.3.4 Consider a 2-category $\mathscr{C}$ as in Section 1.3.1. In order to deal with the monoidal case, we will also consider a sub-2-category $\mathscr{C}^{\otimes}$ of $\mathscr{C}$ such that:

1. the objects of $\mathscr{C}^{\otimes}$ are objects of $\mathscr{C}$ equipped with a symmetric monoidal structure;
2. the 1-morphisms of $\mathscr{C}^{\otimes}$ are exactly the 1-morphisms of $\mathscr{C}$ which are symmetric monoidal as functors;
3. the 2-morphisms of $\mathscr{C}^{\otimes}$ are exactly the 2-morphisms of $\mathscr{C}$ which are symmetric monoidal as natural transformations.

Note that $\mathscr{C}^{\otimes}$ satisfies condition (1) of Section 1.3.1, but it does not satisfy conditions (2) and (3) in general. Instead, we get the following properties:

(2′) If $u : A \longrightarrow B$ is a functor in $\mathscr{C}^{\otimes}$, a right (resp. left) adjoint v is a lax[41] (resp. colax) monoidal functor in $\mathscr{C}$.
(3′) Consider adjunctions

$$u : A \rightleftarrows B : v \text{ and } u' : A \rightleftarrows B : v'$$

in $\mathscr{C}$ (with the left adjoints on the left-hand side). If $u \longrightarrow u'$ (resp. $v \longrightarrow v'$) is a 2-morphism in $\mathscr{C}^{\otimes}$ then $v \longrightarrow v'$ (resp. $u \longrightarrow u'$) is a 2-morphism in $\mathscr{C}$ which is a symmetric monoidal transformation of lax (resp. colax) monoidal functors.

---

[41] For any object a, a' in A, F is lax if there exists a structural map $F(a) \otimes F(a') \xrightarrow{(1)} F(a \otimes a')$ satisfying coherence relations (see [Mac98, XI. 2]). Colax is defined by reversing the arrow (1).

We thus adopt the following definition:

Definition 1.3.5 A $(\mathscr{C},\mathscr{C}^{\otimes})$-*structured monoidal $\mathscr{P}$-fibred category* (resp. a $(\mathscr{C},\mathscr{C}^{\otimes})$-*structured complete monoidal $\mathscr{P}$-fibred category*) is simply a monoidal $\mathscr{P}$-fibred category (resp. a complete monoidal $\mathscr{P}$-fibred category) whose underlying 2-functor $\mathscr{M} : \mathscr{S}^{op} \longrightarrow \mathscr{C}at^{\otimes}$ factors through $\mathscr{C}^{\otimes}$. Morphisms of such objects are defined in the same way.

Note that, with the hypotheses made on $\mathscr{C}$, all the exchange natural transformations defined in the preceding paragraphs lie in $\mathscr{C}$ and satisfy the appropriate coherence property with respect to the monoidal structure.

1.3.b The abelian case

1.3.6 Let $\mathscr{A}b$ be the sub-2-category of $\mathscr{C}at$ comprising the abelian categories, with the additive functors as 1-morphisms, and the natural transformations as 2-morphisms. Obviously, it satisfies the properties of Section 1.3.1. When we apply one of the Definitions 1.3.2, 1.3.3 to the case $\mathscr{C} = \mathscr{A}b$, we will use the simple adjective *abelian* for $\mathscr{A}b$-structured. This allows us to speak of *morphisms of abelian $\mathscr{P}$-fibred categories*.

Let $\mathscr{A}b^{\otimes}$ be the sub-2-category of $\mathscr{A}b$ comprising the abelian monoidal categories, with 1-morphisms the symmetric monoidal additive functors and 2-morphisms the symmetric monoidal natural transformations. It satisfies the hypotheses of Section 1.3.4. When we apply Definition 1.3.5 to the case of $(\mathscr{A}b, \mathscr{A}b^{\otimes})$, we will use the simple expression *abelian monoidal* for $(\mathscr{A}b, \mathscr{A}b^{\otimes})$-structured monoidal. This allows us to speak of *morphisms of abelian monoidal $\mathscr{P}$-fibred categories*.

Lemma 1.3.7 *Consider an abelian $\mathscr{P}$-fibred category $\mathscr{A}$ such that for any object S of $\mathscr{S}$, $\mathscr{A}(S)$ is a Grothendieck abelian category. Then the following conditions are equivalent:*

(i) $\mathscr{A}$ is complete.
(ii) For any morphism $f : T \longrightarrow S$ in $\mathscr{S}$, f^* commutes with sums.

If in addition, $\mathscr{A}$ is monoidal, the following conditions are equivalent:

(i') $\mathscr{A}$ is monoidal complete.
(ii') (a) For any morphism $f : T \longrightarrow S$ in $\mathscr{S}$, f^* is right exact.
(b) For any object S of $\mathscr{S}$, the bifunctor $\otimes_S$ is right exact.

In view of this lemma, we adopt the following definition:

Definition 1.3.8 A Grothendieck abelian (resp. Grothendieck abelian monoidal) $\mathscr{P}$-fibred category $\mathscr{A}$ over $\mathscr{S}$ is an abelian $\mathscr{P}$-fibred category which is complete (resp. complete monoidal) and such that for any scheme S, $\mathscr{A}(S)$ is a Grothendieck abelian category.

Remark 1.3.9 Let $\mathscr{A}$ be a Grothendieck abelian monoidal $\mathscr{P}$-fibred category. Conventionally, we will denote its geometric sections by $M_S(-,\mathscr{A})$. Note that if $\mathscr{A}$ is τ-twisted, then any object of $\mathscr{A}$ is a quotient of a direct sum of objects of shape $M_S(X,\mathscr{A})\{i\}$ for a $\mathscr{P}$-object X/S and a twist $i \in \tau$.

1.3.10 Consider an abelian category $\mathscr{A}$ which admits small sums. Recall the following definition:

An object X of $\mathscr{T}$ is *finitely presented* if the functor $\mathrm{Hom}_{\mathscr{T}}(X,-)$ commutes with small filtering colimits. An essentially small family $\mathscr{G}$ of objects of $\mathscr{A}$ is called generating if for any object A of $\mathscr{A}$ there exists an epimorphism of the form:

$$\bigoplus_{i \in I} G_i \longrightarrow A$$

where $(G_i)_{i \in I}$ is a family of objects in $\mathscr{G}$.

Definition 1.3.11 Let $\mathscr{A}$ be an abelian $\mathscr{P}$-fibred category over $\mathscr{S}$.

Given a set of twists τ of $\mathscr{A}$, we say $\mathscr{A}$ is *finitely τ-presented* if for any object S of $\mathscr{S}$, for any $\mathscr{P}$-object X/S and any twist $i \in \tau$, the object $M_S(X)\{i\}$ is finitely presented and the class of such objects form an essentially small generating family of $\mathscr{A}(S)$.

1.3.c The triangulated case

1.3.12 Let $\mathscr{T}ri$ be the sub-2-category of $\mathscr{C}at$ comprising the triangulated categories, with the triangulated functors as 1-morphisms, and the triangulated natural transformations as 2-morphisms. Then $\mathscr{T}ri$ satisfies the properties of Section 1.3.1 (property (2) can be found for instance in [Ayo07a, Lemma 2.1.23], and we leave property (3) as an exercise for the reader). When we apply one of the Definitions 1.3.2, 1.3.3 to the case $\mathscr{C} = \mathscr{T}ri$, we will use the simple adjective *triangulated* for $\mathscr{T}ri$-structured. This allows us to speak of *morphisms of triangulated $\mathscr{P}$-fibred categories*.

Let $\mathscr{T}ri^{\otimes}$ be the sub-2-category of $\mathscr{T}ri$ comprising the triangulated monoidal categories, with 1-morphisms the symmetric monoidal triangulated functors and 2-morphisms the symmetric monoidal natural transformations. It satisfies the hypotheses of Section 1.3.4. When we apply Definition 1.3.5 to the case of $(\mathscr{T}ri, \mathscr{T}ri^{\otimes})$, we will use the expression *triangulated monoidal* for $(\mathscr{T}ri, \mathscr{T}ri^{\otimes})$-structured monoidal. This allows us to speak of *morphisms of triangulated monoidal $\mathscr{P}$-fibred categories*.

Convention 1.3.13 The set of twists of a triangulated monoidal $\mathscr{P}$-fibred category $\mathscr{T}$ will always be of the form $\mathbf{Z} \times \tau$, by which we mean that τ is a set of twists, while $\mathbf{Z} \times \tau$ is the closure of τ by suspension functors $[n]$, $n \in \mathbf{Z}$. We will often abuse the notation by only indicating τ. In particular, the expression $\mathscr{T}$ is τ-generated will mean conventionally that $\mathscr{T}$ is $(\mathbf{Z} \times \tau)$-generated in the sense of Definition 1.1.41.

1.3.14 Consider a triangulated category $\mathscr{T}$ which admits small sums. Recall the following definitions:

An object X of $\mathscr{T}$ is called *compact* if the functor $\mathrm{Hom}_{\mathscr{T}}(X,-)$ commutes with small sums. A class $\mathscr{G}$ of objects of $\mathscr{T}$ is called generating if the family of functors $\mathrm{Hom}_{\mathscr{T}}(X[n],-)$, $X \in \mathscr{G}$, $n \in \mathbf{Z}$, is conservative.

The triangulated category $\mathscr{T}$ is called *compactly generated* if there exists a generating set $\mathscr{G}$ of compact objects of $\mathscr{T}$. This property of being compact has been generalized by A. Neeman to the property of being α-*small* for some cardinal α (cf. [Nee01, 4.1.1]) — recall compact=$\aleph_0$-small. The property of being compactly generated has been generalized by Neeman to the property of being *well generated*; see [Kra01] for a convenient characterization of well generated triangulated categories.

Definition 1.3.15 Let $\mathscr{T}$ be a triangulated $\mathscr{P}$-fibred category over $\mathscr{S}$. We say that $\mathscr{T}$ is *compactly generated* (resp. *well generated*) if for any object S of $\mathscr{S}$, $\mathscr{T}(S)$ admits small sums and is compactly generated (resp. well generated).

Given a set of twists τ for $\mathscr{T}$, we say $\mathscr{T}$ is *compactly τ-generated* if it is compactly generated in the above sense and for any $\mathscr{P}$-object X/S, any twist $i \in \tau$, $M_S(X)\{i\}$ is compact.

1.3.16 For a triangulated category $\mathscr{T}$ which has small sums, given a family $\mathscr{G}$ of objects of $\mathscr{T}$, we denote by $\langle \mathscr{G} \rangle$ the localizing subcategory of $\mathscr{T}$ generated by $\mathscr{G}$, i.e. $\langle \mathscr{G} \rangle$ is the smallest triangulated full subcategory of $\mathscr{T}$ which is stable under small sums and which contains all the objects in $\mathscr{G}$. Recall that, in the case when $\mathscr{T}$ is well generated (e.g. if $\mathscr{T}$ compactly generated), then the family $\mathscr{G}$ generates $\mathscr{T}$ (in the sense that the family of functors $\{\mathrm{Hom}_{\mathscr{T}}(X,-)\}_{X \in \mathscr{G}}$ is conservative) if and only if $\mathscr{T} = \langle \mathscr{G} \rangle$. The following lemma is a consequence of [Nee01]:

Lemma 1.3.17 *Let $\mathscr{T}$ be a triangulated monoidal $\mathscr{P}$-fibred category over $\mathscr{S}$ with geometric sections M. Assume $\mathscr{T}$ is τ-generated.*

If $\mathscr{T}$ is well generated, then for any object S of $\mathscr{S}$,

$$\mathscr{T}(S) = \langle M_S(X)\{i\}; X/S \text{ a } \mathscr{P}\text{-object}, i \in \tau \rangle.$$

Moreover, there exists a regular cardinal α such that all the objects of shape $M_S(X)\{i\}$ are α-compact.

Note finally that the Brown representability theorem of Neeman (cf. [Nee01]) gives the following lemma (an analog of Lemma 1.3.7):

Lemma 1.3.18 *Consider a well generated triangulated $\mathscr{P}$-fibred category $\mathscr{T}$. Then the following conditions are equivalent:*

(i) $\mathscr{T}$ is complete.
(ii) For any morphism $f : T \longrightarrow S$ in $\mathscr{S}$, f^* commutes with sums.

If in addition, $\mathscr{T}$ is monoidal, the following conditions are equivalent:

(i') $\mathscr{T}$ is monoidal complete.
(ii') (a) For any morphism $f : T \longrightarrow S$ in $\mathscr{S}$, f^* is right exact.

(b) *For any object S of $\mathscr{S}$, the bifunctor $\otimes_S$ is right exact.*

We finish this section with a proposition which will constitute a useful trick:

Proposition 1.3.19 *Consider an adjunction of triangulated categories*
$$a : \mathscr{T} \rightleftarrows \mathscr{T}' : b.$$
Assume that $\mathscr{T}$ admits a set of compact generators $\mathscr{G}$ such that any object in $a(\mathscr{G})$ is compact in $\mathscr{T}'$. Then b commutes with direct sums. If in addition $\mathscr{T}'$ is well generated then b admits a right adjoint.

Proof The second assertion follows from the first one according to a corollary of the Brown representability theorem of Neeman (*cf.* [Nee01, 8.4.4]).

For the first assertion, we consider a family $(X_i)_{i \in I}$ of objects of $\mathscr{T}'$ and prove that the canonical morphism
$$\oplus_{i \in I} b(X_i) \longrightarrow b(\oplus_{i \in I} X_i)$$
is an isomorphism in $\mathscr{T}$. To prove this, it is sufficient to apply the functor $\mathrm{Hom}_{\mathscr{T}}(G, -)$ for any object G of $\mathscr{G}$. Then the result is obvious from the assumptions. □

We shall often use the following standard argument to produce equivalences of triangulated categories.

Corollary 1.3.20 *Let $a : \mathscr{T} \longrightarrow \mathscr{T}'$ be a triangulated functor between triangulated categories. Assume that the functor a preserves small sums, and that $\mathscr{T}$ admits a small set of compact generators $\mathscr{G}$, such that $a(\mathscr{G})$ form a family of compact objects in $\mathscr{T}'$. Then a is fully faithful if and only if, for any pair of objects G and G' in $\mathscr{G}$, the map*
$$\mathrm{Hom}_{\mathscr{T}}(G, G'[n]) \longrightarrow \mathrm{Hom}_{\mathscr{T}'}(a(G), a(G')[n])$$
is bijective for any integer n. If a is fully faithful, then a is an equivalence of categories if and only if $a(\mathscr{G})$ is a generating family in $\mathscr{T}'$.

Proof Let us prove that this is a sufficient condition. As $\mathscr{T}$ is in particular well generated, by the Brown representability theorem, the functor b admits a right adjoint $b : \mathscr{T}' \longrightarrow \mathscr{T}$. By virtue of the preceding proposition, the functor b preserves small sums. Let us prove that a is fully faithful. We have to check that, for any object M of $\mathscr{T}$, the map $M \longrightarrow b(a(M))$ is invertible. As a and b are triangulated and preserve small sums, it is sufficient to check this when M runs over a generating family of objects of $\mathscr{T}$ (e.g. $\mathscr{G}$). As $\mathscr{G}$ is generating, it is sufficient to prove that the map
$$\mathrm{Hom}_{\mathscr{T}}(G, M[n]) \longrightarrow \mathrm{Hom}_{\mathscr{T}'}(a(G), a(M)[n]) = \mathrm{Hom}_{\mathscr{T}'}(a(G), b(a(M))[n])$$
is bijective for any integer n, which holds by assumption. The functor a thus identifies $\mathscr{T}$ with the localizing subcategory of $\mathscr{T}'$ generated by $a(\mathscr{G})$; if moreover $a(\mathscr{G})$ is a generating family in $\mathscr{T}'$, then $\mathscr{T}' = \langle a(\mathscr{G}) \rangle$, which also proves the last assertion. □

1.3.d The model category case

1.3.21 We shall use Hovey's book [Hov99] for a general reference to the theory of model categories. Note that, following *loc. cit.*, all the model categories we shall consider will have small limits and small colimits.

Let $\mathscr{M}$ be the sub-2-category of $\mathscr{C}at$ comprising the model categories, with 1-morphisms the left Quillen functors and 2-morphisms the natural transformations. When we apply Definition 1.3.2 (resp. Definition 1.3.3) to $\mathscr{C} = \mathscr{M}$, we will speak of a $\mathscr{P}$-*fibred model category* for a $\mathscr{M}$-structured $\mathscr{P}$-fibred category $\mathscr{M}$ (resp. morphism of $\mathscr{P}$-fibred model categories). Note that according to the definition of left Quillen functors, $\mathscr{M}$ is then automatically complete.

Given a property (P) of model categories (like being cofibrantly generated, left and/or right proper, combinatorial, stable, etc), we will say that a $\mathscr{P}$-fibred model category $\mathscr{M}$ over $\mathscr{S}$ has the property (P) if, for any object S of $\mathscr{S}$, the model category $\mathscr{M}(S)$ has the property (P).

For the monoidal case, we let $\mathscr{M}^\otimes$ be the sub-2-categories of $\mathscr{M}$ comprising the symmetric monoidal model categories (see [Hov99, Definition 4.2.6]), with 1-morphisms the symmetric monoidal left Quillen functors and 2-morphisms the symmetric monoidal natural transformations, following the conditions of Section 1.3.4. When we apply Definition 1.3.5 to the case of $(\mathscr{M}, \mathscr{M}^\otimes)$, we will speak simply of a *monoidal $\mathscr{P}$-fibred model category* (resp. *morphism of monoidal $\mathscr{P}$-fibred model categories*) for a (resp. morphism of) $(\mathscr{M}, \mathscr{M}^\otimes)$-structured monoidal $\mathscr{P}$-fibred category $\mathscr{M}$. Again, $\mathscr{M}$ is then monoidal complete.

Remark 1.3.22 Let $\mathscr{M}$ be a $\mathscr{P}$-fibred model category over $\mathscr{S}$. Then for any $\mathscr{P}$-morphism $p : X \longrightarrow Y$, the inverse image functor $p^* : \mathscr{M}(Y) \longrightarrow \mathscr{M}(X)$ has very strong exactness properties: it preserves small limits and colimits (having both a left and a right adjoint), and it preserves weak equivalences, cofibrations, and fibrations. The only (completely) non-trivial assertion here concerns the preservation of weak equivalences. For this, one notices first that it preserves trivial cofibrations and trivial fibrations (being both a left Quillen functor and a right Quillen functor). In particular, by virtue of Ken Brown's Lemma [Hov99, Lemma 1.1.12], it preserves weak equivalences between cofibrant (resp. fibrant) objects. Given a weak equivalence $u : M \longrightarrow N$ in $\mathscr{M}(Y)$, we can find a commutative square

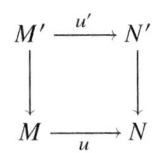

$$\begin{array}{ccc} M' & \xrightarrow{u'} & N' \\ \downarrow & & \downarrow \\ M & \xrightarrow{u} & N \end{array}$$

in which the two vertical maps are trivial fibrations, and where u' is a weak equivalence between cofibrant objects, from which we deduce easily that $p^*(u)$ is a weak equivalence in $\mathscr{M}(X)$.

1 General definitions and axioms

1.3.23 Consider a $\mathscr{P}$-fibred model category $\mathscr{M}$ over $\mathscr{S}$. By assumption, we get the following pairs of adjoint functors:

(a) For any morphism $f : X \longrightarrow S$ of $\mathscr{S}$,

$$\mathbf{L}f^* : \mathrm{Ho}(\mathscr{M}(S)) \rightleftarrows \mathrm{Ho}(\mathscr{M}(X)) : \mathbf{R}f_*.$$

(b) For any $\mathscr{P}$-morphism $p : T \longrightarrow S$, the pullback functor

$$\mathbf{L}p_\sharp : \mathrm{Ho}(\mathscr{M}(S)) \rightleftarrows \mathrm{Ho}(\mathscr{M}(T)) : \mathbf{L}p^* = p^* = \mathbf{R}p^*.$$

Moreover, the canonical isomorphism of shape $(fg)^* \simeq g^* f^*$ induces a canonical isomorphism $\mathbf{R}(fg)^* \simeq \mathbf{R}g^* \mathbf{R}f^*$. In the situation of the $\mathscr{P}$-base change formula 1.1.9, we also obtain that the base change map

$$\mathbf{L}q_\sharp \mathbf{L}g^* \longrightarrow \mathbf{L}f^* \mathbf{L}p_\sharp$$

is an isomorphism from the equivalent property of $\mathscr{M}$. Thus, we have defined a complete $\mathscr{P}$-fibred category whose fiber over S is $\mathrm{Ho}(\mathscr{M}(S))$.

Definition 1.3.24 Given a $\mathscr{P}$-fibred model category $\mathscr{M}$ as above, the complete $\mathscr{P}$-fibred category defined above will be denoted by $\mathrm{Ho}(\mathscr{M})$ and called the *homotopy $\mathscr{P}$-fibred category* associated with $\mathscr{M}$.

1.3.25 Assume that $\mathscr{M}$ is a monoidal $\mathscr{P}$-fibred model category over $\mathscr{S}$. Then, for any object S of $\mathscr{S}$, $\mathrm{Ho}(\mathscr{M})(S)$ has the structure of a symmetric closed monoidal category; see [Hov99, Theorem 4.3.2]. The (derived) tensor product of $\mathrm{Ho}(\mathscr{M})(S)$ will be denoted by $M \otimes_S^{\mathbf{L}} N$, and the (derived) internal Hom will be written $\mathbf{R}Hom_S(M, N)$, while the unit object will be written $\mathbb{1}_S$.

For any morphism $f : T \longrightarrow S$ in $\mathscr{S}$, the derived functor $\mathbf{L}f^*$ is symmetric monoidal, as follows from the equivalent property of its counterpart f^*.

Moreover, for any $\mathscr{P}$-morphism $p : T \longrightarrow S$ and for any object M in $\mathrm{Ho}(\mathscr{M})(T)$ and any object N in $\mathrm{Ho}(\mathscr{M})(S)$, the exchange map of Section 1.1.24

$$\mathbf{L}p_\sharp(M \otimes^{\mathbf{L}} p^*(N)) \longrightarrow \mathbf{L}p_\sharp(M) \otimes^{\mathbf{L}} N$$

is an isomorphism.

Definition 1.3.26 Given a monoidal $\mathscr{P}$-fibred model category $\mathscr{M}$ as above, the complete monoidal $\mathscr{P}$-fibred category defined above will be denoted by $\mathrm{Ho}(\mathscr{M})$ and called the *homotopy monoidal $\mathscr{P}$-fibred category* associated with $\mathscr{M}$.

1.4 Premotivic categories

In the present section, we will focus on a particular type of $\mathscr{P}$-fibred category.

1.4.1 Let $\mathscr{S}$ be a scheme. Assume $\mathscr{S}$ is a full subcategory of the category of $\mathscr{S}$-schemes. In most of this work, we will denote by $\mathscr{S}^{ft}$ the class of morphisms of finite type in $\mathscr{S}$ and by Sm the class of smooth morphisms of finite type in $\mathscr{S}$. There is an exception to this rule: throughout Part III, $\mathscr{S}^{ft}$ will be the class of separated morphisms of finite type in $\mathscr{S}$ and Sm will be the class of separated smooth morphisms of finite type in $\mathscr{S}$. However, the axioms which we will present in the sequel can be applied identically in each case, so that the reader can freely use the restriction that all morphisms of Sm and $\mathscr{S}^{ft}$ are separated.

In any case, the classes Sm and $\mathscr{S}^{ft}$ are admissible in $\mathscr{S}$ in the sense of Section 1.0.1 (this is automatic, for instance, if $\mathscr{S}$ is stable under pullbacks).

Definition 1.4.2 Let $\mathscr{P}$ be an admissible class of morphisms in $\mathscr{S}$.

A $\mathscr{P}$-*premotivic category over* $\mathscr{S}$ — or simply $\mathscr{P}$-*premotivic category* when $\mathscr{S}$ is clear — is a complete monoidal $\mathscr{P}$-fibred category over $\mathscr{S}$. A *morphism of $\mathscr{P}$-premotivic categories* is a morphism of complete monoidal $\mathscr{P}$-fibred categories over $\mathscr{S}$.

As a particular case, when $\mathscr{C}$ is the 2-category $\mathscr{T}ri$ of triangulated categories (resp. $\mathscr{A}b$ of abelian categories), a $\mathscr{P}$-*premotivic triangulated (resp. abelian) category over* $\mathscr{S}$ is a $(\mathscr{C}, \mathscr{C}^\otimes)$-structured complete monoidal $\mathscr{P}$-fibred category over $\mathscr{S}$ (Definition 1.3.5). Morphisms of $\mathscr{P}$-premotivic triangulated (resp. abelian) categories are defined accordingly.

We will also say *premotivic* for Sm-premotivic and *generalized premotivic* for $\mathscr{S}^{ft}$-premotivic.

The sections of a $\mathscr{P}$-premotivic category will be called *premotives*.

Example 1.4.3 Let $\mathscr{S}$ be the category of noetherian schemes of finite dimension.

For such a scheme S, recall $\mathscr{H}_\bullet(S)$ is the pointed homotopy category of Morel and Voevodsky; cf. Examples 1.1.5, 1.1.14, 1.1.30. Then, according to the fact recalled in these examples the 2-functor $\mathscr{H}_\bullet$ is a geometrically generated premotivic category (recall Definition 1.1.41).

For such a scheme S, consider the stable homotopy category $SH(S)$ of Morel and Voevodsky (see [Jar00, Ayo07b]). According to [Ayo07b], it defines a triangulated premotivic category, denoted by SH. Moreover, it is compactly $(\mathbf{Z} \times \mathbf{Z})$-generated in the sense of Definition 1.1.41 where the first factor refers to the suspension and the second one refers to the Tate twist (*i.e.* as a triangulated premotivic category, it is compactly generated by the Tate twists).

1.4.4 Let $\mathscr{T}$ be a $\mathscr{P}$-premotivic triangulated category with geometric sections M and τ be a set of twists for $\mathscr{T}$ (Definition 1.1.39).

Recall from Convention 1.3.13 (resp. and Definition 1.3.15) that $\mathscr{T}$ is said to be τ-generated (resp. compactly τ-generated) if for any scheme S, the family of isomorphism of classes of premotives of the form $M_S(X)\{i\}$ for a $\mathscr{P}$-scheme X

1 General definitions and axioms

over S and a twist $i \in \tau$ is a set of generators (resp. compact generators) for the triangulated category $\mathscr{T}(S)$ (in the respective case, we also assume $\mathscr{T}(S)$ admits small sums).

Let E be a premotive over S and X be a $\mathscr{P}$-scheme over S. For any $(n,i) \in \mathbf{Z} \times \tau$, we define the cohomology of X in degree n and twist i with coefficients in E as:

$$H^{n,i}_{\mathscr{T}}(X,E) = \operatorname{Hom}_{\mathscr{T}(S)}\left(M_S(X), E\{i\}(n)\right).$$

The fact that $\mathscr{T}$ is τ-generated amounts to saying that any such premotive E is determined by its cohomology.

Example 1.4.5 All the known triangulated premotivic categories are τ-generated for a given set of twist τ. In fact, one defines as usual the *Tate twist* $\mathbb{1}_S(1)$ in such a premotivic triangulated category $\mathscr{T}$ by the formula:

$$M_S(\mathbf{P}^1_S) = \mathbb{1} \oplus \mathbb{1}(1)[2].$$

Then $\mathbb{1}(1) = (\mathbb{1}_S(1))_{S \in \mathscr{S}}$ is a cartesian section of $\mathscr{T}$. We will say that $\mathscr{T}$ is *generated by Tate twists* if it is $\mathbf{Z}$-generated, where $\mathbf{Z}$ refers to the set of twists $(\mathbb{1}(n))_{n \in \mathbf{Z}}$.

The premotivic triangulated category SH of the previous example is compactly generated by Tate twists. Similarly, the stable $\mathbf{A}^1$-derived category $D_{\mathbf{A}^1,\Lambda}$ (cf. Example 5.3.31), the category of Voevodsky motives DM (cf. Definition 11.1.1), the category of KGL-modules (cf. Definition 13.3.3) and the category of Beilinson motives DM_B (cf. Definition 14.2.1) are all compactly generated by Tate twists.

Definition 1.4.6 Let $\mathscr{M}$ and $\mathscr{M}'$ be $\mathscr{P}$-premotivic categories.

A morphism of $\mathscr{P}$-premotivic categories (or simply a *premotivic morphism*) is a morphism $\varphi^* : \mathscr{M} \longrightarrow \mathscr{M}'$ of complete monoidal $\mathscr{P}$-fibred categories. We shall also say that

$$\varphi^* : \mathscr{M} \rightleftarrows \mathscr{M}' : \varphi_*$$

is a *premotivic adjunction*. When moreover $\mathscr{M}$ and $\mathscr{M}'$ are $\mathscr{P}$-premotivic triangulated (resp. abelian) categories, we will ask that φ^* is compatible with the triangulated (resp. additive) structure – as in Definition 1.3.3.

If we assume that $\mathscr{M}$ (resp. $\mathscr{M}'$) is τ-twisted (resp. τ'-twisted), we will say as in Definition 1.2.11 that φ^* is *compatible with twists* if for any $i \in \tau$, $\varphi^*(i)$ belongs up to isomorphism to τ'. We say φ^* is *strictly compatible with twists* if it is compatible with twists and if any element of τ' is isomorphic to the image of an element of τ.

Usually, premotivic categories come equipped with canonical twists (especially the Tate twist, see the above example) and premotivic morphisms are compatible with twists.

Example 1.4.7 With the hypotheses and notations of Example 1.4.3, we get a premotivic adjunction

$$\Sigma^\infty : \mathscr{H}_\bullet \rightleftarrows SH : \Omega^\infty$$

induced by the infinite suspension functor according to [Jaroo].

1.4.8 Let $\mathscr{T}$ (resp. $\mathscr{A}$) be a triangulated $\mathscr{P}$-premotivic category with geometric sections M and a set of twists τ. For any scheme S, we let $\mathscr{T}_{\tau,c}(S)$ be the smallest triangulated thick[42] subcategory of $\mathscr{T}(S)$ which contains premotives of shape $M_S(S)\{i\}$ (resp. $M_S(X,\mathscr{A})\{i\}$) for a $\mathscr{P}$-scheme X/S and a twist $i \in \tau$. This subcategory is stable under the operations f^*, $p_\sharp$ and $\otimes$. In particular, $\mathscr{T}_{\tau,c}$ defines a *not necessarily complete* triangulated (resp. abelian) $\mathscr{P}$-fibred category over $\mathscr{S}$. We also obtain a morphism of triangulated (resp. abelian) monoidal $\mathscr{P}$-fibred categories, fully faithful as a functor,
$$\iota : \mathscr{T}_{\tau,c} \longrightarrow \mathscr{T}.$$

Definition 1.4.9 Consider the notations introduced above. We will call $\mathscr{T}_{\tau,c}$ the *τ-constructible part of $\mathscr{T}$*. For any scheme S, the objects of $\mathscr{T}_{\tau,c}(S)$ will be called *τ-constructible*.

When τ is clear from the context, we will put $\mathscr{T}_c := \mathscr{T}_{\tau,c}$ and use the terminology *constructible*.

Remark 1.4.10 The condition of τ-constructibility is a good categorical notion of finiteness which extends the notion of *geometric motives* as introduced by Voevodsky. In the triangulated motivic case, it will be studied thoroughly in Section 4.

Proposition 1.4.11 *Let $\mathscr{T}$ be a τ-twisted $\mathscr{P}$-premotivic triangulated category. Let S be a scheme such that:*

1. *The category $\mathscr{T}(S)$ admits small sums.*
2. *For any $\mathscr{P}$-scheme X over S, and any twist $i \in \tau$, the premotive $M_S(X)\{i\}$ is compact.*

Then, a premotive M over S is τ-constructible if and only if it is compact.

Proof In any triangulated category $\mathscr{D}$, one easily obtains that the property of being compact is stable under extensions and retracts. In particular, the thick triangulated subcategory of $\mathscr{D}$ generated by compact objects consists precisely of the compact objects of $\mathscr{D}$. Moreover, if $\mathscr{D}$ admits small sums and is generated by a family of compact objects G, then the thick triangulated subcategory of $\mathscr{D}$ generated by G contains all compact objects, and is therefore equal to the full subcategory of compact objects (see [Nee92, Lem. 2.2]).

Coming back to the definition of being τ-constructible, this general fact finishes the proof. □

Thus, when the conditions of this proposition are fulfilled, the category $\mathscr{T}_{\tau,c}(S)$ does not depend on the particular choice of τ. This will often be the case in practice (see Corollary 5.1.33, Corollary 5.2.39 and Corollary 5.3.42).

Remark 1.4.12 The notion of compact objects in a triangulated category was heavily developed by A. Neeman. Its relation with finiteness conditions is particularly emphasized when considering the derived category of complexes of quasi-coherent sheaves over a quasi-compact separated scheme: in this triangulated category, being compact is equivalent to being perfect ([Nee96, Cor. 4.3]).

---

[42] *i.e.* stable under direct factors.

Definition 1.4.13 Consider a τ-generated premotivic category $\mathscr{M}$.

An *enlargement* of $\mathscr{M}$ is the data of a τ'-twisted generalized premotivic category $\underline{\mathscr{M}}$ together with a premotivic adjunction

$$\rho_\sharp : \mathscr{M} \longrightarrow \underline{\mathscr{M}} : \rho^*$$

(where $\underline{\mathscr{M}}$ is considered as a premotivic category in the obvious way), satisfying the following properties:

(a) For any scheme S in $\mathscr{S}$, the functor $\rho_{\sharp,S} : \mathscr{M}(S) \longrightarrow \underline{\mathscr{M}}(S)$ is fully faithful and its right adjoint $\rho_S^* : \underline{\mathscr{M}}(S) \longrightarrow \mathscr{M}(S)$ commutes with sums.
(b) $\rho_\sharp$ is strictly compatible with twists.

Again, this notion is defined similarly for a $\mathscr{C}$-structured $\mathscr{P}$-premotivic category.

Note that for any smooth S-scheme X, we get in the context of an enlargement as above the following identifications:

$$\rho_{\sharp,S}(M_S(X)) \simeq \underline{M}_S(X),$$
$$\rho_S^*(\underline{M}_S(X)) \simeq M_S(X),$$

where M (resp. $\underline{M}$) denote the geometric sections of $\mathscr{M}$ (resp. $\underline{\mathscr{M}}$).

Remember also that for any morphism of schemes f and any smooth morphism p, $\rho_\sharp$ commutes with f^* and $p_\sharp$, while ρ^* commutes with f_* and p^*.

2 Triangulated $\mathscr{P}$-fibred categories in algebraic geometry

2.0.1 In this entire section, we fix a base scheme $\mathscr{S}$, assumed to be noetherian, and a full subcategory $\mathscr{S}$ of the category of noetherian $\mathscr{S}$-schemes satisfying the following properties:

(a) $\mathscr{S}$ is closed under finite sums and pullback along morphisms of finite type.
(b) For any scheme S in $\mathscr{S}$, any quasi-projective S-scheme belongs to $\mathscr{S}$.

In Sections 2.2 and 2.4, we will add the following assumption on $\mathscr{S}$:

(c) Any separated morphism $f : Y \longrightarrow X$ in $\mathscr{S}$ admits a compactification in $\mathscr{S}$ in the sense of [AGV73, 3.2.5], i.e. admits a factorization of the form

$$Y \xrightarrow{j} \bar{Y} \xrightarrow{p} X,$$

where j is an open immersion, p is proper, and $\bar{Y}$ belongs to $\mathscr{S}$. Furthermore, if f is quasi-projective, then p can be chosen to be projective.
(d) Chow's lemma holds in $\mathscr{S}$ (i.e., for any proper morphism $Y \longrightarrow X$ in $\mathscr{S}$, there exists a projective birational morphism $p : Y_0 \longrightarrow Y$ in $\mathscr{S}$ such that fp is projective as well).

For future reference, a category $\mathscr{S}$ satisfying all these properties will be called *adequate*.[43]

We also fix an admissible class $\mathscr{P}$ of morphisms in $\mathscr{S}$ and a complete triangulated $\mathscr{P}$-fibred category $\mathscr{T}$. We will add the following assumptions:

(d) In Sections 2.2 and 2.3, $\mathscr{P}$ contains the open immersions.
(e) In Section 2.4, $\mathscr{P}$ contains the smooth morphisms of $\mathscr{S}$.

In the case when $\mathscr{T}$ is monoidal, we denote its geometric sections by

$$M : \mathscr{P}/? \longrightarrow \mathscr{T}.$$

According to the convention of Definition 1.4.2, we will speak of the *premotivic case* when $\mathscr{P}$ is the class of smooth morphisms of finite type[44] in $\mathscr{S}$ and $\mathscr{T}$ is a premotivic triangulated category.

2.1 Elementary properties

Definition 2.1.1 We say that $\mathscr{T}$ is *additive* if for any finite family $(S_i)_{i \in \mathscr{I}}$ of schemes in $\mathscr{S}$, the canonical map

$$\mathscr{T}\left(\bigsqcup_i S_i\right) \longrightarrow \prod_i \mathscr{T}(S_i)$$

is an equivalence.

Recall this property implies in particular that $\mathscr{T}(\varnothing) = 0$.

Lemma 2.1.2 *Let S be a scheme and $p : \mathbf{A}_S^1 \longrightarrow S$ be the canonical projection. The following conditions are equivalent:*

(i) *The functor $p^* : \mathscr{T}(S) \longrightarrow \mathscr{T}(\mathbf{A}_S^1)$ is fully faithful.*
(ii) *The counit adjunction morphism $1 \longrightarrow p_*p^*$ is an isomorphism.*

In the premotivic case, these conditions are equivalent to the following:

(iii) *The unit adjunction morphism $p_\sharp p^* \longrightarrow 1$ is an isomorphism.*

(iv) *The morphism $M_S(\mathbf{A}_S^1) \xrightarrow{p_*} \mathbb{1}_S$ induced by p is an isomorphism.*

(iv') *For any smooth S-scheme X, the morphism $M_S(\mathbf{A}_X^1) \xrightarrow{(1_X \times p)_*} M_S(X)$ is an isomorphism.*

---

[43] For instance, the scheme $\mathscr{S}$ can be the spectrum of a prime field or of a Dedekind domain. The category $\mathscr{S}$ might be the category of all noetherian $\mathscr{S}$-schemes of finite dimension or simply the category of quasi-projective $\mathscr{S}$-schemes. In all these cases, property (c) is ensured by Nagata's theorem (see [Con07]) and property (d) by Chow's lemma (see [GD61, 5.6.1]).
[44] or smooth separated morphisms of finite type when applying this section in Part III.

The only thing to recall is that in the premotivic case, $p_\sharp p^*(M) = M_S(\mathbf{A}_S^1) \otimes M$ and $p_* p^*(M) = Hom_S(M_S(\mathbf{A}_S^1), M)$.

Definition 2.1.3 The equivalent conditions of the previous lemma will be called the *homotopy property* for $\mathscr{T}$, denoted by (Htp).

2.1.4 Recall that a *sieve* R of a scheme X is a class of morphisms in $\mathscr{S}/X$ which is stable under composition on the right by any morphism of schemes (see [AGV73, I.4]).

Given such a sieve R, we will say that $\mathscr{T}$ is *R-separated* if the class of functors f^* for $f \in R$ is conservative. Given two sieves R, R' of X, the following properties are immediate:

(a) If $R \subset R'$ then $\mathscr{T}$ is R-separated implies $\mathscr{T}$ is R'-separated.
(b) If $\mathscr{T}$ is both R-separated and R'-separated then $\mathscr{T}$ is $(R \cup R')$-separated.

A family of morphisms $(f_i : X_i \longrightarrow X)_{i \in I}$ of schemes defines a sieve $R = \langle f_i, i \in I \rangle$ such that f is in R if and only if there exists an $i \in I$ such that f can be factored through f_i. Obviously,

(c) $\mathscr{T}$ is R-separated if and only if the family of functors $(f_i^*)_{i \in I}$ is conservative.

Recall that a topology on $\mathscr{S}$ is the data for any scheme X of a set of sieves of X satisfying certain stability conditions (*cf.* [AGV73, II, 1.1]), called *t-covering sieves*. A pre-topology t_0 on $\mathscr{S}$ is the data for any scheme X of a set of families of morphisms of shape $(f_i : X_i \longrightarrow X)_{i \in I}$ satisfying certain stability conditions (*cf.* [AGV73, II, 1.3]), called t_0-covers. A pre-topology t_0 generates a unique topology t.

Definition 2.1.5 Let t be a Grothendieck topology on $\mathscr{S}$. We say that $\mathscr{T}$ is *t-separated* if the following property holds:

(t-sep) For any t-covering sieve R, $\mathscr{T}$ is R-separated in the sense defined above.

Obviously, given two topologies t and t' on $\mathscr{S}$ such that t' is finer than t, if $\mathscr{T}$ is t-separated then it is t'-separated.

If the topology t on $\mathscr{S}$ is generated by a pre-topology t_0 then $\mathscr{T}$ is t-separated if and only if for any t_0-covers $(f_i)_{i \in I}$, the family of functors $(f_i^*)_{i \in I}$ is conservative – use [AGV73, 1.4] and Section 2.1.4(a)+(c).

2.1.6 Recall that a morphism of schemes $f : T \longrightarrow S$ is *radicial* if it is injective and for any point t of T, the residual extension induced by f at t is radicial (*cf.* [GD60, 3.5.4, 3.5.8]).[45] The following definition is inspired by [Ayo07a, Def. 2.1.160].

Definition 2.1.7 We say that $\mathscr{T}$ is *separated* (resp. *semi-separated*) if $\mathscr{T}$ is separated for the topology generated by surjective families of morphisms of finite type (resp. finite radicial morphisms) in $\mathscr{S}$. We also denote this property by (Sep) (resp. (sSep)).

---

[45] This is equivalent to asking that f is universally injective. When f is surjective, this is equivalent to asking that f is a universal homeomorphism.

Remark 2.1.8 If $\mathscr{T}$ is additive, property (Sep) (resp. (sSep)) is equivalent to asking that for any surjective morphism of finite type (resp. finite surjective radicial morphism) $f : T \longrightarrow S$ in $\mathscr{S}$, the functor f^* is conservative.

Proposition 2.1.9 *Assume $\mathscr{T}$ is semi-separated and satisfies the transversality property with respect to finite surjective radicial morphisms.*
Then for any finite surjective radicial morphism $f : Y \longrightarrow X$, the functor

$$f^* : \mathscr{T}(X) \longrightarrow \mathscr{T}(Y)$$

is an equivalence of categories.

Proof We first consider the case when $f = i$ is in addition a closed immersion. In this case, we can consider the pullback square below.

$$\begin{array}{ccc} Y & = & Y \\ \| & & \downarrow i \\ Y & \xrightarrow{i} & Z \end{array}$$

Using the transversality property with respect to i, we see that the counit $i^* i_* \longrightarrow 1$ is an isomorphism. It thus remains to prove that the unit map $1 \longrightarrow i_* i^*$ is an isomorphism. As i^* is conservative by semi-separability, it is sufficient to check that

$$i^* \longrightarrow i^* i_* i^*(M)$$

is an isomorphism. But this is a section of the map $i^* i_* i^*(M) \longrightarrow i^*(M)$, which is already known to be an isomorphism.

Consider now the general case of a finite radicial extension f. We introduce the pullback square

$$\begin{array}{ccc} Y \times_X Y & \xrightarrow{p} & Y \\ q \downarrow & & \downarrow f \\ Y & \xrightarrow{f} & X. \end{array}$$

Consider the diagonal immersion $i : Y \longrightarrow Y \times_X Y$. Because Y is noetherian and p is separable, i is finite (cf. [GD61, 6.1.5]) thus a closed immersion. As p is a universal homeomorphism, the same is true for its section i. The preceding case thus implies that i^* is an equivalence of categories. Moreover, as $pi = qi = 1_Y$, we see that p^* and q^* are both quasi-inverses to i^*, which implies that they are isomorphic equivalences of categories. More precisely, we get canonical isomorphisms of functors

$$i^* \simeq p_* \simeq q_* \quad \text{and} \quad i_* \simeq p^* \simeq q^*.$$

We check that the unit map $1 \longrightarrow f_* f^*$ is an isomorphism. Indeed, by semi-separability, it is sufficient to prove this after applying the functor f^*, and we get, using the transversality property for f:

2 Triangulated $\mathscr{P}$-fibred categories in algebraic geometry

$$f^* \simeq i^*p^*f^* \simeq q_*p^*f^* \simeq f^*f_*f^*.$$

We then check that the counit map $f^*f_* \longrightarrow 1$ is an isomorphism as well. In fact, using again the transversality property for f, we have isomorphisms

$$f^*f_*(M) \simeq q_*p^*(M) \simeq i^*i_*(M) \simeq M.$$

2.1.10 Recall from [Voe10b] that a *cd-structure* on $\mathscr{S}$ is a collection P of commutative squares of schemes

$$\begin{array}{ccc} B & \longrightarrow & Y \\ \downarrow & Q & \downarrow f \\ A & \xrightarrow{e} & X \end{array}$$

which is closed under isomorphisms. We will say that a square Q in P is P-*distinguished*.

Voevodsky associates to P a topology t_P, the smallest topology such that:

- for any P-distinguished square Q as above, the sieve generated by $\{f : A \longrightarrow X, e : Y \longrightarrow X\}$ is t_P-covering on X,
- the empty sieve covers the empty scheme.

Example 2.1.11 A *Nisnevich distinguished square* is a square Q as above such that Q is cartesian, f is étale, e is an open embedding with reduced complement Z and the induced map $f^{-1}(Z) \longrightarrow Z$ is an isomorphism. The corresponding cd-structure is called the *upper cd-structure* (see Section 2 of [Voe10c]). Because we work with noetherian schemes, the corresponding topology is the *Nisnevich topology* (see Proposition 2.16 of *loc. cit.*).

A *proper cdh-distinguished square* is a square Q as above such that Q is Cartesian, f is proper, e is a closed embedding with open complement U and the induced map $f^{-1}(U) \longrightarrow U$ is an isomorphism. The corresponding cd-structure is called the *lower cd-structure*. The topology associated with the lower cd-structure is called the *proper cdh-topology*.

The topology generated by the lower and upper cd-structures is by definition (according to the preceding remark on Nisnevich topology) the *cdh-topology*.

All three of these examples are complete cd-structures in the sense of [Voe10b, 2.3].

Lemma 2.1.12 *Let P be a complete cd-structure (see [Voe10b, Def. 2.3]) on $\mathscr{S}$ and t_P be the associated topology. The following conditions are equivalent:*

(i) $\mathscr{T}$ is t_P-separated.
(ii) For any distinguished square Q for P of the above form, the pair of functors (e^*, f^*) is conservative.

Proof This follows from the definition of a complete cd-structure and Section 2.1.4(a). □

Remark 2.1.13 If we assume that $\mathscr{S}$ is stable under arbitrary pullback then any cd-structure P on $\mathscr{S}$ such that P-distinguished squares are stable under pullback is complete (see [Voe10b, 2.4]).

2.2 Exceptional functors, following Deligne

2.2.a The support axiom

2.2.1 Consider an open immersion $j : U \longrightarrow S$. Applying Section 1.1.15 to the cartesian square

$$\begin{array}{ccc} U & =\!=\!= & U \\ \| & & \downarrow j \\ U & \xrightarrow{j} & S \end{array}$$

we get a canonical natural transformation

$$\gamma_j : j_\sharp = j_\sharp 1_* \xrightarrow{Ex(\Delta_{\sharp*})} j_* 1_\sharp = j_*.$$

Recall that the functors $j_\sharp$ and j_* are fully faithful (see Corollary 1.1.20).

Note that according to Remark 1.1.7, this natural transformation is compatible with identifications of the kind $(jk)_\sharp = j_\sharp k_\sharp$ and $(jk)_* = j_* k_*$.

Lemma 2.2.2 *Let S be a scheme and U and V be subschemes such that $S = U \sqcup V$. We let $h : U \longrightarrow S$ (resp. $k : V \longrightarrow S$) be the canonical open immersions.*

Assume that the functor $(h^*, k^*) : \mathscr{T}(S) \longrightarrow \mathscr{T}(U) \times \mathscr{T}(V)$ is conservative and that $\mathscr{T}(\varnothing) = 0$. Then the natural transformation γ_h (resp. γ_k) is an isomorphism. Moreover, the functor (h^*, k^*) is then an equivalence of categories.

Proof As $h_\sharp$ and h_* are fully faithful, we have $h^* h_\sharp \simeq h^* h_*$. By $\mathscr{P}$-base change, we also get $k^* h_\sharp \simeq k^* h_* \simeq 0$. It remains to prove the last assertion. The functor $R = (h^*, k^*)$ has a left adjoint L defined by $L = h_\sharp \oplus k_\sharp$:

$$L(M, N) = h_\sharp(M) \oplus k_\sharp(N).$$

The natural transformation $LR \longrightarrow 1$ is an isomorphism: to see this, is it sufficient to evaluate at h^* and k^*, which gives an isomorphism in $\mathscr{T}(U)$ and $\mathscr{T}(V)$, respectively. The natural transformation $1 \longrightarrow RL$ is also an isomorphism because $h_\sharp$ and $k_\sharp$ are fully faithful. $\square$

Remark 2.2.3 Assume $\mathscr{T}$ is Zariski separated (Definition 2.1.5). Then, as a corollary of this lemma, $\mathscr{T}$ is additive (Definition 2.1.1) if and only if $\mathscr{T}(\varnothing) = 0$.

2.2.4 *Exchange structures V.–* Assume $\mathscr{T}$ is additive. We consider a commutative square of schemes

(2.2.4.1)
$$\begin{array}{ccc} V & \xrightarrow{k} & T \\ q \downarrow & \Delta & \downarrow p \\ U & \xrightarrow{j} & S \end{array}$$

such that j, k are open immersions and p, q are proper morphisms.

This diagram can be factored into the following commutative diagram:

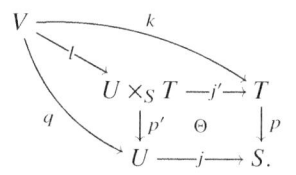

$$\begin{array}{ccc} V & \xrightarrow{k} & \\ {\scriptstyle l}\searrow & & \\ & U \times_S T \xrightarrow{j'} T & \\ q\searrow & \downarrow p' \quad \Theta \quad \downarrow p & \\ & U \xrightarrow{j} S. & \end{array}$$

Then l is an open and closed immersion so that the previous lemma implies the canonical morphism $\gamma_l : l_\sharp \longrightarrow l_*$ is an isomorphism. As a consequence, we get a natural exchange transformation

$$Ex(\Delta_{\sharp *}) : j_\sharp q_* = j_\sharp p'_* l_* \xrightarrow{Ex(\Theta_{\sharp *})} p_* j'_\sharp l_* \xrightarrow{\gamma_l^{-1}} p_* j'_\sharp l_\sharp = p_* k_\sharp$$

using the exchange of Section 1.1.15. Note that, with the notations introduced in Section 2.2.1, the following diagram is commutative.

(2.2.4.2)
$$\begin{array}{ccc} j_\sharp q_* & \xrightarrow{Ex(\Delta_{\sharp *})} & p_* k_\sharp \\ {\scriptstyle \gamma_j q_*}\downarrow & & \downarrow {\scriptstyle p_* \gamma_k} \\ j_* q_* & \xrightarrow{\sim} (jq)_* = (pk)_* \xleftarrow{\sim} & p_* k_*. \end{array}$$

Indeed one sees first that it is sufficient to treat the case where Δ is cartesian. Then, as $j_\sharp$ is a fully faithful left adjoint to j^* it is sufficient to check that (2.2.4.2) commutes after having applied j^*. Using the cotransversality property with respect to open immersions, one sees then that this consists of verifying the commutativity of (2.2.4.2) when j is the identity, in which case it is trivial.

Definition 2.2.5 Let $p : T \longrightarrow S$ be a proper morphism in $\mathscr{S}$.

We say that the triangulated $\mathscr{P}$-fibred category $\mathscr{T}$ satisfies the *support property* with respect to p, denoted by (Supp$_p$), if it is additive and for any commutative square of shape (2.2.4.1) the exchange transformation $Ex(\Delta_{\sharp *}) : j_\sharp q_* \longrightarrow p_* k_\sharp$ defined above is an isomorphism.

We say that $\mathscr{T}$ satisfies the *support property*, also denoted by (Supp), if it satisfies (Supp$_p$) for all proper morphisms p in $\mathscr{S}$.

By definition, it is sufficient to check the last property of property (Supp) in the case where Δ is cartesian.

2.2.b Exceptional direct image

2.2.6 We denote by $\mathscr{S}^{sep}$ (resp. $\mathscr{S}^{open}$, $\mathscr{S}^{prop}$) the sub-category of the category $\mathscr{S}$ with the same objects but where the morphisms are separated morphisms of finite type (resp. open immersions, proper morphisms). We denote by

$$\mathscr{T}_* : \mathscr{S} \longrightarrow \mathscr{T}ri^{\otimes}$$
$$\text{resp. } \mathscr{T}_\sharp : \mathscr{S}^{open} \longrightarrow \mathscr{T}ri^{\otimes}$$

the 2-functor defined respectively by morphisms of type f_* and $j_\sharp$ (f any morphism of schemes). The proposition below is essentially based on a result of Deligne [AGV73, XVII, 3.3.2]:

Proposition 2.2.7 *Assume $\mathscr{T}$ is a monoidal $\mathscr{P}$-fibred category and satisfies property (Supp).*

Then there exists a unique 2-functor

$$\mathscr{T}_! : \mathscr{S}^{sep} \longrightarrow \mathscr{T}ri^{\otimes}$$

with the property that

$$\mathscr{T}_!|_{\mathscr{S}^{prop}} = \mathscr{T}_*|_{\mathscr{S}^{prop}}, \quad \mathscr{T}_!|_{\mathscr{S}^{open}} = \mathscr{T}_\sharp$$

and for any commutative square Δ of shape (2.2.4.1) with p and q proper, the composition of the structural isomorphisms

$$j_\sharp q_* = j_! q_! \simeq (jq)_! = (pk)_! \simeq p_! k_! = p_* k_\sharp$$

is equal to the exchange transformation $Ex(\Delta_{\sharp *})$.

2.2.8 Under the assumptions of the proposition, for any separated morphism of finite type $f : Y \longrightarrow X$, we will denote by $f_! : \mathscr{T}(Y) \longrightarrow \mathscr{T}(X)$ the functor $\mathscr{T}_!(f)$. The functor $f_!$ is called the *direct image functor with compact support* or the *left exceptional functor* associated with f.

Proof We recall the principle of Deligne's proof. Let $f : Y \longrightarrow X$ be a separated morphism of finite type in $\mathscr{S}$.

Let $\mathscr{C}_f$ be the category of compactifications of f in $\mathscr{S}$, i.e. of factorizations of f of the form

(2.2.8.1) $$Y \xrightarrow{j} \bar{Y} \xrightarrow{p} X,$$

where j is an open immersion, p is proper, and $\bar{Y}$ belongs to $\mathscr{S}$. Morphisms of $\mathscr{C}_f$ are given by commutative diagrams of the form

2 Triangulated $\mathscr{P}$-fibred categories in algebraic geometry

(2.2.8.2)
$$Y \xleftarrow{j'} \bar{Y}' \xrightarrow{p'} X$$
$$\downarrow \pi$$
$$Y \xleftarrow{j} \bar{Y} \xrightarrow{p} X$$

in $\mathscr{S}$. To any compactification of f of shape (2.2.8.1), we associate the functor $p_* j_\sharp$. To any morphism of compactifications (2.2.8.2), we associate a natural isomorphism

$$p'_* j'_\sharp = p_* \pi_* j'_\sharp \xrightarrow{Ex(\Delta_{\sharp *})^{-1}} p_* j_\sharp 1_* = p_* j_\sharp,$$

where Δ stands for the commutative square formed by removing π in the diagram (2.2.8.2), and $Ex(\Delta_{\sharp *})$ is the corresponding natural transformation (see Section 2.2.4). The compatibility of $Ex(\Delta_{\sharp *})$ with composition of morphisms of schemes shows that we have defined a functor

$$\Gamma_f : \mathscr{C}_f^{op} \longrightarrow Hom(\mathscr{T}(Y), \mathscr{T}(X))$$

which sends all the maps of $\mathscr{C}_f$ to isomorphisms (by the support property).

The category $\mathscr{C}_f$ is non-empty by the assumption 2.0.1(c) on $\mathscr{S}$, and it is in fact left filtering; see [AGV73, XVII, 3.2.6(ii)]. This defines a canonical functor $f_! : \mathscr{T}(Y) \longrightarrow \mathscr{T}(X)$, independent of any choice compactification of f, defined in the category of functors $Hom(\mathscr{T}(Y), \mathscr{T}(X))$ by the formula

$$f_! = \varinjlim_{\mathscr{C}_f^{op}} \Gamma_f .$$

If $f = p$ is proper, then the compactification

$$Y \xrightarrow{=} Y \xrightarrow{p} X$$

is an initial object of $\mathscr{C}_f$, which gives a canonical identification $p_! = p_*$. Similarly, if $f = j$ is an open immersion, then the compactification

$$Y \xrightarrow{j} X \xrightarrow{=} X$$

is a terminal object of $\mathscr{C}_j$, so that we get a canonical identification $j_! = j_\sharp$.

This construction is compatible with composition of morphisms. Let $g : Z \longrightarrow Y$ and $f : Y \longrightarrow X$ be two separated morphisms of finite type in $\mathscr{S}$. For any two compactifications

$$Z \xrightarrow{k} \bar{Z} \xrightarrow{q} Y \text{ and } Y \xrightarrow{j} \bar{Y} \xrightarrow{p} X$$

of f and g respectively, we can choose a compactification

$$\bar{Z} \xrightarrow{h} T \xrightarrow{r} \bar{Y}$$

of jq, and we get a canonical isomorphism

$$f_! \, g_! \simeq p_* \, j_\sharp \, q_* \, k_\sharp \simeq p_* \, r_* \, h_\sharp \, k_\sharp \simeq (pr)_* \, (hk)_\sharp \simeq (fg)_! \, .$$

The independence of these isomorphisms with respect to the choices of compactification follows from [AGV73, XVII, 3.2.6(iii)]. The cocycle conditions (i.e. the associativity) also follows formally from [AGV73, XVII, 3.2.6]. The uniqueness statement is obvious. □

2.2.9 This construction is functorial in the following sense.

Define a *2-functor with support on* $\mathscr{T}$ to be a triple $(\mathscr{D}, a, b)$, where:

(i) $\mathscr{D} : \mathscr{S}^{sep} \longrightarrow \mathscr{T}ri$ is a 2-functor (we shall write the structural coherence isomorphisms as $c_{g,f} : \mathscr{D}(gf) \xrightarrow{\sim} \mathscr{D}(g)\mathscr{D}(f)$ for composable arrows f and g in $\mathscr{S}^{sep}$);

(ii) $a : \mathscr{T}_*|_{\mathscr{S}^{prop}} \longrightarrow \mathscr{D}|_{\mathscr{S}^{prop}}$ and $b : \mathscr{T}_\sharp \longrightarrow \mathscr{D}|_{\mathscr{S}^{open}}$ are morphisms of 2-functors which agree on objects, i.e. such that for any scheme S in $\mathscr{S}$, we have

$$\psi_S = a_S = b_S : \mathscr{T}(S) \longrightarrow \mathscr{D}(S);$$

(iii) for any commutative square of shape (2.2.4.1) in which j and k are open immersions, while p and q are proper morphisms, the diagram below commutes:

$$\begin{array}{ccc}
\psi_S \, j_\sharp q_* & \xrightarrow{\psi_S \, Ex(\Delta_{\sharp*})} & \psi_S \, p_* k_\sharp \\
{\scriptstyle bq_*} \downarrow & & \downarrow {\scriptstyle ak_\sharp} \\
\mathscr{D}(j)\psi_U q_* & & \mathscr{D}(p)\psi_T k_\sharp \\
{\scriptstyle \mathscr{D}(j)a} \downarrow & & \downarrow {\scriptstyle \mathscr{D}(p)b} \\
\mathscr{D}(j)\mathscr{D}(q)\psi_V & \xrightarrow{c_{j,q}^{-1}} \mathscr{D}(jq) = \mathscr{D}(pk)\psi_V \xleftarrow{c_{p,k}^{-1}} & \mathscr{D}(p)\mathscr{D}(k)\psi_V \, .
\end{array}$$

Morphisms of 2-functors with support on $\mathscr{T}$

$$(\mathscr{D}, a, b) \longrightarrow (\mathscr{D}', a', b')$$

are defined in the obvious way: these are morphisms of 2-functors $\mathscr{D} \longrightarrow \mathscr{D}'$ which preserve all the structure on the nose.

Using the arguments of the proof of Proposition 2.2.7, one easily checks that the category of 2-functors with support has an initial object, which is nothing else but the 2-functor $\mathscr{T}_!$ together with the identities of $\mathscr{T}_*|_{\mathscr{S}^{prop}}$ and of $\mathscr{T}_\sharp$ respectively. In particular, for any 2-functor $\mathscr{D} : \mathscr{S}^{sep} \longrightarrow \mathscr{T}ri$, a morphism of 2-functors $\mathscr{T}_! \longrightarrow \mathscr{D}$ is completely determined by its restrictions to $\mathscr{S}^{prop}$ and $\mathscr{S}^{open}$, and by its compatibility with the exchange isomorphisms of type $Ex(\Delta_{\sharp*})$ in the sense described in condition (iii) above.

Proposition 2.2.10 *Assume that $\mathscr{T}$ satisfies the support property and consider the notations of Proposition 2.2.7. For any separated morphism of finite type f in $\mathscr{S}$, there exists a canonical natural transformation*

$$\alpha_f : f_! \longrightarrow f_*.$$

The collection of maps α_f defines a morphism of 2-functors

$$\alpha : \mathscr{T}_! \longrightarrow \mathscr{T}_*|_{\mathscr{S}^{sep}}, \quad f \mapsto (\alpha_f : f_! \longrightarrow f_*)$$

whose restrictions to $\mathscr{S}^{prop}$ and $\mathscr{S}^{open}$ are respectively the identity and the morphism of 2-functors $\gamma : \mathscr{T}_\sharp \longrightarrow \mathscr{T}_*|_{\mathscr{S}^{open}}$ defined in Section 2.2.1.

Proof The identities $f_* = f_*$ for f proper (resp. projective) and the exchange natural transformations of type $Ex(\Delta_{\sharp*})$ turns $\mathscr{T}_*|_{\mathscr{S}^{sep}}$ into a 2-functor with support (resp. restricted support) on $\mathscr{T}$ (property (iii) of Section 2.2.9 is expressed by the commutative square (2.2.4.2)). □

Proposition 2.2.11 *Let $\mathscr{T}'$ be another triangulated complete $\mathscr{P}$-fibred category over $\mathscr{S}$. Assume that $\mathscr{T}$ and $\mathscr{T}'$ both have the support property, and consider given a triangulated morphism of $\mathscr{P}$-fibred categories $\varphi^* : \mathscr{T} \longrightarrow \mathscr{T}'$ (recall Definition 1.2.2).*

Then, there is a canonical family of natural transformations

$$Ex(\varphi^*, f_!) : \varphi_X^* f_! \longrightarrow f_! \varphi_Y^*$$

for each separated morphism of finite type $f : Y \longrightarrow X$ in $\mathscr{S}$, which is functorial with respect to composition in $\mathscr{S}$ (i.e. defines a morphism of 2-functors) and such that the following conditions are satisfied:

(a) if f is proper, then, under the identification $f_! = f_*$, the map $Ex(\varphi^*, f_!)$ is the exchange transformation $Ex(\varphi^*, f_*) : \varphi_X^* f_* \longrightarrow f_* \varphi_Y^*$ defined in Section 1.2.5;

(b) if f is an open immersion, then, under the identification $f_! = f_\sharp$, the map $Ex(\varphi^*, f_!)$ is the inverse of the exchange isomorphism $Ex(f_\sharp, \varphi^*) : f_\sharp \varphi_Y^* \longrightarrow \varphi_X^* f_\sharp$ defined in Section 1.2.1.

Proof The exchange maps of type $Ex(\varphi^*, f_*)$ define a morphism of 2-functors

$$a : \mathscr{T}_*|_{\mathscr{S}^{prop}} \longrightarrow \mathscr{T}'_*|_{\mathscr{S}^{prop}} = \mathscr{T}'_!|_{\mathscr{S}^{prop}}$$

while the inverse of the exchange isomorphisms of type $Ex(f_\sharp, \varphi^*)$ define a morphism of 2-functors

$$b : \mathscr{T}_\sharp \longrightarrow \mathscr{T}'_\sharp = \mathscr{T}'_!|_{\mathscr{S}^{open}}$$

in such a way that the triple $(\mathscr{T}'_!, a, b)$ is a 2-functor with support on $\mathscr{T}$. □

Corollary 2.2.12 *Suppose $\mathscr{T}$ satisfies the support property and consider the notations of Proposition 2.2.7.*

1. For any cartesian square

$$\begin{array}{ccc} Y' & \xrightarrow{f'} & X' \\ g' \downarrow & \Delta & \downarrow g \\ Y & \xrightarrow{f} & X \end{array}$$

such that f is separated of finite type, there exists a canonical natural transformation

$$Ex(\Delta_!^*) : g^* f_! \longrightarrow f'_! g'^*$$

compatible with horizontal and vertical compositions of squares, and satisfying the following identifications in $\mathscr{T}(X')$

(a) f proper:

$$\begin{array}{ccc} g^* f_! & \xrightarrow{Ex(\Delta_!^*)} & f'_! g'^* \\ \| & & \| \\ g^* f_* & \xrightarrow{Ex(\Delta_*^*)} & f'_* g'^*, \end{array}$$

(b) f open immersion:

$$\begin{array}{ccc} g^* f_! & \xrightarrow{Ex(\Delta_!^*)} & f_! g'^* \\ \| & & \| \\ g^* f_\sharp & \xrightarrow{Ex(\Delta_\sharp^*)^{-1}} & f'_\sharp g'^*. \end{array}$$

Moreover, when g is a $\mathscr{P}$-morphism, $Ex(\Delta_!^*)$ is an isomorphism.

2. For any cartesian square Δ as in (1), assuming f is separated of finite type and g is a $\mathscr{P}$-morphism, there exists a canonical natural transformation

$$Ex(\Delta_{\sharp !}) : g_\sharp f'_! \longrightarrow f_! g'_\sharp$$

compatible with horizontal and vertical compositions of squares, and satisfying the following identifications in $\mathscr{T}(X')$

(a) f proper:

$$\begin{array}{ccc} g_\sharp f'_! & \xrightarrow{Ex(\Delta_{\sharp !})} & f_! g'_\sharp \\ \| & & \| \\ g_\sharp f'^* & \xrightarrow{Ex(\Delta_{\sharp *})} & f_* g'_\sharp, \end{array}$$

(b) f open immersion:

$$\begin{array}{ccc} g_\sharp f'_! & \xrightarrow{Ex(\Delta_{\sharp !})} & f_! g'_\sharp \\ \| & & \| \\ g_\sharp f'_\sharp & =\!=\!=\!= & f_\sharp g'_\sharp. \end{array}$$

3. If furthermore $\mathscr{T}$ is monoidal then for any separated morphism of finite type $f : Y \longrightarrow X$, there is a natural transformation

$$Ex(f_!^*, \otimes) : (f_! K) \otimes L \longrightarrow f_!(K \otimes f^* L)$$

which is compatible with respect to composition in $\mathscr{S}$, and such that, in each of the following cases, we have the following identifications:

(a) f proper:

$$\begin{array}{ccc} (f_! K) \otimes L & \xrightarrow{Ex(f_!^*, \otimes)} & f_!(K \otimes f^* L) \\ \| & & \| \\ (f_* K) \otimes L & \xrightarrow{Ex(f_*^*, \otimes)} & f_*(K \otimes f^* L), \end{array}$$

(b) f open immersion:

$$\begin{array}{ccc} (f_! K) \otimes L & \xrightarrow{Ex(f_!^*, \otimes)} & f_!(K \otimes f^* L) \\ \| & & \| \\ (f_\sharp K) \otimes L & \xrightarrow{Ex(f_\sharp^*, \otimes)^{-1}} & f_\sharp(K \otimes f^* L). \end{array}$$

As in the previous analogous cases, the natural transformations $Ex(\Delta_!^*)$, $Ex(\Delta_{\#,!})$ and $Ex(f_!^*, \otimes)$ will be called *exchange transformations*.

Proof To prove (1), consider a fixed map $g : X' \longrightarrow X$ in $\mathscr{S}$. We consider the triangulated $\mathscr{P}/X$-fibred categories $\mathscr{T}'$ and $\mathscr{T}''$ over $\mathscr{S}/X$ defined by $\mathscr{T}'(Y) = \mathscr{T}(Y)$ and $\mathscr{T}''(Y) = \mathscr{T}(Y')$ for any X-scheme Y (in $\mathscr{S}$), with $g' : Y' = Y \times_X X' \longrightarrow Y$ the map obtained from $Y \longrightarrow X$ by pullback along g. The collection of functors

$$g'^* : \mathscr{T}(Y) \longrightarrow \mathscr{T}(Y')$$

define an exact morphism of triangulated $\mathscr{P}/X$-fibred categories over $\mathscr{S}/X$ (by the $\mathscr{P}$-base change formula):

$$\varphi^* : \mathscr{T}' \longrightarrow \mathscr{T}''.$$

Applying the preceding proposition to the latter gives (1). The fact that we get an isomorphism whenever g is a $\mathscr{P}$-morphism follows from the $\mathscr{P}$-base change formula and from Section 1.1.15.

For point (2), we consider the notations above assuming that g is a $\mathscr{P}$-morphism. The collection of functors

$$g'_\sharp : \mathscr{T}(Y') \longrightarrow \mathscr{T}(Y)$$

associated with an X-scheme Y, $g' : Y' = Y \times_X X' \longrightarrow Y$ obtained from g as above, define an exact morphism of triangulated $\mathscr{P}/X$-fibred categories over $\mathscr{S}/X$ (applying again the $\mathscr{P}$-base change formula):

$$\varphi^* : \mathscr{T}'' \longrightarrow \mathscr{T}'.$$

Applying the preceding proposition to the latter gives (2).

The proof of (3) is similar: fix a scheme X in $\mathscr{S}$, as well as an object L in $\mathscr{T}(X)$. Let $\mathscr{T}'$ be the restriction of $\mathscr{T}$ to $\mathscr{S}/X$ as above. We can consider L as a cartesian section of $\mathscr{T}'$, and by the $\mathscr{P}$-projection formula, we then have an exact morphism of triangulated $\mathscr{P}/X$-fibred categories over $\mathscr{S}/X$:

$$L \otimes (-) : \mathscr{T}' \longrightarrow \mathscr{T}'.$$

Here again, we can apply the preceding proposition and conclude the result.

2.2.c Further properties

We will be particularly interested in the following properties of the triangulated $\mathscr{P}$-fibred category $\mathscr{T}$.

Definition 2.2.13 Let $f : Y \longrightarrow X$ be a morphism in $\mathscr{S}$. We introduce the following properties for $\mathscr{T}$, assuming in the third case that $\mathscr{T}$ is monoidal:

(Adj$_f$) The functor f_* admits a right adjoint. Under this assumption, we denote by $f^!$ the right adjoint of f_*.

(BC$_f$) Any cartesian square of $\mathscr{S}$ of the form

$$\begin{array}{ccc} Y' & \xrightarrow{f'} & X' \\ g' \downarrow & \Delta & \downarrow g \\ Y & \xrightarrow{f} & X, \end{array}$$

is $\mathscr{T}$-transversal (Definition 1.1.17) – *i.e.* the exchange transformation

$$Ex(\Delta_*^*) : g^* f_* \longrightarrow f'_* g'^*$$

associated with Δ is an isomorphism.

(PF$_f$) For any object premotive M over Y, and N over X, the exchange transformation (see Section 1.1.31)

$$Ex(f_*^*, \otimes_X) : (f_* M) \otimes_X N \longrightarrow f_*(M \otimes_Y f^* N)$$

is an isomorphism.

We denote by (Adj) (resp. (BC), (PF)) the property (Adj$_f$) (resp. (BC$_f$), (PF$_f$)) for any *proper* morphism f in $\mathscr{S}$ and call it the *adjoint property* (resp. *proper base change property*, *projection formula*).

We can summarize the construction and properties introduced in this section as follows:

Theorem 2.2.14 *Assume $\mathscr{T}$ satisfies the properties (Supp) and (Adj).*

Then for any separated morphism of finite type $f : Y \longrightarrow X$ in $\mathscr{S}$, there exists an essentially unique pair of adjoint functors

$$f_! : \mathscr{T}(Y) \rightleftarrows \mathscr{T}(X) : f^!$$

called the exceptional functors, *such that:*

1. *There exists a structure of a covariant (resp. contravariant) 2-functor on $f \mapsto f_!$ (resp. $f \mapsto f^!$).*
2. *There exists a natural transformation $\alpha_f : f_! \longrightarrow f_*$ compatible with composition in f which is an isomorphism when f is proper.*
3. *For any open immersion j, $j_! = j_\sharp$ and $j^! = j^*$.*
4. *For any cartesian square*

$$\begin{array}{ccc} Y' & \xrightarrow{f'} & X' \\ g' \downarrow & \Delta & \downarrow g \\ Y & \xrightarrow{f} & X, \end{array}$$

in which f is separated and of finite type, there exist natural transformations

$$Ex(\Delta_!^*) : g^* f_! \longrightarrow f'_! g'^*,$$
$$Ex(\Delta_*^!) : g'_* f'^! \longrightarrow f^! g_*$$

which are isomorphisms in the following three cases:

- *f is an open immersion.*
- *g is a $\mathscr{P}$-morphism.*
- *$\mathscr{T}$ satisfies the proper base change property (BC).*

Assume that $\mathscr{T}$ is in addition monoidal. Then the following property holds:

(5) *For any separated morphism of finite type $f : Y \longrightarrow X$ in $\mathscr{S}$, there exist natural transformations*

$$Ex(f_1^*, \otimes) : (f_! K) \otimes_X L \longrightarrow f_!(K \otimes_Y f^* L),$$
$$Hom_X(f_!(L), K) \longrightarrow f_* Hom_Y(L, f^!(K)),$$
$$f^! Hom_X(L, M) \longrightarrow Hom_Y(f^*(L), f^!(M))$$

which are isomorphisms in the following cases:

- *f is an open immersion.*
- *$\mathscr{T}$ satisfies the projection formula (PF).*

Indeed the existence of $f_!$ follows from Proposition 2.2.7 while that of $f^!$ follows directly from assumption (Adj). Assertions (1) and (3) follow from the construction, (2) is Proposition 2.2.10, and (4) (resp. (5)) follows from Corollary 2.2.12 and the definition of (BC) (resp. (PF)). Note also that the second and third isomorphisms in (5) are obtained by transposition from $Ex(f_!, \otimes)$.

2.2.15 While the properties (BC$_f$) and (PF$_f$) are only reasonable in practice for proper morphisms, this is not the case for the property (Adj$_f$). Recall that an exact functor between well generated triangulated categories admits a right adjoint if and only if it commutes with small sums: this is an immediate consequence of the *Brown representability theorem* proved by Neeman (*cf.* [Nee01, 8.4.4]).

Proposition 2.2.16 *Assume that $\mathscr{T}$ is a compactly τ-generated triangulated premotivic category over $\mathscr{S}$. Then, for any morphism of schemes $f : T \longrightarrow S$, the functor $f_* : \mathscr{T}(T) \longrightarrow \mathscr{T}(S)$ admits a right adjoint.*

Proof This follows directly from Proposition 1.3.19. □

2.3 The localization property

2.3.a Definition

2.3.1 Consider a closed immersion $i : Z \longrightarrow S$ in $\mathscr{S}$. Let $U = S - Z$ be the complement open subscheme of S and $j : U \longrightarrow S$ the canonical immersion. We will use the following consequence of the triangulated $\mathscr{P}$-fibred structure on $\mathscr{T}$:

(a) The unit $1 \longrightarrow j^* j_\sharp$ is an isomorphism.
(b) The counit $j^* j_* \longrightarrow 1$ is an isomorphism.
(c) $i^* j_\sharp = 0$.
(d) $j^* i_* = 0$.
(e) The composite map $j_\sharp j^* \xrightarrow{ad'(j_\sharp, j^*)} 1 \xrightarrow{ad(i^*, i_*)} i_* i^*$ is zero.

In fact, the first four relations all follow from the base change property ($\mathscr{P}$-BC). Relation (e) is a consequence of (d) once we have noticed that the following square is commutative

$$\begin{array}{ccc} j_\sharp j^* & \longrightarrow & 1 \\ \downarrow & & \downarrow \\ j_\sharp j^* i_* i^* & \longrightarrow & i_* i^*. \end{array}$$

For the closed immersion i and the triangulated category $\mathscr{T}$, we introduce the property (Loc$_i$) comprising the following assumptions:

(a) The pair of functors (j^*, i^*) is conservative.
(b) The counit $i^* i_* \xrightarrow{ad'(i^*, i_*)} 1$ is an isomorphism.

Definition 2.3.2 We say that $\mathscr{T}$ satisfies the *localization property*, denoted by (Loc), if:

1. $\mathscr{T}(\varnothing) = 0$.
2. For any closed immersion i in $\mathscr{S}$, (Loc$_i$) is satisfied.

The main consequence of the localization axiom is that it leads to the situation of the six gluing functor (cf. [BBD82, prop. 1.4.5]):

Proposition 2.3.3 *Let $i : Z \longrightarrow S$ be a closed immersion with complementary open immersion $j : U \longrightarrow S$ such that (Loc$_i$) is satisfied.*

1. *The functor i_* admits a right adjoint $i^!$.*
2. *For any K in $\mathscr{T}(S)$, there exists a unique map $\partial_{i,K} : i_* i^* K \longrightarrow j_\sharp j^* K[1]$ such that the triangle*

$$j_\sharp j^* K \xrightarrow{ad'(j_\sharp, j^*)} K \xrightarrow{ad(i^*, i_*)} i_* i^* K \xrightarrow{\partial_{i,K}} j_\sharp j^* K[1]$$

is distinguished. The map $\partial_{i,K}$ is functorial in K.

2 Triangulated $\mathscr{P}$-fibred categories in algebraic geometry

3. *For any K in $\mathscr{T}(S)$, there exists a unique map $\partial'_{i,K} : j_*j^*K \longrightarrow i_*i^!K[1]$ such that the triangle*

$$i_*i^!K \xrightarrow{ad'(i_*,i^!)} K \xrightarrow{ad(j^*,j_*)} j_*j^*K \xrightarrow{\partial'_{i,K}} i_*i^!K[1]$$

is distinguished. The map $\partial'_{i,K}$ is functorial in K.

Under the property (Loc$_i$), the canonical triangles appearing in (2) and (3) above are called the *localization triangles* associated with i.

Proof We first consider point (2). For the existence, we consider a distinguished triangle

$$j_\sharp j^*K \xrightarrow{ad'(j_\sharp,j^*)} K \xrightarrow{\pi} C \xrightarrow{+1}$$

Applying 2.3.1(e), we obtain a factorization

$$K \xrightarrow{ad(i^*,i_*)} i_*i^*K$$
$$\pi \searrow \quad \nearrow w$$
$$C$$

We prove w is an isomorphism. According to the above triangle, $j^*C = 0$. From 2.3.1(d), $j^*i_*i^*K = 0$ so that j^*w is an isomorphism. Applying i^* to the above distinguished triangle, we obtain from 2.3.1(c) that $i^*\pi$ is an isomorphism. Thus, applying i^* to the above commutative diagram together with (Loc$_i$) (b), we obtain that i^*w is an isomorphism, which concludes the proof.

Consider a map $K \xrightarrow{u} L$ in $\mathscr{T}(S)$ and suppose we have chosen maps a and b in the diagram:

$$\begin{array}{ccccccc}
j_\sharp j^*K & \xrightarrow{ad'(j_\sharp,j^*)} & K & \xrightarrow{ad(i^*,i_*)} & i_*i^*K & \xrightarrow{a} & j_\sharp j^*K[1] \\
u\downarrow & & u\downarrow & & & & \downarrow u \\
j_\sharp j^*L & \xrightarrow{ad'(j_\sharp,j^*)} & L & \xrightarrow{ad(i^*,i_*)} & i_*i^*L & \xrightarrow{b} & j_\sharp j^*L[1]
\end{array}$$

such that the horizontal lines are distinguished triangles. We can find a map $h : i_*i^*K \longrightarrow i_*i^*L$ completing the previous diagram into a morphism of triangles. Then the map $w = h - i_*i^*(u)$ satisfies the relation $w \circ ad(i^*,i_*) = 0$. Thus it can be lifted to a map in $\mathrm{Hom}(j_\sharp j^*K[1], i_*i^*L)$. But this is zero by adjunction and the relation 2.3.1(d). This proves both the naturality of $\partial_{i,K}$ and its uniqueness.

For point (1) and (3), for any object K of $\mathscr{T}(S)$, we consider a distinguished triangle

$$D \longrightarrow K \xrightarrow{ad(j^*,j_*)} j_*j^*K \xrightarrow{+1} .$$

According to 2.3.1(b), $j^*D = 0$. Thus according to the triangle of point (2) applied to D, we obtain $D = i_*i^*D$. Arguing as for point (2), we thus obtain that D is unique and depends functorially on K so that, if we put $i^!K = i^*D$, point (1) and (3) follows. $\square$

Remark 2.3.4 Consider the hypotheses and notations of the previous proposition.
1. By transposition from 2.3.1(d), we deduce that $i^! j_* = 0$.
2. Assume that i is a $\mathscr{P}$-morphism. Then the $\mathscr{P}$-base change formula implies that $i^* j_\sharp = 0$. Dually, we get that $i^! j_\sharp = 0$. By adjunction, we thus obtain $\partial_{i,K} = 0$ and $\partial'_{i,K} = 0$ for any object K so that both localization triangles are split. In that case, we get that $\mathscr{T}(S) = \mathscr{T}(Z) \times \mathscr{T}(U)$.[46]

The preceding proposition admits the following reciprocal statement:

Lemma 2.3.5 *Consider a closed immersion $i : Z \longrightarrow S$ in $\mathscr{S}$ with complementary open immersion $j : U \longrightarrow S$. Then the following properties are equivalent:*

(i) $\mathscr{T}$ satisfies (Loc_i).
(ii) (a) The functor i_* is conservative.
  (b) For any object K of $\mathscr{T}(S)$, there exists a map $i_* i^*(K) \longrightarrow j_\sharp j^*(K)[1]$ which fits into a distinguished triangle

$$j_\sharp j^*(K) \xrightarrow{ad'(j_\sharp, j^*)} K \xrightarrow{ad(i^*, i_*)} i_* i^*(K) \longrightarrow j_\sharp j^*(K)[1].$$

Proof The fact that (i) implies (ii) follows from Proposition 2.3.3. Conversely, (ii)(b) implies that the pair (i^*, j^*) is conservative and it remains to prove (Loc_i) (b). Let K be an object of $\mathscr{T}(S)$. Consider the distinguished triangle given by (ii)(b):

$$j_\sharp j^*(K) \xrightarrow{ad'(j_\sharp, j^*)} K \xrightarrow{ad(i^*, i_*)} i_* i^*(K) \longrightarrow j_\sharp j^*(K)[1].$$

If we apply i_* on the left to this triangle, we get using 2.3.1(d) that the morphism

$$i_*(K) \xrightarrow{ad(i^*, i_*).i_*} i_* i^* i_*(K)$$

is an isomorphism. Hence, by the zig-zag equation, the morphism

$$i_* i^* i_*(K) \xrightarrow{i_*.ad'(i^*, i_*)} i_*(K)$$

is an isomorphism. Property (ii)(a) thus implies that $i^* i_*(K) \simeq K$. □

2.3.b First consequences of localization

The following statement is straightforward.

Proposition 2.3.6 *Assume $\mathscr{T}$ satisfies the localization property and consider a scheme S in $\mathscr{S}$.*

---
[46] This remark explains why the localization property is too strong for generalized premotivic categories.

1. Let S_{red} be the reduced scheme associated with S. The canonical immersion $S_{red} \xrightarrow{\nu} S$ induces an equivalence of categories:
$$\nu^* : \mathscr{T}(S) \longrightarrow \mathscr{T}(S_{red}).$$

2. For any partition$(S_i \xrightarrow{\nu_i} S)_{i \in I}$ of S by locally closed subsets, the family of functors $(\nu_i^*)_{i \in I}$ is conservative (S_i is considered with its canonical structure of a reduced subscheme of S).

Lemma 2.3.7 *If $\mathscr{T}$ satisfies the localization property (Loc) then it is additive.*

Proof Note that, by assumption, $\mathscr{T}(\varnothing) = 0$. Then the assertion follows directly from Lemma 2.2.2. □

Proposition 2.3.8 *If $\mathscr{T}$ satisfies the localization property then it satisfies the cdh-separation property.*

Proof Consider a cartesian square of schemes

$$\begin{array}{ccc} B & \longrightarrow & Y \\ \downarrow & Q & \downarrow p \\ A & \xrightarrow{e} & X. \end{array}$$

According to Lemma 2.1.12, we have only to check that the pair of functors (e^*, p^*) is conservative when Q is a Nisnevich (or respectively a proper cdh) distinguished square. Let $\nu : A' \longrightarrow X$ be the complementary closed (resp. open) immersion to e, where A' has the induced reduced subscheme (resp. induced subscheme) structure. Consider the cartesian square

$$\begin{array}{ccc} Y & \longleftarrow & B' \\ p \downarrow & & \downarrow q \\ X & \xleftarrow{\nu} & A'. \end{array}$$

By assumption on Q, q is an isomorphism. According to (Loc) (ii), (e^*, ν^*) is conservative. This concludes the proof. □

The following proposition can be found in a slightly less precise and general form in [Ayo07a, 2.1.162].[47]

Proposition 2.3.9 *Assume $\mathscr{T}$ satisfies the localization property. Then the following conditions are equivalent:*

(i) $\mathscr{T}$ is separated.
(ii) For a morphism $f : T \longrightarrow S$ in $\mathscr{S}$, $f^* : \mathscr{T}(S) \longrightarrow \mathscr{T}(T)$ is conservative whenever f is:

  (a) a finite étale cover;

---

[47] A warning: the proof in *loc. cit.* seems to require that the schemes are excellent.

(b) finite, faithfully flat and radicial.

Proof Only $(ii) \Rightarrow (i)$ requires a proof. Consider a surjective morphism of finite type $f : T \longrightarrow S$ in $\mathscr{S}$. According to [GD67, 17.16.4], there exists a partition $(S_i)_{i \in I}$ of S by (affine) subschemes and a family of maps of the form

$$S_i'' \xrightarrow{g_i} S_i' \xrightarrow{h_i} S_i$$

such that g_i (resp. h_i) satisfies assumption (a) (resp. (b)) above and such that for any $i \in I$, $f \times_S S_i''$ admits a section. Thus, Proposition 2.3.6 concludes the proof. □

2.3.c Localization and exchange properties

2.3.10 Consider a morphism of complete triangulated $\mathscr{P}$-fibred categories over $\mathscr{S}$:

$$\varphi^* : \mathscr{T} \longrightarrow \mathscr{T}'.$$

Recall that for any morphism $f : Y \longrightarrow X$, there is an exchange transformation (1.2.5.1):

$$Ex(\varphi^*, f_*) : \varphi_X^* f_* \longrightarrow f_* \varphi_Y^*.$$

If $\mathscr{T}$ and $\mathscr{T}'$ satisfies the support axiom and f is separated of finite type, we have constructed (Proposition 2.2.11) another exchange transformation:

$$Ex(\varphi^*, f_!) : \varphi_X^* f_! \longrightarrow f_! \varphi_Y^*.$$

Proposition 2.3.11 *Consider a morphism $\varphi^* : \mathscr{T} \longrightarrow \mathscr{T}'$ as above.*

1. *Let $i : Z \longrightarrow X$ be a closed immersion such that $\mathscr{T}$ and $\mathscr{T}'$ satisfy property (Loc_i). Then the exchange $Ex(\varphi^*, i_*) : \varphi_X^* i_* \longrightarrow i_* \varphi_Z^*$ is an isomorphism.*
2. *Assume $\mathscr{T}$ and $\mathscr{T}'$ satisfy property (Loc). Then the following conditions are equivalent:*

   (i) *For any integer $n > 0$ and any scheme X in $\mathscr{S}$, the exchange $Ex(\varphi^*, p_{n*})$ is an isomorphism, where $p_n : \mathbf{P}_X^n \longrightarrow X$ is the canonical projection.*

   (ii) *For any proper morphism $f : Y \longrightarrow X$, the exchange $Ex(\varphi^*, f_*)$ is an isomorphism.*

3. *Assume $\mathscr{T}$ and $\mathscr{T}'$ satisfy properties (Loc) and (Supp). Then conditions (i) and (ii) above are equivalent to the following:*

   (iii) *For any separated morphism $f : Y \longrightarrow X$ of finite type, the exchange $Ex(\varphi^*, f_!)$ is an isomorphism.*

Remark 2.3.12 We will simply say that φ^* commutes with $f_!$ when assertion (iii) is fulfilled. For an important case where this happens, see Proposition 2.4.53.

2 Triangulated $\mathscr{P}$-fibred categories in algebraic geometry

Proof Assertion (1) follows easily from the conservativity of (i^*, j^*), where j is the complementary open immersion, and the relations of Section 2.3.1. Assertion (3) is an easy consequence of the definition of $f_!$ and the exchange $Ex(\varphi^*, f_!)$.

Concerning assertion (2), we have to prove that (i) implies (ii). We fix a morphism $f : Y \longrightarrow X$ and prove that the exchange $Ex(\varphi^*, f_*) : \varphi_Y^* f_* \longrightarrow f_* \varphi_X^*$ is an isomorphism.

We first treat the case where f is projective. According to Proposition 2.3.8, $\mathscr{T}'$ satisfies the Zariski separation property. Using the ($\mathscr{P}$-BC) property, we see that the problem is local in X so that we can assume X is affine. Then X admits an ample line bundle and there exists an integer $n > 0$ such that f can be factored ([GD61, (5.5.4)(ii)]) into a closed immersion $i : Y \longrightarrow \mathbf{P}_X^n$ and the projection $p_n : \mathbf{P}_X^n \longrightarrow X$. Thus, assertion (1) and assumption (i) allow us to conclude this special case.

To treat the general case, we argue by noetherian induction on Y, assuming that for any proper closed subscheme T of Y, the result is known for the restriction of f to T. In fact, the case $T = \varnothing$ is obvious because $\mathscr{T}(\varnothing) = 0$.

According to Chow's lemma [GD61, 5.6.2], there exists a morphism $p : Y_0 \longrightarrow Y$ such that:

(a) p and $f \circ p$ are projective morphisms.
(b) There exists a dense open subscheme V_0 of Y over which p is an isomorphism.

Let T be the complement of V in Y equipped with its reduced subscheme structure. Let j and i be the respective immersion of T and V in Y. According to point (3) of Proposition 2.3.3, it is sufficient to prove that the following natural transformations are isomorphisms:

(2.3.12.1) $$\varphi_Y^* f_* i_* \longrightarrow f_* \varphi_X^* i_*,$$
(2.3.12.2) $$\varphi_Y^* f_* j_* \longrightarrow f_* \varphi_X^* j_*.$$

Concerning the first one, we consider the following commutative diagram:

$$\begin{array}{ccccc} \varphi_Y^* f_* i_* & \xrightarrow{Ex(\varphi^*, f_*)} & f_* \varphi_X^* i_* & \xrightarrow{Ex(\varphi^*, i_*)} & f_* i_* \varphi_X^* \\ \| & & & & \| \\ \varphi_Y^* (fi)_* & & \xrightarrow{Ex(\varphi^*, (fi)_*)} & & (fi)_* \varphi_X^*. \end{array}$$

Thus the result follows from assertion (1) and the induction hypothesis.

Concerning the natural transformation (2.3.12.2), we consider the pullback square

$$\begin{array}{ccc} V_0 & \xrightarrow{l} & Y_0 \\ q \downarrow & & \downarrow p \\ V & \xrightarrow{j} & Y. \end{array}$$

Assumption (b) above says that q is an isomorphism which implies the relation: $j_* = p_* l_* q^*$. In particular, it is sufficient to prove that the natural transformation

$\varphi_Y^* f_* p_* \longrightarrow f_* \varphi_X^* p_*$ is an isomorphism. This follows from the commutativity of the following diagram

$$\begin{array}{ccccc} \varphi_Y^* f_* p_* & \xrightarrow{Ex(\varphi^*, f_*)} & f_* \varphi_X^* p_* & \xrightarrow{Ex(\varphi^*, p_*)} & f_* p_* \varphi_X^* \\ \| & & & & \| \\ \varphi_Y^* (fp)_* & & \xrightarrow{Ex(\varphi^*, (fp)_*)} & & (fp)_* \varphi_X^*, \end{array}$$

according to the projective case treated above and assumption (b). The proof is complete. □

Corollary 2.3.13 *In the next statements, we assume $\mathscr{T}$ is monoidal when it is needed.*

1. *Let $i : Z \longrightarrow X$ be a closed immersion such that $\mathscr{T}$ satisfies property (Loc_i). Then $\mathscr{T}$ satisfies property $(Supp_i)$ (resp. (BC_i), (PF_i)).*
2. *Assume $\mathscr{T}$ satisfies the localization property. Then the following properties of $\mathscr{T}$ are equivalent:*

    (i) *For any integer $n > 0$ and any scheme X in $\mathscr{S}$, $p_n : \mathbf{P}_X^n \longrightarrow X$ being the canonical projection, $\mathscr{T}$ satisfies $(Supp_{p_n})$ (resp. (BC_{p_n}), (PF_{p_n})).*
    (ii) *$\mathscr{T}$ satisfies $(Supp)$ (resp. (BC), (PF)).*

3. *Assume $\mathscr{T}$ is well generated and satisfies the localization property. Then the following properties of $\mathscr{T}$ are equivalent:*

    (i') *For any integer $n > 0$ and any scheme X in $\mathscr{S}$, $p_n : \mathbf{P}_X^n \longrightarrow X$ being the canonical projection, $\mathscr{T}$ satisfies (Adj_{p_n}).*
    (ii') *$\mathscr{T}$ satisfies (Adj).*

Proof As in the proof of Corollary 2.2.12, each respective case of assertions (1) and (2) follows from the previous proposition applied to a particular type of morphism $\varphi^* : \mathscr{T}' \longrightarrow \mathscr{T}''$ of complete $\mathscr{P}$-fibred triangulated categories over a subcategory $\mathscr{S}'$ of $\mathscr{S}$.

For property (Supp), we proceed as follows. We fix an open immersion $j : U \longrightarrow X$ and let $\mathscr{S}' = \mathscr{S}/X$. For any Y/X, we let $j_Y = Y \times_X U \longrightarrow Y$ be the pullback of j. We put $\mathscr{T}'(Y) = \mathscr{T}(Y \times_X U)$ and $\mathscr{T}''(Y) = \mathscr{T}(Y)$ and let φ_Y^* be the functor:

$$j_{Y\sharp} : \mathscr{T}(Y \times_X U) \longrightarrow \mathscr{T}(Y).$$

For the property (BC) (resp. (PF)), we refer the reader to the proof of assertion (1) (resp. (2)) in Corollary 2.2.12.

Finally we consider assertion (3). It is sufficient to prove that (i') implies (ii'). According to the Brown representability theorem [Nee01, 8.4.4], the property (Adj_f) for a proper morphism f is equivalent to asking that f_* preserves small sums. Consider an arbitrary set I. For any scheme S, we put $\mathscr{T}^I(S) = \mathscr{T}(S)^I$, that is, the category of families of object of $\mathscr{T}(S)$ indexed by I. Then $\mathscr{T}^I$ is obviously a complete triangulated $\mathscr{P}$-fibred category over $\mathscr{S}$ (limits and colimits are computed termwise). For any scheme S, we consider the functor:

$$\varphi_S^* : \mathscr{T}^I(S) \longrightarrow \mathscr{T}(S), \quad (M_i)_{i \in I} \longmapsto \bigoplus_{i \in I} M_i.$$

Then $\varphi^* : \mathscr{T}^I \longrightarrow \mathscr{T}$ is obviously a morphism of complete $\mathscr{P}$-fibred categories. Thus, given condition (i'), the preceding proposition applied to φ^* shows that for any proper morphism f, f_* commutes with sums indexed by I. As this is true for any I, we obtain (ii'). □

2.3.d Localization and monoidal structure

2.3.14 Assume $\mathscr{T}$ is monoidal and let M denote its geometric sections. Fix a closed immersion $i : Z \longrightarrow S$ in $\mathscr{S}$ with complementary open immersion $j : U \longrightarrow S$. We fix an object $M_S(S/S - Z)$ of $\mathscr{T}(S)$ and a distinguished triangle

$$(2.3.14.1) \quad M_S(S - Z) \xrightarrow{j_*} \mathbb{1}_S \xrightarrow{p_i} M_S(S/S - Z) \xrightarrow{d_i} M_S(S - Z)[1].$$

Note that according to 2.3.1(c), the map $i^*(p_i) : \mathbb{1}_Z \longrightarrow i^* M_S(S/S - Z)$ is an isomorphism. Given any object K in $\mathscr{T}(S)$, we thus obtain an isomorphism

$$i^*(M_S(S/S - Z) \otimes_S K) = i^*(M_S(S/S - Z)) \otimes_Z i^*(K) \xrightarrow{(i^* p_i)^{-1}} \mathbb{1}_Z \otimes_Z i^*(K) = i^*(K)$$

which is natural in K. It induces by adjunction a map

$$(2.3.14.2) \quad \psi_{i,K} : M_S(S/S - Z) \otimes_S K \longrightarrow i_* i^*(K)$$

which is natural in K.

For any $\mathscr{P}$-scheme X/S, we put $M_S(X/X - X_Z) = M_S(S/S - Z) \otimes_S M_S(X)$ so that we get from (2.3.14.1) a canonical distinguished triangle:

$$M_S(X - X_Z) \xrightarrow{j_{X*}} M_S(X) \longrightarrow M_S(X/X - X_Z) \longrightarrow M_S(X - X_Z)[1].$$

The map (2.3.14.2) for $K = M_S(X)$ gives a canonical map

$$(2.3.14.3) \quad \psi_{i,X} : M_S(X/X - X_Z) \longrightarrow i_*(M_Z(X_Z)).$$

Proposition 2.3.15 *Consider the previous hypotheses and notations. Then the following conditions are equivalent:*

(i) $\mathscr{T}$ satisfies the property (Loc_i).
(ii) (a) The functor i_* is conservative.
  (b) The morphism $\psi_{i,S} : M_S(S/S - Z) \longrightarrow i_*(\mathbb{1}_Z)$ is an isomorphism.
  (c) For any object K of $\mathscr{T}(S)$, the exchange transformation

$$Ex(i_*^*, \otimes) : (i_* \mathbb{1}_Z) \otimes_S K \longrightarrow i_* i^* K$$

is an isomorphism.

(iii) (a) *The functor i_* is conservative.*
 (b) *The morphism $\psi_{i,S} : M_S(S/S - Z) \longrightarrow i_*(\mathbb{1}_Z)$ is an isomorphism.*
 (c) *For any objects K and L of $\mathscr{T}(S)$, the exchange transformation*

$$Ex(i_*^*, \otimes) : (i_*K) \otimes_S L \longrightarrow i_*(K \otimes_Z i^*L)$$

is an isomorphism.

Assume in addition that $\mathscr{T}$ is well generated and τ-generated as a triangulated $\mathscr{P}$-fibred category. Then the above conditions are equivalent to the following one:

(iv) (a) *The functor i_* is conservative, commutes with direct sums and with τ-twists.*
 (b) *The morphism $\psi_{i,X} : M_S(X/X - X_Z) \longrightarrow i_*(M_Z(X_Z))$ is an isomorphism for any $\mathscr{P}$-scheme X/S.*

In particular, (Loc$_i$) implies that for any object K of $\mathscr{T}(S)$, the localization triangle of Proposition 2.3.3

$$j_\sharp j^*(K) \longrightarrow K \longrightarrow i_*i^*(K) \xrightarrow{\partial_K} j_\sharp j^*(K)[1]$$

is canonically isomorphic (through exchange transformations) to the triangle (2.3.14.1) tensored with K.

Proof (i) $\Rightarrow$ (iii) : According to (Loc$_i$) (a), we need only to check that the maps in (iii)(b) and (iii)(c) are isomorphisms after applying i^* and j^*. This follows easily from (Loc$_i$) (b).
(iii) $\Rightarrow$ (ii) : Obvious
(ii) $\Rightarrow$ (i) : According to (ii)(b), the distinguished triangle (2.3.14.1) is isomorphic to a triangle of the form

$$j_\sharp j^*(\mathbb{1}_S) \xrightarrow{ad'(j_\sharp, j^*)} \mathbb{1}_S \xrightarrow{ad(i^*, i_*)} i_*i^*(\mathbb{1}_S) \longrightarrow j_\sharp j^*(\mathbb{1}_S).$$

According to (ii)(c), this latter triangle tensored with K is isomorphic through exchange transformations to a triangle of the form

$$j_\sharp j^*(K) \xrightarrow{ad'(j_\sharp, j^*)} K \xrightarrow{ad(i^*, i_*)} i_*i^*(K) \longrightarrow j_\sharp j^*(K).$$

Thus the result follows from Lemma 2.3.5.

To end the proof, we remark by using the equations for the adjunction (i^*, i_*) that for any object M of $\mathscr{T}(S)$, the following diagram is commutative:

$$\begin{array}{ccc}
 & i_*i^*(\mathbb{1}_S) \otimes K = i_*(\mathbb{1}_Z) \otimes K \\
\psi_i \otimes 1_K \nearrow & & \downarrow Ex(i_*^*, \otimes) \\
M_S(S/S - Z) \otimes K & & \\
\psi_{i,K} \searrow & & \\
 & i_*i^*(K) = i_*(\mathbb{1}_Z \otimes i^*i^*(K)).
\end{array}$$

2 Triangulated $\mathscr{P}$-fibred categories in algebraic geometry

Note that (i) implies that i_* is conservative and commutes with direct sums (see Proposition 2.3.3) and (ii)(c) implies it commutes with twists. According to the above diagram, (ii)(b) implies (iv)(b).

We prove that conversely (iv) implies (ii). Because (ii)(b) (resp. (ii)(a)) is a particular case of (iv)(b) (resp. (iv)(a)), we have only to prove (ii)(b). In view of the previous diagram, we are reduced to proving that for any object K of $\mathscr{T}(S)$, the map $\psi_{i,K}$ is an isomorphism. Consider the full subcategory $\mathscr{U}$ of $\mathscr{T}(S)$ comprising the objects K such that $\psi_{i,K}$ is an isomorphism. Then $\mathscr{U}$ is triangulated. Using (iv)(a), $\mathscr{U}$ is stable under small sums and τ-twists. By assumption, it contains the objects of the form $M_S(X)$ for a $\mathscr{P}$-scheme X/S. Thus, because $\mathscr{T}$ is well generated by assumption, the result follows by Lemma 1.3.17. □

Lemma 2.3.16 *Consider a closed immersion $i : Z \longrightarrow S$. We assume the following conditions are satisfied in addition to that of Section 2.0.1:*

- $\mathscr{T}$ *is well generated, τ-generated, and satisfies the Zariski separation property.*
- *For any $\mathscr{P}$-scheme X_0/Z and any point x_0 of X_0, there exists an open neighborhood U_0 of x_0 in X_0 and a $\mathscr{P}$-scheme U/S such that $U_0 = U \times_S Z$.*[48]

Then the functor i_* is conservative.

Proof Consider an object K of $\mathscr{T}(Z)$ such that $i_*(K) = 0$. We prove that $K = 0$. Because $\mathscr{T}$ is τ-generated, it is sufficient to prove that for a $\mathscr{P}$-morphism $p_0 : X_0 \longrightarrow Z$ and a twist $(n,m) \in \mathbf{Z} \times \tau$,

$$\mathrm{Hom}_{\mathscr{T}(Z)}(M_Z(X_0)\{m\}[n], K) = 0.$$

Because $M_Z(X_0) = p_{0\sharp}(\mathbb{1}_{X_0})$, this equivalent to proving that

$$\mathrm{Hom}_{\mathscr{T}(X_0)}(\mathbb{1}_{X_0}\{m\}[n], p_0^*(K)) = 0.$$

Using the Zariski separation property on $\mathscr{T}$, this latter assumption is local in X_0. Thus, according to the assumption on the class $\mathscr{P}$, we can assume there exists a $\mathscr{P}$-scheme X/S such that $X_0 = X \times_S Z$. Thus $M_Z(X_0)\{m\}[n] = i^*(M_S(X)\{m\}[n])$ and the initial assumption on K allows us to conclude the proof. □

Note for future applications the following interesting corollaries:

Corollary 2.3.17 *Assume $\mathscr{T}$ is a premotivic triangulated category which is compactly τ-generated for a group of twists τ (i.e. any twist in τ admits a tensor inverse) and which satisfies the Zariski separation property.*

Then, for any closed immersion i, the functor i_* is conservative and commutes with sums and with twists.

This is a consequence of Lemmas 2.3.16 and 2.2.16. In fact, under these conditions, i_* commutes with arbitrary τ-twists because this is true for its (left) adjoint i^*.

---

[48] This property is trivial when $\mathscr{P}$ is the class of open immersions or the class of morphisms of finite type in $\mathscr{S}$. It is also true when $\mathscr{P}$ is the class of étale morphism or $\mathscr{P} = Sm$ (cf. [GD67, 18.1.1]).

Corollary 2.3.18 *Assume $\mathscr{T}$ satisfies the assumptions of the preceding corollary. Then the following conditions on a closed immersion i are equivalent:*

(i) $\mathscr{T}$ satisfies the property (Loc_i).
(ii) For any scheme S in $\mathscr{S}$ and any smooth S-scheme X, the map (2.3.14.3)

$$\psi_{i,X} : M_S(X/X - X_Z) \longrightarrow i_*M_Z(X_Z)$$

is an isomorphism.

We finish this section with the following useful result:

Proposition 2.3.19 *Assume $\mathscr{T}$ is τ-generated and consider a τ'-generated triangulated $\mathscr{P}$-fibred category $\mathscr{T}'$ and a morphism*

$$\varphi^* : (\mathscr{T}, \tau) \rightleftarrows (\mathscr{T}', \tau') : \varphi_*.$$

We assume the following properties:

(a) the morphism φ^* is strictly compatible with twists;
(b) $\mathscr{T}'$ is well generated.

We consider a closed immersion $i : Z \longrightarrow S$ and further assume the following properties:

(c) $\mathscr{T}$ satisfies the property (Loc_i).
(d) The exchange transformation $Ex(\varphi^*, i_*) : \varphi^* i_* \longrightarrow i_* \varphi^*$ is an isomorphism.
(e) The functor $i_* : \mathscr{T}'(Z) \longrightarrow \mathscr{T}'(S)$ commutes with τ'-twists.[49]

Then $\mathscr{T}'$ satisfies the property (Loc_i).

Proof Note that, under the above assumptions, φ_* is conservative (in fact, for any $\mathscr{P}$-scheme X/S and any twists $i \in \tau'$, the premotive $M_S(X)\{i\}$ is in the essential image of φ^*). Thus, if $i_* : \mathscr{T}(Z) \longrightarrow \mathscr{T}(S)$ is conservative (resp. commutes with sums), then $i_* : \mathscr{T}'(S) \longrightarrow \mathscr{T}'(S)$ is conservative (resp. commutes with sums) using the isomorphism $\varphi_* i_* \simeq i_* \varphi_*$.

Let M (resp. M') be the geometric sections of $\mathscr{T}$ (resp. $\mathscr{T}'$). As in Section 2.3.14, we fix a distinguished triangle

$$M_S(S - Z) \xrightarrow{j_*} \mathbb{1}_S \xrightarrow{p_i} M_S(S/S - Z) \xrightarrow{d_i} M_S(S - Z)[1]$$

and we put $M'_S(S/S - Z) = \varphi^* M_S(S/S - Z)$. According to *loc. cit.*, we thus get for any $\mathscr{P}$-scheme X/S canonical maps

$$\psi_{i,X} : M_S(X/X - X_Z) \longrightarrow i_* M_Z(X_Z),$$
$$\psi'_{i,X} : M'_S(X/X - X_Z) \longrightarrow i_* M'_Z(X_Z).$$

---

[49] This will be satisfied if any τ'-twist is invertible because the left adjoint of i_* commutes with τ'-twists.

2 Triangulated $\mathscr{P}$-fibred categories in algebraic geometry

By construction, the following diagram is commutative:

$$\begin{array}{ccccc} \varphi^* M_S(X/X - X_Z) & \xrightarrow{\varphi^* \psi_{i,X}} & \varphi^* i_* M_Z(X_Z) & \xrightarrow{Ex(\varphi^*, i_*)} & i_* \varphi^* M_Z(X_Z) \\ \| & & & & \| \\ M'_S(X/X - X_Z) & & \xrightarrow{\psi'_{i,X}} & & M'_Z(X_Z). \end{array}$$

Thus, the result follows from Proposition 2.3.15. □

2.4 Purity and the theorem of Voevodsky–Röndigs–Ayoub

Recall we assume $\mathscr{P} = Sm$ in this section.

2.4.a The stability property

The following section is directly inspired by the work of Ayoub in [Ayo07a, §1.5].[50] We claim no originality except for a closer look at the needed axioms.

Definition 2.4.1 A *pointed smooth S-scheme* will be a pair (f, s) of morphisms of $\mathscr{S}$ such that $f : X \longrightarrow S$ is a smooth separated morphism of finite type and $s : S \longrightarrow X$ is a section of f.

We associate with a pointed smooth scheme (f, s) the following endofunctor of $\mathscr{T}(S)$

$$Th(f, s) := f_\sharp s_*,$$

called the associated *Thom transformation*.

If $\mathscr{T}$ satisfies (Adj$_s$) (recall: s_* admits a right adjoint denoted by $s^!$), we put

$$Th'(f, s) := s^! f^*$$

and call it the associated *adjoint Thom transformation*.

Remark 2.4.2 Note that because f is separated, s is a closed immersion.

Example 2.4.3

1. Let $p : E \longrightarrow X$ be a vector bundle and s_0 be its zero section. Following [Ayo07a], we put $Th(E) := Th(p, s_0)$ and simply call it the Thom transformation associated with E/X.
2. Consider a pointed smooth S-scheme (f, s) such that f is étale. Then s is an open and closed immersion. Thus, if $\mathscr{T}$ is additive, $s_* = s_\sharp$ according to Lemma 2.2.2. In particular, $Th(f, s) = \text{Id}_S$.

---
[50] See also [Del01, §5].

Definition 2.4.4 We will say that $\mathscr{T}$ satisfies the *stability property*, denoted by (Stab), if for any point smooth scheme (f,s), the Thom transformation $Th(f,s)$ is an equivalence of categories.

2.4.5 Consider a commutative diagram in $\mathscr{S}$ of the form

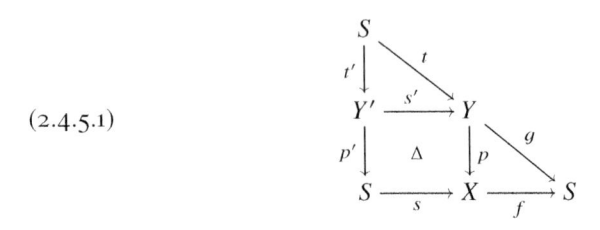

(2.4.5.1)
$$\begin{array}{ccc}
& S & \\
{}^{t'}\downarrow & \searrow^{t} & \\
Y' & \xrightarrow{s'} Y & \\
{}^{p'}\downarrow & \Delta \quad \downarrow^{p} & \searrow^{g} \\
S & \xrightarrow{s} X & \xrightarrow{f} S
\end{array}$$

such that Δ is a cartesian square, (f,s), (g,t) are smooth pointed schemes and g is a smooth separated morphism of finite type. Then we get a canonical exchange morphism:

(2.4.5.2) $\quad Th(g,t) = f_\sharp p_\sharp s'_* t'_* \xrightarrow{Ex(\Delta_{\sharp *})} f_\sharp s_* p'_\sharp t'_* = Th(f,s)Th(p',t').$

This is an isomorphism as soon as $Ex(\Delta_{\sharp *})$ is an isomorphism. The following lemma gives a sufficient condition for this to happen.

Lemma 2.4.6 *Consider the above notations. If $\mathscr{T}$ satisfies (Loc_S) then the natural transformations $Ex(\Delta_{\sharp *})$ is an isomorphism for any square Δ as above.*

This lemma follows easily from the definition of (Loc_S), the relations of Section 2.3.1 and the $\mathscr{P}$-base change formula ($\mathscr{P}$-BC). It motivates the next definition:

Definition 2.4.7 We say that $\mathscr{T}$ satisfies the *weak localization property* (wLoc) if it satisfies (Loc_S) for any closed immersion s which admits a smooth retraction.

Proposition 2.4.8 *Assume that $\mathscr{T}$ satisfies the Nisnevich separation property. Then the following conditions are equivalent:*

(i) $\mathscr{T}$ satisfies (wLoc).
(ii) For any scheme S and any closed immersion $i : Z \longrightarrow X$ between smooth S-schemes, $\mathscr{T}$ satisfies (Loc_i).

Proof Of course, (ii) implies (i). We prove the converse. The Nisnevich separation property says that for any Nisnevich cover $f : X' \longrightarrow X$, the functor f^* is conservative. We deduce that the properties (Loc_i) (a) and (Loc_i) (b) are local in X with respect to the Nisnevich topology – for (b), one also uses the smooth projection formula. Thus, we can conclude as locally for the Nisnevich topology, i admits a smooth retraction (see for example [Dég07, 4.5.11]). $\square$

Applying the second point of Example 2.4.3, we easily deduce from that construction the following kind of excision property:

2 Triangulated $\mathscr{P}$-fibred categories in algebraic geometry

Lemma 2.4.9 *Assume that $\mathscr{T}$ satisfies (wLoc).*

Then, given any diagram (2.4.5.1) satisfying the assumption as above and such that p is étale, the natural transformation (2.4.5.2) gives an isomorphism:

$$Th(g,t) \xrightarrow{\sim} Th(f,s).$$

2.4.10 To any short exact sequence of vector bundles over a scheme S

(σ) $\qquad\qquad 0 \longrightarrow E' \xrightarrow{\nu} E \xrightarrow{\pi} E'' \longrightarrow 0,$

we can associate a commutative diagram

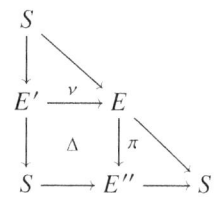

$$\begin{array}{ccc} S & & \\ \downarrow & \searrow & \\ E' & \xrightarrow{\nu} & E \\ \downarrow & \Delta & \downarrow \pi \quad \searrow \\ S & \longrightarrow & E'' \longrightarrow S \end{array}$$

where the unlabeled maps are either the canonical projections or the zero sections of the relevant vector bundles, and Δ is cartesian. Using the notation of Example 2.4.3, the exchange transformation (2.4.5.2) associated with this diagram has the following form:

$$Th(\sigma) : Th(E) \longrightarrow Th(E'') \circ Th(E').$$

Recall from the above that this natural transformation is an isomorphism as soon as $\mathscr{T}$ satisfies (wLoc).

Proposition 2.4.11 *Assume $\mathscr{T}$ satisfies (wLoc) and (Zar-sep). Then the following conditions are equivalent:*

(i) The complete triangulated Sm-fibred category $\mathscr{T}$ satisfies the stability property.
(ii) For any scheme S, the Thom transformation $Th(\mathbf{A}_S^1)$ is an equivalence of categories.

Proof We have to prove that (ii) implies (i). Note that according to the above section, we already know that for any scheme S and any integer $n \geq 0$, $Th(\mathbf{A}_S^n) \simeq Th(\mathbf{A}_S^1)^{\circ,n}$ is an equivalence.

We consider a smooth pointed scheme $(f : X \longrightarrow S, s)$ and we prove that $Th(f,s)$ is an equivalence.

Recall that (Loc_s) implies (Adj) s (first point of Proposition 2.3.3). In particular, $Th(f,s)$ admits a right adjoint $Th'(f,s)$ and we have to prove that the adjunction morphisms are isomorphisms.

Consider an open immersion $j : U \longrightarrow S$ and let (f_0, s_0) be the restriction of the smooth S-point (f,s) over U. Property (Loc_s) implies (BC_s) (Corollary 2.3.13). Thus, using also property $(\mathscr{P}\text{-BC})$, we obtain a canonical isomorphism:

$$j^*Th(f,s) \xrightarrow{\sim} Th(f_0,s_0)j^*.$$

Recall also that (Loc$_s$) implies (Supp$_s$) (again Corollary 2.3.13). Thus we get a canonical isomorphism:

$$j_\sharp Th(f_0, s_0) \xrightarrow{\sim} Th(f,s) j_\sharp$$

which gives by adjunction an isomorphism:

$$Th'(f_0, s_0) j^* \xrightarrow{\sim} j^* Th'(f,s).$$

Thus, (*Zar*-sep) shows that the property of $Th(f,s)$ being an equivalence is Zariski local in S.

Consider a point $a \in S$, $x = s(a)$. As X is smooth over S, there exists an open subscheme $U \subset X$, an integer $n \geq 0$ and an étale S-morphism $\pi : U \longrightarrow \mathbf{A}_S^n$ which fits into the following cartesian square:

$$\begin{array}{ccc} S_0 & \longrightarrow & U \\ \downarrow & & \downarrow \pi \\ S & \xrightarrow{v} & \mathbf{A}_S^n, \end{array}$$

where v is the zero section (*cf.* [GD67, 17.12.2]). Note that the scheme $S_0 = s^{-1}(U)$ is an open neighborhood of a in S. Let us put $X_0 = f^{-1}(S_0)$ and $U_0 = U \cap X_0$. Then we get the following commutative diagram:

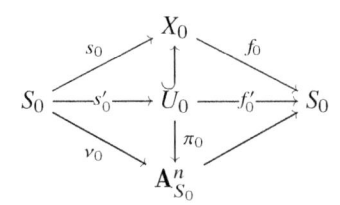

$$\begin{array}{c} X_0 \\ {}^{s_0}\nearrow \quad \uparrow \quad \searrow^{f_0} \\ S_0 \xrightarrow{s_0'} U_0 \xrightarrow{f_0'} S_0 \\ {}_{v_0}\searrow \quad \downarrow^{\pi_0} \\ \mathbf{A}_{S_0}^n \end{array}$$

where π_0 is the restriction of π above S_0 and v_0 is again the zero section. According to Lemma 2.4.9, we get isomorphisms

$$Th(f_0, s_0) \simeq Th(f_0', s_0') \simeq Th(\mathbf{A}_S^n).$$

Thus, according to the beginning of the proof, $Th(f_0, s_0)$ is an equivalence. This finishes the proof because S_0 is an open neighborhood of a in S. □

Definition 2.4.12 Assume that $\mathscr{T}$ is monoidal.

1. For any smooth pointed scheme $(f : X \longrightarrow S, s)$, we put $M_S(X/X - s(S)) := f_\sharp s_*(\mathbb{1}_S)$.
2. For any vector bundle E/S with projection f and zero section s, we define the *Thom premotive* associated with E over S as $MTh_S(E) = f_\sharp s_*(\mathbb{1}_S)$.

2 Triangulated $\mathscr{P}$-fibred categories in algebraic geometry

2.4.13 We assume $\mathscr{T}$ is monoidal and satisfies properties (wLoc) and (Zar-sep).

In each case of the previous definition, if we apply $f_\sharp$ to the distinguished triangle obtained from point (2) of Proposition 2.3.3 applied to s, we get the following canonical distinguished triangles:

$$M_S(X - s(S)) \longrightarrow M_S(X) \longrightarrow M_S(X/X - s(S)) \xrightarrow{+1}$$

$$M_S(E^\times) \longrightarrow M_S(E) \longrightarrow MTh_S(E) \xrightarrow{+1}$$

where the first map is induced by the obvious open immersion.

Moreover, property (Loc$_s$) implies (PF$_s$) (see Corollary 2.3.13). Thus for any premotive K over S, the following composite map is an isomorphism:

(2.4.13.1)

$$Th(f,s).K = f_\sharp s_*(K) = f_\sharp s_*(\mathbb{1}_S \otimes_S s^* f^*(K)) \xrightarrow{Ex(s_*^*, \otimes)^{-1}} f_\sharp(s_*(\mathbb{1}_S) \otimes_X f^*(K))$$
$$\xrightarrow{Ex(f_\sharp^*, \otimes)} (f_\sharp s_*(\mathbb{1}_S)) \otimes_S K = M_S(X/X - s(S)) \otimes_S K.$$

Similarly, in the case of a vector bundle E/S, we get a canonical isomorphism:

$$Th(E).K \xrightarrow{\sim} MTh_S(E) \otimes_S K.$$

From these isomorphisms, we easily deduce the following corollary of the previous proposition:

Corollary 2.4.14 *Consider the above notations and assumptions. Then the following properties are equivalent:*

(i) $\mathscr{T}$ satisfies the stability property.
(ii) For any smooth pointed scheme $(X \longrightarrow S, s)$, the premotive $M_S(X/X - s(S))$ is $\otimes$-invertible.
(iii) For any vector bundle E/S the Thom premotive $MTh_S(E)$ is $\otimes$-invertible.
(iv) For any scheme S, the premotive $MTh_S(\mathbf{A}_S^1)$ is $\otimes$-invertible.

Remark 2.4.15 Assume that $\mathscr{T}$ satisfies the assumptions and the equivalent conditions of the previous corollary. Then, under the notations of Section 2.4.10, we associate with the exact sequence (σ) a canonical isomorphism

(2.4.15.1) $\qquad Th_S(\sigma) : MTh_S(E) \longrightarrow MTh_S(E'') \otimes_S MTh_S(E').$

Recall that Deligne introduced in [Del87, 4.12] the Picard category $\underline{K}(S)$ of *virtual vector bundles* over a scheme S.

Then, it follows from the above isomorphism and the universal properties of $\underline{K}(S)$ (see [Del87, 4.3]) that the functor MTh_S can be extended uniquely to a symmetric monoidal functor:

$$MTh_S : \underline{K}(S) \longrightarrow \mathscr{T}(S).$$

The reader is referred to [Ayo07a, th. 1.5.18] for a detailed argument.

2.4.16 Assume $\mathscr{T}$ is monoidal. For any scheme S, the canonical projection $p : \mathbf{P}_S^1 \longrightarrow S$ is a split epimorphism. A splitting is given by the inclusion of the infinite point $v : S \longrightarrow \mathbf{P}_S^1$. The induced map $p_* : M_S(\mathbf{P}_S^1) \longrightarrow \mathbb{1}_S$ is a split epimorphism. Thus it admits a kernel K in the triangulated category $\mathscr{T}(S)$.

Definition 2.4.17 Under the above assumption and notations, we define the *Tate premotive* over S as the object $\mathbb{1}_S(1) = K[-2]$ of $\mathscr{T}(S)$.

The monoid generated by the cartesian section $(\mathbb{1}_S)_S$ defines a canonical **N**-twist on $\mathscr{T}$ called the *Tate twist*. The n-th Tate twist of an object K is denoted by $K(n)$.

2.4.18 Consider again the assumption of Section 2.4.13.

According to Lemma 2.4.9, we get a canonical isomorphism

$$MTh_S(\mathbf{A}_S^1) = M_S(\mathbf{A}_S^1/\mathbf{A}_S^1 - \{0\}) \longrightarrow M_S(\mathbf{P}_S^1/\mathbf{P}_S^1 - \{0\}).$$

On the other hand, $\mathbb{1}_S(1)[2]$ is by definition the cokernel of the monomorphism $v_* : \mathbb{1}_S \longrightarrow M_S(\mathbf{P}_S^1)$. Thus we get a canonical morphism:

(2.4.18.1) $\qquad \mathbb{1}_S(1)[2] \longrightarrow M_S(\mathbf{P}_S^1/\mathbf{P}_S^1 - \{0\}) \xrightarrow{\sim} MTh_S(\mathbf{A}_S^1).$

From this definition and Corollary 2.4.14 the following result is obvious:

Corollary 2.4.19 *Consider the above assumption and notations. Then the following conditions are equivalent:*

(i) $\mathscr{T}$ satisfies the homotopy property.
(ii) For any scheme S, the arrow (2.4.18.1) is an isomorphism.

When these equivalent assertions are satisfied, the following conditions are equivalent:

(iii) $\mathscr{T}$ satisfies the stability property.
(iv) For any scheme S, the Tate premotive $\mathbb{1}_S(1)$ is $\otimes$-invertible.

2.4.b The purity property

2.4.20 Let $f : X \longrightarrow S$ be a smooth proper morphism in $\mathscr{S}$. We consider the following cartesian square:

(2.4.20.1)
$$\begin{array}{ccc} X \times_S X & \xrightarrow{f''} & X \\ {\scriptstyle f'}\downarrow & \Delta & \downarrow{\scriptstyle f} \\ X & \xrightarrow{f} & S \end{array}$$

where f' (resp. f'') is the projection on the first (resp. second) factor. Let $\delta : X \longrightarrow X \times_S X$ be the diagonal embedding. Note that (f', δ) is a smooth pointed scheme which depends only on f. We put:

2 Triangulated $\mathscr{P}$-fibred categories in algebraic geometry

$$\Sigma_f := Th(f', \delta) = f'_\sharp \delta_*.$$

We then define a canonical morphism:

$$\mathfrak{p}_f : f_\sharp = f_\sharp f''_* \delta_* \xrightarrow{Ex(\Delta_{\sharp *})} f_* f'_\sharp \delta_* = f_* \circ \Sigma_f$$

using the exchange transformation introduced in Section 1.1.15.

Definition 2.4.21 We say that f is $\mathscr{T}$-*pure*, or simply *pure* when $\mathscr{T}$ is clear, when the following conditions are satisfied:

1. The natural transformation Σ_f is an equivalence.
2. The morphism $\mathfrak{p}_f : f_\sharp \longrightarrow f_* \circ \Sigma_f$ is an isomorphism.

Then $\mathfrak{p}_f$ is called the *purity isomorphism* associated with f. We also say that f is *universally $\mathscr{T}$-pure* if f is pure after any base change along a morphism of $\mathscr{S}$.

We introduce the following properties on $\mathscr{T}$:

- $\mathscr{T}$ satisfies the *purity property* (Pur) if any proper smooth morphism is pure.
- $\mathscr{T}$ satisfies the *weak purity property* (wPur) if for any scheme S and any integer $n > 0$, the canonical projection $p_n : \mathbf{P}^n_S \longrightarrow S$ is pure.

Remark 2.4.22 Consider the above notations and assume f is pure.

Then f_* admits a right adjoint $f^!$ and we deduce by transposition from $\mathfrak{p}_f$ a canonical isomorphism:

$$\mathfrak{p}'_f : f^* \longrightarrow \Sigma_f^{-1} \circ f^!.$$

Recall also that, when δ_* admits a right adjoint $\delta^!$, Σ_f admits as a right adjoint the transformation $\Omega_f := \delta^! f^*$. In particular, $\Omega_f = \Sigma_f^{-1}$.

The following lemma shows the importance of the purity property.

Lemma 2.4.23 *Assume that $\mathscr{T}$ satisfies (wLoc). Let $f : Y \longrightarrow X$ be a proper smooth morphism. If f is universally pure then the following conditions hold:*

1. *$\mathscr{T}$ satisfies $(Supp_f)$ and (BC_f).*
2. *For any cartesian square*

$$\begin{array}{ccc} Z & \xrightarrow{\tilde{f}} & Y \\ h \downarrow & \Delta & \downarrow g \\ X & \xrightarrow{f} & S \end{array}$$

such that g is smooth, the exchange transformation:

$$Ex(\Delta_{\sharp *}) : g_\sharp \tilde{f}_* \longrightarrow f_* h_\sharp$$

is an isomorphism.

3. *If moreover $\mathscr{T}$ is monoidal then $\mathscr{T}$ satisfies (PF_f).*

Proof We first prove condition (2). By assumption, the natural transformations $\Sigma_f = f'_\sharp \delta_*$ and $\Sigma_{\tilde f} = \tilde f' \tilde\delta_*$ are equivalences. Thus, it is sufficient to prove that the natural transformation

$$g_\sharp \tilde f_* \Sigma_{\tilde f} \xrightarrow{Ex(\Delta_{\sharp *})} f_* h_\sharp \Sigma_{\tilde f}$$

is an isomorphism.

For ease of notation, let us also introduce the following cartesian squares:

$$\begin{array}{ccccc} Z & \xrightarrow{\tilde\delta} & Z \times_Y Z & \xrightarrow{\tilde f'} & Z \\ {\scriptstyle h}\downarrow & \Gamma & {\scriptstyle k}\downarrow & \Theta & \downarrow{\scriptstyle h} \\ X & \xrightarrow{\delta} & X \times_S X & \xrightarrow{f'} & X \end{array}$$

using the notations of Section 2.4.20. Thus, by definition: $\Sigma_f = f'_\sharp \delta_*$, $\Sigma_{\tilde f} = \tilde f' \tilde\delta_*$. Then we consider the following diagram of exchange transformations:

$$\begin{array}{ccc} g_\sharp \tilde f_\sharp & \xrightarrow{\mathfrak{p}_{\tilde f}} & g_\sharp \tilde f_* \tilde f'_\sharp \tilde\delta_* \\ \parallel & & \downarrow{\scriptstyle Ex(\Delta_{\sharp *})} \\ f_\sharp h_\sharp \xrightarrow{\mathfrak{p}_f} f_* f'_\sharp \delta_* h_\sharp \xleftarrow{Ex(\Gamma_{\sharp *})} f_* f'_\sharp k_\sharp \tilde\delta_* & = & f_* h_\sharp \tilde f'_\sharp \tilde\delta_*. \end{array}$$

Note that it only involves exchange transformations of type $Ex(?_{\sharp *})$: it is commutative by compatibility of these exchange transformations with composition. By assumption, the transformations $\mathfrak{p}_f$ and $\mathfrak{p}_{\tilde f}$ are isomorphisms. Moreover the property (Loc_δ) is satisfied and it implies $(Supp_\delta)$ according to Corollary 2.3.13. Thus $Ex(\Gamma_{\sharp *})$ is an isomorphism and this concludes the proof of (2).

For condition (1), we note that (2) already implies $(Supp_f)$. Thus we have only to prove (BC_f). We consider a square of shape Δ as in the statement of the lemma without assuming that g is smooth. We have to prove that

$$Ex(\Delta^*_*) : g^* f_* \longrightarrow \tilde f_* h^*$$

is an isomorphism. We proceed as for condition (2). It is sufficient to prove that $Ex(\Delta^*_*)$ is an isomorphism after composition on the right with Σ_f. Then we consider the following commutative diagram of exchange transformations:

$$\begin{array}{ccc} g^* f_\sharp & \xrightarrow{\mathfrak{p}_f} & g^* f_* f'_\sharp \delta_* \\ {\scriptstyle Ex(\Delta^*_\sharp)}\downarrow & & \downarrow{\scriptstyle Ex(\Delta^*_*)} \\ \tilde f_\sharp h^* \xrightarrow{\mathfrak{p}_{\tilde f}} \tilde f_* \tilde f'_\sharp \tilde\delta_* h^* \xleftarrow{Ex(\Gamma^*_*)} \tilde f_* \tilde f'_\sharp k^* \delta_* \xleftarrow{Ex(\Theta^*_\sharp)} \tilde f_* h^* f'_\sharp \delta_*. \end{array}$$

According to ($\mathscr{P}$-BC), $Ex(\Delta_\sharp^*)$ and $Ex(\Theta_\sharp^*)$ are isomorphisms. By assumption, $\mathfrak{p}_f$ and $\mathfrak{p}_{\bar{f}}$ are isomorphisms. Moreover, property (Loc$_\delta$) is satisfied and this implies $Ex(\Gamma_*^*)$ is an isomorphism according to Corollary 2.3.13. Condition (1) is proved.

It remains to prove (3). We consider again the notations of the cartesian diagram (2.4.20.1). For any premotives K over X and L over S, we consider the following commutative diagram of exchange transformations (see Remark 1.1.32):

$$\begin{array}{ccc}
f_\sharp(K \otimes f^*(L)) & \xrightarrow{\mathfrak{p}_f} & f_* f_\sharp' \delta_* (K \otimes \delta^* f'^* f^*(L)) \\
& & \downarrow Ex(\delta_*^*, \otimes) \\
& & f_* f_\sharp' (\delta_*(K) \otimes f'^* f^*(L)) \\
Ex(f_\sharp^*, \otimes) \downarrow & & \downarrow Ex(f_\sharp'^*, \otimes) \\
& & f_* (f_\sharp' \delta_*(K) \otimes f^*(L)) \\
& & \downarrow Ex(f_*^*, \otimes) \\
f_\sharp(K) \otimes L & \xrightarrow{\mathfrak{p}_f} & f_* f_\sharp' \delta_*(K) \otimes L.
\end{array}$$

By definition, the exchanges $Ex(f_\sharp^*, \otimes)$ and $Ex(f_\sharp'^*, \otimes)$ are isomorphisms. By assumption, the arrows labeled $\mathfrak{p}_f$ are isomorphisms. Moreover, the property (Loc$_\delta$) is satisfied: Corollary 2.3.13 implies that $Ex(\delta_*^*, \otimes)$ is an isomorphism. We deduce from this that the arrow $Ex(f_*^*, \otimes)$ is an isomorphism. This concludes the proof of (3) as the functor $\Sigma_f = f_\sharp' \delta_*$ is an equivalence according to the hypothesis on f. □

2.4.24 Assume that $\mathscr{T}$ satisfies the support property (Supp). Then we can extend Definition 2.4.21 to the case of a smooth separated morphism of finite type $f : X \longrightarrow S$. We still consider the cartesian square (2.4.20.1) and the diagonal embedding $\delta : X \longrightarrow X \times_S X$. Again, (f', δ) is a smooth pointed scheme so that we can put

$$\Sigma_f := Th(f', \delta) = f_\sharp' \delta_*$$

and we define a canonical morphism:

(2.4.24.1) $\qquad \mathfrak{p}_f : f_\sharp = f_\sharp f_!'' \delta_! \xrightarrow{Ex(\Delta_\sharp!)} f_! f_\sharp' \delta_! = f_! \circ \Sigma_f$

using the exchange transformation of point (2) in Corollary 2.2.12.

Definition 2.4.25 Using the notations above, we say that f is $\mathscr{T}$-*pure*, or simply *pure* when $\mathscr{T}$ is clear, when the following conditions are satisfied:

1. The natural transformation Σ_f is an equivalence.
2. The morphism $\mathfrak{p}_f : f_\sharp \longrightarrow f_! \circ \Sigma_f$ is an isomorphism.

We can easily deduce from the construction of the exchange transformation $Ex(\Delta_\sharp!)$ that, when $\mathscr{T}$ satisfies properties (Stab) and (Pur), any smooth separated morphism of finite type f is pure. The following theorem is a consequence of the formalism developed previously.

Theorem 2.4.26 *Assume that $\mathscr{T}$ satisfies the localization and weak purity properties. Then the following conditions hold:*

1. *$\mathscr{T}$ satisfies the stability property.*
2. *$\mathscr{T}$ satisfies the support and base change properties.*
   If moreover $\mathscr{T}$ is monoidal, it satisfies the projection formula.
3. *Any smooth separated morphism of finite type is pure.*
4. *For any projective morphism f, the property (Adj_f) holds.*
   If moreover $\mathscr{T}$ is well generated, then the adjoint property holds in general.

Proof We start by proving condition (1). As (Loc) implies (Zar-sep), we can apply Proposition 2.4.11 and we have only to prove that for any scheme S, $Th(\mathbf{A}_S^1)$ is an equivalence. Let $s : S \longrightarrow \mathbf{A}_S^1$ be the zero section and $j : \mathbf{A}_S^1 \longrightarrow \mathbf{P}_S^1$ be the canonical open immersion. Put $t = j \circ s$. According to Lemma 2.4.9, j induces an isomorphism $Th(\mathbf{A}_S^1) \simeq Th(p_1, s)$. Consider now the following cartesian squares:

$$\begin{array}{ccccc} S & \xrightarrow{s} & \mathbf{P}_S^1 & \xrightarrow{p_1} & S \\ s \downarrow & & \downarrow s' & \Delta & \downarrow s \\ \mathbf{P}_S^1 & \xrightarrow{\delta} & \mathbf{P}_S^1 \times_S \mathbf{P}_S^1 & \xrightarrow{p_1'} & \mathbf{P}_S^1 \end{array}$$

where p_1' (resp. δ) is the projection on the first factor (resp. diagonal embedding). The property (Loc_s) implies that $s^* s_* = 1$ and that the exchange transformation $Ex(\Delta_{\#*})$ is an isomorphism according to Corollary 2.3.13. Thus we get an isomorphism of functors:

$$Th(p_1, s) = p_{1\#} s_* = s^* s_* p_{1\#} s_* \xrightarrow{Ex(\Delta_{\#*})^{-1}} s^* p_{1\#}' s'_* s_* = s^* p_{1\#}' \delta_* s_* = s^* \Sigma_{p_1} s_*$$

and this proves (1) because p_1 is pure.

Condition (2) follows simply from Corollary 2.3.13. In fact, for any scheme S, the weak purity assumption on $\mathscr{T}$ implies that $p_n : \mathbf{P}_S^n \longrightarrow S$ is universally pure. Thus, Lemma 2.4.23 implies properties (Supp_{p_n}) and (BC_{p_n}) so that we can apply Corollary 2.3.13 to get (Supp) and (BC). The same argument applies to the property (PF) in the monoidal case.

For condition (3), we consider a smooth separated morphism of finite type $g : Y \longrightarrow S$ and we prove it is pure. According to (1), Σ_g is an equivalence. Thus, by definition of $\mathfrak{p}_g$, it is sufficient to prove that for any cartesian square:

$$\begin{array}{ccc} Z & \xrightarrow{\tilde{f}} & Y \\ h \downarrow & \Delta & \downarrow g \\ X & \xrightarrow{f} & S \end{array}$$

with f separated of finite type, the exchange transformation
$$Ex(\Delta_{\sharp !}^*) : g_\sharp \tilde{f}_! \longrightarrow f_! h_\sharp$$
is an isomorphism.

To do this, we apply Proposition 2.3.11, as in the case of Corollary 2.3.13. We consider the obvious complete Sm-fibred triangulated categories $\mathscr{T}'$ and $\mathscr{T}''$ over $\mathscr{S}/S$ which to an S-scheme Y associates:

- $\mathscr{T}'(Y) = \mathscr{T}(Y \times_S X)$.
- $\mathscr{T}''(Y) = \mathscr{T}(Y)$.

We consider the morphism $\varphi^* : \mathscr{T}' \longrightarrow \mathscr{T}''$ such that for any S-scheme Y, $\varphi_Y^* = (Y \times_S p)_\sharp$. As for any scheme S, $p_n : \mathbf{P}_S^n \longrightarrow S$ is universally pure, Lemma 2.4.23 shows that φ^* satisfies condition (i) of Proposition 2.3.11. According to that Proposition, (i) is equivalent to condition (iii), and (iii) is precisely what we want.

It remains only to prove condition (4). According to property (Pur), any smooth proper morphism f satisfies (Adj$_f$). According to (Loc) and Proposition 2.3.3 any closed immersion i satisfies (Adj$_i$). It follows easily that any projective morphism f satisfies (Adj$_f$). When $\mathscr{T}$ is well generated, we simply apply point (4) of Corollary 2.3.13. □

Remark 2.4.27 In particular, in the assumption of the previous theorem, if $\mathscr{T}$ satisfies properties (Loc), (wPur) and (Adj),[51] we can apply Theorem 2.2.14 to $\mathscr{T}$ so that we get a complete formalism of operations $(f^*, f_*, f_!, f^!)$ satisfying all the desired formulas.

Thus the preceding theorem gives another look at the main result of [Ayo07a, 1.4.2]. In fact, the proof given here is simpler as the assumptions of our theorem are stronger. However, we do not use the homotopy property in our theorem.

We end up this section with a theorem due to Ayoub [Ayo07a, 1.4.2]. The particular case $\mathscr{T}(X) = SH(X)$ was also established by Röndigs in [Rön05], after Voevodsky, with a proof which extends immediately to Ayoub's axiomatic setting. It may be stated in a simpler form, according to Theorem 2.4.26 above:

Theorem 2.4.28 (Voevodsky–Röndigs–Ayoub) *Assume $\mathscr{T}$ satisfies the localization, homotopy and stability properties. Then $\mathscr{T}$ is weakly pure.*

In fact, this theorem is stated explicitly in [Ayo07a, Theorem 1.7.9].

Remark 2.4.29 Recall that Ayoub proves more than just this theorem: indeed he constructs the whole formalism of the six functors for quasi-projective morphisms for his *monoidal homotopy stable functors* — see again [Ayo07a]. Similarly, the fact that one can deduce the proper base change formula from relative purity was also observed by Röndigs [Rön05]. The work we have done here is to isolate the crucial

---
[51] Note that under the assumptions of the previous theorem, we know that for any proper smooth morphism f, f_* admits a right adjoint. The same is true for a proper morphism which can be factorized as a closed immersion followed by a smooth proper morphism according to (Loc).

properties of purity and weak purity. Also, using the construction of Deligne, we see how to avoid the assumption of quasi-projectiveness made by Ayoub. Finally, the interest of Theorem 2.4.26 is to give a possible approach to the *six functors formalism* without requiring the homotopy property.

2.4.c Duality, purity and orientation

2.4.30 This section is concerned with the relation between purity and duality. We will assume that $\mathscr{T}$ is premotivic.

Recall that an object M of a monoidal category $\mathscr{M}$ is called *strongly dualizable* if there exists an object M' such that $(M' \otimes -)$ is both right and left adjoint to $(M \otimes -)$. Then, M' is called the *strong dual* of M.

If $\mathscr{M}$ is closed monoidal, we will say that a morphism of the form

$$\mu : M \otimes M' \longrightarrow \mathbb{1}$$

is a *perfect pairing* if the natural transformation

$$(M \otimes -) \longrightarrow Hom(M', -)$$

obtained from μ by adjunction is an isomorphism. Then M is strongly dualizable with dual M'.

Proposition 2.4.31 *Let $f : X \longrightarrow S$ be a smooth proper morphism. If f is pure then the premotive $M_S(X)$ is strongly dualizable in $\mathscr{T}(S)$ with dual:*

$$f_*(\mathbb{1}_X) \simeq f_\sharp\bigl(\Omega_f(\mathbb{1}_X)\bigr),$$

where Ω_f denotes the inverse of Σ_f.

Proof By assumption, Σ_f is an automorphism of the category $\mathscr{T}(X)$. Moreover, the identification (2.4.13.1) can be rewritten as $\Sigma_f(M) = \Sigma_f(\mathbb{1}_X) \otimes_X M$ for any premotive M over X. The fact that Σ_f is an equivalence means that $\Sigma_f(\mathbb{1}_X)$ is a $\otimes$-invertible object, whose inverse is $T := \Omega_f(\mathbb{1}_S)$. In particular, we get: $\Omega_f(M) = T \otimes M$.

According to the Sm-projection formula, the functor $M_S(X) \otimes .$ is isomorphic to $f_\sharp f^*$. Thus, its right adjoint is $f_* f^*$. As f is pure by assumption, this last functor is isomorphic to $f_\sharp \Omega_f f^*$. Using the observation at the beginning of the proof and the Sm-projection formula again, we obtain:

$$f_\sharp \Omega_f f^*(N) = f_\sharp(T \otimes f^*(N)) = f_\sharp(T) \otimes N.$$

Moreover, the right adjoint of $f_\sharp \Omega_f f^*$ is $f_* \Sigma_f f^*$. Using again the purity isomorphism for f, this last functor can be identified with $f_\sharp f^*$ and this concludes the proof. □

2.4.32 Assume again that the premotivic triangulated category $\mathscr{T}$ satisfies properties (wLoc) and (Nis-sep).

Let S be a scheme. A *smooth closed S-pair* will be a pair (X, Z) of smooth S-schemes such that Z is a closed subscheme of X. We consider the canonical projection $p : X \longrightarrow S$ and the immersion $i : Z \longrightarrow X$ associated with (X, Z). Note that according to Proposition 2.4.8, $\mathscr{T}$ satisfies property (Loc_i). Then we define the premotive of (X, Z) as follows:

(2.4.32.1) $$M_S(X/X - Z) := p_\sharp i_*(\mathbb{1}_Z).$$

According to property (Loc_i), we thus get a canonical distinguished triangle:

(2.4.32.2) $$M_S(X - Z) \xrightarrow{j_*} M_S(X) \longrightarrow M_S(X/X - Z) \xrightarrow{+1}$$

Note that given any smooth morphism $p : S \longrightarrow S_0$, we obviously have:

(2.4.32.3) $$p_\sharp M_S(X/X - Z) = M_{S_0}(X/X - Z).$$

Moreover, given any morphism $f : T \longrightarrow S$, we get an exchange isomorphism:

(2.4.32.4) $$f^* M_S(X/X - Z) \xrightarrow{\sim} M_T(X_T/X_T - Z_T).$$

A morphism of smooth closed S-pairs $(Y, T) \longrightarrow (X, Z)$ will be a pair (f, g) which fits into a commutative diagram

$$\begin{array}{ccc} T & \xrightarrow{k} & Y \\ g \downarrow & \Delta & \downarrow f \\ Z & \xrightarrow{i} & X, \end{array}$$

with i, k the canonical immersions, and such that $T = f^{-1}(Z)$ as a set. We can associate with (f, g) a morphism of premotives:

$$M_S(Y/Y - T) = q_\sharp k_* g^*(\mathbb{1}_Z) \xrightarrow{Ex(\Delta_*^*)^{-1}} q_\sharp f^* i_*(\mathbb{1}_Z) \xrightarrow{Ex_\sharp^*} p_\sharp i_*(\mathbb{1}_Z) = M_S(X/X - Z).$$

Indeed, the exchange map $Ex(\Delta_*^*)$ is an isomorphism according to (Loc_i) and Corollary 2.3.13.

It is easy to check that the triangle (2.4.32.2) is functorial with respect to morphisms of closed S-pairs. Before proving the next theorem, we state the following lemma.

Lemma 2.4.33 *Consider the assumptions and notations above.*
Let $(f, g) : (Y, T) \longrightarrow (X, Z)$ be a morphism of smooth closed S-pairs such that f is étale and g is an isomorphism. Then the induced map $M_S(Y/Y - T) \longrightarrow M_S(X/X - Z)$ is an isomorphism.

Proof According to the identification 2.4.32.3, it is sufficient to treat the case where $X = Z$. Let $U = X - Z$ and $j : U \longrightarrow X$ be the obvious immersion. Then (f, j) is a Nisnevich cover of X. According to (*Nis*-sep), it is sufficient to prove that the pullback of $M_X(Y/Y - T) \longrightarrow M_X(X/X - Z)$ along f and j is an isomorphism. This is obvious using 2.4.32.4. □

2.4.34 We consider again the assumption of the section preceding the above lemma.

Fix a smooth closed S-pair (X, Z). Let $B_Z X$ (resp. $B_Z(\mathbf{A}_X^1)$) be the blow-up of X (resp. $\mathbf{A}_X^1$) with center in Z (resp. $\{0\} \times Z$). We define the deformation space associated with (X, Z) as the S-scheme $D_Z X = B_Z(\mathbf{A}_X^1) - B_Z X$. Note also $D_Z Z = \mathbf{A}_Z^1$ is a closed subscheme of $D_Z X$; the pair $(D_Z X, \mathbf{A}_Z^1)$ is a smooth closed S-pair.

Let $N_Z X$ be the normal bundle of Z in X. The scheme $D_Z X$ is fibred over $\mathbf{A}^1$. Moreover, the 0-fiber of $(D_Z X, \mathbf{A}^1)$ is the closed pair $(N_Z X, Z)$ corresponding to the zero section and the 1-fiber is the closed pair (X, Z). In particular, we get the following morphisms of closed pairs:

(2.4.34.1) $\qquad (X, Z) \xrightarrow{d_1} (D_Z X, \mathbf{A}_Z^1) \xleftarrow{d_0} (N_Z X, Z).$

We are now ready to state the purity theorem for smooth closed pairs in our abstract formalism. Though our assumptions are more general, this theorem follows exactly from the method of Morel and Voevodsky used to prove this result in the homotopy category $\mathcal{H}$ (see [MV99, §3, 2.24]):

Theorem 2.4.35 *Consider the above assumptions and notations and suppose that $\mathcal{T}$ satisfies the homotopy property. Then the morphisms*

$$M_S(X/X - Z) \xrightarrow{d_{1*}} M_S(D_Z X/D_Z X - \mathbf{A}_Z^1) \xleftarrow{d_{0*}} M_S(N_Z X/N_Z^\times X) =: MTh_S(N_Z X)$$

are isomorphisms.

Proof By noetherian induction and the preceding lemma, the statement is local in X for the Nisnevich topology. Thus, because (X, Z) is a smooth closed S-pair, we can assume that there exists an étale map $\pi : X \longrightarrow \mathbf{A}_S^{n+c}$ such that $\pi^{-1}(\mathbf{A}_S^c) = Z$, *cf.* [GD67, 17.12.2]. Consider the pullback square

$$\begin{array}{ccc} X' & \xrightarrow{p} & X \\ q \downarrow & & \downarrow \pi \\ \mathbf{A}^n \times Z & \xrightarrow{1 \times \pi|Z} & \mathbf{A}^n \times \mathbf{A}_S^c. \end{array}$$

There is an obvious closed immersion $Z \longrightarrow X'$ and its image is contained in $q^{-1}(Z)$. As q is étale, Z is a direct factor of $q^{-1}(Z)$. Put $W = q^{-1}(Z) - Z$ and $\Omega = X' - W$. Thus Ω is an open subscheme of X', and the reader can check that p and q induces morphisms of smooth closed S-pairs

$$(X, Z) \longleftarrow (\Omega, Z) \longrightarrow (\mathbf{A}_Z^n, Z).$$

Applying again the preceding lemma, these morphisms induce isomorphisms on the associated premotives. Thus we are reduced to the case of the closed S-pair $(\mathbf{A}^n_Z, Z)$. A direct computation shows that $D_Z(\mathbf{A}^n_Z) \simeq \mathbf{A}^1 \times \mathbf{A}^n_Z$. Under this isomorphism d_0 (resp. d_1) corresponds to the 0-section (resp. 1-section) of $\mathbf{A}^1 \times \mathbf{A}^n_Z$ corresponding to the first factor. Thus, the proof is concluded by using the homotopy property. $\square$

2.4.36 The interest of the previous theorem is to simplify the purity isomorphism. Let us restate the assumptions on the triangulated premotivic category $\mathscr{T}$:

- $\mathscr{T}$ satisfies properties (Nis-sep), (wLoc) and (Htp).

Then applying the above theorem, we get for any smooth closed S-pair (X, Z) a canonical isomorphism

(2.4.36.1) $$\mathfrak{p}_{X,Z} : M_S(X/X - Z) \longrightarrow MTh_S(N_Z X).$$

Corollary 2.4.37 *Consider the assumptions and notations above.*

1. *For any smooth pointed S-scheme (f, s) and any premotive K over S, we get a canonical isomorphism*

$$Th(f, s).K \simeq M_S(X/X - s(S)) \otimes_S K \xrightarrow{\mathfrak{p}_{X,s}} MTh_S(N_s) \otimes_S K,$$

where the first isomorphism is given by the map (2.4.13.1) and N_s is the normal bundle of s.

2. *For any smooth separated morphism of finite type $f : X \longrightarrow S$ with tangent bundle [52] T_f, and any premotive K over X, we get a canonical isomorphism:*

$$\mathfrak{p}_{XX,X} : \Sigma_f(K) \xrightarrow{\sim} MTh_X(T_f) \otimes_X K$$

— here, (XX, X) stands for the closed pair corresponding to the diagonal embedding of X/S.

In the assumption of point (2), we thus get a canonical map:

(2.4.37.1) $$f_\sharp(K) \xrightarrow{\mathfrak{p}_f} f_!(\Sigma_f K) \xrightarrow{\sim} f_!\big(MTh_X(T_f) \otimes_X K\big)$$

that we will still denote by $\mathfrak{p}_f$ and call the *purity isomorphism* associated with f.

Definition 2.4.38 Assume the triangulated premotivic category $\mathscr{T}$ satisfies (wLoc). As usual, $M(1)$ denotes the Tate twist of a premotive M.

An *orientation* $\mathfrak{t}$ of $\mathscr{T}$ will be the data for each smooth scheme X and each vector bundle E/X of rank n of an isomorphism

$$\mathfrak{t}_E : MTh_X(E) \longrightarrow \mathbb{1}_X(n)[2n],$$

called the *Thom isomorphism*, satisfying the following coherence properties:

---
[52] We define T_f as the normal bundle of the diagonal immersion $\delta : X \longrightarrow X \times_S X$.

(a) Given a scheme X and an isomorphism of vector bundles $\varphi : E \longrightarrow F$ of ranks n over X, the following diagram is commutative:

$$\begin{array}{ccc} MTh_X(E) & \xrightarrow{\varphi_*} & MTh_X(F) \\ & \searrow{\scriptstyle t_E} \quad \swarrow{\scriptstyle t_F} & \\ & \mathbb{1}_X(n)[2n] & \end{array}$$

(b) For any morphism $f : Y \longrightarrow X$ of schemes, and any vector bundle E/X of rank n with pullback F over Y, the following diagram commutes:

$$\begin{array}{ccc} f^*(MTh_X(E)) & \xrightarrow{f^*t_E} & f^*(\mathbb{1}_X(n)[2n]) \\ {\scriptstyle \sim}\downarrow & & \downarrow{\scriptstyle \sim} \\ MTh_Y(F) & \xrightarrow{t_F} & \mathbb{1}_Y(n)[2n] \end{array}$$

where the vertical maps are the canonical isomorphisms.

(c) For any scheme X and any exact sequence (σ) of vector bundles over X

$$0 \longrightarrow E' \xrightarrow{\nu} E \xrightarrow{\pi} E'' \longrightarrow 0,$$

if n (resp. m) denotes the rank of the vector bundle E' (resp. E''), the following diagram commutes:

$$\begin{array}{ccc} MTh_X(E) & \xrightarrow{Th_X(\sigma)} & MTh_X(E') \otimes MTh_X(E'') \\ {\scriptstyle t_E}\downarrow & & \downarrow{\scriptstyle t_{E'} \otimes t_{E''}} \\ \mathbb{1}_X(n+m)[2n+2m] & \longrightarrow & \mathbb{1}_X(n)[2n] \otimes \mathbb{1}_X(m)[2m] \end{array}$$

where the map $Th_X(\sigma)$ is the isomorphism (2.4.15.1) associated with (σ) and the bottom vertical one is the obvious identification.

We will also say that $\mathscr{T}$ is *oriented* when the choice of one particular orientation is not essential.

Note that the Thom isomorphism can be viewed as a cohomology class in

$$H^{2n,n}_{\mathscr{T}}(Th_X(E)) := \mathrm{Hom}_{\mathscr{T}(X)}\left(MTh_X(E), \mathbb{1}_S(n)[2n]\right),$$

which in classical homotopy theory is called the *Thom class*.

2.4.39 Suppose the triangulated premotivic category $\mathscr{T}$ satisfies the following properties:

- $\mathscr{T}$ satisfies properties (*Nis*-sep), (wLoc), (Htp).
- $\mathscr{T}$ admits an orientation t.

Consider a smooth closed S-pair (X, Z) of codimension n. Let p (resp. q) be the structural morphism of X/S (resp. Z/S) and $i : Z \longrightarrow X$ the associated immersion. Then we associate with (X, Z) the following form of the purity isomorphism:

$$(2.4.39.1) \qquad \mathfrak{p}^t_{X,Z} : M_S(X/X - Z) \xrightarrow{\mathfrak{p}_{X,Z}} MTh_S(N_Z X) \xrightarrow{q_\sharp(\mathfrak{t}_{N_Z X})} M_S(Z)(n)[2n],$$

where $\mathfrak{p}_{X,Z}$ is the isomorphism (2.4.36.1). For future reference, note that we deduce from this the so-called Gysin morphism:

$$(2.4.39.2) \qquad i^* : M_S(X) \xrightarrow{\pi} M_S(X/X - Z) \xrightarrow{\mathfrak{p}^t_{X,Z}} M_S(Z)(n)[2n],$$

where π is the following map:

$$M_S(X) = p_\sharp(\mathbb{1}_X) \xrightarrow{ad(i^*, i_*)} p_\sharp i_* i^*(\mathbb{1}_X) = M_S(X/X - Z).$$

As a particular case, we get using the notation of Corollary 2.4.37, point (2), an isomorphism:

$$\mathfrak{p}^t_{XX,X} : \Sigma_f(K) \xrightarrow{\mathfrak{p}_{XX,X}} MTh_X(T_f) \otimes K \xrightarrow{\mathfrak{t}_{T_f}} K(d)[2d].$$

In particular, when $\mathscr{T}$ satisfies property (Supp), the purity comparison map associated with f can be rewritten as:

$$(2.4.39.3) \qquad \mathfrak{p}^t_f : f_\sharp \xrightarrow{\mathfrak{p}_f} f_! \circ \Sigma_f \xrightarrow{\mathfrak{p}^t_{XX,X}} f_!(d)[2d].$$

Example 2.4.40 Assume as in the above definition that $\mathscr{T}$ is premotivic and satisfies properties (wLoc) and (Nis-sep).

We suppose the following two additional conditions are fulfilled:

(a') There exists a morphism of triangulated premotivic categories:

$$\varphi^* : \mathrm{SH} \rightleftarrows \mathscr{T} : \varphi_*,$$

where SH is the stable homotopy category of Morel and Voevodsky — see Example 1.4.3.

(b') For any scheme X, let $\mathrm{Pic}(X)$ be the Picard group of X. We assume there exists a map

$$c_1 : \mathrm{Pic}(X) \longrightarrow H^{2,1}_{\mathscr{T}}(X) := \mathrm{Hom}_{\mathscr{T}(X)}(M(X), \mathbb{1}_X(1)[2])$$

which is natural with respect to contravariant functoriality — we do not require c_1 to be a morphism of abelian groups.

Then one can apply the results of [Dég08] to $\mathscr{T}(X)$ for any scheme X. All of the following references will be within *loc. cit.*: according to section 2.3.2, the triangulated category $\mathscr{T}(X)$ satisfies the axioms of Paragraph 2.1.[53] Then the existence of the Thom isomorphism follows from Proposition 4.3 and, more explicitly, from Paragraph 4.4. Property (a) and (b) of the above definition are easy — explicitly, this is a consequence of 4.10 — and Property (c) follows from Lemma 4.30.

To sum up, assumptions (a') and (b') guarantee the existence of a canonical orientation of $\mathscr{T}$ in the sense of the above definition. Moreover, the purity isomorphism (2.4.39.1) as well as the associated Gysin morphism (2.4.39.2) for this particular orientation coincide with the one defined in [Dég08] (see in particular the uniqueness statement of [Dég08, Prop. 4.3]).

Note moreover that assuming $\mathscr{T}$ satisfies all the properties above except (b'), the data of an orientation of $\mathscr{T}$ is equivalent to the data of a map c_1 as in (b'). Indeed, if t is an orientation of $\mathscr{T}$, given any line bundle L/X with zero section s, we put $c_1(L) = \rho(t_L)$ where ρ is the following composite map:

$$H_{\mathscr{T}}^{2,1}(Th_X(L)) \longrightarrow H_{\mathscr{T}}^{2,1}(L) \xrightarrow{s^*} H_{\mathscr{T}}^{2,1}(X),$$

where the first map is induced by the canonical projection $M_X(L) \longrightarrow MTh_X(L)$. Then c_1 depends only on the isomorphism classes of L/X — property (a) of the above definition — and it is compatible with pullbacks — property (c) of the above definition.

2.4.41 We now assume the following conditions on the triangulated premotivic category $\mathscr{T}$:

- $\mathscr{T}$ satisfies properties (Nis-sep), (wLoc), (Htp) and (Stab).
- $\mathscr{T}$ admits an orientation t.

Let $f : X \longrightarrow S$ be a smooth proper morphism of dimension d. Note we do not need $\mathscr{T}$ to satisfy property (Supp) to rewrite the purity comparison map as follows:

(2.4.41.1) $$\mathfrak{p}_f^t : f_\sharp \longrightarrow f_*(d)[2d]$$

(see Section 2.4.39).

Note also that using the Gysin morphism (2.4.39.2) associated with the diagonal immersion $\delta : X \longrightarrow X \times_S X$, we get the following morphism:
(2.4.41.2)
$$\mu_f^t : M_S(X) \otimes M_S(X)(-d)[-2d] = M_S(X \times_S X)(-d)[-2d] \xrightarrow{\delta^*} M_S(X) \xrightarrow{f_*} \mathbb{1}_S.$$

Theorem 2.4.42 *Consider the assumptions and notations above. Then the following conditions are equivalent:*

(i) f is pure: $\mathfrak{p}_f$ is an isomorphism.

---

[53] Note in particular that for any smooth closed S-pair, we obtain a canonical isomorphism in $\mathscr{T}(S)$ of the form:
$$\varphi^*(\Sigma^\infty X/X - Z) \simeq M_S(X/X - Z)$$
where on the left-hand side $X/X - Z$ stands for the homotopy cofiber of the open immersion $(X - Z) \longrightarrow X$ while the right-hand side is defined by Equality (2.4.32.1).

2 Triangulated $\mathscr{P}$-fibred categories in algebraic geometry

(i') The natural transformation $\mathfrak{p}_f.f^*$ is an isomorphism.
(ii) The premotive $M_S(X)$ is strongly dualizable and μ^t_f is a perfect pairing.

Proof In this proof, we put $\tau(K) = K(d)[2d]$. As $\mathscr{T}$ satisfies property (Stab), f_* commutes with Tate twists (Definition 1.1.44). This means the following exchange transformation is an isomorphism:

$$(2.4.42.1) \qquad Ex_\tau : \tau f_* \longrightarrow f_*\tau.$$

We first prove that (i) is equivalent to (i'). One implication is obvious so that we have only to prove that (i') implies (i). Guided by a method of Ayoub (see [Ayo07a, 1.7.14, 1.7.15]), we will construct a right inverse ϕ_1 and a left inverse ϕ_2 to the morphism $\mathfrak{p}^t_f$ as the following composite maps:

$$\phi_1 : f_*\tau \xrightarrow{ad(f^*,f_*)} f_*f^*f_*\tau \xrightarrow{Ex_\tau^{-1}} f_*f^*\tau f_* = f_*\tau f^* f_* \xrightarrow{(\mathfrak{p}^t_f.f^*f_*)^{-1}} f_\sharp f^* f_* \xrightarrow{ad'(f^*,f_*)} f_\sharp$$

$$\phi_2 : f_*\tau \xrightarrow{\beta_f} f_*\tau f^* f_\sharp \xrightarrow{(\mathfrak{p}^t_f.f^*f_\sharp)^{-1}} f_\sharp f^* f_\sharp \xrightarrow{ad'(f_\sharp,f^*)} f_\sharp.$$

Let us check that $\mathfrak{p}^t_f \circ \phi_1 = 1$. To prove this relation, we prove that the following diagram is commutative:

[commutative diagram]

The commutativity of (1) and (2) is obvious and the commutativity of (3) follows from Formula (2.4.42.1) defining Ex_τ. Then the result follows from the usual formula between the unit and counit of an adjunction. The relation $\phi_2 \circ \mathfrak{p}^t_f = 1$ is proved using the same kind of computations.

It remains to prove that (i) and (i') are equivalent to (ii). We already know from Proposition 2.4.31 that (i) implies the premotive $M_S(X)$ is strongly dualizable. Saying that μ^t_f is a perfect pairing amounts to proving that the natural transformation obtained by adjunction

$$d^t_f : (M_S(X) \otimes -) \longrightarrow Hom(M_S(X), -(d)[2d])$$

is an isomorphism. On the other hand, as we have already seen previously, the smooth projection formula implies an identification of functors:

(2.4.42.2)
$$f_\sharp f^* \simeq (M_S(X) \otimes -),$$
$$f_* f^* \simeq Hom(M_S(X), -).$$

Thus, to finish the proof, it will be enough to show that the map

$$f_\sharp f^* \xrightarrow{p_f^! f^*} f_* \tau f^* = f_* f^* \tau$$

is equal to d_f^t through the identifications (2.4.42.2).

Let us consider the following cartesian square

$$\begin{array}{ccc} X \times_S X & \xrightarrow{f''} & X \\ {\scriptstyle f'}\downarrow & \Delta & \downarrow{\scriptstyle f} \\ X & \xrightarrow{f} & S \end{array}$$

and put $g = f \circ f''$. According to the definition of μ_f^t, and notably Formula (2.4.39.2) for the Gysin map δ^*, the natural transformation of functors $(\mu_f^t \otimes -)$ can be described as the following composition:

$$f_\sharp f^* f_\sharp f^* \xrightarrow{Ex(\Delta_\sharp^*)} f_\sharp f'_\sharp f''^* f^* = g_\sharp g^* \xrightarrow{ad(\delta^*,\delta_*)} g_\sharp \delta_* \delta^* g^*$$

$$= f_\sharp f'_\sharp \delta_* f^* \xrightarrow{p_{X\times X,X}^!} f_\sharp \tau f^* = f_\sharp f^* \tau \xrightarrow{ad'(f_\sharp, f^*)} \tau.$$

Note in particular that the base change map $Ex(\Delta_\sharp^*)$ corresponds to the first identification in Formula (2.4.41.2). Thus we have to prove the preceding composite map is equal to the following one, obtained by adjunction from $p_f^!$:

$$f_\sharp f^* f_\sharp f^* = f_\sharp f^* f_\sharp f''_* \delta_* f^* \xrightarrow{Ex(\Delta_{\sharp *})} f_\sharp f^* f_* f'_\sharp \delta_* f^*$$

$$\xrightarrow{p_{X\times X,X}^!} f_\sharp f^* f_* \tau f^* = f_\sharp f^* f_* f^* \tau \xrightarrow{ad'(f^*, f_*)} f_\sharp f^* \tau \xrightarrow{ad'(f_\sharp, f^*)} \tau.$$

This amounts to proving, after some easy cancellation, the commutativity of the following diagram:

$$\begin{array}{ccccc} f^* f_\sharp & = & f^* f_\sharp f''_* \delta_* & \xrightarrow{Ex(\Delta_{\sharp *})} & f^* f_* f'_\sharp \delta_* \\ {\scriptstyle Ex(\Delta_\sharp^*)}\downarrow & & & & \downarrow{\scriptstyle ad'(f^*, f_*)} \\ f'_\sharp f''^* & \xrightarrow{ad(\delta^*,\delta_*)} & f'_\sharp \delta_* \delta^* f''^* & = & f'_\sharp \delta_*. \end{array}$$

According to the definition of the exchange transformation $Ex(\Delta_{\sharp*})$ (cf. Example 1.1.14), we can divide this diagram into the following pieces:

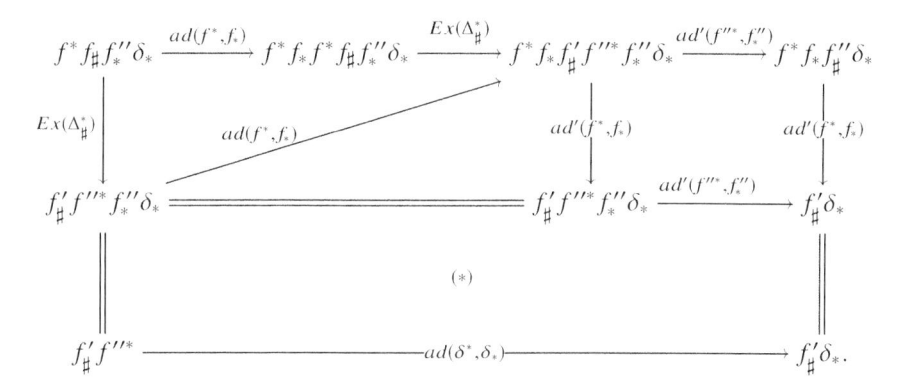

$$\begin{array}{ccccccc}
f^*f_\sharp f''_*\delta_* & \xrightarrow{ad(f^*,f_*)} & f^*f_*f^*f_\sharp f''_*\delta_* & \xrightarrow{Ex(\Delta_\sharp^*)} & f^*f_*f'_\sharp f'''^*f''_*\delta_* & \xrightarrow{ad'(f''^*,f''_*)} & f^*f_*f'_\sharp \delta_* \\
{\scriptstyle Ex(\Delta_\sharp^*)}\downarrow & & {\scriptstyle ad(f^*,f_*)}\nearrow & & \downarrow{\scriptstyle ad'(f^*,f_*)} & & \downarrow{\scriptstyle ad'(f^*,f_*)} \\
f'_\sharp f'''^*f''_*\delta_* & = & = & = & f'_\sharp f'''^*f''_*\delta_* & \xrightarrow{ad'(f'''^*,f''_*)} & f'_\sharp \delta_* \\
\| & & & (*) & & & \| \\
f'_\sharp f'''^* & & & \xrightarrow{ad(\delta^*,\delta_*)} & & & f'_\sharp \delta_*.
\end{array}$$

Every part of this diagram is obviously commutative except for part $(*)$. As $f''\delta = 1$, the axioms of 2-functors (for f^* and f_* say) implies that the unit map

$$\alpha : f'_\sharp f'''^* \longrightarrow f'_\sharp f'''^*(f''\delta)_*(f''\delta)^*$$

is the canonical identification that we get using $1_* = 1$ and $1^* = 1$. We can consider the following diagram:

$$\begin{array}{ccccccc}
f'_\sharp f'''^* & \xrightarrow{\alpha} & f'_\sharp f'''^*(f''\delta)_*(f''\delta)^* & = & f'_\sharp f'''^*f''_*\delta_* \\
\| & & \| & & \| \\
f'_\sharp f'''^* & \xrightarrow{ad(f'''^*,f''_*)} & f'_\sharp f'''^*f''_*f'''^* & \xrightarrow{ad(\delta^*,\delta_*)} & f'_\sharp f'''^*(f''\delta)_*(f''\delta)^* & & {\scriptstyle ad'(f'''^*,f''_*)}\downarrow \\
\| & & \downarrow{\scriptstyle ad'(f'''^*,f''_*)} & & \downarrow{\scriptstyle ad'(f'''^*,f''_*)} & & \\
f'_\sharp f'''^* & = & f'_\sharp f'''^* & \xrightarrow{ad(\delta^*,\delta_*)} & f'_\sharp \delta_*\delta^*f'''^* & = & f'_\sharp \delta_*
\end{array}$$

for which each part is obviously commutative. This concludes the proof. $\square$

As a corollary, together with the results of [Dég08], we get the following:

Corollary 2.4.43 *Let us assume the following conditions on the triangulated premotivic category $\mathscr{T}$:*

(a) $\mathscr{T}$ satisfies properties (Nis-sep), (wLoc), (Htp) and (Stab).
(b) $\mathscr{T}$ admits an orientation t.
(c) There exists a morphism of triangulated premotivic categories:

$$\varphi^* : \mathrm{SH} \rightleftarrows \mathscr{T} : \varphi_* \, .$$

Then any smooth projective morphism is $\mathscr{T}$-pure. In particular, $\mathscr{T}$ is weakly pure.

Proof According to Example 2.4.40, one can apply the results of [Dég08] to the triangulated category $\mathscr{T}(X)$. Then it follows from [Dég08, 5.23] that condition (ii) of the above theorem is satisfied. □

Remark 2.4.44 This theorem should be compared with the result of Ayoub recalled in Theorem 2.4.28. On the one hand, if $\mathscr{T}$ satisfies the localization property, we get another proof of this result under the additional assumption that $\mathscr{T}$ is oriented. On the other hand, the above theorem does not require the assumption that $\mathscr{T}$ satisfies (Loc); this is important as we can only prove (wLoc) for the category DM_Λ introduced in Definition 11.1.1.

2.4.d Motivic categories

This section summarizes the main constructions of this part and draws a conclusive theorem.

Definition 2.4.45 A *motivic triangulated category over $\mathscr{S}$* is a premotivic triangulated category over $\mathscr{S}$ which satisfies the homotopy, stability, localization and adjoint property.

Remark 2.4.46 Without the adjoint property, this definition corresponds to what Ayoub called a *monoidal stable homotopy 2-functor* (cf. [Ayo07a, def. 2.3.1]). We think our shorter terminology fits well in the spirit of the current theory of mixed motives.

Remark 2.4.47 Assume $\mathscr{T}$ is a premotivic triangulated category such that:

1. $\mathscr{T}$ is well generated.
2. $\mathscr{T}$ satisfies the homotopy and stability properties.
3. $\mathscr{T}$ satisfies the localization property.

Then $\mathscr{T}$ is a motivic triangulated category in the above sense. Indeed, property (Adj) is proved under the above assumptions in point (4) of Theorem 2.4.26. Note also that if $\mathscr{T}$ is compactly τ-generated, we simply obtain property (Adj) from Lemma 2.2.16.[54]

Example 2.4.48 According to the previous remark, the premotivic category SH of Example 1.4.3 is a motivic category. In fact, property (1) is proved in [Ayo07a, 4.5.67], property (2) follows by definition and property (3) is proved in [Ayo07a, 4.5.44].

2.4.49 In the next theorem, we summarize what is now called the *Grothendieck six functors formalism*. In fact, this is a consequence of the axioms in the above definition, as a result of the work done in previous sections. More precisely:

---

[54] In our examples, (1) will always be satisfied, (2) will be obtained by construction and (3) will be the hard point.

- We apply Theorem 2.4.26 using the theorem of Ayoub recalled in Theorem 2.4.28, and use the generalized theorem of Morel and Voevodsky, Theorem 2.4.35, to get the form (2.4.37.1) of the purity isomorphism.
- In the case where $\mathscr{T}$ is oriented, we use the form (2.4.41.1) of the purity isomorphism. Recall that, when $\mathscr{T}$ satisfies assumption (c) of Corollary 2.4.43, then we have given a different proof of the Theorem of Ayoub and the theorem below follows from Theorem 2.4.26 and Corollary 2.4.43.

Theorem 2.4.50 Let $\mathscr{T}$ be a motivic triangulated category.
Then, for any separated morphism of finite type $f : Y \longrightarrow X$ in $\mathscr{S}$, there exists a pair of adjoint functors, the exceptional functors,

$$f_! : \mathscr{T}(Y) \rightleftarrows \mathscr{T}(X) : f^!$$

such that:

1. There exists a structure of a covariant (resp. contravariant) 2-functor on $f \mapsto f_!$ (resp. $f \mapsto f^!$).
2. There exists a natural transformation $\alpha_f : f_! \longrightarrow f_*$ which is an isomorphism when f is proper. Moreover, α is a morphism of 2-functors.
3. For any smooth separated morphism of finite type $f : X \longrightarrow S$ in $\mathscr{S}$ with tangent bundle T_f, there are canonical natural isomorphisms

$$\mathfrak{p}_f : f_\sharp \longrightarrow f_!\bigl(MTh_X(T_f) \otimes_X .\bigr)$$
$$\mathfrak{p}'_f : f^* \longrightarrow MTh_X(-T_f) \otimes_X f^!$$

which are dual to each other – the Thom premotive $MTh_X(T_f)$ is $\otimes$-invertible with inverse $MTh_X(-T_f)$.
If $\mathscr{T}$ admits an orientation $\mathfrak{t}$ and f has dimension d then there are canonical natural isomorphisms

$$\mathfrak{p}_f^\mathfrak{t} : f_\sharp \longrightarrow f_!(d)[2d]$$
$$\mathfrak{p}_f'^\mathfrak{t} : f^* \longrightarrow f^!(-d)[-2d]$$

which are dual to each other.
4. For any Cartesian square:

$$\begin{array}{ccc} Y' & \xrightarrow{f'} & X' \\ g' \downarrow & \Delta & \downarrow g \\ Y & \xrightarrow{f} & X \end{array}$$

such that f is separated of finite type, there exist natural isomorphisms

$$Ex(\Delta_!^*) : g^* f_! \xrightarrow{\sim} f'_! g'^*,$$
$$Ex(\Delta_*^!) : g'_* f'^! \xrightarrow{\sim} f^! g_*.$$

5. For any separated morphism of finite type $f : Y \longrightarrow X$ in $\mathscr{S}$, there exist natural isomorphisms

$$Ex(f_!^*, \otimes) : (f_!K) \otimes_X L \xrightarrow{\sim} f_!(K \otimes_Y f^*L),$$

$$Hom_X(f_!(L), K) \xrightarrow{\sim} f_* Hom_Y(L, f^!(K)),$$

$$f^! Hom_X(L, M) \xrightarrow{\sim} Hom_Y(f^*(L), f^!(M)).$$

Remark 2.4.51 It is important to note that in the case where the morphisms in $\mathscr{S}$ are assumed to be quasi-projective, this theorem is proved by Ayoub in [Ayo07a] if we exclude the case where $\mathscr{T}$ is oriented in point (3).[55]

Regarding this theorem, our contribution is to extend Ayoub's result to the non-quasi-projective case and to consider the oriented case — which is crucial in the theory of motives. Recall also we have given another proof of this result in the case where the motivic category $\mathscr{T}$ satisfies in addition the assumptions of Corollary 2.4.43 — which will always be the case for the different categories of motives introduced here.

Remark 2.4.52 The purity isomorphism is compatible with composition. Given smooth separated morphisms of finite type

$$Y \xrightarrow{g} X \xrightarrow{f} S$$

we obtain (*cf.* [GD67, 17.2.3]) an exact sequence of vector bundles over Y

$$(\sigma) \qquad 0 \longrightarrow g^{-1}T_f \longrightarrow T_{fg} \longrightarrow T_g \longrightarrow 0,$$

which according to Remark 2.4.15 induces an isomorphism:

$$\epsilon_\sigma : MTh_Y(T_{fg}) \xrightarrow{MTh_Y(\sigma)} MTh_Y(T_g) \otimes_Y MTh_Y(g^{-1}T_f)$$

$$\xrightarrow{\sim} g^* MTh_X(T_f) \otimes_Y MTh_Y(T_g).$$

One can check the following diagram is commutative:

---

[55] This theorem was first announced by Voevodsky but only notes covering the basic setting were to be found by the time Ayoub wrote the proof.

2 Triangulated $\mathscr{P}$-fibred categories in algebraic geometry

$$\begin{CD}
(fg)_\sharp(K) @= f_\sharp g_\sharp(K) \\
@VV{\mathfrak{p}_{fg}}V @VV{\mathfrak{p}_f \circ \mathfrak{p}_g}V \\
@. f_!\Big(MTh_X(T_f) \otimes_X g_!\big(MTh_Y(T_g) \otimes_Y K\big)\Big) \\
@. @VV{Ex(g_!^*,\otimes)^{-1}}V \\
@. f_! g_!\big(g^* MTh_Y(T_f) \otimes_Y MTh_Y(T_g) \otimes_Y K\big) \\
@. @VV{\epsilon_\sigma^{-1}}V \\
(fg)_!(MTh(T_{fg}) \otimes K) @= f_! g_!(MTh(T_{fg}) \otimes K).
\end{CD}$$

This is not an easy check.[56] In fact, this is one of the key technical point in the proof of Ayoub's main theorem ([Ayo07a, 1.4.2]). We refer the reader to [Ayo07a, 1.5] for details.

Note also that given the commutativity of the above diagram, if $\mathscr{T}$ admits an orientation t, it readily follows from axiom (c) of Definition 2.4.38 that the following diagram is commutative:

$$\begin{CD}
(fg)_\sharp(K) @= f_\sharp g_\sharp(K) \\
@VV{\mathfrak{p}^t_{fg}}V @VV{\mathfrak{p}^t_f \circ \mathfrak{p}^t_g}V \\
(fg)_!(K)(n+m)[2n+2m] @= f_! g_!(K)(n+m)[2n+2m],
\end{CD}$$

where n (resp. m) is the relative dimension of f (resp. g).

Morphisms of triangulated motivic categories are compatible with the six Grothendieck operations in the following sense:

Proposition 2.4.53 *Let $\mathscr{T}$ and $\mathscr{T}'$ be motivic triangulated categories and*

$$\varphi^* : \mathscr{T} \rightleftarrows \mathscr{T}' : \varphi_*$$

be an adjunction of premotivic categories.

Then φ^* (resp. φ_*) commutes with the operations f^* (resp. f_*), for any morphism of schemes f, as well as with the operation $p_!$ (resp. $p^!$), for any separated morphism of finite type p.

Moreover, φ^* is monoidal and for any premotive $M \in \mathscr{T}(S)$, $N \in \mathscr{T}'(S)$, the canonical map

$$Hom(M, \varphi_*(N)) \longrightarrow \varphi_* Hom(\varphi^*(M), N)$$

is an isomorphism.

---

[56] The main point is to check that the isomorphism of Theorem 2.4.35 is compatible with composition (of closed immersions). On that particular point, see [Dég08, Th. 4.32, Cor. 4.33].

Proof The only thing to prove is that φ^* commutes with $p_!$ since the other statements follow either from the definitions or by adjunction. This follows from Proposition 2.3.11, the purity property in $\mathscr{T}$ and $\mathscr{T}'$ (property (3) in the above theorem) and the fact that φ^* commutes with $p_\sharp$ when p is smooth by assumption. □

Remark 2.4.54 With additional assumptions on $\mathscr{T}$ and $\mathscr{T}'$, and over a field, we will see that φ^* commutes with all six operations (see Theorem 4.4.25).

3 Descent in $\mathscr{P}$-fibred model categories

3.0.1 In this section, $\mathscr{S}$ is an abstract category and $\mathscr{P}$ an admissible class of morphisms in $\mathscr{S}$.

In Section 3.3 however, we will consider as in Section 2.0.1 a noetherian base scheme $\mathscr{S}$ and we will assume that $\mathscr{S}$ is an adequate category of $\mathscr{S}$-schemes satisfying the following condition on $\mathscr{S}$:

(a) Any scheme in $\mathscr{S}$ is finite-dimensional.

Moreover, in Sections 3.3.c and 3.3.d, we will even assume:

(a') Any scheme in $\mathscr{S}$ is quasi-excellent and finite-dimensional.

We fix an admissible class $\mathscr{P}$ of morphisms in $\mathscr{S}$ which contains the class of étale morphisms in $\mathscr{S}$ and a stable combinatorial $\mathscr{P}$-fibred model category $\mathscr{M}$ over $\mathscr{S}$ (see Section 1.3.21).

In Section 3.3.d, we will assume furthermore that:

(b) The stable model $\mathscr{P}$-fibred category $\mathscr{M}$ is **Q**-linear (see Section 3.2.14).

3.1 Extension of $\mathscr{P}$-fibred categories to diagrams

3.1.a The general case

3.1.1 Assume given a $\mathscr{P}$-fibered category $\mathscr{M}$ over $\mathscr{S}$. Then $\mathscr{M}$ can be extended to $\mathscr{S}$-diagrams (i.e. functors from a small category to $\mathscr{S}$) as follows. Let I be a small category, and $\mathscr{X}$ a functor from I to $\mathscr{S}$. For an object i of I, we will denote by $\mathscr{X}_i$ the fiber of $\mathscr{X}$ at i (i.e. the evaluation of $\mathscr{X}$ at i), and, for a map $u : i \longrightarrow j$ in I, we will still denote by $u : \mathscr{X}_i \longrightarrow \mathscr{X}_j$ the morphism induced by u. We define the category $\mathscr{M}(\mathscr{X}, I)$ as follows.

An object of $\mathscr{M}(\mathscr{X}, I)$ is a pair (M, a), where M is the data of an object M_i in $\mathscr{M}(\mathscr{X}_i)$ for any object i of I, and a is the data of a morphism $a_u : u^*(M_j) \longrightarrow M_i$ for any morphism $u : i \longrightarrow j$ in I, such that, for any object i of I, the map a_{1_i} is the identity of M_i (we will always assume that 1_i^* is the identity functor), and, for

3 Descent in $\mathscr{P}$-fibred model categories

any composable morphisms $u : i \longrightarrow j$ and $v : j \longrightarrow k$ in I, the following diagram commutes.

$$\begin{array}{ccc} u^*v^*(M_k) & \xrightarrow{\simeq} & (vu)^*(M_k) \\ {\scriptstyle u^*(a_v)}\downarrow & & \downarrow {\scriptstyle a_{vu}} \\ u^*(M_j) & \xrightarrow{a_u} & M_i \end{array}$$

A morphism $p : (M, a) \longrightarrow (N, b)$ is a collection of morphisms

$$p_i : M_i \longrightarrow N_i$$

in $\mathscr{M}(\mathscr{X}_i)$, for each object i in I, such that, for any morphism $u : i \longrightarrow j$ in I, the following diagram commutes.

$$\begin{array}{ccc} u^*(M_j) & \xrightarrow{u^*(p_j)} & u^*(N_j) \\ {\scriptstyle a_u}\downarrow & & \downarrow {\scriptstyle b_u} \\ M_i & \xrightarrow{p_i} & N_i \end{array}$$

In the case where $\mathscr{M}$ is a monoidal $\mathscr{P}$-fibred category, the category $\mathscr{M}(\mathscr{X}, I)$ is naturally endowed with a symmetric monoidal structure. Given two objects (M, a) and (N, b) of $\mathscr{M}(\mathscr{X}, I)$, their tensor product

$$(M, a) \otimes (N, b) = (M \otimes N, a \otimes b)$$

is defined as follows. For any object i of I,

$$(M \otimes N)_i = M_i \otimes N_i \,,$$

and for any map $u : i \longrightarrow j$ in I, the map $(a \otimes b)_u$ is the composition of the isomorphism $u^*(M_j \otimes N_j) \simeq u^*(M_j) \otimes u^*(N_j)$ with the morphism

$$a_u \otimes b_u : u^*(M_j) \otimes u^*(N_j) \longrightarrow M_i \otimes N_i \,.$$

Note finally that if $\mathscr{M}$ is a complete monoidal $\mathscr{P}$-fibred category, then $\mathscr{M}(\mathscr{X}, I)$ admits an internal Hom.

3.1.2 *Evaluation functors.* Assume now that for any S, $\mathscr{M}(S)$ admits small sums. For each object i of I, we have a functor

(3.1.2.1) $$\begin{aligned} i^* : \mathscr{M}(\mathscr{X}, I) &\longrightarrow \mathscr{M}(\mathscr{X}_i) \\ (M, a) &\longmapsto M_i \end{aligned}$$

called the *evaluation functor* associated with i. This functor i^* has a left adjoint

(3.1.2.2) $$i_\sharp : \mathscr{M}(\mathscr{X}_i) \longrightarrow \mathscr{M}(\mathscr{X}, I)$$

defined as follows. If M is an object of $\mathscr{M}(\mathscr{X}_i)$, then $i_\sharp(M)$ is the data (M', a') such that for any object j of I,

$$(i_\sharp(M))_j = M'_j = \coprod_{u \in \mathrm{Hom}_I(j,i)} u^*(M), \tag{3.1.2.3}$$

and, for any morphism $v : k \longrightarrow j$ in I, the map a'_v is the canonical map induced by the collection of maps

$$v^* u^*(M) \simeq (uv)^*(M) \longrightarrow \coprod_{w \in \mathrm{Hom}_I(k,i)} w^*(M) \tag{3.1.2.4}$$

for $u \in \mathrm{Hom}_I(j,i)$.

If we assume that $\mathscr{M}$ is a complete $\mathscr{P}$-fibred category and that $\mathscr{M}(S)$ admits small products for any S, then i^* has a right adjoint

$$i_* : \mathscr{M}(\mathscr{X}_i) \longrightarrow \mathscr{M}(\mathscr{X}, I) \tag{3.1.2.5}$$

given, for any object M of $\mathscr{M}(\mathscr{X}_i)$, by the formula

$$(i_*(M))_j = \prod_{u \in \mathrm{Hom}_I(i,j)} u_*(M), \tag{3.1.2.6}$$

with transition map given by the dual formula of 3.1.2.4.

3.1.3 *Functoriality.* Assume that $\mathscr{M}$ if a $\mathscr{P}$-fibred category such that for any object S of $\mathscr{S}$, $\mathscr{M}(S)$ has small colimits.

Recall that, if $\mathscr{X}$ and $\mathscr{Y}$ are $\mathscr{S}$-diagrams, indexed respectively by small categories I and J, a morphism of $\mathscr{S}$-diagrams $\varphi : (\mathscr{X}, I) \longrightarrow (\mathscr{Y}, J)$ is a pair $\varphi = (\alpha, f)$, where $f : I \longrightarrow J$ is a functor, and $\alpha : \mathscr{X} \longrightarrow f^*(\mathscr{Y})$ is a natural transformation (where $f^*(\mathscr{Y}) = \mathscr{Y} \circ f$). In particular, for any object i of I, we have a morphism

$$\alpha_i : \mathscr{X}_i \longrightarrow \mathscr{Y}_{f(i)}$$

in $\mathscr{S}$. This turns $\mathscr{S}$-diagrams into a strict 2-category: the identity of $(\mathscr{X}, I)$ is the pair $(1_{\mathscr{X}}, 1_I)$, and, if $\varphi = (\alpha, f) : (\mathscr{X}, I) \longrightarrow (\mathscr{Y}, J)$ and $\psi = (\beta, g) : (\mathscr{Y}, J) \longrightarrow (\mathscr{Z}, K)$ are two composable morphisms, the composed morphism $\psi \circ \varphi : (\mathscr{X}, I) \longrightarrow (\mathscr{Z}, K)$ is the pair (gf, γ), where for each object i of I, the map

$$\gamma_i : \mathscr{X}_i \longrightarrow \mathscr{Z}_{g(f(i))}$$

is the composition

$$\mathscr{X}_i \xrightarrow{\alpha_i} \mathscr{Y}_{f(i)} \xrightarrow{\beta_{f(i)}} \mathscr{Z}_{g(f(i))}\,.$$

There is also a notion of natural transformation between morphisms of $\mathscr{S}$-diagrams: if $\varphi = (\alpha, f)$ and $\varphi' = (\alpha', f')$ are two morphisms from $(\mathscr{X}, I)$ to $(\mathscr{Y}, J)$, a natural transformation t from φ to φ' is a natural transformation $t : f \longrightarrow f'$ such that the following diagram of functors commutes.

3 Descent in $\mathscr{P}$-fibred model categories

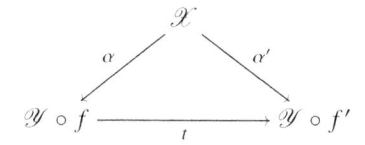

$$\begin{array}{c} & \mathscr{X} & \\ \alpha \swarrow & & \searrow \alpha' \\ \mathscr{Y} \circ f & \xrightarrow{\quad t \quad} & \mathscr{Y} \circ f' \end{array}$$

This makes the category of $\mathscr{S}$-diagrams a (strict) 2-category.

To a morphism of diagrams $\varphi = (\alpha, f) : (\mathscr{X}, I) \longrightarrow (\mathscr{Y}, J)$, we associate a functor

$$\varphi^* : \mathscr{M}(\mathscr{Y}, J) \longrightarrow \mathscr{M}(\mathscr{X}, I)$$

as follows. For an object (M, a) of $\mathscr{M}(\mathscr{Y})$, $\varphi^*(M, a) = (\varphi^*(M), \varphi^*(a))$ is the object of $\mathscr{M}(\mathscr{X})$ defined by $\varphi^*(M)_i = \alpha_i^*(M_{f(i)})$ for i in I, and by the formula

$$\varphi^*(a)_u = \alpha_i^*(a_{f(u)}) : \alpha_i^* f(u)^*(M_{f(j)}) = u^* \alpha_j^*(M_{f(j)}) \longrightarrow \alpha_i^*(M_{f(i)})$$

for $u : i \longrightarrow j$ in I.

We will say that a morphism $\varphi : (\mathscr{X}, I) \longrightarrow (\mathscr{Y}, J)$ is a $\mathscr{P}$-morphism if, for any object i in I, the morphism $\alpha_i : \mathscr{X}_i \longrightarrow \mathscr{Y}_{f(i)}$ is a $\mathscr{P}$-morphism. For such a morphism φ, the functor φ^* has a left adjoint which we denote by

$$\varphi_\sharp : \mathscr{M}(\mathscr{X}, I) \longrightarrow \mathscr{M}(\mathscr{Y}, J).$$

For instance, given a $\mathscr{S}$-diagram $\mathscr{X}$ indexed by a small category I, each object i of I defines a $\mathscr{P}$-morphism of diagrams $i : \mathscr{X}_i \longrightarrow (\mathscr{X}, I)$ (where $\mathscr{X}_i$ is indexed by the terminal category), so that the corresponding functor $i_\sharp$ corresponds precisely to (3.1.2.2).

Assume that $\mathscr{M}$ is a complete $\mathscr{P}$-fibred category such that $\mathscr{M}(S)$ has small limits for any object S of $\mathscr{S}$. Then the functor φ^* has a right adjoint which we denote by

$$\varphi_* : \mathscr{M}(\mathscr{X}, I) \longrightarrow \mathscr{M}(\mathscr{Y}, J).$$

In the case where φ is the morphism $i : \mathscr{X}_i \longrightarrow (\mathscr{X}, I)$ defined by an object i of I, i_* corresponds precisely to (3.1.2.5).

Remark 3.1.4 This construction can be applied in particular to any Grothendieck abelian (monoidal) $\mathscr{P}$-fibred category (cf. Definition 1.3.8). The triangulated case cannot be treated in general without assuming a thorough structure – this is the purpose of the next section.

3.1.b The model category case

3.1.5 Let $\mathscr{M}$ be a $\mathscr{P}$-fibred model category over $\mathscr{S}$ (cf. Section 1.3.21). Given a $\mathscr{S}$-diagram $\mathscr{X}$ indexed by a small category I, we will say that a morphism of $\mathscr{M}(\mathscr{X}, I)$ is a *termwise weak equivalence* (resp. a *termwise fibration*, resp. a *termwise cofibration*) if, for any object i of I, its image under the functor i^* is a weak equivalence (resp. a fibration, resp. a cofibration) in $\mathscr{M}(\mathscr{X}_i)$.

Proposition 3.1.6 *If $\mathscr{M}$ is a cofibrantly generated $\mathscr{P}$-fibred model category over $\mathscr{S}$, then, for any $\mathscr{S}$-diagram $\mathscr{X}$ indexed by a small category I, the category $\mathscr{M}(\mathscr{X}, I)$ is a cofibrantly generated model category whose weak equivalences (resp. fibrations) are the termwise weak equivalences (resp. the termwise fibrations). This model category structure on $\mathscr{M}(\mathscr{X}, I)$ will be called the* projective model structure.

Moreover, any cofibration of $\mathscr{M}(\mathscr{X}, I)$ is a termwise cofibration, and the family of functors
$$i^* : \mathrm{Ho}(\mathscr{M})(\mathscr{X}, I) \longrightarrow \mathrm{Ho}(\mathscr{M})(\mathscr{X}_i), \quad i \in \mathrm{Ob}(I),$$
is conservative.

If $\mathscr{M}$ is left proper (resp. right proper, resp. combinatorial, resp. stable), then so is the projective model category structure on $\mathscr{M}(\mathscr{X})$.

Proof Let $\mathscr{X}^\delta$ be the $\mathscr{S}$-diagram indexed by the set of objects of I (seen as a discrete category), whose fiber at i is $\mathscr{X}_i$. Let $\varphi : (\mathscr{X}^\delta, \mathrm{Ob}\, I) \longrightarrow (\mathscr{X}, I)$ be the inclusion (i.e. the map which is the identity on objects and which is the identity on each fiber). As φ is clearly a $\mathscr{P}$-morphism, we have an adjunction

$$\varphi_\sharp : \mathscr{M}(\mathscr{X}^\delta, \mathrm{Ob}\, I) \simeq \prod_i \mathscr{M}(\mathscr{X}_i) \rightleftarrows \mathscr{M}(\mathscr{X}, I) : \varphi^*.$$

The functor $\varphi_\sharp$ can be made explicit: it sends a family of objects $(M_i)_i$ (with M_i in $\mathscr{M}(\mathscr{X}_i)$) to the sum of the $i_\sharp(M_i)$'s indexed by the set of objects of I. Note also that this proposition is trivially true whenever $\mathscr{X}^\delta = \mathscr{X}$. Using the explicit formula for $i_\sharp$ given in Section 3.1.2, it is then straightforward to check that the adjunction $(\varphi_\sharp, \varphi^*)$ satisfies the assumptions of [Cra95, Theorem 3.3], which proves the existence of the projective model structure on $\mathscr{M}(\mathscr{X}, I)$. Furthermore, the generating cofibrations (resp. trivial cofibrations of $\mathscr{M}(\mathscr{X}, I)$) can be described as follows. For each object i of I, let A_i (resp. B_i) be a generating set of cofibrations (resp. of trivial cofibrations in $\mathscr{M}(\mathscr{X}_i)$. The class of termwise trivial fibrations (resp. of termwise fibrations) of $\mathscr{M}(\mathscr{X}, I)$ is the class of maps which have the right lifting property with respect to the set $A = \cup_{i \in I}\, i_\sharp(A_i)$ (resp. to the set $B = \cup_{i \in I}\, i_\sharp(B_i)$). Hence, the set A (resp. B) generates the class of cofibrations (resp. of trivial cofibrations). In particular, as any element of A is a termwise cofibration (which follows immediately from the explicit formula for $i_\sharp$ given in Section 3.1.2), and as termwise cofibrations are stable under pushouts, transfinite compositions and retracts, any cofibration is a termwise cofibration (by the small object argument).

As any fibration (resp. cofibration) of $\mathscr{M}(\mathscr{X}, I)$ is a termwise fibration (resp. a termwise cofibration), it is clear that, whenever the model categories $\mathscr{M}(\mathscr{X}_i)$ are right (resp. left) proper, the model category $\mathscr{M}(\mathscr{X}, I)$ has the same property.

The functor φ^* preserves fibrations and cofibrations, while it also preserves and detects weak equivalences (by definition). This implies that the induced functor

$$\varphi^* : \mathrm{Ho}(\mathscr{M})(\mathscr{X}, I) \longrightarrow \mathrm{Ho}(\mathscr{M})(\mathscr{X}^\delta, \mathrm{Ob}\, I) \simeq \prod_i \mathrm{Ho}(\mathscr{M})(\mathscr{X}_i)$$

is conservative (using the facts that the set of maps from a cofibrant object to a fibrant object in the homotopy category of a model category is the set of homotopy classes of maps, and that a morphism of a model category is a weak equivalence if and only if it induces an isomorphism in the homotopy category). As φ^* commutes with limits and colimits, this implies that it commutes with homotopy limits and homotopy colimits (up to weak equivalences). Using the conservativity property, this implies that a commutative square of $\mathcal{M}(\mathcal{X},I)$ is a homotopy pushout (resp. a homotopy pullback) if and only if it is so in $\mathcal{M}(\mathcal{X}^\delta, Ob\, I)$. Recall that stable model categories are characterized as those in which a commutative square is a homotopy pullback square if and only if it is a homotopy pushout square. As a consequence, if all the model categories $\mathcal{M}(\mathcal{X}_i)$ are stable, as $\mathcal{M}(\mathcal{X}^\delta, Ob\, I)$ is then obviously stable as well, the model category $\mathcal{M}(\mathcal{X},I)$ has the same property.

It remains to prove that if $\mathcal{M}(X,I)$ is a combinatorial model category for any object X of $\mathcal{S}$, then $\mathcal{M}(\mathcal{X},I)$ is combinatorial as well. For each object i in I, let G_i be a set of accessible generators of $\mathcal{M}(\mathcal{X}_i)$. Note that, for any object i of I, the functor $i_\sharp$ has a left adjoint i^* which commutes to colimits (having itself a right adjoint i_*). It is then easy to check that the set of objects of shape $i_\sharp(M)$, for M in G_i and i in I, is a small set of accessible generators of $\mathcal{M}(\mathcal{X},I)$. This implies that $\mathcal{M}(\mathcal{X},I)$ is accessible and ends the proof. $\square$

Proposition 3.1.7 *Let $\mathcal{M}$ be a combinatorial $\mathcal{P}$-fibred model category over $\mathcal{S}$. Then, for any $\mathcal{S}$-diagram $\mathcal{X}$ indexed by a small category I, the category $\mathcal{M}(\mathcal{X},I)$ is a combinatorial model category whose weak equivalences (resp. cofibrations) are the termwise weak equivalences (resp. the termwise cofibrations). This model category structure on $\mathcal{M}(\mathcal{X},I)$ will be called the* injective model structure.[57] *Moreover, any fibration of the injective model structure on $\mathcal{M}(\mathcal{X},I)$ is a termwise fibration.*

If $\mathcal{M}$ is left proper (resp. right proper, resp. stable), then so is the injective model category structure on $\mathcal{M}(\mathcal{X},I)$.

Proof See [Bar10, Theorem 2.28] for the existence of such a model structure (if, for any object X in $\mathcal{S}$, all the cofibrations of $\mathcal{M}(X)$ are monomorphisms, this can also be done following mutatis mutandis the proof of [Ayo07a, Proposition 4.5.9]). Any trivial cofibration of the projective model structure being a termwise trivial cofibration, any fibration of the injective model structure is a fibration of the projective model structure, hence a termwise fibration.

The assertions about properness follow from their analogs for the projective model structure and from [Cis06, Corollary 1.5.21] (or can be proved directly; see [Bar10, Proposition 2.31]). Similarly, the assertion on stability follows from their analogs for the projective model structure. $\square$

3.1.8 From now on, we assume that a combinatorial $\mathcal{P}$-fibred model category $\mathcal{M}$ over $\mathcal{S}$ is given. Then, for any $\mathcal{S}$-diagram $(\mathcal{X},I)$, we have two model category structures on $\mathcal{M}(\mathcal{X},I)$, and the identity defines a left Quillen equivalence from the

---

[57] Quite unfortunately, this corresponds to the 'semi-projective' model structure introduced in [Ayo07a, Def. 4.5.8].

projective model structure to the injective model structure. This fact will be used to understand the functorialities coming from morphisms of diagrams of S-schemes.

3.1.9 The category of $\mathscr{S}$-diagrams admits small sums. If $\{(\mathscr{Y}_j, I_j)\}_{j \in J}$ is a small family of $\mathscr{S}$-diagrams, then their sum is the $\mathscr{S}$-diagram $(\mathscr{X}, I)$, where

$$I = \coprod_{j \in J} I_j,$$

and $\mathscr{X}$ is the functor from I to $\mathscr{S}$ defined by

$$\mathscr{X}_i = \mathscr{Y}_j \quad \text{whenever } i \in I_j.$$

Proposition 3.1.10 *For any small family of $\mathscr{S}$-diagrams $\{(\mathscr{Y}_j, I_j)\}_{j \in J}$, the canonical functor*

$$\mathrm{Ho}(\mathscr{M})\left(\coprod_{j \in J} \mathscr{Y}_j\right) \longrightarrow \prod_{j \in J} \mathrm{Ho}(\mathscr{M})(\mathscr{Y}_j)$$

is an equivalence of categories.

Proof The functor

$$\mathscr{M}\left(\coprod_{j \in J} \mathscr{Y}_j\right) \longrightarrow \prod_{j \in J} \mathscr{M}(\mathscr{Y}_j)$$

is an equivalence of categories. It thus remains an equivalence after localization. To conclude, it is sufficient to see that the homotopy category of a product of model categories is the product of their homotopy categories, which follows rather easily from the explicit description of the homotopy category of a model category; see e.g. [Hov99, Theorem 1.2.10]. □

Proposition 3.1.11 *Let $\varphi = (\alpha, f) : (\mathscr{X}, I) \longrightarrow (\mathscr{Y}, J)$ be a morphism of $\mathscr{S}$-diagrams.*

(i) *The adjunction $\varphi^* : \mathscr{M}(\mathscr{Y}, J) \rightleftarrows \mathscr{M}(\mathscr{X}, I) : \varphi_*$ is a Quillen adjunction with respect to the injective model structures. In particular, it induces a derived adjunction*

$$\mathbf{L}\varphi^* : \mathrm{Ho}(\mathscr{M})(\mathscr{Y}, J) \rightleftarrows \mathrm{Ho}(\mathscr{M})(\mathscr{X}, I) : \mathbf{R}\varphi_*.$$

(ii) *If φ is a $\mathscr{P}$-morphism, then the adjunction $\varphi_\sharp : \mathscr{M}(\mathscr{X}, I) \rightleftarrows \mathscr{M}(\mathscr{Y}, J) : \varphi^*$ is a Quillen adjunction with respect to the projective model structures, and the functor φ^* preserves weak equivalences. In particular, we get a derived adjunction*

$$\mathbf{L}\varphi_\sharp : \mathrm{Ho}(\mathscr{M})(\mathscr{X}, I) \rightleftarrows \mathrm{Ho}(\mathscr{M})(\mathscr{Y}, J) : \mathbf{L}\varphi^* = \varphi^* = \mathbf{R}\varphi^*.$$

Proof The functor φ^* obviously preserves termwise cofibrations and termwise trivial cofibrations (we reduce to the case of a morphism of $\mathscr{S}$ using the explicit description of φ^* given in Section 3.1.3), which proves the first assertion. Similarly, the second assertion follows from the fact that, under the assumption that φ is a $\mathscr{P}$-morphism,

3 Descent in $\mathscr{P}$-fibred model categories

the functor φ^* preserves termwise weak equivalences (see Remark 1.3.22), as well as termwise fibrations. □

3.1.12 The computation of the (derived) functors $\mathbf{R}\varphi_*$ (and $\mathbf{L}\varphi_\sharp$ whenever it makes sense) given by Proposition 3.1.11 has to do with homotopy limits (and homotopy colimits). It is easier to first understand this in the non-derived version as follows.

Consider first the trivial case of a constant $\mathscr{S}$-diagram: let X be an object of $\mathscr{S}$, and I a small category. Then, regarding X as the constant functor $I \longrightarrow \mathscr{S}$ with value X, we have a projection map $p_I : (X, I) \longrightarrow X$. From the very definition, the category $\mathscr{M}(X, I)$ is simply the category of functors on I with values in $\mathscr{M}(X)$, so that the inverse image functor

$$(3.1.12.1) \qquad p_I^* : \mathscr{M}(X) \longrightarrow \mathscr{M}(X, I) = \mathscr{M}(X)^{I^{op}}$$

is the 'constant diagram functor', while its right adjoint

$$(3.1.12.2) \qquad \varprojlim_{I^{op}} = p_{I,*} : \mathscr{M}(X, I) \longrightarrow \mathscr{M}(X)$$

is the limit functor, and its left adjoint,

$$(3.1.12.3) \qquad \varinjlim_{I^{op}} = p_{I,\sharp} : \mathscr{M}(X, I) \longrightarrow \mathscr{M}(X)$$

is the colimit functor.

Let S be an object of $\mathscr{S}$. A $\mathscr{S}$-*diagram over* S is the data of a $\mathscr{S}$-diagram $(\mathscr{X}, I)$, together with a morphism of $\mathscr{S}$-diagrams $p : (\mathscr{X}, I) \longrightarrow S$ (i.e. it is a $\mathscr{S}/S$-diagram). Such a map p factors as

$$(3.1.12.4) \qquad (\mathscr{X}, I) \xrightarrow{\pi} (S, I) \xrightarrow{p_I} S,$$

where $\pi = (p, 1_I)$. Then one easily checks that, for any object M of $\mathscr{M}(\mathscr{X}, I)$, and for any object i of I, one has

$$(3.1.12.5) \qquad \pi_*(M)_i \simeq p_{i,*}(M_i),$$

where $p_i : \mathscr{X}_i \longrightarrow S$ is the structural map, from which we deduce the formula

$$(3.1.12.6) \qquad p_*(M) \simeq \varprojlim_{i \in I^{op}} \pi_*(M)_i \simeq \varprojlim_{i \in I^{op}} p_{i,*}(M_i).$$

Note that, if I is a small category with a terminal object ω, then any $\mathscr{S}$-diagram $\mathscr{X}$ indexed by I is a $\mathscr{S}$-diagram over $\mathscr{X}_\omega$, and we deduce from the computations above that, if $p : (\mathscr{X}, I) \longrightarrow \mathscr{X}_\omega$ denotes the canonical map, then, for any object M of $\mathscr{M}(\mathscr{X}, I)$,

$$(3.1.12.7) \qquad p_*(M) \simeq M_\omega.$$

Consider now a morphism of $\mathscr{S}$-diagrams $\varphi = (\alpha, f) : (\mathscr{X}, I) \longrightarrow (\mathscr{Y}, J)$. For each object j, we can form the following pullback square of categories.

(3.1.12.8)
$$\begin{array}{ccc} I/j & \xrightarrow{u_j} & I \\ {\scriptstyle f/j}\downarrow & & \downarrow{\scriptstyle f} \\ J/j & \xrightarrow{v_j} & J \end{array}$$

in which J/j is the category of objects of J over j (which has a terminal object, namely $(j, 1_j)$, and v_j is the canonical projection; the category I/j is thus the category of pairs (i, a), where i is an object of I, and $a : f(i) \longrightarrow j$ a morphism in J. From this, we can form the following pullback of $\mathscr{S}$-diagrams

(3.1.12.9)
$$\begin{array}{ccc} (\mathscr{X}/j, I/j) & \xrightarrow{\mu_j} & (\mathscr{X}, I) \\ {\scriptstyle \varphi/j}\downarrow & & \downarrow{\scriptstyle \varphi} \\ (\mathscr{Y}/j, J/j) & \xrightarrow{v_j} & (\mathscr{Y}, J) \end{array}$$

in which $\mathscr{X}/j = \mathscr{X} \circ u_j$, $\mathscr{Y}/j = \mathscr{Y} \circ v_j$, and the maps μ_j and v_j are induced by u_j and v_j respectively. For an object M of $\mathscr{M}(\mathscr{X}, I)$ (resp. an object N of $\mathscr{M}(\mathscr{Y}, J)$), we define M/j (resp. N/j) as the object of $\mathscr{M}(\mathscr{X}/j, I/j)$ (resp. of $\mathscr{M}(\mathscr{Y}/j, J/j)$) obtained as $M/j = \mu_j^*(M)$ (resp. $N/j = v_j^*(N)$). With these conventions, for any object M of $\mathscr{M}(\mathscr{X}, I)$ and any object j of the indexing category J, one gets the formula

(3.1.12.10) $$\varphi_*(M)_j \simeq (\varphi/j)_*(M/j)_{(j,1_j)} \simeq \varprojlim_{(i,a)\in I/j^{op}} \alpha_{i,*}(M_i).$$

This implies that the natural map

(3.1.12.11) $$\varphi_*(M)/j = v_j^*\, \varphi_*(M) \longrightarrow (\varphi/j)_*\, \mu_j^*(M) = (\varphi/j)_*(M/j)$$

is an isomorphism: to prove this, it is sufficient to obtain an isomorphism from (3.1.12.11) after evaluating by any object $(j', a : j' \longrightarrow j)$ of J/j, which follows readily from (3.1.12.10) and from the obvious fact that $(I/j)/(j', a)$ is canonically isomorphic to I/j'.

In order to deduce from the computations above their derived versions, we need two lemmata.

Lemma 3.1.13 *Let $\mathscr{X}$ be a $\mathscr{S}$-diagram indexed by a small category I, and i an object of I. Then the evaluation functor*

$$i^* : \mathscr{M}(\mathscr{X}, I) \longrightarrow \mathscr{M}(\mathscr{X}_i)$$

3 Descent in $\mathscr{P}$-fibred model categories

is a right Quillen functor with respect to the injective model structure, and it preserves weak equivalences.

Proof Proving that the functor i^* is a right Quillen functor is equivalent to proving that its left adjoint (3.1.2.2) is a left Quillen functor with respect to the injective model structure, which follows immediately from its computation (3.1.2.3), as, in any model category, cofibrations and trivial cofibrations are stable under small sums. The last assertion is obvious from the very definition of the weak equivalences in $\mathscr{M}(\mathscr{X}, I)$. □

Lemma 3.1.14 *For any pullback square of $\mathscr{S}$-diagrams of shape (3.1.12.9), the functors*

$$\mu_j^* : \mathscr{M}(\mathscr{X}, I) \longrightarrow \mathscr{M}(\mathscr{X}/j, I/j), \quad M \mapsto M/j,$$
$$\nu_j^* : \mathscr{M}(\mathscr{Y}, I) \longrightarrow \mathscr{M}(\mathscr{Y}/j, J/j), \quad N \mapsto N/j$$

are right Quillen functors with respect to the injective model structure, and they preserve weak equivalences.

Proof It is sufficient to prove this for the functor μ_j^* (as ν_j^* is simply the special case where $I = J$ and f is the identity). The fact that μ_j^* preserves weak equivalences is obvious, so that it remains to prove that it is a right Quillen functor. We thus have to prove that left adjoint of μ_j^*,

$$\mu_{j,\sharp} : \mathscr{M}(\mathscr{X}/j, I/j) \longrightarrow \mathscr{M}(\mathscr{X}, I),$$

is a left Quillen functor. In other words, we have to prove that, for any object i of I, the functor

$$i^* \mu_{j,\sharp} : \mathscr{M}(\mathscr{X}, I) \longrightarrow \mathscr{M}(\mathscr{X})$$

is a left Quillen functor. For any object M of $\mathscr{M}(\mathscr{X}, I)$, we have a natural isomorphism

$$i^* \mu_{j,\sharp}(M) \simeq \coprod_{a \in \mathrm{Hom}_J(f(i), j)} (i, a)_\sharp (M_i).$$

But we know that the functors $(i, a)_\sharp$ are left Quillen functors, so that the stability of cofibrations and trivial cofibrations under small sums and this description of the functor $i^* \mu_{j,\sharp}$ completes the proof. □

Proposition 3.1.15 *Let S be an object of $\mathscr{S}$, and $p : (\mathscr{X}, I) \longrightarrow S$ a $\mathscr{S}$-diagram over S, and consider the canonical factorization (3.1.12.4). For any object M of* $\mathrm{Ho}(\mathscr{M})(\mathscr{X}, I)$, *there are canonical isomorphisms and* $\mathrm{Ho}(\mathscr{M})(S)$:

$$\mathbf{R}\pi_*(M)_i \simeq \mathbf{R}p_{i,*}(M_i) \quad \text{and} \quad \mathbf{R}p_*(M) \simeq \mathbf{R}\varprojlim_{i \in I^\mathrm{op}} \mathbf{R}p_{i,*}(M_i).$$

In particular, if furthermore the category I has a terminal object ω, then

$$\mathbf{R}p_*(M) \simeq \mathbf{R}p_{\omega,*}(M_\omega).$$

Proof This follows immediately from Formulas (3.1.12.5), (3.1.12.6) and from the fact that deriving (right) Quillen functors is compatible with composition. □

Proposition 3.1.16 *We consider the pullback square of $\mathscr{S}$-diagrams (3.1.12.9) (as well as the notations thereof). For any object M of $\mathrm{Ho}(\mathscr{M})(\mathscr{X}, I)$, and any object j of J, we have natural isomorphisms*

$$\mathbf{R}\varphi_*(M)_j \simeq \varprojlim_{(i,a)\in I/j^{op}} \mathbf{R}\alpha_{i,*}(M_i) \quad \text{and} \quad \mathbf{R}\varphi_*(M)/j \simeq \mathbf{R}(\varphi/j)_*(M/j)$$

in $\mathrm{Ho}(\mathscr{M})(\mathscr{Y}_j)$ and in $\mathrm{Ho}(\mathscr{M})(\mathscr{Y}/j, J/j)$, respectively.

Proof Using again the fact that deriving right Quillen functors is compatible with composition, by virtue of Lemma 3.1.13 and Lemma 3.1.14, this is a direct translation of (3.1.12.10) and (3.1.12.11). □

Proposition 3.1.17 *Let $u : T \longrightarrow S$ be a $\mathscr{P}$-morphism of $\mathscr{S}$, and $p : (\mathscr{X}, I) \longrightarrow S$ a $\mathscr{S}$-diagram over S. Consider the pullback square of $\mathscr{S}$-diagrams*

$$\begin{array}{ccc} (\mathscr{Y}, I) & \xrightarrow{\varphi} & (\mathscr{X}, I) \\ q \downarrow & & \downarrow p \\ T & \xrightarrow{u} & S \end{array}$$

(i.e. $\mathscr{Y}_i = T \times_S \mathscr{X}_i$ for any object i of I). Then, for any object M of $\mathrm{Ho}(\mathscr{M})(\mathscr{X}, I)$, the canonical map

$$\mathbf{L}u^* \mathbf{R}p_*(M) \longrightarrow \mathbf{R}q_* \mathbf{L}v^*(M)$$

is an isomorphism in $\mathrm{Ho}(\mathscr{M})(T)$.

Proof By Remark 1.3.22, the functor v^* is both a left and a right Quillen functor which preserves weak equivalences, so that the functor $\mathbf{L}v^* = v^* = \mathbf{R}v^*$ preserves homotopy limits. Hence, by Proposition 3.1.15, one reduces to the case where I is the terminal category, i.e. to the transposition of the isomorphism given by the $\mathscr{P}$-base change formula ($\mathscr{P}$-BC) for the homotopy $\mathscr{P}$-fibred category $\mathrm{Ho}(\mathscr{M})$ (see Proposition 1.1.19). □

3.1.18 A morphism of $\mathscr{S}$-diagrams $\nu = (\alpha, f) : (\mathscr{Y}', J') \longrightarrow (\mathscr{Y}, J)$ is *cartesian* if, for any arrow $i \longrightarrow j$ in J', the induced commutative square

$$\begin{array}{ccc} \mathscr{Y}'_i & \longrightarrow & \mathscr{Y}'_j \\ \alpha_i \downarrow & & \downarrow \alpha_j \\ \mathscr{Y}_{f(i)} & \longrightarrow & \mathscr{Y}_{f(j)} \end{array}$$

is cartesian.

A morphism of $\mathscr{S}$-diagrams $\nu = (\alpha, f) : (\mathscr{Y}', J') \longrightarrow (\mathscr{Y}, J)$ is *reduced* if $J = J'$ and $f = 1_J$.

3 Descent in $\mathscr{P}$-fibred model categories

Proposition 3.1.19 *Let* $v : (\mathscr{Y}', J) \longrightarrow (\mathscr{Y}, J)$ *be a reduced cartesian $\mathscr{P}$-morphism of $\mathscr{S}$-diagrams, and* $\varphi = (\alpha, f) : (\mathscr{X}, I) \longrightarrow (\mathscr{Y}, J)$ *a morphism of $\mathscr{S}$-diagrams. Consider the pullback square of $\mathscr{S}$-diagrams*

$$\begin{CD} (\mathscr{X}', I) @>\mu>> (\mathscr{X}, I) \\ @V\psi VV @VV\varphi V \\ (\mathscr{Y}', J) @>>v> (\mathscr{Y}, J) \end{CD}$$

(i.e. $\mathscr{X}'_i = \mathscr{Y}'_{f(i)} \times_{\mathscr{Y}_{f(i)}} \mathscr{X}_i$ *for any object i of I). Then, for any object M of* $\mathrm{Ho}(\mathscr{M})(\mathscr{X}, I)$, *the canonical map*

$$\mathbf{L}v^* \, \mathbf{R}\varphi_*(M) \longrightarrow \mathbf{R}\psi_* \, \mathbf{L}\mu^*(M)$$

is an isomorphism in $\mathrm{Ho}(\mathscr{M})(\mathscr{Y}', J)$.

Proof By virtue of Proposition 3.1.6, it is sufficient to prove that the map

$$j^* \mathbf{L}v^* \, \mathbf{R}\varphi_*(M) \longrightarrow j^* \mathbf{R}\psi_* \, \mathbf{L}\mu^*(M)$$

is an isomorphism for any object j of J. Let $p : (\mathscr{X}/j, I/j) \longrightarrow \mathscr{Y}_j$ and $q : (\mathscr{X}'/j, J, j) \longrightarrow \mathscr{Y}'_j$ be the canonical maps. As v is cartesian, we have a pullback square of $\mathscr{S}$-diagrams

$$\begin{CD} (\mathscr{X}'/j, I/j) @>\mu/j>> (\mathscr{X}/j, I/j) \\ @VqVV @VVpV \\ \mathscr{Y}'_j @>>v_j> \mathscr{Y}_j. \end{CD}$$

But v_j being a $\mathscr{P}$-morphism, by virtue of Proposition 3.1.17, we thus have an isomorphism

$$\mathbf{L}v_j^* \, \mathbf{R}p_*(M/j) \simeq \mathbf{R}q_* \, \mathbf{L}(\mu/j)^*(M/j) = \mathbf{R}q_*(\mathbf{L}\mu^*(M)/j).$$

Applying Proposition 3.1.16 and the last assertion of Proposition 3.1.15 twice, we also have canonical isomorphisms

$$j^* \mathbf{R}\varphi_*(M) \simeq \mathbf{R}p_*(M/j) \quad \text{and} \quad j^* \mathbf{R}\psi_* \, \mathbf{L}\mu^*(M) \simeq \mathbf{R}q_*(\mathbf{L}\mu^*(M)/j).$$

The obvious identity $j^* \mathbf{L}v^* = \mathbf{L}v_j^* j^*$ completes the proof. □

Corollary 3.1.20 *Under the assumptions of Proposition 3.1.19, for any object N of the category $\operatorname{Ho}(\mathscr{M})(\mathscr{Y}', j)$, the canonical map*

$$\mathbf{L}\mu_\sharp \mathbf{L}\psi^*(N) \longrightarrow \mathbf{L}\varphi^* \mathbf{L}\nu_\sharp(N)$$

is an isomorphism in $\operatorname{Ho}(\mathscr{M})(\mathscr{X}, I)$.

Remark 3.1.21 The class of cartesian $\mathscr{P}$-morphisms form an admissible class of morphisms in the category of $\mathscr{S}$-diagrams, which we denote by $\mathscr{P}_{cart}$. Proposition 3.1.11 and the preceding corollary thus asserts that $\operatorname{Ho}(\mathscr{M})$ is a $\mathscr{P}_{cart}$-fibred category over the category of $\mathscr{S}$-diagrams.

3.1.22 We shall sometimes deal with diagrams of $\mathscr{S}$-diagrams. Let I be a small category, and $\mathscr{F}$ a functor from I to the category of $\mathscr{S}$-diagrams. For each object i of I, we have a $\mathscr{S}$-diagram $(\mathscr{F}(i), J_i)$, and, for each map $u : i \longrightarrow i'$, we have a functor $f_u : J_i \longrightarrow J_{i'}$ as well as a natural transformation $\alpha_u : \mathscr{F}(i) \longrightarrow \mathscr{F}(i') \circ f_u$, subject to coherence identities. In particular, the correspondence $i \mapsto J_i$ defines a functor from I to the category of small categories. Let $I_{\mathscr{F}}$ be the cofibred category over I associated to it; see [Gro03, Exp. VI]. Explicitly, $I_{\mathscr{F}}$ is described as follows. The objects are the pairs (i, x), where i is an object of I, and x is an object of J_i. A morphism $(i, x) \longrightarrow (i', x')$ is a pair (u, v), where $u : i \longrightarrow i'$ is a morphism of I, and $v : f_u(x) \longrightarrow x'$ is a morphism of $J_{i'}$. The identity of (i, x) is the pair $(1_i, 1_x)$, and, for two morphisms $(u, v) : (i, x) \longrightarrow (i', x')$ and $(u', v') : (i', x') \longrightarrow (i'', x'')$, their composition $(u'', v'') : (i, x) \longrightarrow (i'', x'')$ is defined by $u'' = u' \circ u$, while v'' is the composition

$$f_{u''}(x) = f_{u'}(f_u(x)) \xrightarrow{f_{u'}(v)} f_{u'}(x') \xrightarrow{v'} x''.$$

The functor $p : I_{\mathscr{F}} \longrightarrow I$ is simply the projection $(i, x) \mapsto i$. For each object i of I, we get a canonical pullback square of categories

(3.1.22.1)
$$\begin{array}{ccc} J_i & \xrightarrow{\ell_i} & I_{\mathscr{F}} \\ q \downarrow & & \downarrow p \\ e & \xrightarrow{i} & I \end{array}$$

in which i is the functor from the terminal category e which corresponds to the object i, and ℓ_i is the functor defined by $\ell_i(x) = (i, x)$.

The functor $\mathscr{F}$ defines a $\mathscr{S}$-diagram $(\int \mathscr{F}, I_{\mathscr{F}})$: for an object (i, x) of $I_{\mathscr{F}}$, $(\int \mathscr{F})_{(i,x)} = \mathscr{F}(i)_x$, and for a morphism $(u, v) : (i, x) \longrightarrow (i', x')$, the map

$$(u, v) : (\textstyle\int \mathscr{F})_{(i,x)} = \mathscr{F}(i)_x \longrightarrow (\textstyle\int \mathscr{F})_{(i',x')} = \mathscr{F}(i')_{x'}$$

is simply the morphism induced by α_u and v. For each object i of I, there is a natural morphism of $\mathscr{S}$-diagrams

(3.1.22.2) $\qquad\qquad \lambda_i : (\mathscr{F}(i), J_i) \longrightarrow (\textstyle\int \mathscr{F}, I_{\mathscr{F}}),$

3 Descent in $\mathscr{P}$-fibred model categories

given by $\lambda_i = (1_{\mathscr{F}(i)}, \ell_i)$

Proposition 3.1.23 *Let X be an object of $\mathscr{S}$, and $f : \mathscr{F} \longrightarrow X$ a morphism of functors (where X is considered as the constant functor from I to $\mathscr{S}$-diagrams with value the functor from e to $\mathscr{S}$ defined by X). Then, for each object i of I, we have a canonical pullback square of $\mathscr{S}$-diagrams*

$$\begin{array}{ccc} (\mathscr{F}(i), J_i) & \xrightarrow{\lambda_i} & (\int \mathscr{F}, I_{\mathscr{F}}) \\ \varphi_i \downarrow & & \downarrow \varphi \\ X & \xrightarrow{i} & (X, I) \end{array}$$

in which φ and φ_i are the obvious morphisms induced by f (where, this time, (X, I) is seen as the constant functor from I to $\mathscr{S}$ with value X).

Moreover, for any object M of $\mathrm{Ho}(\mathscr{M})(\int \mathscr{F}, I_{\mathscr{F}})$, the natural map

$$i^* \mathbf{R}\varphi_*(M) = \mathbf{R}\varphi_*(M)_i \longrightarrow \mathbf{R}\varphi_{i,*} \lambda_i^*(M)$$

is an isomorphism. In particular, if we also write by abuse of notation f for the induced map of $\mathscr{S}$-diagrams from $(\int \mathscr{F}, I_{\mathscr{F}})$ to X, we have a natural isomorphism

$$\mathbf{R}f_*(M) \simeq \mathbf{R}\varprojlim_{i \in I^{op}} \mathbf{R}\varphi_{i,*} \lambda_i^*(M).$$

Proof This pullback square is the one induced by (3.1.22.1). We shall prove first that the map

$$i^* \mathbf{R}\varphi_*(M) = \mathbf{R}\varphi_*(M)_i \longrightarrow \mathbf{R}\varphi_{i,*} \lambda_i^*(M)$$

is an isomorphism in the particular case where I has a terminal object ω and $i = \omega$. By virtue of Propositions 3.1.15 and 3.1.16, we have isomorphisms

(3.1.23.1) $\qquad \omega^* \mathbf{R}\varphi_*(M) \simeq \mathbf{R}\varprojlim_{i \in I^{op}} \mathbf{R}\varphi_*(M)_i \simeq \mathbf{R}\varprojlim_{(i,x) \in I_{\mathscr{F}}^{op}} \mathbf{R}\varphi_{i,x,*}(M_{(i,x)}),$

where $\varphi_{i,x} : \mathscr{F}(i)_x \longrightarrow X$ denotes the map induced by f. We are thus reduced to proving that the canonical map
(3.1.23.2)
$$\mathbf{R}\varprojlim_{(i,x) \in I_{\mathscr{F}}^{op}} \mathbf{R}\varphi_{i,x,*}(M_{(i,x)}) \longrightarrow \mathbf{R}\varprojlim_{x \in J_\omega^{op}} \mathbf{R}\varphi_{\omega,x,*}(M_{(\omega,x)}) \simeq \mathbf{R}\varphi_{\omega,*} \lambda_\omega^*(M)$$

is an isomorphism. As $I_{\mathscr{F}}$ is cofibred over I, and as ω is a terminal object of I, the inclusion functor $\ell_\omega : J_\omega \longrightarrow I_{\mathscr{F}}$ has a left adjoint, whence is coaspherical in any weak basic localizer (i.e. is homotopy cofinal); see [Malo5, 1.1.9, 1.1.16 and 1.1.25]. As any model category defines a Grothendieck derivator ([Ciso3, Thm. 6.11]), it follows from [Ciso3, Cor. 1.15] that the map (3.1.23.2) is an isomorphism.

To prove the general case, we proceed as follows. Let $\mathscr{F}/i$ be the functor obtained by composing $\mathscr{F}$ with the canonical functor $v_i : I/i \longrightarrow I$. Then, keeping track of the conventions adopted in Section 3.1.12, we easily check that $(I/i)_{\mathscr{F}/i} = (I_{\mathscr{F}})/i$ and that $\int (\mathscr{F}/i) = (\int \mathscr{F})/i$. Moreover, the pullback square (3.1.22.1) is the composition of the following pullback squares of categories:

$$\begin{array}{ccccc} J_i & \xrightarrow{a_i} & I_{\mathscr{F}}/i & \xrightarrow{u_i} & I_{\mathscr{F}} \\ q \downarrow & & p/i \downarrow & & \downarrow p \\ e & \xrightarrow[(i,1_i)]{} & I/i & \xrightarrow[v_i]{} & I. \end{array}$$

The pullback square of the proposition is thus the composition of the following pullback squares:

$$\begin{array}{ccccc} (\mathscr{F}(i), J_i) & \xrightarrow{\alpha_i} & (\int \mathscr{F}/i, I_{\mathscr{F}}/i) & \xrightarrow{\mu_i} & (\int \mathscr{F}, I_{\mathscr{F}}) \\ \varphi_i \downarrow & & \varphi/i \downarrow & & \downarrow \varphi \\ X & \xrightarrow[(i,1_i)]{} & (X, I/i) & \xrightarrow[v_i]{} & (X, I). \end{array}$$

The natural transformations

$$(i, 1_i)^* \, \mathbf{R}(\varphi/i)_* \longrightarrow \mathbf{R}\varphi_{i,*} \, \alpha_i^* \quad \text{and} \quad v_i^* \, \mathbf{R}\varphi_* \longrightarrow \mathbf{R}(\varphi/i)_* \, \mu_i^*$$

are both isomorphisms: the first one comes from the fact that $(i, 1_i)$ is a terminal object of I/i, and the second one from Proposition 3.1.16. We thus get:

$$\begin{aligned} i^* \, \mathbf{R}\varphi_*(M) &\simeq (i, 1_i)^* \, v_i^* \, \mathbf{R}\varphi_*(M) \\ &\simeq (i, 1_i)^* \, \mathbf{R}(\varphi/i)_* \, \mu_i^*(M) \\ &\simeq \mathbf{R}\varphi_{i,*} \, \alpha_i^* \, \mu_i^*(M) \\ &\simeq \mathbf{R}\varphi_{i,*} \, \lambda_i^*(M) \, . \end{aligned}$$

The last assertion of the proposition is then a straightforward application of Proposition 3.1.15. $\square$

Proposition 3.1.24 *If $\mathscr{M}$ is a monoidal $\mathscr{P}$-fibred combinatorial model category over $\mathscr{S}$, then, for any $\mathscr{S}$-diagram $\mathscr{X}$ indexed by a small category I, the injective model structure turns $\mathscr{M}(\mathscr{X}, I)$ into a symmetric monoidal model category. In particular, the categories $\mathrm{Ho}(\mathscr{M})(\mathscr{X}, I)$ are canonically endowed with a closed symmetric monoidal structure, in such a way that, for any morphism of $\mathscr{S}$-diagrams $\varphi : (\mathscr{X}, I) \longrightarrow (\mathscr{Y}, J)$, the functor $\mathbf{L}\varphi^* : \mathrm{Ho}(\mathscr{M})(\mathscr{Y}, J) \longrightarrow \mathrm{Ho}(\mathscr{M})(\mathscr{X}, I)$ is symmetric monoidal.*

Proof This is obvious from the definition of a symmetric monoidal model category, as the tensor product of $\mathscr{M}(\mathscr{X}, I)$ is defined termwise, as well as the cofibrations and the trivial cofibrations. $\square$

3 Descent in $\mathscr{P}$-fibred model categories

Proposition 3.1.25 *Assume that $\mathscr{M}$ is a monoidal $\mathscr{P}$-fibred combinatorial model category over $\mathscr{S}$, and consider a reduced cartesian $\mathscr{P}$-morphism $\varphi = (\alpha, f)$: $(\mathscr{X}, I) \longrightarrow (\mathscr{Y}, I)$. Then, for any object M in $\mathrm{Ho}(\mathscr{M})(\mathscr{X}, I)$ and any object N in $\mathrm{Ho}(\mathscr{M})(\mathscr{Y}, I)$, the canonical map*

$$\mathbf{L}\varphi_\sharp(M \otimes^{\mathbf{L}} \varphi^*(N)) \longrightarrow \mathbf{L}\varphi_\sharp(M) \otimes^{\mathbf{L}} N$$

is an isomorphism.

Proof Let i be an object of I. It is sufficient to prove that the map

$$i^*\mathbf{L}\varphi_\sharp(M \otimes^{\mathbf{L}} \varphi^*(N)) \longrightarrow i^*\mathbf{L}\varphi_\sharp(M) \otimes^{\mathbf{L}} N$$

is an isomorphism in $\mathrm{Ho}(\mathscr{M})(\mathscr{X}_i)$. Using Corollary 3.1.20, we see that this map can be identified with the map

$$\mathbf{L}\varphi_{i,\sharp}(M_i \otimes^{\mathbf{L}} \varphi_i^*(N_i)) \longrightarrow \mathbf{L}\varphi_{i,\sharp}(M_i) \otimes^{\mathbf{L}} N_i,$$

which is an isomorphism according to the $\mathscr{P}$-projection formula for the homotopy $\mathscr{P}$-fibred category $\mathrm{Ho}(\mathscr{M})$. □

3.1.26 Let $(\mathscr{X}, I)$ be a $\mathscr{S}$-diagram. An object M of $\mathscr{M}(\mathscr{X}, I)$ is *homotopy cartesian* if, for any map $u : i \longrightarrow j$ in I, the structural map $u^*(M_j) \longrightarrow M_i$ induces an isomorphism

$$\mathbf{L}u^*(M_i) \simeq M_j$$

in $\mathrm{Ho}(\mathscr{M})(\mathscr{X}, I)$ (i.e. if there exists a weak equivalence $M'_j \longrightarrow M_j$ with M'_j cofibrant in $\mathscr{M}(\mathscr{X}_j)$ such that the map $u^*(M'_j) \longrightarrow M_i$ is a weak equivalence in $\mathscr{M}(\mathscr{X}_i)$).

We denote by $\mathrm{Ho}(\mathscr{M})(\mathscr{X}, I)_{hcart}$ the full subcategory of $\mathrm{Ho}(\mathscr{M})(\mathscr{X}, I)$ spanned by homotopy cartesian sections.

Definition 3.1.27 A cofibrantly generated model category $\mathscr{V}$ is *tractable* if there exist sets I and J of cofibrations between cofibrant objects which generate the class of cofibrations and the class of trivial cofibrations respectively.

Remark 3.1.28 If $\mathscr{M}$ is a combinatorial and tractable $\mathscr{P}$-fibred model category over $\mathscr{S}$, then so are the projective and the injective model structures on $\mathscr{M}(\mathscr{X}, I)$; see [Bar10, Thm. 2.28 and 2.30].

Proposition 3.1.29 *If $\mathscr{M}$ is tractable, then the inclusion functor*

$$\mathrm{Ho}(\mathscr{M})(\mathscr{X}, I)_{hcart} \longrightarrow \mathrm{Ho}(\mathscr{M})(\mathscr{X}, I)$$

admits a right adjoint.

Proof This follows from the fact that the cofibrant homotopy cartesian sections are the cofibrant objects of a right Bousfield localization of the injective model structure on $\mathscr{M}(\mathscr{X}, I)$; see [Bar10, Theorem 5.25]. □

Definition 3.1.30 Let $\mathcal{M}$ and $\mathcal{M}'$ be two $\mathscr{P}$-fibred model categories over $\mathscr{S}$. A *Quillen morphism* γ from $\mathcal{M}$ to $\mathcal{M}'$ is a morphism of $\mathscr{P}$-fibred categories $\gamma : \mathcal{M} \longrightarrow \mathcal{M}'$ such that $\gamma^* : \mathcal{M}(X) \longrightarrow \mathcal{M}'(X)$ is a left Quillen functor for any object X of $\mathscr{S}$.

Remark 3.1.31 If $\gamma : \mathcal{M} \longrightarrow \mathcal{M}'$ is a Quillen morphism between $\mathscr{P}$-fibred combinatorial model categories, then, for any $\mathscr{S}$-diagram $(\mathcal{X}, I)$, we get a Quillen adjunction

$$\gamma^* : \mathcal{M}(\mathcal{X}, I) \rightleftarrows \mathcal{M}'(\mathcal{X}, I) : \gamma_*$$

(with the injective model structures as well as with the projective model structures).

Proposition 3.1.32 *For any Quillen morphism* $\gamma : \mathcal{M} \longrightarrow \mathcal{M}'$, *the derived adjunction*

$$\mathbf{L}\gamma^* : \mathrm{Ho}(\mathcal{M})(X) \rightleftarrows \mathrm{Ho}(\mathcal{M}')(X) : \mathbf{R}\gamma_*$$

defines a morphism of $\mathscr{P}$-fibred categories $\mathrm{Ho}(\mathcal{M}) \longrightarrow \mathrm{Ho}(\mathcal{M}')$ *over $\mathscr{S}$. If moreover $\mathcal{M}$ and $\mathcal{M}'$ are combinatorial, then the morphism* $\mathrm{Ho}(\mathcal{M}) \longrightarrow \mathrm{Ho}(\mathcal{M}')$ *extends to a morphism of $\mathscr{P}_{cart}$-fibred categories over the category of $\mathscr{S}$-diagrams.*

Proof This follows immediately from [Hov99, Theorem 1.4.3]. □

3.2 Hypercovers, descent, and derived global sections

3.2.1 Let $\mathscr{S}$ be an essentially small category, and $\mathscr{P}$ an admissible class of morphisms in $\mathscr{S}$. We assume that a Grothendieck topology t on $\mathscr{S}$ is given. We shall write $\mathscr{S}^{\amalg}$ for the full subcategory of the category of $\mathscr{S}$-diagrams whose objects are the small families $X = \{X_i\}_{i \in I}$ of objects of $\mathscr{S}$ (regarded as functors from a discrete category to $\mathscr{S}$). The category $\mathscr{S}^{\amalg}$ is equivalent to the full subcategory of the category of presheaves of sets on $\mathscr{S}$ spanned by sums of representable presheaves. In particular, small sums are representable in $\mathscr{S}^{\amalg}$ (but note that the functor from $\mathscr{S}$ to $\mathscr{S}^{\amalg}$ does not preserve sums). Finally, we remark that the topology t extends naturally to a Grothendieck topology on $\mathscr{S}^{\amalg}$ such that the topology t on $\mathscr{S}$ is the topology induced from the inclusion $\mathscr{S} \subset \mathscr{S}^{\amalg}$. The covering maps for this topology on $\mathscr{S}^{\amalg}$ will be called *t-covers* (note that the inclusion $\mathscr{S} \subset \mathscr{S}^{\amalg}$ is continuous and induces an equivalence between the topos of t-sheaves on $\mathscr{S}$ and the topos of t-sheaves on $\mathscr{S}^{\amalg}$).

Let Δ be the category of non-empty finite ordinals. Recall that a simplicial object of $\mathscr{S}^{\amalg}$ is a presheaf on Δ with values in $\mathscr{S}^{\amalg}$. For a simplicial set K and an object X of $\mathscr{S}^{\amalg}$, we denote by $K \times X$ the simplicial object of $\mathscr{S}^{\amalg}$ defined by

$$(K \times X)_n = \coprod_{x \in K_n} X \quad , \qquad n \geq 0 \, .$$

We write Δ^n for the standard combinatorial simplex of dimension n, and $i_n : \partial \Delta^n \longrightarrow \Delta^n$ for its boundary inclusion.

3 Descent in $\mathscr{P}$-fibred model categories   99

A morphism $p : \mathscr{X} \longrightarrow \mathscr{Y}$ between simplicial objects of $\mathscr{S}^{II}$ is a *t-hypercover* if, locally for the *t*-topology, it has the right lifting property with respect to boundary inclusions of standard simplices, which, in a more precise way, means that, for any integer $n \geq 0$, any object U of $\mathscr{S}^{II}$, and any commutative square

$$\begin{array}{ccc} \partial\Delta^n \times U & \xrightarrow{x} & \mathscr{X} \\ {\scriptstyle i_n \times 1} \downarrow & & \downarrow {\scriptstyle p} \\ \Delta^n \times U & \xrightarrow{y} & \mathscr{Y}, \end{array}$$

there exists a *t*-covering $q : V \longrightarrow U$, and a morphism of simplicial objects $z : \Delta^n \times V \longrightarrow \mathscr{X}$, such that the diagram below commutes.

$$\begin{array}{ccc} \partial\Delta^n \times V & \xrightarrow{x(1\times q)} & \mathscr{X} \\ {\scriptstyle i_n \times 1} \downarrow & {\scriptstyle z} \nearrow & \downarrow {\scriptstyle p} \\ \Delta^n \times V & \xrightarrow{y(1\times q)} & \mathscr{Y}. \end{array}$$

A *t-hypercover* of an object X of $\mathscr{S}^{II}$ is a *t*-hypercover $p : \mathscr{X} \longrightarrow X$ (where X is considered as a constant simplicial object).

Remark 3.2.2 This definition of *t*-hypercover is equivalent to the one given in [AGV73, Exp. V, 7.3.1.4].

3.2.3 Let $\mathscr{X}$ be a simplicial object of $\mathscr{S}^{II}$. It is in particular a functor from the category Δ^{op} to the category of $\mathscr{S}$-diagrams, so that the constructions and considerations of Section 3.1.22 apply to $\mathscr{X}$. In particular, there is a $\mathscr{S}$-diagram $\tilde{\mathscr{X}}$ associated to $\mathscr{X}$, namely $\tilde{\mathscr{X}} = (\int \mathscr{X}, (\Delta^{op})_{\mathscr{X}})$. More explicitly, for each integer $n \geq 0$, there is a family $\{\mathscr{X}_{n,x}\}_{x \in K_n}$ of objects of $\mathscr{S}$ such that

(3.2.3.1) $$\mathscr{X}_n = \coprod_{x \in K_n} \mathscr{X}_{n,x}.$$

In fact, the sets K_n form a simplicial set K, and the category $(\Delta^{op})_{\mathscr{X}}$ can be identified over Δ^{op} with the category $(\Delta/K)^{op}$, where Δ/K is the fibred category over Δ whose fiber over n is the set K_n (seen as a discrete category), i.e. the category of simplices of K. We shall call K the *underlying simplicial set of* $\mathscr{X}$, while the decomposition (3.2.3.1) will be called the *local presentation of* $\mathscr{X}$. The construction $\mathscr{X} \mapsto \tilde{\mathscr{X}}$ is functorial. If $p : \mathscr{X} \longrightarrow \mathscr{Y}$ is a morphism of simplicial objects of $\mathscr{S}^{II}$, we shall still denote by $p : \tilde{\mathscr{X}} \longrightarrow \tilde{\mathscr{Y}}$ the induced morphism of $\mathscr{S}$-diagrams. In particular, for a morphism of $p : \mathscr{X} \longrightarrow X$, where X is an object of $\mathscr{S}^{II}$, $p : \tilde{\mathscr{X}} \longrightarrow X$ denotes the corresponding morphism of $\mathscr{S}$-diagrams.

Let $\mathscr{M}$ be a $\mathscr{P}$-fibred combinatorial model category over $\mathscr{S}$. Given a simplicial object $\mathscr{X}$ of $\mathscr{S}^{II}$, we define the category $\text{Ho}(\mathscr{M})(\mathscr{X})$ by the formula:

(3.2.3.2) $$\mathrm{Ho}(\mathscr{M})(\mathscr{X}) = \mathrm{Ho}(\mathscr{M})(\int \mathscr{X}, (\Delta^{op})_{\mathscr{X}}).$$

Given an object X of $\mathscr{S}^{\amalg}$ and a morphism $p : \mathscr{X} \longrightarrow X$, we have a derived adjunction

(3.2.3.3) $$\mathbf{L}p^* : \mathrm{Ho}(\mathscr{M})(X) \rightleftarrows \mathrm{Ho}(\mathscr{M})(\mathscr{X}) : \mathbf{R}p_*.$$

Proposition 3.2.4 *Consider an object X of $\mathscr{S}$, a simplicial object $\mathscr{X}$ of $\mathscr{S}^{\amalg}$, as well as a morphism $p : \mathscr{X} \longrightarrow X$. Denote by K the underlying simplicial set of $\mathscr{X}$, and for each integer $n \geq 0$ and each simplex $x \in K_n$, write $p_{n,x} : \mathscr{X}_{n,x} \longrightarrow X$ for the morphism of $\mathscr{S}^{\amalg}$ induced by the local presentation of $\mathscr{X}$ (3.2.3.1). Then, for any object M of $\mathrm{Ho}(\mathscr{M})(X)$, there are canonical isomorphisms*

$$\mathbf{R}p_*\mathbf{R}p^*(M) \simeq \mathbf{R}\varprojlim_{n \in \Delta} \mathbf{R}p_{n,*}\mathbf{L}p_n^*(M) \simeq \mathbf{R}\varprojlim_{n \in \Delta}\Big(\prod_{x \in K_n} \mathbf{R}p_{n,x,*}\mathbf{L}p_{n,x}^*(M)\Big).$$

Proof The first isomorphism is a direct application of the last assertion of Proposition 3.1.23 for $\mathscr{F} = \mathscr{X}$, while the second one follows from the first by Proposition 3.1.10. □

Definition 3.2.5 Given an object Y of $\mathscr{S}^{\amalg}$, an object M of $\mathrm{Ho}(\mathscr{M})(Y)$ will be said to satisfy *t-descent* if it has the following property: for any morphism $f : X \longrightarrow Y$ and any *t*-hypercover $p : \mathscr{X} \longrightarrow X$, the map

$$\mathbf{R}f_*\mathbf{L}f^*(M) \longrightarrow \mathbf{R}f_*\mathbf{R}p_*\mathbf{L}p^*\mathbf{L}f^*(M)$$

is an isomorphism in $\mathrm{Ho}(\mathscr{M})(Y)$.

We shall say that $\mathscr{M}$ (or by abuse, that $\mathrm{Ho}(\mathscr{M})$) satisfies *t-descent* if, for any object Y of $\mathscr{S}^{\amalg}$, any object of $\mathrm{Ho}(\mathscr{M})(Y)$ satisfies *t*-descent.

Proposition 3.2.6 *If $Y = \{Y_i\}_{i \in I}$ is a small family of objects of $\mathscr{S}$ (regarded as an object of $\mathscr{S}^{\amalg}$), then an object M of $\mathrm{Ho}(\mathscr{M})(Y)$ satisfies t-descent if and only if, for any $i \in I$, any morphism $f : X \longrightarrow Y_i$ of $\mathscr{S}$, and any t-hypercover $p : \mathscr{X} \longrightarrow X$, the map*

$$\mathbf{R}f_*\mathbf{L}f^*(M_i) \longrightarrow \mathbf{R}f_*\mathbf{R}p_*\mathbf{L}p^*\mathbf{L}f^*(M_i)$$

is an isomorphism in $\mathrm{Ho}(\mathscr{M})(Y_i)$.

Proof This follows from the definition and from Proposition 3.1.10. □

Corollary 3.2.7 *The $\mathscr{P}$-fibred model category $\mathscr{M}$ satisfies t-descent if and only if, for any object X of $\mathscr{S}$, and any t-hypercover $p : \mathscr{X} \longrightarrow X$, the functor*

$$\mathbf{L}p^* : \mathrm{Ho}(\mathscr{M})(X) \longrightarrow \mathrm{Ho}(\mathscr{M})(\mathscr{X})$$

is fully faithful.

Proposition 3.2.8 *If $\mathscr{M}$ satisfies t-descent, then, for any t-cover $f : Y \longrightarrow X$, the functor*

$$\mathbf{L}f^* : \mathrm{Ho}(\mathscr{M})(X) \longrightarrow \mathrm{Ho}(\mathscr{M})(Y)$$

is conservative.

3 Descent in $\mathscr{P}$-fibred model categories

Proof Let $f : Y \longrightarrow X$ be a t-cover, and $u : M \longrightarrow M'$ a morphism of $\text{Ho}(\mathscr{M})(X)$ whose image under $\mathbf{L}f^*$ is an isomorphism. We can consider the Čech t-hypercover associated to f, that is, the simplicial object $\mathscr{Y}$ over X defined by

$$\mathscr{Y}_n = \underbrace{Y \times_X Y \times_X \cdots \times_X Y}_{n+1 \text{ times}}.$$

Let $p : \mathscr{Y} \longrightarrow X$ be the canonical map. For each $n \geq 0$, the map $p_n : \mathscr{Y}_n \longrightarrow X$ factor through f, from which we deduce that the functor

$$\mathbf{L}p_n^* : \text{Ho}(\mathscr{M})(X) \longrightarrow \text{Ho}(\mathscr{M})(\mathscr{Y}_n)$$

sends u to an isomorphism. This implies that the functor

$$\mathbf{L}p^* : \text{Ho}(\mathscr{M})(X) \longrightarrow \text{Ho}(\mathscr{M})(\mathscr{Y})$$

sends u to an isomorphism as well. Since $\mathscr{Y}$ is a t-hypercover of X, the functor $\mathbf{L}p^*$ is fully faithful, from which we deduce that u is an isomorphism by the Yoneda Lemma. □

3.2.9 Let $\mathscr{V}$ be a complete and cocomplete category. For an object X of $\mathscr{S}$, define $\text{PSh}(\mathscr{S}/X, \mathscr{V})$ as the category of presheaves on $\mathscr{S}/X$ with values in $\mathscr{V}$. Then $\text{PSh}(C/-, \mathscr{V})$ is a $\mathscr{P}$-fibred category (where, by abuse of notation, $\mathscr{S}$ also denotes the class of all maps in $\mathscr{S}$): this is a special case of the constructions described in Section 3.1.2 applied to $\mathscr{V}$, viewed as a fibred category over the terminal category. To be more explicit, for each object X of $\mathscr{S}^{\text{II}}$, we have a $\mathscr{V}$-enriched Yoneda embedding

(3.2.9.1) $\qquad \mathscr{S}^{\text{II}}/X \times \mathscr{V} \longrightarrow \text{PSh}(\mathscr{S}/X, \mathscr{V}) \quad , \quad (U, M\} \longmapsto U \otimes M,$

where, if $U = \{U_i\}_{i \in I}$ is a small family of objects of $\mathscr{S}/X$, $U \otimes M$ is the presheaf

(3.2.9.2) $\qquad V \longmapsto \coprod_{i \in I} \coprod_{a \in \text{Hom}_{\mathscr{S}/S}(V, U_i)} M.$

For a morphism $f : X \longrightarrow Y$ in $\mathscr{S}$, the functor

$$f^* : \text{PSh}(\mathscr{S}/Y, \mathscr{V}) \longrightarrow \text{PSh}(\mathscr{S}/X, \mathscr{V})$$

is the functor defined by composition with the corresponding functor $\mathscr{S}/X \longrightarrow \mathscr{S}/Y$. The functor f^* always has a left adjoint

$$f_\sharp : \text{PSh}(\mathscr{S}/X, \mathscr{V}) \longrightarrow \text{PSh}(\mathscr{S}/Y, \mathscr{V}),$$

which is the unique colimit preserving functor defined by

$$f_\sharp(U \otimes M) = U \otimes M,$$

where, on the left-hand side U is considered as an object over X, while, on the right-hand side, U is considered as an object over Y by composition with f. Similarly, if all the pullbacks by f are representable in $\mathscr{S}$ (e.g. if f is a $\mathscr{P}$-morphism), the functor f^* can be described as the colimit preserving functor defined by the formula

$$f^*(U \otimes M) = (X \times_Y U) \otimes M.$$

If $\mathscr{V}$ is a cofibrantly generated model category, then, for each object X of $\mathscr{S}$, the category $\mathrm{PSh}(\mathscr{S}/X, \mathscr{V})$ is naturally endowed with the *projective model category structure*, i.e. with the cofibrantly generated model category structure whose weak equivalences and fibrations are defined termwise (this is Proposition 3.1.6 applied to $\mathscr{V}$, viewed as a fibred category over the terminal category). The cofibrations of the projective model category structure on $\mathrm{PSh}(\mathscr{S}/X, \mathscr{V})$ will be called the *projective cofibrations*. If moreover $\mathscr{V}$ is combinatorial (resp. left proper, resp. right proper, resp. stable), so is $\mathrm{PSh}(\mathscr{S}/X, \mathscr{V})$. In particular, if $\mathscr{V}$ is a combinatorial model category, then $\mathrm{PSh}(\mathscr{S}/-, \mathscr{V})$ is a $\mathscr{P}$-fibred combinatorial model category over $\mathscr{S}$.

According to Definition 3.2.5, it thus makes sense to speak of t-descent in $\mathrm{PSh}(\mathscr{S}/-, \mathscr{V})$.

If $U = \{U_i\}_{i \in I}$ is a small family of objects of $\mathscr{S}$ over X, and if F is a presheaf over $\mathscr{S}/X$, we define

(3.2.9.3) $$F(U) = \prod_{i \in I} F(U_i).$$

The functor $F \mapsto F(U)$ is a right adjoint to the functor $E \mapsto U \otimes E$.

We remark that a termwise fibrant presheaf F on $\mathscr{S}/X$ satisfies t-descent if and only if, for any object Y of $\mathscr{S}^{\mathrm{II}}$, and any t-hypercover $\mathscr{Y} \longrightarrow Y$ over X, the map

$$F(Y) \longrightarrow \mathbf{R}\varprojlim_{n \in \Delta} F(\mathscr{Y}_n)$$

is an isomorphism in $\mathrm{Ho}(\mathscr{V})$.

Proposition 3.2.10 *If $\mathscr{V}$ is combinatorial and left proper, then the category of presheaves $\mathrm{PSh}(\mathscr{S}/X, \mathscr{V})$ admits a combinatorial model category structure whose cofibrations are the projective cofibrations, and whose fibrant objects are the termwise fibrant objects which satisfy t-descent. This model category structure will be called the t-local model category structure, and the corresponding homotopy category will be denoted by $\mathrm{Ho}_t(\mathrm{PSh}(\mathscr{S}/X, \mathscr{V}))$.*

Moreover, any termwise weak equivalence is a weak equivalence for the t-local model structure, and the induced functor

$$a^* : \mathrm{Ho}(\mathrm{PSh}(\mathscr{S}/X, \mathscr{V})) \longrightarrow \mathrm{Ho}_t(\mathrm{PSh}(\mathscr{S}/X, \mathscr{V}))$$

admits a fully faithful right adjoint

$$a_* : \mathrm{Ho}_t(\mathrm{PSh}(\mathscr{S}/X, \mathscr{V})) \longrightarrow \mathrm{Ho}(\mathrm{PSh}(\mathscr{S}/X, \mathscr{V}))$$

3 Descent in $\mathscr{P}$-fibred model categories

whose essential image consists precisely of the full subcategory of $\mathrm{Ho}(\mathrm{PSh}\,(\mathscr{S}/X,\mathscr{V}))$ spanned by the presheaves which satisfy t-descent.

Proof Let H be the class of maps of shape

(3.2.10.1) $$\underset{n\in\Lambda^{op}}{\mathrm{hocolim}}\,\mathscr{U}_n\otimes E \longrightarrow Y\otimes E,$$

where Y is an object of $\mathscr{S}^{\mathrm{II}}$ over X, $\mathscr{U}\longrightarrow Y$ is a t-hypercover, and E is a cofibrant replacement of an object which is either a source or a target of a generating cofibration of $\mathscr{V}$. Define the t-local model category structure as the left Bousfield localization of $Pr(\mathscr{S}/X,\mathscr{V})$ by H; see [Bar10, Theorem 4.7]. We shall call t-*local weak equivalences* the weak equivalences of the t-local model category structure. For each object Y over X, the functor $Y\otimes(-)$ is a left Quillen functor from $\mathscr{V}$ to $Pr(\mathscr{S}/X,\mathscr{V})$. We thus get a total left derived functor

$$Y\otimes^{\mathbf{L}}(-):\mathrm{Ho}(\mathscr{V})\longrightarrow \mathrm{Ho}_t(\mathrm{PSh}\,(\mathscr{S}/X,\mathscr{V}))$$

whose right adjoint is the evaluation at Y. For any object E of $\mathscr{V}$ and any t-local fibrant presheaf F on $\mathscr{S}/X$ with values in $\mathscr{V}$, we thus have natural bijections

(3.2.10.2) $$\mathrm{Hom}(E,F(Y))\simeq\mathrm{Hom}(Y\otimes^{\mathbf{L}}E,F),$$

and, for any simplicial object $\mathscr{U}$ of $\mathscr{S}/X$, identifications

(3.2.10.3) $$\mathrm{Hom}(E,\mathbf{R}\varprojlim_{n\in\Lambda} F(\mathscr{U}_n))\simeq\mathrm{Hom}(\mathbf{L}\varinjlim_{n\in\Lambda}\mathscr{U}_n\otimes^{\mathbf{L}}E,F).$$

One easily sees that, for any t-hypercover $\mathscr{U}\longrightarrow Y$ and any cofibrant object E of $\mathscr{V}$, the map

(3.2.10.4) $$\mathbf{L}\varinjlim_{n\in\Lambda}\mathscr{U}_n\otimes^{\mathbf{L}}E\longrightarrow Y\otimes^{\mathbf{L}}E$$

is an isomorphism in the t-local homotopy category $\mathrm{Ho}_t(\mathrm{PSh}\,(\mathscr{S}/X,\mathscr{V}))$: by the small object argument, the smallest full subcategory of $\mathrm{Ho}(\mathrm{PSh}\,(\mathscr{S}/X,\mathscr{V}))$ which is stable by homotopy colimits and which contains the source and the targets of the generating cofibrations is $\mathrm{Ho}_t(\mathrm{PSh}\,(\mathscr{S}/X,\mathscr{V}))$ itself, and the class of objects E of $\mathscr{V}$ such that the map (3.2.10.4) is an isomorphism in $\mathrm{Ho}(\mathscr{V})$ is stable under homotopy colimits. Similarly, we see that, for any object E, the functor $(-)\otimes^{\mathbf{L}}E$ preserves sums. As a consequence, we get from (3.2.10.2) and (3.2.10.3) that the fibrant objects of the t-local model category structure are precisely the termwise fibrant objects F of the projective model structure which satisfy t-descent. The last part of the proposition follows from the general yoga of left Bousfield localizations. □

3.2.11 Let $\mathscr{M}$ be a $\mathscr{P}$-fibred combinatorial model category over $\mathscr{S}$, and S an object of $\mathscr{S}$. Denote by
$$\mathscr{S} : \mathscr{S}/S \longrightarrow \mathscr{S}$$
the canonical forgetful functor. Then there is a canonical morphism of $\mathscr{S}$-diagrams

(3.2.11.1) $$\sigma : (\mathscr{S}, \mathscr{S}/S) \longrightarrow (S, \mathscr{S}/S)$$

(where $(S, \mathscr{S}/S)$ stands for the constant diagram with value S). This defines a functor
(3.2.11.2)
$$\mathbf{R}\sigma_* : \mathrm{Ho}(\mathscr{M})(\mathscr{S}, \mathscr{S}/S) \longrightarrow \mathrm{Ho}(\mathscr{M})(S, \mathscr{S}/S) = \mathrm{Ho}(\mathrm{PSh}\,(\mathscr{S}/S, \mathscr{M}(S)))\,.$$

For an object M of $\mathrm{Ho}(\mathscr{M})(S)$, one defines the presheaf of *geometric derived global sections of M over S* by the formula

(3.2.11.3) $$\mathbf{R}\Gamma_{geom}(-, M) = \mathbf{R}\sigma_* \, \mathbf{L}\sigma^*(M)\,.$$

This is a presheaf on $\mathscr{S}/S$ with values in $\mathscr{M}(S)$ whose evaluation on a morphism $f : X \longrightarrow S$ is, by virtue of Propositions 3.1.15 and 3.1.16,

(3.2.11.4) $$\mathbf{R}\Gamma_{geom}(X, M) \simeq \mathbf{R}f_* \, \mathbf{L}f^*(M)\,.$$

Proposition 3.2.12 *For an object M of $\mathrm{Ho}(\mathscr{M})(S)$, the following conditions are equivalent.*

(a) The object M satisfies t-descent.
(b) The presheaf $\mathbf{R}\Gamma_{geom}(-, M)$ satisfies t-descent.

Proof For any morphism $f : X \longrightarrow S$ and any t-hypercover $p : \mathscr{X} \longrightarrow X$ over S, we have, by Proposition 3.2.4 and formula (3.2.11.4), an isomorphism

$$\mathbf{R}f_* \mathbf{R}p_* \, \mathbf{L}p^* \, \mathbf{L}f^*(M) \simeq \mathbf{R}\varprojlim_{n \in \Delta} \mathbf{R}\Gamma_{geom}(\mathscr{X}_n, M)\,.$$

From there, we see easily that conditions (a) and (b) are equivalent. $\square$

3.2.13 The preceding proposition allows us to reduce descent problems in a fibred model category to descent problems in a category of presheaves with values in a model category. On can even go further and reduce the problem to a category of presheaves with values in an 'elementary model category' as follows.

Consider a model category $\mathscr{V}$. Then one can associate to $\mathscr{V}$ its corresponding *prederivator* $\mathbf{Ho}(\mathscr{V})$, that is, the strict 2-functor from the 2-category of small categories to the 2-category of categories, defined by

(3.2.13.1) $$\mathbf{Ho}(\mathscr{V})(I) = \mathrm{Ho}(\mathscr{V}^{I^{op}}) = \mathrm{Ho}(\mathrm{PSh}\,(I, \mathscr{V}))$$

for any small category I. More explictly: for any functor $u : I \longrightarrow J$, one gets a functor
$$u^* : \mathbf{Ho}(\mathscr{V})(J) \longrightarrow \mathbf{Ho}(\mathscr{V})(I)$$

3 Descent in $\mathscr{P}$-fibred model categories

(induced by the composition with u), and for any morphism of functors

$$I \underset{v}{\overset{u}{\rightrightarrows}} J \;, \quad \Downarrow \alpha$$

one has a morphism of functors

$$\mathbf{Ho}(\mathscr{V})(I) \underset{v^*}{\overset{u^*}{\rightleftarrows}} \mathbf{Ho}(\mathscr{V})(J) \;, \quad \alpha^* \Uparrow$$

Moreover, the prederivator $\mathbf{Ho}(\mathscr{V})$ is then a Grothendieck derivator; see [Cis03, Thm. 6.11]. This means in particular that, for any functor between small categories $u : I \longrightarrow J$, the functor u^* has a left adjoint

(3.2.13.2) $$\mathbf{L}u_\sharp : \mathbf{Ho}(\mathscr{V})(I) \longrightarrow \mathbf{Ho}(\mathscr{V})(J)$$

as well as a right adjoint

(3.2.13.3) $$\mathbf{R}u_* : \mathbf{Ho}(\mathscr{V})(I) \longrightarrow \mathbf{Ho}(\mathscr{V})(J)$$

(in the case where $J = e$ is the terminal category, $\mathbf{L}u_\sharp$ is the homotopy colimit functor, while $\mathbf{R}u_*$ is the homotopy limit functor).

If $\mathscr{V}$ and $\mathscr{V}'$ are two model categories, a *morphism of derivators*

$$\Phi : \mathbf{Ho}(\mathscr{V}) \longrightarrow \mathbf{Ho}(\mathscr{V}')$$

is simply a morphism of 2-functors, that is, the data of functors

$$\Phi_I : \mathbf{Ho}(\mathscr{V})(I) \longrightarrow \mathbf{Ho}(\mathscr{V}')(I)$$

together with coherent isomorphisms

$$u^*(\Phi_J(F)) \simeq \Phi_I(u^*(F))$$

for any functor $u : I \longrightarrow J$ and any presheaf F on J with values in $\mathscr{V}$ (see [Cis03, p. 210] for a precise definition).

Such a morphism Φ is said to be *continuous* if, for any functor $u : I \longrightarrow J$, and any object F of $\mathbf{Ho}(\mathscr{V})(I)$, the canonical map

(3.2.13.4) $$\Phi_J \, \mathbf{R}u_*(F) \longrightarrow \mathbf{R}u_* \, \Phi_I(F)$$

is an isomorphism. One can check that a morphism of derivators Φ is continuous if and only if it commutes with homotopy limits (i.e. if and only if the maps (3.2.13.4) are isomorphisms in the case where $J = e$ is the terminal category); see [Cis08,

Prop. 2.6]. For instance, the total right derived functor of any right Quillen functor defines a continuous morphism of derivators; see [Cis03, Prop. 6.12].

Dually a morphism Φ of derivators is *cocontinuous* if, for any functor $u : I \longrightarrow J$, and any object F of $\mathbf{Ho}(\mathscr{V})(I)$, the canonical map

(3.2.13.5) $$\mathbf{L}u_! \, \Phi_I(F) \longrightarrow \Phi_J \, \mathbf{L}u_!(F)$$

is an isomorphism.

3.2.14 We shall say that a stable model category $\mathscr{V}$ is **Q**-*linear* if all the objects of the triangulated category $\mathrm{Ho}(\mathscr{V})$ are uniquely divisible.

Theorem 3.2.15 *Let $\mathscr{V}$ be a model category (resp. a stable model category, resp. a **Q**-linear stable model category), and denote by $\mathscr{S}$ the model category of simplicial sets (resp. the stable model category of S^1-spectra, resp. the **Q**-linear stable model category of complexes of **Q**-vector spaces). Denote by $\mathbb{1}$ the unit object of the closed symmetric monoidal category $\mathrm{Ho}(\mathscr{S})$. Then, for each object E of $\mathrm{Ho}(\mathscr{V})$, there exists a unique continuous morphism of derivators*

$$\mathbf{R}\mathrm{Hom}(E,-) : \mathbf{Ho}(\mathscr{V}) \longrightarrow \mathbf{Ho}(\mathscr{S})$$

such that, for any object F of $\mathrm{Ho}(\mathscr{V})$, there is a functorial bijection

$$\mathrm{Hom}_{\mathrm{Ho}(\mathscr{S})}(\mathbb{1}, \mathbf{R}\mathrm{Hom}(E,F)) \simeq \mathrm{Hom}_{\mathrm{Ho}(\mathscr{V})}(E,F)).$$

Proof Note that the stable **Q**-linear case follows from the stable case and from the fact that the derivator of complexes of **Q**-vector spaces is (equivalent to) the full subderivator of the derivator of S^1-spectra spanned by uniquely divisible objects.

It thus remains to prove the theorem in the case where $\mathscr{V}$ is a model category (resp. a stable model category) and $\mathscr{S}$ is the model category of simplicial sets (resp. the stable model category of S^1-spectra). The existence of $\mathbf{R}\mathrm{Hom}(E,-)$ then follows from [Cis03, Prop. 6.13] (resp. [CT11, Lemma A.6]).

For the unicity, as we don't really need it here, we shall only sketch the proof (the case of simplicial sets is covered in [Cis03, Rem. 6.14]). One uses the universal property of the derivator $\mathbf{Ho}(\mathscr{S})$: by virtue of [Cis08, Cor. 3.26] (resp. of [CT11, Thm. A.5]), for any model category (resp. stable model category) $\mathscr{V}'$ there is a canonical equivalence of categories between the category of cocontinuous morphisms from $\mathbf{Ho}(\mathscr{S})$ to $\mathbf{Ho}(\mathscr{V}')$ and the homotopy category $\mathrm{Ho}(\mathscr{V})$. As a consequence, the derivator $\mathbf{Ho}(\mathscr{S})$ admits a unique closed symmetric monoidal structure, and any derivator (resp. triangulated derivator) is naturally and uniquely enriched in $\mathbf{Ho}(\mathscr{S})$; see [Cis08, Thm. 5.22]. More concretely, this universal property gives, for any object E in $\mathrm{Ho}(\mathscr{V}')$, a unique cocontinuous morphism of derivators

$$\mathbf{Ho}(\mathscr{S}) \longrightarrow \mathbf{Ho}(\mathscr{V}'), \qquad K \longmapsto K \otimes E$$

such that $\mathbb{1} \otimes E = E$. For a fixed K in $\mathbf{Ho}(\mathscr{S})(I)$, this defines a cocontinuous morphism of derivators

3 Descent in $\mathscr{P}$-fibred model categories

$$\mathbf{Ho}(\mathscr{V}') \longrightarrow \mathbf{Ho}(\mathscr{V}'^{I^{op}}) \quad , \quad E \longmapsto K \otimes E$$

which has a right adjoint

$$\mathbf{Ho}(\mathscr{V}'^{I^{op}}) \longrightarrow \mathbf{Ho}(\mathscr{V}'), \quad F \longmapsto F^K.$$

Let

$$\mathbf{R}\operatorname{Hom}(E, -) : \mathbf{Ho}(\mathscr{V}) \longrightarrow \mathbf{Ho}(\mathscr{S})$$

be a continuous morphism such that, for any object F of $\mathscr{V}$, there is a functorial bijection

$$i_F : \operatorname{Hom}_{\mathbf{Ho}(\mathscr{S})}(\mathbb{1}, \mathbf{R}\operatorname{Hom}(E, F)) \simeq \operatorname{Hom}_{\mathbf{Ho}(\mathscr{V})}(E, F)).$$

Then, for any object K of $\mathbf{Ho}(\mathscr{S})(I)$, and any object F of $\mathbf{Ho}(\mathscr{V})(I)$, there is a canonical isomorphism

$$\mathbf{R}\operatorname{Hom}(E, F^K) \simeq \mathbf{R}\operatorname{Hom}(E, F)^K$$

which is completely determined by being the identity for $K = \mathbb{1}$ (this requires the full universal property of $\mathbf{Ho}(\mathscr{S})$ given by [Cis08, Thm. 3.24] (resp. by the dual version of [CT11, Thm. A.5])). We thus get from the functorial bijections i_F the natural bijections:

$$\operatorname{Hom}_{\mathbf{Ho}(\mathscr{S})(I)}(K, \mathbf{R}\operatorname{Hom}(E, F)) \simeq \operatorname{Hom}_{\mathbf{Ho}(\mathscr{S})}(\mathbb{1}, \mathbf{R}\operatorname{Hom}(E, F)^K)$$
$$\simeq \operatorname{Hom}_{\mathbf{Ho}(\mathscr{S})}(\mathbb{1}, \mathbf{R}\operatorname{Hom}(E, F^K))$$
$$\simeq \operatorname{Hom}_{\mathbf{Ho}(\mathscr{V})}(E, F^K)$$
$$\simeq \operatorname{Hom}_{\mathbf{Ho}(\mathscr{V})(I)}(K \otimes E, F).$$

In other words, $\mathbf{R}\operatorname{Hom}(E, -)$ has to be a right adjoint to $(-) \otimes E$. □

Remark 3.2.16 The preceding theorem mostly holds for abstract derivators. The only problem is for the existence of the morphism $\mathbf{R}\operatorname{Hom}(E, -)$ (the unicity is always clear). However, this problem disappears for derivators which have a Quillen model (as we have seen above), as well as for triangulated derivators (see [CT11, Lemma A.6]). Hence Theorem 3.2.15 holds in fact for any triangulated Grothendieck derivator.

In the case when $\mathscr{V}$ is a combinatorial model category (which, in practice, will essentially always be the case), the enrichment over simplicial sets (resp. in the stable case, over spectra) can be constructed via Quillen functors by Dugger's presentation theorems [Dug01] (resp. [Dug06]).

Corollary 3.2.17 *Let $\mathscr{M}$ be a $\mathscr{P}$-fibred combinatorial model category (resp. a stable $\mathscr{P}$-fibred combinatorial model category, resp. a $\mathbf{Q}$-linear stable $\mathscr{P}$-fibred combinatorial model category) over $\mathscr{S}$, and $\mathscr{S}$ the model category of simplicial sets (resp. the stable model category of S^1-spectra, resp. the $\mathbf{Q}$-linear stable model category of complexes of $\mathbf{Q}$-vector spaces).*

Consider an object S of $\mathscr{S}$, a morphism $f : X \longrightarrow S$, and a morphism of $\mathscr{S}$-diagrams $p : (\mathscr{X}, I) \longrightarrow X$ over S. Then, for an object M of $\operatorname{Ho}(\mathscr{M})(S)$, the following conditions are equivalent.

(a) The map
$$\mathbf{R} f_* \, \mathbf{L} f^*(M) \longrightarrow \mathbf{R} f_* \, \mathbf{R} p_* \, \mathbf{L} p^* \, \mathbf{L} f^*(M)$$
is an isomorphism in $\operatorname{Ho}(\mathscr{M})(S)$.
(b) The map
$$\mathbf{R} \Gamma_{geom}(X, M) \longrightarrow \mathbf{R} \varprojlim_{i \in I^{op}} \mathbf{R} \Gamma_{geom}(\mathscr{X}_i, M)$$
is an isomorphism in $\operatorname{Ho}(\mathscr{M})(S)$.
(c) For any object E of $\operatorname{Ho}(\mathscr{M})(S)$, the map
$$\mathbf{R} \operatorname{Hom}(E, \mathbf{R} \Gamma_{geom}(X, M)) \longrightarrow \mathbf{R} \varprojlim_{i \in I^{op}} \mathbf{R} \operatorname{Hom}(E, \mathbf{R} \Gamma_{geom}(\mathscr{X}_i, M))$$
is an isomorphism in $\operatorname{Ho}(\mathscr{S})$.

Proof The equivalence between (a) and (b) follows from Propositions 3.1.15 and 3.1.16, which give the formula
$$\mathbf{R} f_* \, \mathbf{R} p_* \, \mathbf{L} p^* \, \mathbf{L} f^*(M) \simeq \mathbf{R} \varprojlim_{i \in I^{op}} \mathbf{R} \Gamma_{geom}(\mathscr{X}_i, M).$$

The identification
$$\operatorname{Hom}_{\operatorname{Ho}(\mathscr{S})}(\mathbb{1}, \mathbf{R} \operatorname{Hom}(E, F)) \simeq \operatorname{Hom}_{\operatorname{Ho}(\mathscr{M})(S)}(E, F)$$

and the Yoneda Lemma show that a map in $\operatorname{Ho}(\mathscr{M})(S)$ is an isomorphism if and only its image under $\mathbf{R} \operatorname{Hom}(E, -)$ is an isomorphism for any object E of $\operatorname{Ho}(\mathscr{M})(S)$. Moreover, as $\mathbf{R} \operatorname{Hom}(E, -)$ is continuous, for any small category I and any presheaf F on I with values in $\mathscr{M}(S)$, there is a canonical isomorphism
$$\mathbf{R} \operatorname{Hom}(E, \mathbf{R} \varprojlim_{i \in I^{op}} F_i) \simeq \mathbf{R} \varprojlim_{i \in I^{op}} \mathbf{R} \operatorname{Hom}(E, F_i).$$

This proves the equivalence between conditions (b) and (c). □

Corollary 3.2.18 *Under the assumptions of Corollary 3.2.17, given an object S of $\mathscr{S}$, an object M of $\operatorname{Ho}(\mathscr{M})(S)$ satisfies t-descent if and only if, for any object E of $\operatorname{Ho}(\mathscr{M})(S)$ the presheaf of simplicial sets (resp. of S^1-spectra, resp. of complexes of $\mathbb{Q}$-vector spaces)*
$$\mathbf{R} \operatorname{Hom}(E, \mathbf{R} \Gamma_{geom}(-, M))$$
satisfies t-descent over $\mathscr{S}/S$.

Proof This follows from the preceding corollary, using the formula given by Proposition 3.2.4. □

3 Descent in $\mathscr{P}$-fibred model categories

Remark 3.2.19 We need the category $\mathscr{S}$ to be small in some sense to apply the two preceding corollaries because we need to make sense of the projective model category structure of Proposition 3.2.10. However, we can use these corollaries even if the site $\mathscr{S}$ is not small: we can either use the theory of universes, or apply these corollaries to all the adequate small subsites of $\mathscr{S}$. As a consequence, we shall feel free to use Corollaries 3.2.17 and 3.2.18 for not necessarily small sites $\mathscr{S}$, leaving to the reader the task of avoiding set-theoretic difficulties according to her/his taste.

Definition 3.2.20 For an S^1-spectrum E and an integer n, we define its nth cohomology group $H^n(E)$ by the formula

$$H^n(E) = \pi_{-n}(E),$$

where π_i stands for the ith stable homotopy group functor.

Let $\mathscr{M}$ be a monoidal $\mathscr{P}$-fibred stable combinatorial model category over $\mathscr{S}$. Given an object S of $\mathscr{S}$ as well as an object M of $\mathrm{Ho}(\mathscr{M})(S)$, we define the presheaf of *absolute derived global sections of M over S* by the formula

$$\mathbf{R}\Gamma(-, M) = \mathbf{R}\operatorname{Hom}(\mathbb{1}_S, \mathbf{R}\Gamma_{geom}(-, M)).$$

For a map $X \longrightarrow S$ of $\mathscr{S}$, we thus have the *absolute cohomology of X with coefficients in M*, $\mathbf{R}\Gamma(X, M)$, as well as the *cohomology groups of X with coefficients in M*:

$$H^n(X, M) = H^n(\mathbf{R}\Gamma(X, M)).$$

We have canonical isomorphisms of abelian groups

$$\begin{aligned} H^n(X, M) &\simeq \operatorname{Hom}_{\mathrm{Ho}(\mathscr{M})(S)}(\mathbb{1}_S, \mathbf{R}f_* \mathbf{L}f^*(M)) \\ &\simeq \operatorname{Hom}_{\mathrm{Ho}(\mathscr{M})(X)}(\mathbb{1}_X, \mathbf{L}f^*(M)). \end{aligned}$$

Note that, if moreover $\mathscr{M}$ is $\mathbf{Q}$-linear, the presheaf $\mathbf{R}\Gamma(-, M)$ can be considered as a presheaf of complexes of $\mathbf{Q}$-vector spaces on $\mathscr{S}/S$.

3.3 Descent over schemes

The aim of this section is to give natural sufficient conditions for $\mathscr{M}$ to satisfy descent with respect to various Grothendieck topologies.[58]

---

[58] In fact, using Remark 3.2.16, all of this section (results and proofs) holds for an abstract algebraic prederivator in the sense of Ayoub [Ayo07a, Def. 2.4.13] without any changes (note that the results of Section 3.1.b are in fact a proof that (stable) combinatorial fibred model categories over $\mathscr{S}$ give rise to algebraic prederivators). The only interest in considering a fibred model category over $\mathscr{S}$ is that it allows one to formulate things in a slightly more naive way. Of course, the optimal setting in which to formulate descent theory is that of ∞-categories. However, restricting to presentable ∞-categories, using Dugger's presentation theorem [Dug01], as well as rectification theorems such

3.3.a Localization and Nisnevich descent

3.3.1 Recall from Example 2.1.11 that a *Nisnevich distinguished square* is a pullback square of schemes

(3.3.1.1)
$$\begin{array}{ccc} V & \xrightarrow{l} & Y \\ g \downarrow & & \downarrow f \\ U & \xrightarrow{j} & X \end{array}$$

in which f is étale, j is an open immersion with reduced complement Z and the induced morphism $f^{-1}(Z) \longrightarrow Z$ is an isomorphism.

For any scheme X in $\mathscr{S}$, we denote by X_{Nis} the small Nisnevich site of X.

Theorem 3.3.2 (Morel–Voevodsky) *Let $\mathscr{V}$ be a (combinatorial) model category and T a scheme in $\mathscr{S}$. For a presheaf F on T_{Nis} with values in $\mathscr{V}$, the following conditions are equivalent.*

(i) $F(\varnothing)$ is a terminal object in $\mathrm{Ho}(\mathscr{V})$, and for any Nisnevich distinguished square (3.3.1.1) in T_{Nis}, the square

$$\begin{array}{ccc} F(X) & \longrightarrow & F(Y) \\ \downarrow & & \downarrow \\ F(U) & \longrightarrow & F(V) \end{array}$$

is a homotopy pullback square in $\mathscr{V}$.
(ii) The presheaf F satisfies Nisnevich descent on T_{Nis}.

Proof By virtue of Corollaries 3.2.17 and 3.2.18, it is sufficient to prove this in the case where $\mathscr{V}$ is the usual model category of simplicial sets, in which case this is precisely Morel and Voevodsky's theorem; see [MV99, Voe10b, Voe10c]. □

3.3.3 Consider a Nisnevich distinguished square (3.3.1.1) and put $a = jg = fl$. According to our general assumption 3.0.1, the maps a, j and f are $\mathscr{P}$-morphisms. For any object M of $\mathscr{M}(X)$, we obtain a commutative square in $\mathscr{M}$ (which is well defined as an object in the homotopy of commutative squares in $\mathscr{M}(X)$):

(3.3.3.1)
$$\begin{array}{ccc} \mathbf{L}a_\sharp a^*M & \longrightarrow & \mathbf{L}f_\sharp f^*(M) \\ \downarrow & & \downarrow \\ \mathbf{L}j_\sharp j^*(M) & \longrightarrow & M. \end{array}$$

---

as [Cis19, Thm. 7.5.30 and 7.9.8] as well as those from [Bal19], we can see that the case of model categories remains meaningful.

3 Descent in $\mathscr{P}$-fibred model categories

We also obtain another commutative square in $\mathscr{M}$ by applying the functor $\mathbf{R}Hom_X(-, \mathbb{1}_X)$:

(3.3.3.2)
$$\begin{array}{ccc} M & \longrightarrow & \mathbf{R}f_* f^*(M) \\ \downarrow & & \downarrow \\ \mathbf{R}j_* j^*(M) & \longrightarrow & \mathbf{R}a_* a^*(M). \end{array}$$

Proposition 3.3.4 *If the category* $\mathrm{Ho}(\mathscr{M})$ *has the localization property, then for any Nisnevich distinguished square* (3.3.1.1) *and any object M of* $\mathrm{Ho}(\mathscr{M})(X)$, *the squares* (3.3.3.1) *and* (3.3.3.2) *are homotopy cartesians.*

Proof Let $i : Z \longrightarrow X$ be the complement of the open immersion j (Z being endowed with the reduced structure) and $p : f^{-1}(Z) \longrightarrow Z$ the map induced by f.

We have only to prove that one of the squares (3.3.3.1), (3.3.3.2) is cartesian. We choose the square (3.3.3.1).

Because the pair of functors $(\mathbf{L}i^*, j^*)$ is conservative on $\mathrm{Ho}(\mathscr{M})(X)$, we have only to check that the pullback of (3.3.3.1) along j^* or $\mathbf{L}i^*$ is homotopy cartesian. But, using the $\mathscr{P}$-base change property, we see that the image of (3.3.3.1) under j^* is (canonically isomorphic to) the commutative square

$$\begin{array}{ccc} \mathbf{L}g_\sharp a^*(M) & =\!=\!= & \mathbf{L}g_\sharp a^*(M) \\ \downarrow & & \downarrow \\ j^*(M) & =\!=\!= & j^*(M), \end{array}$$

which is obviously homotopy cartesian.

Using again the $\mathscr{P}$-base change property, we obtain that the image of (3.3.3.1) under $\mathbf{L}i^*$ is isomorphic in $\mathrm{Ho}(\mathscr{M})$ to the square

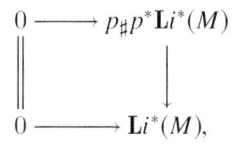

$$\begin{array}{ccc} 0 & \longrightarrow & p_\sharp p^* \mathbf{L}i^*(M) \\ \| & & \downarrow \\ 0 & \longrightarrow & \mathbf{L}i^*(M), \end{array}$$

which is again obviously homotopy cartesian because p is an isomorphism (note for this last reason, $p_\sharp = \mathbf{L}p_\sharp$). $\square$

Corollary 3.3.5 *If* $\mathrm{Ho}(\mathscr{M})$ *has the localization property then it satisfies Nisnevich descent.*

Proof This corollary thus follows immediately from Corollary 3.2.17, Theorem 3.3.2 and Proposition 3.3.4. $\square$

Remark 3.3.6 Note that using Theorem 3.3.2, if we assume only that $\mathrm{Ho}(\mathcal{M})$ satisfies Nisnevich descent, then the squares (3.3.3.1) and (3.3.3.2) are homotopy cartesian for any Nisnevich distinguished square (3.3.1.1).

Assume that $\mathcal{M}$ is monoidal with geometric sections M. Let S be a base scheme and consider a Nisnevich distinguished square (3.3.1.1) of smooth S-schemes. Then the fact that the square (3.3.3.1) is homotopy cartesian implies there exists a *canonical distinguished triangle*:

$$M_S(V) \xrightarrow{g_*+l_*} M_S(U) \oplus M_S(Y) \xrightarrow{f_*+j_*} M_S(X) \longrightarrow M_S(V)[1].$$

It is called the *Mayer–Vietoris triangle* associated with the square (3.3.1.1).

3.3.b Proper base change isomorphism and descent by blow-ups

3.3.7 Recall from Example 2.1.11 that a *cdh-distinguished square* is a pullback square of schemes

(3.3.7.1)
$$\begin{array}{ccc} T & \xrightarrow{k} & Y \\ g \downarrow & & \downarrow f \\ Z & \xrightarrow{i} & X \end{array}$$

in which f is proper surjective, i a closed immersion and the induced map $f^{-1}(X - Z) \longrightarrow X - Z$ is an isomorphism.

Recall from Example 2.1.11 the *cdh-topology* is the Grothendieck topology on the category of schemes generated by Nisnevich coverings and by coverings of shape $\{Z \longrightarrow X, Y \longrightarrow X\}$ for any *cdh*-distinguished square (3.3.7.1).

Theorem 3.3.8 (Voevodsky) *Let $\mathcal{V}$ be a (combinatorial) model category. For a presheaf F on $\mathcal{S}$ with values in $\mathcal{V}$, the following conditions are equivalent.*

(i) The presheaf F satisfies cdh-descent on $\mathcal{S}$.
(ii) The presheaf F satisfies Nisnevich descent and, for any cdh-distinguished square (3.3.7.1) of $\mathcal{S}$, the square

$$\begin{array}{ccc} F(X) & \longrightarrow & F(Y) \\ \downarrow & & \downarrow \\ F(Z) & \longrightarrow & F(T) \end{array}$$

is a homotopy pullback square in $\mathcal{V}$.

Proof It is sufficient to prove this in the case where $\mathcal{V}$ is the usual model category of simplicial sets; see Corollaries 3.2.17 and 3.2.18. As the distinguished *cdh*-squares

3 Descent in $\mathscr{P}$-fibred model categories

define a bounded regular and reduced cd-structure on $\mathscr{S}$, the equivalence between (i) and (ii) follows from Voevodsky's theorems on descent with respect to topologies defined by cd-structures [Voe10b, Voe10c]. □

3.3.9 Consider a cdh-distinguished square (3.3.7.1) and put $a = ig = fk$. For any object M of $\mathscr{M}(X)$, we obtain a commutative square in $\mathscr{M}$ (which is well defined as an object in the homotopy of commutative squares in $\mathscr{M}(X)$):

(3.3.9.1)
$$\begin{array}{ccc} M & \longrightarrow & \mathrm{R}f_* \, \mathrm{L}f^*(M) \\ \downarrow & & \downarrow \\ \mathrm{R}i_* \, \mathrm{L}i^*(M) & \longrightarrow & \mathrm{R}a_* \, \mathrm{L}a^*(M). \end{array}$$

Proposition 3.3.10 *Assume* $\mathrm{Ho}(\mathscr{M})$ *satisfies the localization property and the transversality property with respect to proper morphisms. Then the following conditions hold:*

(i) For any cdh-distinguished square (3.3.7.1), and any object M of $\mathrm{Ho}(\mathscr{M})(X)$ the commutative square (3.3.9.1) is homotopy cartesian.
(ii) The $\mathscr{P}$-fibred model category $\mathrm{Ho}(\mathscr{M})$ satisfies cdh-descent.

Proof We first prove (i). Consider a cdh-distinguished square (3.3.7.1) and let $j : U \longrightarrow X$ be the complement open immersion of i. As the pair of functors $(\mathrm{L}i^*, j^*)$ is conservative on $\mathrm{Ho}(\mathscr{M})(X)$, we have only to check that the image of (3.3.9.1) under $\mathrm{L}i^*$ and j^* is homotopy cartesian.

Using projective transversality, we see that the image of (3.3.9.1) under the functor $\mathrm{L}i^*$ is (isomorphic to) the homotopy pullback square

$$\begin{array}{ccc} \mathrm{L}i^*(M) & \longrightarrow & \mathrm{R}g_* \, \mathrm{L}g^* \, \mathrm{L}i^*(M) \\ \| & & \| \\ \mathrm{L}i^*(M) & \longrightarrow & \mathrm{R}g_* \, \mathrm{L}g^* \, \mathrm{L}i^*(M) \, . \end{array}$$

Let $h : f^{-1}(U) \longrightarrow U$ be the pullback of f over U. As j is an open immersion, it is by assumption a $\mathscr{P}$-morphism and the $\mathscr{P}$-base change formula implies that the image of (3.3.9.1) under j^* is (isomorphic to) the commutative square

$$\begin{array}{ccc} \mathrm{L}j^*(M) & \longrightarrow & \mathrm{R}h_* \mathrm{L}h^* \mathrm{L}j^*(M) \\ \downarrow & & \downarrow \\ 0 & =\!=\!=\!=\!= & 0, \end{array}$$

which is obviously homotopy cartesian because h is an isomorphism.

We then prove (ii). We already know that $\mathscr{M}$ satisfies Nisnevich descent (Corollary 3.3.5). Thus, by virtue of the equivalence between conditions (i) and (ii) of Theorem

3.3.8, the computation above, together with Corollaries 3.2.17 and 3.2.18, imply that $\mathscr{M}$ satisfies cdh-descent. □

3.3.11 To any cdh-distinguished square (3.3.7.1), one associates a diagram of schemes $\mathscr{Y}$ over X as follows. Let Γ be the category freely generated by the oriented graph

(3.3.11.1)
$$\begin{array}{c} a \longrightarrow b \\ \downarrow \\ c \end{array}$$

Then $\mathscr{Y}$ is the functor from Γ to $\mathscr{S}/X$ defined by the following diagram.

(3.3.11.2)
$$\begin{array}{ccc} T & \xrightarrow{k} & Y \\ g \downarrow & & \\ Z & & \end{array}$$

We then have a canonical map $\varphi : \mathscr{Y} \longrightarrow X$, and the second assertion of Theorem 3.3.10 can be reformulated by saying that the adjunction map

$$M \longrightarrow \mathbf{R}\varphi_* \, \mathbf{L}\varphi^*(M)$$

is an isomorphism for any object M of $\mathrm{Ho}(\mathscr{M})(X)$: indeed, by virtue of Proposition 3.1.15, $\mathbf{R}\varphi_* \, \mathbf{L}\varphi^*(M)$ is the homotopy limit of the diagram

$$\begin{array}{c} \mathbf{R}f_* \, \mathbf{L}f^*(M) \\ \downarrow \\ \mathbf{R}i_* \, \mathbf{L}i^*(M) \longrightarrow \mathbf{R}a_* \, \mathbf{L}a^*(M) \end{array}$$

in $\mathrm{Ho}(\mathscr{M})(X)$. In other words, if $\mathscr{M}$ has the properties of localization and of projective transversality, then the functor

$$\mathbf{L}\varphi^* : \mathrm{Ho}(\mathscr{M})(X) \longrightarrow \mathrm{Ho}(\mathscr{M})(\mathscr{Y}, \Gamma)$$

is fully faithful.

3.3.c Proper descent with rational coefficients I: Galois excision

From now on, we assume that any scheme in $\mathscr{S}$ is quasi-excellent[59] (in fact, we shall only use the fact that the normalization of a quasi-excellent schemes gives rise to a finite surjective morphism, so that, in fact, universally Japanese schemes would

---

[59] See Section 4.1.1 below for a reminder on quasi-excellent schemes.

3 Descent in $\mathscr{P}$-fibred model categories

be enough). We fix a scheme S in $\mathscr{S}$, and we shall work with S-schemes in $\mathscr{S}$ (assuming these form an essentially small category).

3.3.12 The h-*topology* (resp. the qfh-*topology*) is the Grothendieck topology on the category of schemes associated to the pretopology whose coverings are the universal topological epimorphisms (resp. the quasi-finite universal topological epimorphisms). This topology was introduced and studied by Voevodsky in [Voe96].

The h-topology is finer than the cdh-topology and, of course, finer than the qfh-topology. The qfh-topology is in turn finer than the étale topology. An interesting feature of the h-topology (resp. of the qfh-topology) is that any proper (resp. finite) surjective map is an h-cover. In fact, the h-topology (resp. the qfh-topology) can be described as the topology generated by the Nisnevich coverings and by the proper (resp. finite) surjective maps; see Lemma 3.3.28 (resp. Lemma 3.3.27) below for a precise statement.

3.3.13 Consider a morphism of schemes $f : Y \longrightarrow X$. Consider the group of automorphisms $G = \mathrm{Aut}_Y(X)$ of the X-scheme Y.

Assuming X is connected, we say according to [Gro03, exp. V] that f is a *Galois cover* if it is finite étale (thus surjective) and G operates transitively and faithfully on any (or simply one) of the geometric fibers of Y/X. Then G is called the *Galois group* of Y/X.[60]

When X is not connected, we will still say that f is a *Galois cover* if it is so over any connected component of X. Then G will be called the *Galois group* of X. If $(X_i)_{i \in I}$ is the family connected components of X, then G is the product of the Galois groups G_i of $f \times_X X_i$ for each $i \in I$. The group G_i is equal to the Galois group of any residual extension over a generic point of X_i.

The following definition is an extension of definition 5.5 of [SV00b]:

Definition 3.3.14 A *pseudo-Galois cover* is a finite surjective morphism of schemes $f : Y \longrightarrow X$ which can be factored as

$$Y \xrightarrow{f'} X' \xrightarrow{p} X,$$

where f' is a Galois cover and p is radicial[61] (such a p is automatically finite and surjective).

Note that the group G defined by the Galois cover f' is independent of the choice of the factorization. In fact, if $\bar{X}$ denotes the semi-localization of X at its generic points, considering the cartesian squares

$$\begin{array}{ccc} \bar{Y} & \longrightarrow \bar{X}' & \longrightarrow \bar{X} \\ \downarrow & \downarrow & \downarrow \\ Y & \xrightarrow{f'} X' & \xrightarrow{p} X \end{array}$$

---

[60] The map f induces a one to one correspondence between the generic points of Y and those of X. For any generic point $y \in Y$, $x = f(y)$, the residual extension κ_y/κ_x is a Galois extension with Galois group G.

[61] See Section 2.1.6 for a reminder on radicial morphisms.

then $G = \mathrm{Aut}_{\bar{X}}(\bar{Y})$ – for any point $y \in \bar{Y}$, $x' = f'(y)$, $x = f(y)$, $\kappa_{x'}/\kappa_x$ is the maximal radicial sub-extension of the normal extension κ_y/κ_x. It will be called the *Galois group* of Y/X.

Note also that Y is a G-torsor over X locally for the *qfh*-topology (i.e. it is a Galois object of group G in the *qfh*-topos of X): this comes from the fact that finite radicial epimorphisms are isomorphisms locally for the *qfh*-topology (any universal homeomorphism has this property by [Voe96, prop. 3.2.5]).

Let $f : Y \longrightarrow X$ be a finite morphism, and G a finite group acting on Y over X. Note that, as Y is affine on X, the scheme-theoretic quotient Y/G exists; see [Gro03, Exp. V, Cor. 1.8]. Such scheme-theoretic quotients are stable under flat pullbacks; see [Gro03, Exp. V, Prop. 1.9].

Definition 3.3.15 Let G be finite group. A *qfh-distinguished square of group G* is a pullback square of S-schemes of shape

(3.3.15.1)
$$\begin{array}{ccc} T & \xrightarrow{h} & Y \\ g \downarrow & & \downarrow f \\ Z & \xrightarrow{i} & X \end{array}$$

in which Y is endowed with an action of G over X, and satisfying the following three conditions.

(a) The morphism f is finite and surjective.
(b) The induced morphism $f^{-1}(X - Z) \longrightarrow f^{-1}(X - Z)/G$ is flat.
(c) The morphism $f^{-1}(X - Z)/G \longrightarrow X - Z$ is radicial.

Immediate examples of *qfh*-distinguished squares of trivial group are the following. The scheme Y might be the normalization of X, and Z is a nowhere dense closed subscheme out of which f is an isomorphism; or Y is dense open subscheme of X which is the disjoint union of its irreducible components; or Y is a closed subscheme of X inducing an isomorphism $Y_{red} \simeq X_{red}$.

A *qfh*-distinguished square of group G (3.3.15.1) will be said to be *pseudo-Galois* if Z is nowhere dense in X and if the map $f^{-1}(X - Z) \longrightarrow X - Z$ is a pseudo-Galois cover of group G.

The main examples of pseudo-Galois *qfh*-distinguished squares will come from the following situation.

Proposition 3.3.16 *Consider an irreducible normal scheme X, and a finite extension L of its field of functions $k(X)$. Let K be the inseparable closure of $k(X)$ in L, and assume that L/K is a Galois extension of group G. Denote by Y the normalization of X in L. Then the action of G on $k(Y) = L$ extends naturally to an action on Y over X. Furthermore, there exists a closed subscheme Z of X such that the pullback square*

3 Descent in $\mathscr{P}$-fibred model categories

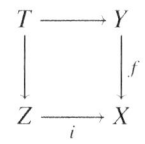

$$\begin{array}{ccc} T & \longrightarrow & Y \\ \downarrow & & \downarrow f \\ Z & \xrightarrow{i} & X \end{array}$$

is a pseudo-Galois qfh-distinguished square of group G.

Proof The action of G on L extends naturally to an action on Y over X by functoriality. Furthermore, Y/G is the normalization of X in K, so that $Y/G \longrightarrow X$ is finite radicial and surjective (see [Voe96, Lemma 3.1.7] or [Bou98, V, §2, n° 3, lem. 4]). By construction, Y is generically a Galois cover over Y/G, which implies the result (see [GD67, Cor. 18.2.4]). □

3.3.17 For a given S-scheme T, we shall denote by $L(T)$ the corresponding representable *qfh*-sheaf of sets (recall that the *qfh*-topology is not subcanonical, so that $L(T)$ has to be distinguished from T itself). Beware that, in general, there is no reason that, given a finite group G acting on T, the scheme-theoretic quotient $L(T/G)$ (whenever defined) and the *qfh*-sheaf-theoretic quotient $L(T)/G$ would coincide.

Lemma 3.3.18 *Let $f : Y \longrightarrow X$ be a separated morphism, G a finite group acting on Y over X, and Z a closed subscheme of X such that f is finite and surjective over $X - Z$, and such that the quotient map $f^{-1}(X - Z) \longrightarrow f^{-1}(X - Z)/G$ is flat, while the map $f^{-1}(X - Z)/G \longrightarrow X - Z$ is radicial. For $g \in G$, write $g : Y \longrightarrow Y$ for the corresponding automorphism of Y, and define Y_g as the image of the diagonal $Y \longrightarrow Y \times_X Y$ composed with the automorphism $1_Y \times_X g : Y \times_X Y \longrightarrow Y \times_X Y$. Then, if $T = Z \times_X Y$, we get a qfh-cover of $Y \times_X Y$ by closed subschemes:*

$$Y \times_X Y = (T \times_Z T) \cup \bigcup_{g \in G} Y_g \, .$$

Proof Note that, as f is separated, the diagonal $Y \longrightarrow Y \times_X Y$ is a closed embedding, so that the Y_g's are closed subschemes of $Y \times_X Y$. As the map $Y \times_{Y/G} Y \longrightarrow Y \times_X Y$ is a universal homeomorphism, we may assume that $Y/G = X$. It is sufficient to prove that, if y and y' are two geometric points of Y whose images coincide in X and do not belong to Z, there exists an element g of G such that $y' = gy$ (which means that the pair (y, y') belongs to Y_g). For this purpose, we may assume, without loss of generality, that $Z = \emptyset$. Then, by assumption, Y is flat over X, from which we get the identification $(Y \times_X Y)/G \simeq Y \times_X (Y/G) \simeq Y$ (where the action of G on $Y \times_X Y$ is trivial on the first factor and is induced by the action on Y on the second factor). This completes the proof. □

Proposition 3.3.19 *For any qfh-distinguished square of group G (3.3.15.1), the commutative square*

$$\begin{array}{ccc} L(T)/G & \longrightarrow & L(Y)/G \\ \downarrow & & \downarrow \\ L(Z) & \longrightarrow & L(X) \end{array}$$

is a pullback and a pushout in the category of qfh-sheaves. Moreover, if X is normal and if Z is nowhere dense in X, then the canonical map $L(Y)/G \longrightarrow L(Y/G) \simeq L(X)$ is an isomorphism of qfh-sheaves (which implies that $L(T)/G \longrightarrow L(Z)$ is an isomorphism as well).

Proof Note that this commutative square is a pullback because it was so before taking the quotients by G (as colimits are universal in any topos). As f is a qfh-cover, it is sufficient to prove that

$$\begin{array}{ccc} L(T) \times_{L(Z)} L(T)/G & \longrightarrow & L(Y) \times_{L(X)} L(Y)/G \\ \downarrow & & \downarrow \\ L(T) & \longrightarrow & L(Y) \end{array}$$

is a pushout square. This latter square fits into the following commutative diagram

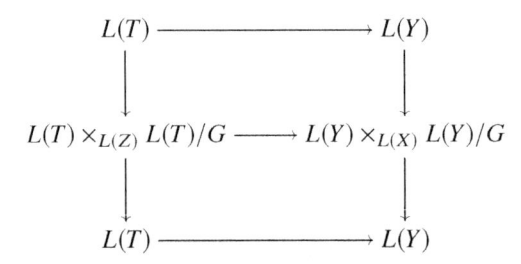

$$\begin{array}{ccc} L(T) & \longrightarrow & L(Y) \\ \downarrow & & \downarrow \\ L(T) \times_{L(Z)} L(T)/G & \longrightarrow & L(Y) \times_{L(X)} L(Y)/G \\ \downarrow & & \downarrow \\ L(T) & \longrightarrow & L(Y) \end{array}$$

in which the two vertical composed maps are identities (the vertical maps of the upper commutative square are obtained from the diagonals by taking the quotients under the natural action of G on the right component). It is thus sufficient to prove that the upper square is a pushout. As the lower square is a pullback, the upper one shares the same property; moreover, all the maps in the upper commutative square are monomorphisms of qfh-sheaves, so that it is sufficient to prove that the map $(L(T) \times_{L(Z)} L(T)/G) \amalg L(Y) \longrightarrow L(Y) \times_{L(X)} L(Y)/G$ is an epimorphism of qfh-sheaves. According to Lemma 3.3.18, this follows from the commutativity of the diagram

$$\begin{array}{ccc} L(T \times_Z T) \amalg \left(\coprod_{g \in G} L(Y_g) \right) & \longrightarrow & L(Y \times_X Y) \\ \downarrow & & \downarrow \\ (L(T) \times_{L(Z)} L(T)/G) \amalg L(Y) & \longrightarrow & L(Y) \times_{L(X)} L(Y)/G \end{array}$$

in which the vertical maps are obviously epimorphic.

Assume now that X is normal and that Z is nowhere dense in X, and let us prove that the canonical map $L(Y)/G \longrightarrow L(X)$ is an isomorphism of qfh-sheaves. This is equivalent to proving that, for any qfh-sheaf of sets F, the map $f^* : F(X) \longrightarrow F(Y)$ induces a bijection

3 Descent in $\mathscr{P}$-fibred model categories

$$F(X) \simeq F(Y)^G.$$

Let F be a qfh-sheaf. The map $f^* : F(X) \longrightarrow F(Y)$ is injective because f is a qfh-cover, and it is clear that the image of f^* lies in $F(Y)^G$.

Let a be a section of F over Y which is invariant under the action of G. Denote by $pr_1, pr_2 : Y \times_X Y \longrightarrow Y$ the two canonical projections. With the notations introduced in Lemma 3.3.18, we have

$$pr_1^*(a)|_{Y_g} = a = a.g = pr_2^*(a)|_{Y_g}$$

for every element g in G. As Z does not contain any generic point of X, the scheme $T \times_Z T$ does not contain any generic point of $Y \times_X Y$ either: as any irreducible component of Y dominates an irreducible component of X, and, as X is normal, the finite map $Y \longrightarrow X$ is universally open; in particular, the projection $pr_1 : Y \times_X Y \longrightarrow Y$ is universally open, which implies that any generic point of $Y \times_X Y$ lies over a generic point of Y. By virtue of [Voe96, prop. 3.1.4], Lemma 3.3.18 thus gives a qfh-cover of $Y \times_X Y$ by closed subschemes of shape

$$Y \times_X Y = \bigcup_{g \in G} Y_g.$$

This implies that
$$pr_1^*(a) = pr_2^*(a).$$

The morphism $Y \longrightarrow X$ being a qfh-cover and F a qfh-sheaf, we deduce that the section a lies in the image of f^*. □

Corollary 3.3.20 *For any qfh-distinguished square of group G (3.3.15.1), we get a bicartesian square of qfh-sheaves of abelian groups*

$$\begin{array}{ccc} \mathbf{Z}_{qfh}(T)_G & \longrightarrow & \mathbf{Z}_{qfh}(Y)_G \\ \downarrow & & \downarrow \\ \mathbf{Z}_{qfh}(Z) & \longrightarrow & \mathbf{Z}_{qfh}(X) \end{array}$$

(where the subscript G stands for the coinvariants under the action of G). In other words, there is a canonical short exact sequence of sheaves of abelian groups

$$0 \longrightarrow \mathbf{Z}_{qfh}(T)_G \longrightarrow \mathbf{Z}_{qfh}(Z) \oplus \mathbf{Z}_{qfh}(Y)_G \longrightarrow \mathbf{Z}_{qfh}(X) \longrightarrow 0.$$

Proof As the abelianization functor preserves colimits and monomorphisms, the preceding proposition implies formally that we have a short exact sequence of shape

$$\mathbf{Z}_{qfh}(T)_G \longrightarrow \mathbf{Z}_{qfh}(Z) \oplus \mathbf{Z}_{qfh}(Y)_G \longrightarrow \mathbf{Z}_{qfh}(X) \longrightarrow 0,$$

while the left exactness follows from the fact that $Z \longrightarrow X$ being a monomorphism, the map obtained by pullback, $L(T)/G \longrightarrow L(Y)/G$, is a monomorphism as well. □

3.3.21 Let $\mathscr{V}$ be a **Q**-linear stable model category (see Section 3.2.14).

Consider a finite group G, and an object E of $\mathscr{V}$, endowed with an action of G. By viewing G as a category with one object we can see E as functor from G to $\mathscr{V}$ and take its homotopy limit in $\mathrm{Ho}(\mathscr{V})$, which we denote by E^{hG} (in the literature, E^{hG} is called the *object of homotopy fixed points* under the action of G on E). One the other hand, the category $\mathrm{Ho}(\mathscr{V})$ is, by assumption, a **Q**-linear triangulated category with small sums, and, in particular, a **Q**-linear pseudo-abelian category so that we can define E^G as the object of $\mathrm{Ho}(\mathscr{V})$ defined by

$$(3.3.21.1) \qquad E^G = \mathrm{Im}\, p,$$

where $p : E \longrightarrow E$ is the projector defined in $\mathrm{Ho}(\mathscr{V})$ by the formula

$$(3.3.21.2) \qquad p(x) = \frac{1}{\#G} \sum_{g \in G} g.x\,.$$

The inclusion $E^G \longrightarrow E$ induces a canonical isomorphism

$$(3.3.21.3) \qquad E^G \xrightarrow{\sim} E^{hG}$$

in $\mathrm{Ho}(\mathscr{V})$: to see this, by virtue of Theorem 3.2.15, we can assume that $\mathscr{V}$ is the model category of complexes of **Q**-vector spaces, in which case it is obvious.

Corollary 3.3.22 *Let C be a presheaf of complexes of **Q**-vector spaces on the category of S-schemes. Then, for any qfh-distinguished square of group G* (3.3.15.1), *the commutative square*

$$\begin{array}{ccc} \mathrm{R}\Gamma_{qfh}(X, C_{qfh}) & \longrightarrow & \mathrm{R}\Gamma_{qfh}(Y, C_{qfh})^G \\ \downarrow & & \downarrow \\ \mathrm{R}\Gamma_{qfh}(Z, C_{qfh}) & \longrightarrow & \mathrm{R}\Gamma_{qfh}(T, C_{qfh})^G \end{array}$$

*is a homotopy pullback square in the derived category of **Q**-vector spaces. In particular, we get a long exact sequence of shape*

$$\cdots \longrightarrow H^n_{qfh}(X, C_{qfh}) \longrightarrow H^n_{qfh}(Z, C_{qfh}) \oplus H^n_{qfh}(Y, C_{qfh})^G \longrightarrow H^n_{qfh}(T, C_{qfh})^G \longrightarrow \cdots$$

If furthermore X is normal and Z is nowhere dense in X, then the maps

$$H^n_{qfh}(X, C_{qfh}) \longrightarrow H^n_{qfh}(Y, C_{qfh})^G \quad \text{and} \quad H^n_{qfh}(Z, C_{qfh}) \longrightarrow H^n_{qfh}(T, C_{qfh})^G$$

are isomorphisms for any integer n.

Proof Let $C_{qfh} \longrightarrow C'$ be a fibrant resolution in the *qfh*-local injective model category structure on the category of *qfh*-sheaves of complexes of **Q**-vector spaces; see for instance [Ayo07a, Cor. 4.4.42]. Then for $U = Y, T$, we have a natural isomorphism of complexes

3 Descent in $\mathscr{P}$-fibred model categories

$$\mathrm{Hom}(\mathbf{Q}_{qfh}(U)_G, C') = C'(U)^G$$

which gives an isomorphism

$$\mathbf{R}\,\mathrm{Hom}(\mathbf{Q}_{qfh}(U)_G, C_{qfh}) \simeq \mathbf{R}\Gamma_{qfh}(U, C_{qfh})^G$$

in the derived category of the abelian category of $\mathbf{Q}$-vector spaces. This corollary thus follows formally from Corollary 3.3.20 by evaluating at the derived functor $\mathbf{R}\,\mathrm{Hom}(-, C_{qfh})$.

If furthermore X is normal, then one deduces the isomorphism $H^n_{qfh}(X, C_{qfh}) \simeq H^n_{qfh}(Y, C_{qfh})^G$ from the fact that $L(Y)/G \simeq L(Y/G) \simeq X$ (Proposition 3.3.19), which implies that $\mathbf{Z}_{qfh}(Y)_G \simeq \mathbf{Z}_{qfh}(X)$. The isomorphism $H^n_{qfh}(Z, C_{qfh}) \simeq H^n_{qfh}(T, C_{qfh})^G$ then comes as a byproduct of the long exact sequence above. $\square$

Theorem 3.3.23 *Let X be a scheme, and C be a presheaf of complexes of $\mathbf{Q}$-vector spaces on the small étale site of X. Then C satisfies étale descent if and only if it has the following properties.*

(a) The complex C satisfies Nisnevich descent.
(b) For any étale X-scheme U and any Galois cover $V \longrightarrow U$ of group G, the map $C(U) \longrightarrow C(V)^G$ is a quasi-isomorphism.

Proof These are certainly necessary conditions. To prove that they are sufficient, note that the Nisnevich cohomological dimension and the rational étale cohomological dimension of a noetherian scheme are bounded by the dimension; see [MV99, proposition 1.8, page 98] and [Voe96, Lemma 3.4.7]. By virtue of [SV00a, Theorem 0.3], for $\tau = Nis, \acute{e}t$, we have strongly convergent spectral sequences

$$E^{p,q}_2 = H^p_\tau(U, H^q(C)_\tau) \Rightarrow H^{p+q}_\tau(U, C_\tau)\,.$$

Condition (a) gives isomorphisms $H^{p+q}(C(U)) \simeq H^{p+q}_{Nis}(U, C_{Nis})$, so that it is sufficient to prove that, for each of the cohomology presheaves $F = H^q(C)$, we have

$$H^p_{Nis}(U, F_{Nis}) \simeq H^p_{\acute{e}t}(U, F_{\acute{e}t})\,.$$

As the rational étale cohomology of any henselian scheme is trivial in non-zero degrees, it is sufficient to prove that, for any local henselian scheme U (obtained as the henselisation of an étale X-scheme at some point), $F_{Nis}(U) \simeq F_{\acute{e}t}(U)$. Let G be the absolute Galois group of the closed point of U. Then we have

$$F_{Nis}(U) = F(U) \quad \text{and} \quad F_{\acute{e}t}(U) = \varinjlim_\alpha F(U_\alpha)^{G_\alpha}\,,$$

where the U_α's run over all the Galois covers of U corresponding to the finite quotients $G \longrightarrow G_\alpha$. It follows from (b) that $F(U) \simeq F(U_\alpha)^{G_\alpha}$ for any α, so that $F_{Nis}(U) \simeq F_{\acute{e}t}(U)$. $\square$

Lemma 3.3.24 *Any qfh-cover admits a refinement of the form $Z \longrightarrow Y \longrightarrow X$, where $Z \longrightarrow Y$ is a finite surjective morphism, and $Y \longrightarrow X$ is an étale cover.*

Proof This property being clearly local on X with respect to the étale topology, we can assume that X is strictly henselian, in which case this follows from [Voe96, Lemma 3.4.2]. □

Theorem 3.3.25 *A presheaf of complexes of* $\mathbf{Q}$*-vector spaces C on the category of S-schemes satisfies qfh-descent if and only if it has the following two properties:*

(a) the complex C satisfies Nisnevich descent;
(b) for any pseudo-Galois qfh-distinguished square of group G (3.3.15.1), the commutative square

$$\begin{array}{ccc} C(X) & \longrightarrow & C(Y)^G \\ \downarrow & & \downarrow \\ C(Z) & \longrightarrow & C(T)^G \end{array}$$

is a homotopy pullback square in the derived category of $\mathbf{Q}$*-vector spaces.*

Proof Any complex of presheaves of $\mathbf{Q}$-vector spaces satisfying qfh-descent satisfies properties (a) and (b): property (a) follows from the fact that the qfh-topology is finer than the étale topology; property (b) is Corollary 3.3.22.

Assume now that C satisfies these two properties. Let $\varphi : C \longrightarrow C'$ be a morphism of presheaves of complexes of $\mathbf{Q}$-vector spaces which is a quasi-isomorphism locally for the qfh-topology, and such that C' satisfies qfh-descent (such a morphism exists thanks to the qfh-local model category structure on the category of presheaves of complexes of $\mathbf{Q}$-vector spaces; see Proposition 3.2.10). Then the cone of φ also satisfies conditions (a) and (b). Hence it is sufficient to prove the theorem in the case where C is acyclic locally for the qfh-topology.

Assume from now on that C_{qfh} is an acyclic complex of qfh-sheaves, and denote by $H^n(C)$ the nth cohomology presheaf associated to C. We know that the associated qfh-sheaves vanish, and we want to deduce that $H^n(C) = 0$.

We shall prove by induction on d that, for any S-scheme X of dimension d and for any integer n, the group $H^n(C)(X) = H^n(C(X))$ vanishes. The case where $d < 0$ follows from the fact, that by (a), the presheaves $H^n(C)$ send finite sums to finite direct sums, so that, in particular, $H^n(C)(\varnothing) = 0$. Before going further, notice that condition (b) implies $H^n(C)(X_{red}) = H^n(C)(X)$ for any S-scheme X (consider the case where, in the diagram (3.3.15.1), $Z = Y = T = X_{red}$), so that it is always harmless to replace X by its reduction. Assume now that $d \geq 0$, and that the vanishing of $H^n(C)(X)$ is known whenever X is of dimension $< d$ and for any integer n. Under this inductive assumption, we have the following reduction principle.

Consider a pseudo-Galois qfh-distinguished square of group G (3.3.15.1). If Z and T are of dimension $< d$, then by condition (b), the map $H^n(C)(X) \longrightarrow H^n(C)(Y)^G$ is an isomorphism: indeed, we have an exact sequence of shape

$$H^{n-1}(C)(T)^G \longrightarrow H^n(C)(X) \longrightarrow H^n(C)(Z) \oplus H^n(C)(Y)^G \longrightarrow H^n(C)(T)^G,$$

which implies our assertion by induction on d.

3 Descent in $\mathscr{P}$-fibred model categories

We shall prove now the vanishing of $H^n(C)(T)$ for normal S-schemes T of dimension d. Let a be a section of $H^n(C)$ over such a T. As $H^n(C)_{qfh}(T) = 0$, there exists a qfh-cover $g : Y \longrightarrow T$ such that $g^*(a) = 0$. By virtue of Lemma 3.3.24, we can assume g is the composition of a finite surjective morphism $f : Y \longrightarrow X$ and of an étale cover $e : X \longrightarrow T$. We claim that $e^*(a) = 0$. To prove it, since, by (a), the presheaf $H^n(C)$ sends finite sums to finite direct sums, we can assume that X is normal and connected. Refining f further, we can assume that Y is the normalization of X in a finite extension of $k(X)$, and that $k(Y)$ is a Galois extension of group G over the inseparable closure of $k(X)$ in $k(Y)$. By virtue of Proposition 3.3.16, we get by the reduction principle the identification $H^n(C)(X) = H^n(C)(Y)^G$, whence $e^*(a) = 0$. As a consequence, the restriction of the presheaf of complexes C to the category of normal S-schemes of dimension $\leq d$ is acyclic locally for the étale topology (note that this is quite meaningful, as any étale scheme over a normal scheme is normal; see [GD67, Prop. 18.10.7]). But C satisfies étale descent (by virtue of Theorem 3.3.23 this follows formally from property (a) and from property (b) for $Z = \varnothing$), so that $H^n(C)(T) = H^n_{\acute{e}t}(T, C_{\acute{e}t}) = 0$ for any normal S-scheme T of dimension $\leq d$ and any integer n.

Consider now a reduced S-scheme X of dimension $\leq d$. Let $p : T \longrightarrow X$ be the normalization of X. As p is birational (see [GD61, Cor. 6.3.8]) and finite surjective (because X is quasi-excellent), we can apply the reduction principle and see that the pullback map $p^* : H^n(C)(X) \longrightarrow H^n(C)(T) = 0$ is an isomorphism for any integer n, which achieves the induction and the proof. □

Lemma 3.3.26 *Étale coverings are finite étale coverings locally for the Nisnevich topology: any étale cover admits a refinement of the form $Z \longrightarrow Y \longrightarrow X$, where $Z \longrightarrow Y$ is a finite étale cover and $Y \longrightarrow X$ is a Nisnevich cover.*

Proof This property being local on X for the Nisnevich topology, it is sufficient to prove this in the case where X is local henselian. Then, by virtue of [GD67, Cor. 18.5.12 and Prop. 18.5.15], we can even assume that X is the spectrum of a field, in which case this is obvious. □

Lemma 3.3.27 *Any qfh-cover admits a refinement of the form $Z \longrightarrow Y \longrightarrow X$, where $Z \longrightarrow Y$ is a finite surjective morphism, and $Y \longrightarrow X$ is a Nisnevich cover.*

Proof As finite surjective morphisms are stable under pullback and composition, this follows immediately from Lemmas 3.3.24 and 3.3.26. □

Lemma 3.3.28 *Any h-cover of an integral scheme X admits a refinement of the form*

$$U \longrightarrow Z \longrightarrow Y \longrightarrow X,$$

where $U \longrightarrow Z$ is a finite surjective morphism, $Z \longrightarrow Y$ is a Nisnevich cover, $Y \longrightarrow X$ is a proper surjective birational map, and Y is normal.

Proof By virtue of [Voe96, Theorem 3.1.9], any h-cover admits a refinement of shape

$$W \longrightarrow V \longrightarrow X,$$

where $W \longrightarrow V$ is a qfh-cover, and $V \longrightarrow X$ is a proper surjective birational map. By replacing V by its normalization Y, we get a refinement of shape

$$W \times_V Y \longrightarrow Y \longrightarrow X,$$

where $W \times_V Y \longrightarrow Y$ is a qfh-cover, and $Y \longrightarrow X$ is proper surjective birational map. The result follows by Lemma 3.3.27. □

Lemma 3.3.29 *Let C be a presheaf of complexes of $\mathbf{Q}$-vector spaces on the category of S-schemes satisfying qfh-descent. Then, for any finite surjective morphism $f : Y \longrightarrow X$ with X normal, the map $f^* : H^n(C)(X) \longrightarrow H^n(C)(Y)$ is a monomorphism.*

Proof It is clearly sufficient to prove this when X is connected. Then, up to refinement, we can assume that f is a map as in Proposition 3.3.16. In this case, by virtue of Corollary 3.3.22, the $\mathbf{Q}$-vector space $H^n(C)(X) \simeq H^n(C)(Y)^G$ is a direct factor of $H^n(C)(Y)$. □

Theorem 3.3.30 *A presheaf of complexes of $\mathbf{Q}$-vector spaces on the category of S-schemes satisfies h-descent if and only if it satisfies qfh-descent and cdh-descent.*

Proof This is certainly a necessary condition, as the h-topology is finer than the qfh-topology and the cdh-topology. For the converse, as in the proof of Theorem 3.3.25, it is sufficient to prove that any presheaf of complexes of $\mathbf{Q}$-vector spaces C on the category of S-schemes satisfying qfh-descent and cdh-descent, and which is acyclic locally for the h-topology, is acyclic. We shall prove by noetherian induction that, given such a complex C, for any integer n, and any S-scheme X, for any section a of $H^n(C)$ over X, there exists a cdh-cover $X' \longrightarrow X$ on which a vanishes. In other words, we shall get that C is acyclic locally for the cdh-topology, and, as C satisfies cdh-descent, this will imply that $H^n(C)(X) = H^n_{cdh}(X, C_{cdh}) = 0$ for any integer n and any S-scheme X. Note that the presheaves $H^n(C)$ send finite sums to finite direct sums (which follows, for instance, from the fact that C satisfies Nisnevich descent). In particular, $H^n(C)(\varnothing) = 0$ for any integer n.

Let X be an S-scheme, and $a \in H^n(C)(X)$. We have a cdh-cover of X of shape $X' \amalg X'' \longrightarrow X$, where X' is the sum of the irreducible components of X_{red} and X'' is a nowhere dense closed subscheme of X, so that we can assume X is integral. Let a be a section of the presheaf $H^n(C)$ over X. As $H^n(C)_h = 0$, by virtue of Lemma 3.3.28, there exists a proper surjective birational map $p : Y \longrightarrow X$ with Y normal, a Nisnevich cover $q : Z \longrightarrow Y$, and a surjective finite morphism $r : U \longrightarrow Z$ such that $r^*(q^*(p^*(a))) = 0$ in $H^n(C)(U)$. Then Z is normal as well (see [GD67, Prop. 18.10.7]), so that, by Lemma 3.3.29, we have $q^*(p^*(a)) = 0$ in $H^n(C)(Z)$. Let T be a nowhere dense closed subscheme of X such that p is an isomorphism over $X - T$. By noetherian induction, there exists a cdh-cover $T' \longrightarrow T$ such that $a|_{T'}$ vanishes. Hence the section a vanishes on the cdh-cover $T' \amalg Z \longrightarrow X$. □

3.3.d Proper descent with rational coefficients II: separation

From now on, we assume that $\mathrm{Ho}(\mathscr{M})$ is $\mathbf{Q}$-linear.

Proposition 3.3.31 *Let* $f : Y \longrightarrow X$ *be a morphism of schemes in* $\mathscr{S}$, *and* G *a finite group acting on* Y *over* X. *Denote by* $\mathscr{Y}$ *the scheme* Y *considered as a functor from* G *to the category of* S-*schemes, and denote by* $\varphi : (\mathscr{Y}, G) \longrightarrow X$ *the morphism induced by* f. *Then, for any object* M *of* $\mathrm{Ho}(\mathscr{M})(X)$, *there are canonical isomorphisms*

$$(\mathbf{R}f_* \mathbf{L}f^*(M))^G \simeq (\mathbf{R}f_* \mathbf{L}f^*(M))^{hG} \simeq \mathbf{R}\varphi_* \mathbf{L}\varphi^*(M)$$

(where G *acts by functoriality of the construction* $\mathbf{R}f_* \mathbf{L}f^*$, *as expressed by formulas (3.2.11.3) and (3.2.11.4)).*

Proof The second isomorphism comes from Proposition 3.1.15, and the first, from (3.3.21.3). □

Theorem 3.3.32 *If* $\mathrm{Ho}(\mathscr{M})$ *satisfies Nisnevich descent, the following conditions are equivalent:*

(i) $\mathrm{Ho}(\mathscr{M})$ *satisfies étale descent;*
(ii) for any finite étale cover $f : Y \longrightarrow X$, *the functor*

$$\mathbf{L}f^* : \mathrm{Ho}(\mathscr{M})(X) \longrightarrow \mathrm{Ho}(\mathscr{M})(Y)$$

is conservative;
(iii) for any finite Galois cover $f : Y \longrightarrow X$ *of group* G, *and for any object* M *of* $\mathrm{Ho}(\mathscr{M})(X)$, *the canonical map*

$$M \longrightarrow (\mathbf{R}f_* \mathbf{L}f^*(M))^G$$

is an isomorphism.

Proof The equivalence between (i) and (iii) follows from Theorem 3.3.23 by Corollaries 3.2.17 and 3.2.18, and Proposition 3.2.8 shows that (i) implies (ii). It is thus sufficient to prove that (ii) implies (iii). Let $f : Y \longrightarrow X$ be a finite Galois cover of group G. As the functor $f^* = \mathbf{L}f^*$ is conservative by assumption, it is sufficient to check that the map $M \longrightarrow (\mathbf{R}f_* \mathbf{L}f^*(M))^G$ becomes an isomorphism after applying f^*. By virtue of Proposition 3.1.17, this just means that it is sufficient to prove (iii) when f has a section, i.e. when Y is isomorphic to the trivial G-torsor over X. In this case, we have the (equivariant) identification $\bigoplus_{g \in G} M \simeq \mathbf{R}f_* \mathbf{L}f^*(M)$, where G acts on the left term by permuting the factors. Hence $M \simeq (\mathbf{R}f_* \mathbf{L}f^*(M))^G$. □

Proposition 3.3.33 *Assume that* $\mathrm{Ho}(\mathscr{M})$ *has the localization property. The following conditions are equivalent:*

(i) $\mathrm{Ho}(\mathscr{M})$ *is separated.*
(ii) $\mathrm{Ho}(\mathscr{M})$ *is semi-separated and satisfies étale descent.*

Proof This follows from Proposition 2.3.9 and Theorem 3.3.32. □

Corollary 3.3.34 *Assume that all the residue fields of $\mathscr{S}$ are of characteristic zero, and that* $\mathrm{Ho}(\mathscr{M})$ *has the property of localization. Then the following conditions are equivalent:*

(i) $\mathrm{Ho}(\mathscr{M})$ *is separated.*
(ii) $\mathrm{Ho}(\mathscr{M})$ *satisfies étale descent.*

Proof Consider a radicial finite surjective morphism $f : Y \longrightarrow X$ in $\mathscr{S}$. To prove that the functor $\mathbf{L}f^*$ is conservative, as $\mathrm{Ho}(\mathscr{M})$ has the property of localization, by noetherian induction, we may replace X by any dense open subscheme U (and Y by $U \times_X Y$). The residue fields of X being of characteristic zero, this means that we may assume that f induces an isomorphism after reduction $Y_{red} \simeq X_{red}$. It is clear that, by the localization property, such a morphism f induces an equivalence of categories $\mathbf{L}f^*$, so that $\mathrm{Ho}(\mathscr{M})$ is automatically semi-separated. We conclude by Proposition 3.3.33. □

Proposition 3.3.35 *Assume that* $\mathrm{Ho}(\mathscr{M})$ *is separated, satisfies the localization property and the proper transversality property. Then, for any pseudo-Galois cover* $f : Y \longrightarrow X$ *of group G, and for any object M of* $\mathrm{Ho}(\mathscr{M})(X)$, *the canonical map*

$$M \longrightarrow (\mathbf{R}f_* \mathbf{L}f^*(M))^G$$

is an isomorphism.

Proof By Proposition 3.3.33, this is an easy consequence of Proposition 2.1.9 and of condition (iii) of Theorem 3.3.32. □

3.3.36 From now on, we assume furthermore that any scheme in $\mathscr{S}$ is quasi-excellent.

Theorem 3.3.37 *Assume that* $\mathrm{Ho}(\mathscr{M})$ *satisfies the localization and proper transversality properties. Then the following conditions are equivalent:*

(i) $\mathrm{Ho}(\mathscr{M})$ *is separated;*
(ii) $\mathrm{Ho}(\mathscr{M})$ *satisfies h-descent;*
(iii) $\mathrm{Ho}(\mathscr{M})$ *satisfies qfh-descent;*
(iv) for any qfh-distinguished square (3.3.15.1) of group G, if we write $a = fh = ig : T \longrightarrow X$ for the composed map, then, for any object M of $\mathrm{Ho}(\mathscr{M})(X)$, the commutative square

(3.3.37.1)
$$\begin{array}{ccc} M & \longrightarrow & (\mathbf{R}f_* \mathbf{L}f^*(M))^G \\ \downarrow & & \downarrow \\ \mathbf{R}i_* \mathbf{L}i^*(M) & \longrightarrow & (\mathbf{R}a_* \mathbf{L}a^*(M))^G \end{array}$$

is homotopy cartesian;

3 Descent in $\mathscr{P}$-fibred model categories

(v) the same as condition (iv), but only for pseudo-Galois qfh-distinguished squares.

Proof As $\mathscr{M}$ satisfies cdh-descent (Theorem 3.3.10), the equivalence between conditions (ii) and (iii) follows from Theorem 3.3.30 by Corollary 3.2.18. Similarly, Theorem 3.3.25 and Corollaries 3.3.22, 3.2.17 and 3.2.18 show that conditions (iii), (iv) and (v) are equivalent. As étale surjective morphisms as well as finite radicial epimorphisms are qfh-coverings, it follows from Proposition 3.2.8, Theorem 3.3.32 and Proposition 3.3.33, that condition (iii) implies condition (i). It thus remains to prove that condition (i) implies condition (v). So let us consider a pseudo-Galois qfh-distinguished square (3.3.15.1) of group G, and prove that (3.3.37.1) is homotopy cartesian. Using proper transversality, we see that the image of (3.3.37.1) under the functor $\mathbf{L}i^*$ is (isomorphic to) the homotopy pullback square

$$\begin{array}{ccc} \mathbf{L}i^*(M) & \longrightarrow & (\mathbf{R}g_* \mathbf{L}g^* \mathbf{L}i^*(M))^G \\ \| & & \| \\ \mathbf{L}i^*(M) & \longrightarrow & (\mathbf{R}g_* \mathbf{L}g^* \mathbf{L}i^*(M))^G \,. \end{array}$$

Write $j : U \longrightarrow X$ for the complement open immersion of i, and $b : f^{-1}(U) \longrightarrow U$ for the map induced by f. As j is étale, we see, using Proposition 3.1.17, that the image of (3.3.9.1) under $j^* = \mathbf{L}j^*$ is (isomorphic to) the square

$$\begin{array}{ccc} j^*(M) & \longrightarrow & (\mathbf{R}b_* \mathbf{L}b^* j^*(M))^G \\ \downarrow & & \downarrow \\ 0 & =\!=\!= & 0 \,, \end{array}$$

in which the upper horizontal map is an isomorphism by Proposition 3.3.35. Hence it is a homotopy pullback square. Thus, because the pair of functors $(\mathbf{L}i^*, j^*)$ is conservative on $\mathrm{Ho}(\mathscr{M})(X)$, the square (3.3.37.1) is homotopy cartesian. □

Corollary 3.3.38 *Assume that all the residue fields of $\mathscr{S}$ are of characteristic zero, and that $\mathrm{Ho}(\mathscr{M})$ has the localization and proper transversality properties. Then $\mathrm{Ho}(\mathscr{M})$ satisfies h-descent if and only if it satisfies étale descent.*

Proof This follows from Corollary 3.3.34 and Theorem 3.3.37. □

Corollary 3.3.39 *Assume that $\mathrm{Ho}(\mathscr{M})$ is separated and has the localization and proper transversality properties. Let $f : Y \longrightarrow X$ be a finite surjective morphism, with X normal, and G a group acting on Y over X, such that the map $Y/G \longrightarrow X$ is generically radicial (i.e. radicial over a dense open subscheme of X). Lastly, consider a pullback square of the following shape.*

$$\begin{array}{ccc} Y' & \longrightarrow & Y \\ f' \downarrow & & \downarrow f \\ X' & \longrightarrow & X \,. \end{array}$$

Then, for any object M of $\mathrm{Ho}(\mathcal{M})(X')$, the natural map

$$M \longrightarrow (\mathbf{R}f'_* \mathbf{L}f'^*(M))^G$$

is an isomorphism.

Proof For any presheaf C of complexes of $\mathbf{Q}$-vector spaces on $\mathcal{S}/X$, one has an isomorphism

$$\mathbf{R}\Gamma_{qfh}(X', C_{qfh}) \simeq \mathbf{R}\Gamma_{qfh}(Y', C_{qfh})^G.$$

This follows from the fact that we have an isomorphism of qfh-sheaves of sets $L(Y)/G \simeq L(X)$ (the map $Y \longrightarrow Y/G$ being generically flat, this is Proposition 3.3.19), which implies that the map $L(Y')/G \longrightarrow L(X')$ is an isomorphism of qfh-sheaves (by the universality of colimits in topoi), and implies this assertion (as in the proof of Corollary 3.3.22).

By virtue of Theorem 3.3.37, $\mathrm{Ho}(\mathcal{M})$ satisfies qfh-descent, so that the preceding computations imply the result by Corollaries 3.2.17 and 3.2.18. □

Corollary 3.3.40 *Assume that $\mathrm{Ho}(\mathcal{M})$ is separated and has the localization and proper transversality properties. Then for any finite surjective morphism $f : Y \longrightarrow X$ with X normal, the morphism*

$$M \longrightarrow \mathbf{R}f_* \mathbf{L}f^*(M)$$

is a monomorphism and admits a functorial splitting in $\mathrm{Ho}(\mathcal{M})(X)$. Furthermore, this remains true after base change by any map $X' \longrightarrow X$.

Proof It is sufficient to treat the case where X is connected. We may replace Y by a normalization of X in a suitable finite extension of its field of functions, and assume that a finite group G acts on Y over X, so that the properties described in the preceding corollary are fulfilled (see Proposition 3.3.16). □

Remark 3.3.41 The condition (iv) of Theorem 3.3.37 can be reformulated in a more global way as follows (this won't be used in these notes, but this might be useful for the reader who might want to formulate all this in terms of (pre-)algebraic derivators [Ayo07a, Def. 2.4.13]). Given a qfh-distinguished square (3.3.15.1) of group G, we can form a functor $\mathscr{F}$ from category $I = \Gamma$ (3.3.11.1) to the category of diagrams of S-schemes corresponding to the diagram of diagrams of S-schemes

$$(\mathscr{T}, G) \xrightarrow{(h, 1_G)} (\mathscr{Y}, G)$$
$$\downarrow g$$
$$Z$$

in which $\mathscr{T}$ and $\mathscr{Y}$ correspond to T and Y respectively, viewed as functors from G to $\mathscr{S}/X$. The construction of Section 3.1.22 gives a diagram of X-schemes $(\int \mathscr{F}, I_{\mathscr{F}})$ which can be described explicitly as follows. The category $I_{\mathscr{F}}$ is the cofibred category

3 Descent in $\mathscr{P}$-fibred model categories

over Γ associated to the functor from Γ to the category of small categories defined by the diagram

$$G \xrightarrow{1_G} G$$
$$\downarrow$$
$$e$$

in which e stands for the terminal category, and G for the category with one object associated to G. It thus has three objects a, b, c (see (3.3.11.1)), and the morphisms are determined by

$$\mathrm{Hom}_{I_{\mathscr{F}}}(x, y) = \begin{cases} * & \text{if } y = c; \\ \varnothing & \text{if } x \neq y \text{ and } x = b, c; \\ G & \text{otherwise.} \end{cases}$$

The functor $\mathscr{F}$ sends a, b, c to T, Y, Z respectively, and simply encodes the fact that the diagram

$$T \xrightarrow{h} Y$$
$$g \downarrow$$
$$Z$$

is G-equivariant, the action on Z being trivial. Now, by Propositions 3.1.23 and 3.3.31, if $\varphi : (\mathscr{F}, I_{\mathscr{F}}) \longrightarrow (X, \Gamma)$ denotes the canonical map, for any object M of $\mathrm{Ho}(\mathscr{M})(X)$, the object $\mathbf{R}\varphi_* \mathbf{L}\varphi^*(M)$ is the functor from $\lrcorner = \Gamma^{op}$ to $\mathscr{M}(X)$ corresponding to the diagram below (of course, this is well defined only in the homotopy category of the category of functors from $\lrcorner$ to $\mathscr{M}(X)$).

$$(\mathbf{R}f_* \mathbf{L}f^*(M))^G$$
$$\downarrow$$
$$\mathbf{R}i_* \mathbf{L}i^*(M) \longrightarrow (\mathbf{R}a_* \mathbf{L}a^*(M))^G$$

As a consequence, if $\psi : (\int \mathscr{F}, I_{\mathscr{F}}) \longrightarrow X$ denotes the structural map, the object $\mathbf{R}\psi_* \mathbf{L}\psi^*(M)$ is simply the homotopy limit of the diagram of $\mathscr{M}(X)$ above, so that condition (iv) of Theorem 3.3.37 can now be reformulated by saying that the map

$$M \longrightarrow \mathbf{R}\psi_* \mathbf{L}\psi^*(M)$$

is an isomorphism, i.e. that the functor

$$\mathbf{L}\psi^* : \mathrm{Ho}(\mathscr{M})(X) \longrightarrow \mathrm{Ho}(\mathscr{M})(\int \mathscr{F}, I_{\mathscr{F}})$$

is fully faithful.

4 Constructible motives

4.0.1 Consider as in Section 2.0.1 a base scheme $\mathscr{S}$ and a sub-category $\mathscr{S}$ of the category of $\mathscr{S}$-schemes. In Section 4.4, and for the main theorem of Section 4.2, we will assume:

(a) Any scheme in $\mathscr{S}$ is quasi-excellent.[62]

Apart from Definition 4.3.2 and the subsequent proposition, where we will consider an abstract situation, we will be concerned with the study of a fixed premotivic triangulated category $\mathscr{T}$ over $\mathscr{S}$ (recall Definition 2.4.45) such that:

(b) $\mathscr{T}$ is motivic (see Definition 2.4.45).
(c) $\mathscr{T}$ is endowed with a set of twists τ (see Section 1.4.4) which is stable under Tate twists $\mathbb{1}(p)[q]$, for $p, q \in \mathbf{Z}$.
(d) $\mathscr{T}$ is the homotopy category associated with a stable combinatorial Sm-fibred model category $\mathscr{M}$ over $\mathscr{S}$.[63]

As usual, the geometric sections of $\mathscr{T}$ will be denoted by M.

Unless explicitly referring to the underlying model category $\mathscr{M}$, we will not indicate in the notation of the six operations that the functors are derived functors.

4.1 Resolution of singularities

The aim of this subsection is to gather the results from the theory of resolution of singularities that will be used subsequently.

4.1.1 In [GD67, IV, 7.8.2], Grothendieck defined the notion of an *excellent ring*. Matsumura introduced in [Mat70] the weaker notion of a *quasi-excellent* ring. Recall a ring A is quasi-excellent if the following conditions hold:

1. A is noetherian.
2. For any prime ideal $\mathfrak{p}$, $\hat{A}_\mathfrak{p}$ being the completion of A at $\mathfrak{p}$, the canonical morphism $A \longrightarrow \hat{A}_\mathfrak{p}$ is regular (see Section 4.1.4 below).
3. For any A-algebra B of finite type, the regular locus of $\mathrm{Spec}\,(B)$ is open.

Then a ring A is excellent if it is quasi-excellent and universally catenary. Following Gabber, we say a scheme X is *quasi-excellent* (*excellent*) if it admits an open cover by affine schemes whose rings are quasi-excellent (excellent, respectively).

Theorem 4.1.2 (**Gabber's weak local uniformization**) *Let X be a quasi-excellent scheme, and $Z \subset X$ a nowhere dense closed subscheme. Then there exists a finite h-cover $\{f_i : Y_i \longrightarrow X\}_{i \in I}$ such that for all i in I, f_i is a morphism of finite type, the*

---

[62] See Section 4.1.1. The reader can safely restrict his attention to the more classical notion of an excellent scheme ([GD67, IV, 7.8.5]).

[63] We need this assumption to apply descent theory as described in Section 3.3.

4 Constructible motives

scheme Y_i is regular, and $f_i^{-1}(Z)$ is either empty or the support of a strict normal crossing divisor in Y_i.

See [ILO14] for a proof. Note that, if we are only interested in schemes of finite type over $\operatorname{Spec}(R)$, for R a field, a complete discrete valuation ring, or a Dedekind domain whose field of functions is a global field, this is an immediate consequence of de Jong's resolution of singularities by alterations; see [dJ96]. One can also deduce the case of schemes of finite type over an excellent noetherian scheme of dimension less than or equal to 2 from [dJ97]; see Theorem 4.1.10 and Corollary 4.1.11 below for a precise statement.

Remark 4.1.3 This theorem will be used in the proof of Lemma 4.2.14, which is the key point for the proof of Theorem 4.2.16.

4.1.4 Recall that a morphism of rings $u : A \longrightarrow B$ is *regular* if it is flat, and if, for any prime ideal $\mathfrak{p}$ in A, with residue field $\kappa(\mathfrak{p})$, the $\kappa(\mathfrak{p})$-algebra $\kappa(\mathfrak{p}) \otimes_A B$ is geometrically regular (equivalently, this means that, for any prime ideal $\mathfrak{q}$ of B, the A-algebra $B_\mathfrak{q}$ is formally smooth in the $\mathfrak{q}$-adic topology). We recall the following great generalization of Neron's desingularization theorem:

Theorem 4.1.5 (Popescu–Spivakovsky) *A morphism of noetherian rings $u : A \longrightarrow B$ is regular if and only if B is a filtered colimit of smooth A-algebras of finite type.*

Proof See [Spi99, theorems 1.1 and 1.2]. □

4.1.6 Recall that an *alteration* is a proper surjective morphism $p : X' \longrightarrow X$ which is generically finite, i.e. such that there exists a dense open subscheme $U \subset X$ over which p is finite.

Definition 4.1.7 (de Jong) Let X be a noetherian scheme endowed with an action of a finite group G. A *Galois alteration* of the pair (X, G) is the data of a finite group G', of a surjective morphism of groups $G' \longrightarrow G$, of an alteration $X' \longrightarrow X$, and of an action of G' on X', such that:

(i) the map $X' \longrightarrow X$ is G'-equivariant;
(ii) for any irreducible component T of X, there exists a unique irreducible component T' of X' over T, and the corresponding finite field extension

$$k(T)^G \subset k(T')^{G'}$$

is purely inseparable.

In practice, we shall keep the morphism of groups $G' \longrightarrow G$ implicit, and we shall say that $(X' \longrightarrow X, G')$ is a Galois alteration of (X, G).

Given a noetherian scheme X, a *Galois alteration* of X is a Galois alteration $(X' \longrightarrow X, G)$ of (X, e), where e denotes the trivial group. In this case, we shall say that $X' \longrightarrow X$ is a *Galois alteration of X of group G*.

Remark 4.1.8 If $p : X' \longrightarrow X$ is a Galois alteration of group G over X, then, if X and X' are normal, irreducible and quasi-excellent, p can be factored as a radicial finite surjective morphism $X'' \longrightarrow X$, followed by a Galois alteration $X' \longrightarrow X''$ of group G, such that $k(X'') = k(X')^G$ (just define X'' as the normalization of X in $k(X')^G$). In other words, up to a radicial finite surjective morphism, X is generically the quotient of X' under the action of G.

Definition 4.1.9 A noetherian scheme S admits *canonical dominant resolution of singularities up to quotient singularities* if, for any Galois alteration $S' \longrightarrow S$ of group G, and for any G-equivariant nowhere dense closed subscheme $Z' \subset S'$, there exists a Galois alteration $(p : S'' \longrightarrow S', G')$ of (S', G), such that S'' is regular and projective over S, and such that the inverse image of Z' in S'' is contained in a G'-equivariant strict normal crossing divisor (i.e. a strict normal crossing divisor whose irreducible components are stable under the action of G').

A noetherian scheme S admits *canonical resolution of singularities up to quotient singularities* if any integral closed subscheme of S admits canonical dominant resolution of singularities up to quotient singularities.

A noetherian scheme S admits *wide resolution of singularities up to quotient singularities* if, for any separated S-scheme of finite type X, and any nowhere dense closed subscheme $Z \subset X$, there exists a projective Galois alteration $p : X' \longrightarrow X$ of group G, with X' regular, such that, in each connected component of X', $Z' = p^{-1}(Z)$ is either empty or the support of a strict normal crossing divisor.

Theorem 4.1.10 (de Jong) *If an excellent noetherian scheme of finite dimension S admits canonical resolution of singularities up to quotient singularities, then any separated S-scheme of finite type admits canonical resolution of singularities up to quotient singularities.*

Proof Let X be an integral separated S-scheme of finite type. There exists a finite morphism $S' \longrightarrow S$, with S' integral, an integral dominant S'-scheme X' and a radicial extension $X' \longrightarrow X$ over S, such that X' has a geometrically irreducible generic fiber over S'. It then follows from (the proof of) [dJ97, theorem 5.13] that X' admits canonical dominant resolution of singularities up to quotient singularities, which implies that X has the same property. □

Corollary 4.1.11 (de Jong) *Let S be an excellent noetherian scheme of dimension less than or equal to 2. Then any separated scheme of finite type over S admits canonical resolution of singularities up to quotient singularities. In particular, S admits wide resolution of singularities up to quotient singularities.*

Proof See [dJ97, corollary 5.15]. □

4.2 Finiteness theorems

The aim of this section is to study the notion of τ-constructibility in the triangulated motivic case and to study its stability properties under Grothendieck's six operations. Recall the following particular case of Definition 1.4.9:

Definition 4.2.1 For a scheme X in $\mathscr{S}$, we denote by $\mathscr{T}_c(X)$ the thick triangulated sub-category of $\mathscr{T}(X)$ generated by premotives of the form $M_X(Y)\{i\}$ for a smooth X-scheme Y and a twist $i \in \tau$. We will say that a premotive in $\mathscr{T}_c(X)$ is τ-*constructible*, or, simply, *constructible*.

Remark 4.2.2 Let us mention that our main examples:

- the stable homotopy category SH (cf. Example 1.4.3),
- the category of Voevodsky motives DM (cf. Definition 11.1.1),
- the category of Beilinson motives DM_B (cf. Definition 14.2.1)

are all generated by Tate twists (*i.e.* $\tau = \mathbf{Z}$). Recall also Proposition 1.4.11: it applies to all these examples so that constructible premotives coincide with compact objects.[64]

Proposition 4.2.3 *If M and N are constructible in $\mathscr{T}(X)$, so is $M \otimes_X N$.*

Proof For a fixed M, the full subcategory of $\mathscr{T}(X)$ spanned by objects such that $M \otimes_X N$ is constructible is a thick triangulated subcategory of $\mathscr{T}(X)$. If M is of shape $M_X(Y)\{n\}$ for Y smooth over X and $n \in \tau$, this proves that $M \otimes_X N$ is constructible whenever N is. By the same argument, using the symmetry of the tensor product, we get to the general case. □

Similarly, one has the following conservation property.

Proposition 4.2.4 *For any morphism $f : X \longrightarrow Y$ of schemes, the functor*

$$f^* : \mathscr{T}(Y) \longrightarrow \mathscr{T}(X)$$

preserves constructible objects. If moreover f is smooth, the functor

$$f_\sharp : \mathscr{T}(X) \longrightarrow \mathscr{T}(Y)$$

also preserves constructible objects.

Corollary 4.2.5 *The categories $\mathscr{T}_c(X)$ form a thick triangulated monoidal Sm-fibred subcategory of $\mathscr{T}$.*

Proposition 4.2.6 *Let X be a scheme, and $X = \bigcup_{i \in I} U_i$ a cover of X by open subschemes. An object M of $\mathscr{T}(X)$ is constructible if and only if its restriction to U_i is constructible in $\mathscr{T}(U_i)$ for all $i \in I$.*

---

[64] Notice, however, this fact is not true for étale motivic complexes.

Proof This is a necessary condition by Proposition 4.2.4. For the converse, as X is noetherian, it is sufficient to treat the case where I is finite. Proceeding by induction on the cardinality of I it is sufficient to treat the case of a cover by two open subschemes $X = U \cup V$. For an open immersion $j : W \longrightarrow X$, write $M_W = j_\sharp j^*(M)$. If the restrictions of M to U and V are constructible, then so is its restriction to $U \cap V$. According to Proposition 3.3.4, we get a distinguished triangle

$$M_{U \cap V} \longrightarrow M_U \oplus M_V \longrightarrow M \longrightarrow M_{U \cap V}[1]$$

in which M_W is constructible for $W = U, V, U \cap V$ (using Proposition 4.2.4 again). Thus the premotive M is constructible. □

Corollary 4.2.7 *For any scheme X and any vector bundle E over X, the functors $Th(E)$ and $Th(-E)$ preserve constructible objects in $\mathscr{T}(X)$.*

Proof To prove that $Th(E)$ and $Th(-E)$ preserves constructible objects, by virtue of the preceding proposition, we may assume that E is trivial of rank r. It is thus sufficient to prove that $M(r)$ is constructible whenever M is so for any integer r. Since we may assume that $M = \mathbb{1}_X\{n\}$ for some $n \in \tau$ (using Proposition 4.2.4), this is true by assumption on τ; see 4.0.1(c). □

Corollary 4.2.8 *Let $f : X \longrightarrow Y$ be a morphism of finite type. The property that the functor*

$$f_* : \mathscr{T}(X) \longrightarrow \mathscr{T}(Y)$$

preserves constructible objects is local on Y with respect to the Zariski topology.

Proof Consider a finite Zariski cover $\{v_i : Y_i \longrightarrow Y\}_{i \in I}$, and write $f_i : X_i \longrightarrow Y_i$ for the pullback of f along v_i for each i in I. Assume that the functors $f_{i,*}$ preserve constructible objects; we shall prove that f_* has the same property. Let M be a constructible object in $\mathscr{T}(X)$. Then for $i \in I$, using the smooth base change isomorphism (for open immersions), we see that the restriction of $f_*(M)$ to Y_i is isomorphic to the image under $f_{i,*}$ of the restriction of M to X_i, hence is constructible. The preceding proposition thus implies that $f_*(M)$ is constructible. □

Proposition 4.2.9 *For any closed immersion $i : Z \longrightarrow X$, the functor*

$$i_* : \mathscr{T}(Z) \longrightarrow \mathscr{T}(X)$$

preserves constructible objects.

Proof It is sufficient to prove that for any smooth Z-scheme Y_0 and any twist $n \in \tau$, the premotive $i_*(M_Z(Y_0)\{n\})$ is constructible in $\mathscr{T}(X)$. According to the Mayer–Vietoris triangle (see Remark 3.3.6), this assertion is local in X. Thus we can assume there exists a smooth X-scheme Y such that $Y_0 = Y \times_X Z$ (apply [GD67, 18.1.1]). Put $U = X - Z$ and let $j : U \longrightarrow X$ be the obvious open immersion. From the localization property, we get a distinguished triangle

$$M_X(Y \times_X U)\{n\} \longrightarrow M_X(Y)\{n\} \longrightarrow i_*(M_Z(Y_0)\{n\}) \longrightarrow M_X(Y \times_X U)\{n\}[1]$$

4 Constructible motives   135

and this concludes the proof.   □

Corollary 4.2.10 *Let* $i : Z \longrightarrow X$ *be a closed immersion with open complement* $j : U \longrightarrow X$. *Then an object M of $\mathscr{T}(X)$ is constructible if and only if $j^*(M)$ and $i^*(M)$ are constructible in $\mathscr{T}(U)$ and $\mathscr{T}(Z)$, respectively.*

Proof We have a distinguished triangle

$$j_\sharp j^*(M) \longrightarrow M \longrightarrow i_* i^*(M) \longrightarrow j_\sharp j^*(M)[1].$$

Hence this assertion follows from Propositions 4.2.4 and 4.2.9.   □

Proposition 4.2.11 *If $f : X \longrightarrow Y$ is proper, then the functor*

$$f_* : \mathscr{T}(X) \longrightarrow \mathscr{T}(Y)$$

preserves constructible objects.

Proof We shall first consider the case where f is projective. As this property is local on Y (Corollary 4.2.8), we may assume that f factors as a closed immersion $i : X \longrightarrow \mathbf{P}^n_Y$ followed by the canonical projection $p : \mathbf{P}^n_Y \longrightarrow Y$. By virtue of Proposition 4.2.9, we can assume that $f = p$. In this case, the functor p_* is isomorphic to $p_\sharp$ composed with the quasi-inverse of the Thom endofunctor associated to the cotangent bundle of p; see Theorem 2.4.50 (3). Therefore, the functor p_* preserves constructible objects by virtue of Proposition 4.2.4 and of Corollary 4.2.7. The case where f is proper follows easily from the projective case, using Chow's lemma and *cdh*-descent (the homotopy pullback squares (3.3.9.1)), by induction on the dimension of X.   □

Corollary 4.2.12 *If $f : X \longrightarrow Y$ is separated of finite type, then the functor*

$$f_! : \mathscr{T}(X) \longrightarrow \mathscr{T}(Y)$$

preserves constructible objects.

Proof It is sufficient to treat the case where f is either an open immersion or a proper morphism, which follows respectively from Proposition 4.2.4 and Proposition 4.2.11.   □

Proposition 4.2.13 *Let X be a scheme. The category of constructible objects in $\mathscr{T}(X)$ is the smallest thick triangulated subcategory which contains the objects of shape $f_*(\mathbb{1}_{X'}\{n\})$, where $f : X' \longrightarrow X$ is a (strictly) projective morphism, and $n \in \tau$.*

Proof Let $\mathscr{T}_p(X)$ be the smallest thick triangulated subcategory which contains the objects of shape $f_*(\mathbb{1}_{X'}\{n\})$, where $f : X' \longrightarrow X$ is a (strictly) projective morphism, and $n \in \tau$. Proposition 4.2.11 shows that $\mathscr{T}_p(X) \subset \mathscr{T}_c(X)$, so that it is sufficient to prove the reverse inclusion. Note that, for any separated smooth morphism f, locally for the Zariski topology, $f_\sharp$ coincides with $f_!$ up to a Tate twist. In other words, it is sufficient to prove that, for any separated morphism of finite type $f : Y \longrightarrow X$, $f_!(\mathbb{1}_Y)$

belongs to $\mathscr{T}_p(X)$. If we factor f into an open immersion $j : Y \longrightarrow X'$ followed by a proper morphism $p : X' \longrightarrow X$, we see that it is sufficient to prove that $j_\sharp(\mathbb{1}_Y)$ belongs to $\mathscr{T}_p(X')$. This follows straight away from the localization property. □

The following lemma is the key geometrical point for the finiteness Theorem 4.2.16

Lemma 4.2.14 *Let* $j : U \longrightarrow X$ *be a dense open immersion such that* X *is quasi-excellent. Then, there exists the following data:*

(i) a finite h-cover $\{f_i : Y_i \longrightarrow X\}_{i \in I}$ *such that for all i in I, f_i is a morphism of finite type, the scheme Y_i is regular, and $f_i^{-1}(U)$ is either Y_i itself or the complement of a strict normal crossing divisor in Y_i; we shall write*

$$f : Y = \coprod_{i \in I} Y_i \longrightarrow X$$

for the induced global h-cover;
(ii) a commutative diagram

(4.2.14.1)
$$\begin{array}{ccc} X''' & \xrightarrow{g} & Y \\ {\scriptstyle q}\downarrow & & \downarrow{\scriptstyle f} \\ X'' & \xrightarrow{u} X' \xrightarrow{p} & X \end{array}$$

in which: p is a proper birational morphism, X' is normal, u is a Nisnevich cover, and q is a finite surjective morphism.

Let T (resp. T') be a closed subscheme of X (resp. X') and assume that for any irreducible component T_0 of T, the following inequality is satisfied:

(4.2.14.2) $$\operatorname{codim}_{X'}(T') \geq \operatorname{codim}_X(T_0).$$

Then, possibly after shrinking X to an open neighborhood of the generic points of T in X, one can replace X'' by an open cover and X''' by its pullback along this cover, in such a way that we have in addition the following properties:

(iii) $p(T') \subset T$ and the induced map $T' \longrightarrow T$ is finite and pseudo-dominant; [65]
(iv) if we write $T'' = u^{-1}(T')$, the induced map $T'' \longrightarrow T'$ is an isomorphism.

Proof The existence of $f : Y \longrightarrow X$ as in (i) follows from Gabber's weak uniformization theorem (see Theorem 4.1.2), while the commutative diagram (4.2.14.1) satisfying property (ii) is ensured by Lemma 3.3.28.

Consider moreover closed subschemes $T \subset X$ and $T' \subset X'$ satisfying (4.2.14.2).

We first show that, by shrinking X to an open neighborhood of the generic points of T and by replacing the diagram (4.2.14.1) by its pullback over this neighborhood,

---

[65] Recall from Section 8.1.3 that this means that any irreducible component of T' dominates an irreducible component of T.

4 Constructible motives

we can assume that condition (iii) is satisfied. Note that shrinking X in this way does not change the condition (4.2.14.2) because $\operatorname{codim}_X(T_0)$ does not change and $\operatorname{codim}_{X'}(T')$ can only increase.[66]

Note first that, by shrinking X, we can assume that any irreducible component T'_0 of T' dominates an irreducible component T_0 of T. In fact, given an irreducible component T'_0 which does not satisfy this condition, $p(T'_0)$ is a closed subscheme of X disjoint from the set of generic points of T and replacing X by $X - f(T'_0)$, we can throw out T'_0.

Further, shrinking X again, we can assume that for any pair (T'_0, T_0) as in the preceding paragraph, $p(T'_0) \subset T_0$. In fact, in any case, as $p(T'_0)$ is closed we get that $T_0 \subset p(T'_0)$. Let Z be the closure of $p(T'_0) - T_0$ in X. Then Z does not contain any generic points of T (because $p(T'_0)$ is irreducible), and $p(T'_0) \cap (X - Z) \subset T_0$. Thus it is sufficient to replace X by $X - Z$ to ensure this assumption.

Consider again a pair (T'_0, T_0) as in the two preceding paragraphs and the induced commutative square:

(4.2.14.3)
$$\begin{array}{ccc} T'_0 & \longrightarrow & X' \\ {\scriptstyle p_0}\downarrow & & \downarrow{\scriptstyle p} \\ T_0 & \longrightarrow & X. \end{array}$$

We show that the map p_0 is generically finite. In fact, this will conclude the first step, because if it is true for any irreducible component T'_0 of T', we can shrink X again so that the dominant morphism $p_0 : T'_0 \longrightarrow T_0$ becomes finite.

Denote by c' (resp. c) the codimension of T_0 in X (resp. T'_0 in X'). Note that (4.2.14.2) gives the inequality $c' \geq c$. Let t_0 be the generic point of T_0, Ω the localization of X at t_0 and consider the pullback of (4.2.14.3):

(4.2.14.4)
$$\begin{array}{ccc} W' & \longrightarrow & \Omega' \\ {\scriptstyle q_0}\downarrow & & \downarrow{\scriptstyle q} \\ \{t_0\} & \longrightarrow & \Omega. \end{array}$$

We have to prove that $\dim(W') = 0$. Consider an irreducible component Ω'_0 of Ω' containing W'. As q is still proper birational, Ω'_0 corresponds to a unique irreducible component Ω_0 of Ω such that q induces a proper birational map $\Omega'_0 \longrightarrow \Omega_0$. According to [GD67, 5.6.6], we get the inequality

$$\dim(\Omega'_0) \leq \dim(\Omega_0).$$

Thus, we obtain the following inequalities:

---

[66] Recall that for any scheme X, $\operatorname{codim}_X(\emptyset) = +\infty$.

$$\dim(W') \leq \dim(\Omega'_0) - \operatorname{codim}_{\Omega'_0}(W')$$
$$\leq \dim(\Omega_0) - \operatorname{codim}_{\Omega'_0}(W')$$
$$\leq \dim(\Omega) - \operatorname{codim}_{\Omega'_0}(W').$$

As this is true for any irreducible component Ω'_0 of Ω', we finally obtain:

$$\dim(W') \leq \dim(\Omega) - \operatorname{codim}_{\Omega'}(W') \leq c - c'$$

and this concludes the first step.

Keeping T' and T as above, as the map from T'' to T' is a Nisnevich cover, it is a split epimorphism in a neighborhood of the generic points of T' in X'. Hence, as the map $X' \longrightarrow X$ is proper and birational, we can find a neighborhood of the generic points of T in X over which the map $T'' \longrightarrow T'$ admits a section $s : T' \longrightarrow T''$. Let S be a closed subset of X'' such that $T'' = s(T') \amalg S$ (which exists because $X'' \longrightarrow X'$ is étale). The map $(X''-T'') \amalg (X''-S) \longrightarrow X'$ is then a Nisnevich cover. Replacing X'' by $(X''-T'') \amalg (X''-S)$ (and X''' by the pullback of $X''' \longrightarrow X''$ along $(X''-T'') \amalg (X''-S) \longrightarrow X'$), we may assume that the induced map $T'' \longrightarrow T'$ is an isomorphism, without modifying further the data f, p, T and T'. This gives property (iv) and ends the proof the lemma. □

4.2.15 Let $\mathscr{T}_0$ be a full *Open*-fibred subcategory of $\mathscr{T}$ (where *Open* stands for the class of open immersions). We assume that $\mathscr{T}_0$ has the following properties:

(a) for any scheme X in $\mathscr{S}$, $\mathscr{T}_0(X)$ is a thick subcategory of $\mathscr{T}(X)$ which contains the objects of the form $\mathbb{1}_X\{n\}$, $n \in \tau$;
(b) for any separated morphism of finite type $f : X \longrightarrow Y$ in $\mathscr{S}$, $\mathscr{T}_0$ is stable under $f_!$;
(c) for any dense open immersion $j : U \longrightarrow X$, with X regular, which is the complement of a strict normal crossing divisor, $j_*(\mathbb{1}_U\{n\})$ is in $\mathscr{T}_0(U)$ for any $n \in \tau$.

Properties (a) and (b) have the following consequences: any constructible object belongs to $\mathscr{T}_0$; given a closed immersion $i : Z \longrightarrow X$ with complement open immersion $j : U \longrightarrow X$, an object M of $\mathscr{T}(X)$ belongs to $\mathscr{T}_0(X)$ if and only if $j^*(M)$ and $i^*(M)$ belongs to $\mathscr{T}_0(U)$ and $\mathscr{T}_0(Z)$, respectively; for any scheme X in $\mathscr{S}$, the condition that an object of $\mathscr{T}(X)$ belongs to $\mathscr{T}_0(X)$ is local on X for the Zariski topology.

Theorem 4.2.16 *Consider the above hypotheses and assume that $\mathscr{T}$ is **Q**-linear and separated. Let Y be a quasi-excellent scheme and $f : X \longrightarrow Y$ be a morphism of finite type. Then for any constructible object M of $\mathscr{T}(X)$, the object $f_*(M)$ belongs to $\mathscr{T}_0(Y)$.*

Proof It is sufficient to prove that, for any dense open immersion $j : U \longrightarrow X$, and for any $n \in \tau$, the object $j_*(\mathbb{1}_U\{n\})$ is in $\mathscr{T}_0$. Indeed, assume this is known. We want to prove that $f_*(M)$ is in $\mathscr{T}_0(Y)$ whenever M is constructible. We deduce from property (b) of Section 4.2.15 and from Proposition 4.2.13 that it is sufficient to consider the case where $M = \mathbb{1}_X\{n\}$, with $n \in \tau$. Then, as this property is assumed to be known for dense open immersions, by an easy Mayer–Vietoris argument, we see that

4 Constructible motives

the condition that $f_*(\mathbb{1}_X\{n\})$ belongs to $\mathcal{T}_0$ is local on X with respect to the Zariski topology. Therefore, we may assume that f is separated. Consider a compactification of f, i.e. a commutative diagram

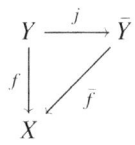

$$\begin{array}{ccc} Y & \xrightarrow{j} & \bar{Y} \\ {\scriptstyle f}\downarrow & \swarrow {\scriptstyle \bar{f}} & \\ X & & \end{array}$$

with j a dense open immersion, and $\bar{f}$ proper. By property (b) of Section 4.2.15, we may assume that $f = j$ is a dense open immersion.

Let $j : U \longrightarrow X$ be a dense open immersion. We shall prove by induction on the dimension of X that, for any $n \in \tau$, the object $j_*(\mathbb{1}_U\{n\})$ is in $\mathcal{T}_0$. The case where X is of dimension ≤ 0 follows from the fact that the map j is then an isomorphism, which implies that $j_\sharp \simeq j_*$, and allows to conclude the result (because $\mathcal{T}_0$ is $Open$-fibred).

Assume that $\dim X > 0$. Following an argument used by Gabber [ILO14] in the context of ℓ-adic sheaves, we shall prove by induction on $c \geq 0$ that there exists a closed subscheme $T \subset X$ of codimension $> c$ such that, for any $n \in \tau$, the restriction of $j_*(\mathbb{1}_U\{n\})$ to $X - T$ is in $\mathcal{T}_0(X - T)$. As X is of finite dimension, this will obviously prove Theorem 4.2.16.

The case where $c = 0$ is clear: we can choose T such that $X - T = U$. If $c > 0$, we choose a closed subscheme T of X, of codimension $> c - 1$, such that the restriction of $j_*(\mathbb{1}_U\{n\})$ to $X - T$ is in $\mathcal{T}_0$. It is then sufficient to find a dense open subscheme V of X, which contains all the generic points of T, and such that the restriction of $j_*(\mathbb{1}_U\{n\})$ to V is in $\mathcal{T}_0$: for such a V, we shall obtain that the restriction of $j_*(\mathbb{1}_U\{n\})$ to $V \cup (X - T)$ is in $\mathcal{T}_0$, the complement of $V \cup (X - T)$ being the support of a closed subscheme of codimension $> c$ in X. In particular, using the smooth base change isomorphism (for open immersions), we can always replace X by a generic neighborhood of T. It is sufficient to prove that, possibly after shrinking X as above, the pullback of $j_*(\mathbb{1}_U\{n\})$ along $T \longrightarrow X$ is in $\mathcal{T}_0$ (as we already know that its restriction to $X - T$ is in $\mathcal{T}_0$).

We may assume that T is purely of codimension c. We may assume that we have data as in points (i) and (ii) of Lemma 4.2.14. We let $j' : U' \longrightarrow X'$ denote the pullback of j along $p : X' \longrightarrow X$. Then, we can find, by induction on c, a closed subscheme T' in X', of codimension $> c - 1$, such that the restriction of $j'_*(\mathbb{1}_{U'}\{n\})$ to $X' - T'$ is in $\mathcal{T}_0$. By shrinking X, we may assume that conditions (iii) and (iv) of Lemma 4.2.14 are fulfilled as well.

For an X-scheme $w : W \longrightarrow X$ and a closed subscheme $Z \subset W$, we shall write

$$\varphi(W, Z) = w_* i_* i^* j_{W,*} j_W^*(\mathbb{1}_W\{n\}),$$

where $i : Z \longrightarrow W$ denotes the inclusion, and $j_W : W_U \longrightarrow W$ stands for the pullback of j along w. This construction is functorial with respect to morphisms of pairs of X-schemes: if $W' \longrightarrow W$ is a morphism of X-schemes, with Z' and Z two closed subschemes of W' and W respectively, such that Z' is sent to Z, then we get a natural map $\varphi(W,Z) \longrightarrow \varphi(W',Z')$. Remember that we want to prove that $\varphi(X,T)$ is in $\mathscr{T}_0$. This will be done via the following lemmas (which hold assuming all the conditions stated in Lemma 4.2.14 as well as our inductive assumptions).

Lemma 4.2.17 *The cone of the map $\varphi(X,T) \longrightarrow \varphi(X',T')$ is in $\mathscr{T}_0$.* □

The map $\varphi(X,T) \longrightarrow \varphi(X',T')$ factors as

$$\varphi(X,T) \longrightarrow \varphi(X',p^{-1}(T)) \longrightarrow \varphi(X',T').$$

By the octahedral axiom, it is sufficient to prove that each of these two maps has a cone in $\mathscr{T}_0$.

We shall prove first that the cone of the map $\varphi(X',p^{-1}(T)) \longrightarrow \varphi(X',T')$ is in $\mathscr{T}_0$. Given an immersion $a : S \longrightarrow X'$, we shall write

$$M_S = a_! \, a^*(M).$$

We then have distinguished triangles

$$M_{p^{-1}(T)-T'} \longrightarrow M_{p^{-1}(T)} \longrightarrow M_{T'} \longrightarrow M_{p^{-1}(T)-T'}[1].$$

For $M = j'_*(\mathbb{1}_{U'}\{n\})$ (recall j' is the pullback of j along p) the image of this triangle under p_* gives a distinguished triangle

$$p_*(M_{p^{-1}(T)-T'}) \longrightarrow \varphi(X',p^{-1}(T)) \longrightarrow \varphi(X',T') \longrightarrow p_*(M_{p^{-1}(T)-T'})[1].$$

As the restriction of $M = j'_*(\mathbb{1}_{U'}\{n\})$ to $X' - T'$ is in $\mathscr{T}_0$ by assumption on T', the object $M_{p^{-1}(T)-T'}$ is in $\mathscr{T}_0$ as well (by property (b) of Section 4.2.15 and because $\mathscr{T}_0$ is $Open$-fibred), from which we deduce that $p_*(M_{p^{-1}(T)-T'})$ is in $\mathscr{T}_0$ (using condition (iii) of Lemma 4.2.14 and property (b) of Section 4.2.15).

Let V be a dense open subscheme of X such that $p^{-1}(V) \longrightarrow V$ is an isomorphism. We may assume that $V \subset U$, and write $i : Z \longrightarrow U$ for the complement closed immersion. Let $p_U : U' = p^{-1}(U) \longrightarrow U$ be the pullback of p along j, and let $\bar{Z}$ be the reduced closure of Z in X. We thus get the commutative squares of immersions below,

$$\begin{array}{ccc} Z & \xrightarrow{k} & \bar{Z} \\ {\scriptstyle i}\downarrow & & \downarrow{\scriptstyle l} \\ U & \xrightarrow{j} & X \end{array} \quad \text{and} \quad \begin{array}{ccc} Z' & \xrightarrow{k'} & \bar{Z}' \\ {\scriptstyle i'}\downarrow & & \downarrow{\scriptstyle l'} \\ U' & \xrightarrow{j'} & X', \end{array}$$

where the square on the right is obtained from the one on the left by pulling back along $p : X' \longrightarrow X$. As p is an isomorphism over V, we get by cdh-descent (Proposition 3.3.10) the homotopy pullback square below:

4 Constructible motives 141

$$\begin{array}{ccc} \mathbb{1}_U\{n\} & \longrightarrow & p_{U,*}(\mathbb{1}_{U'}\{n\}) \\ \downarrow & & \downarrow \\ i_* i^*(\mathbb{1}_Z\{n\}) & \longrightarrow & i_* i^* p_{U,*}(\mathbb{1}_{U'}\{n\}). \end{array}$$

If $a : T \longrightarrow X$ denotes the inclusion, applying the functor $a_* \, a^* \, j_*$ to the commutative square above, we see from the proper base change formula and from the identification $j_* \, i_* \simeq l_* \, k_*$ that we get a commutative square isomorphic to the following one

$$\begin{array}{ccc} \varphi(X,T) & \longrightarrow & \varphi(X',p^{-1}(T)) \\ \downarrow & & \downarrow \\ \varphi(\bar{Z}, \bar{Z} \cap T) & \longrightarrow & \varphi(\bar{Z}', p^{-1}(\bar{Z} \cap T)), \end{array}$$

which is thus homotopy cartesian as well. It is sufficient to prove that the two objects $\varphi(\bar{Z}, \bar{Z} \cap T)$ and $\varphi(\bar{Z}', p^{-1}(\bar{Z} \cap T))$ are in $\mathcal{T}_0$. It follows from the proper base change formula that the object $\varphi(\bar{Z}, \bar{Z} \cap T)$ is canonically isomorphic to the restriction to T of $l_* \, k_*(\mathbb{1}_Z\{n\})$. As $\dim \bar{Z} < \dim X$, we know that the object $k_*(\mathbb{1}_Z\{n\})$ is in $\mathcal{T}_0$. By property (b) of Section 4.2.15, we obtain that $\varphi(\bar{Z}, \bar{Z} \cap T)$ is in $\mathcal{T}_0$. Similarly, the object $\varphi(\bar{Z}', p^{-1}(\bar{Z} \cap T))$ is canonically isomorphic to the restriction of $p_* \, l'_* \, k'_*(\mathbb{1}_{Z'}\{n\})$ to T, and, as $\dim \bar{Z}' < \dim X'$ (because, p being an isomorphism over the dense open subscheme V of X, $\bar{Z}'$ does not contain any generic points of X'), $k'_*(\mathbb{1}_{Z'}\{n\})$ is in $\mathcal{T}_0$. We deduce again from property (b) of Section 4.2.15 that $\varphi(\bar{Z}', p^{-1}(\bar{Z} \cap T))$ is in $\mathcal{T}_0$ as well, which completes the proof of the lemma.

Lemma 4.2.18 *The map* $\varphi(X',T') \longrightarrow \varphi(X'',T'')$ *is an isomorphism in* $\mathcal{T}(X)$. $\square$

Condition (iv) of Lemma 4.2.14 can be reformulated by saying that we have the Nisnevich distinguished square below:

$$\begin{array}{ccc} X'' - T'' & \longrightarrow & X'' \\ \downarrow & & \downarrow v \\ X' - T' & \longrightarrow & X'. \end{array}$$

This lemma then follows by Nisnevich excision (Proposition 3.3.4) and smooth base change (for étale maps).

Lemma 4.2.19 *Let T''' be the pullback of T'' along the finite surjective morphism $X''' \longrightarrow X''$. The map $\varphi(X'',T'') \longrightarrow \varphi(X''',T''')$ is a split monomorphism in $\mathcal{T}(X)$.* $\square$

We have the following pullback squares

$$\begin{array}{ccccc} T''' & \xrightarrow{t} & X''' & \xleftarrow{j'''} & U''' \\ {\scriptstyle r}\downarrow & & {\scriptstyle q}\downarrow & & \downarrow{\scriptstyle q_U} \\ T'' & \xrightarrow{s} & X'' & \xleftarrow{j''} & U' \end{array}$$

in which j'' and j''' denote the pullback of j along pu and puq, respectively, while s and t are the inclusions. By the proper base change formula applied to the left-hand square, we see that the map $\varphi(X'',T'') \longrightarrow \varphi(X''',T''')$ is isomorphic to the image of the map

$$j''_*(\mathbb{1}_{U''}\{n\}) \longrightarrow q_* q^* j''_*(\mathbb{1}_{U''}\{n\}) \longrightarrow q_* j'''_*(\mathbb{1}_{U'''}\{n\})$$

under $f_* s^*$, where $f : T'' \longrightarrow T$ is the map induced by p (note that f is proper as $T'' \simeq T'$ by assumption). As $q_* j'''_* \simeq j''_* q_{U,*}$, we are thus reduced to proving that the unit map

$$\mathbb{1}_{U''}\{n\} \longrightarrow q_{U,*}(\mathbb{1}_{U'''}\{n\})$$

is a split monomorphism. As X'' is normal (because X' is so by assumption, while $X'' \longrightarrow X'$ is étale), this follows immediately from Corollary 3.3.40.

Now, we can finish the proof of Theorem 4.2.16. Consider the Verdier quotient

$$D = \mathscr{T}(X)/\mathscr{T}_0(X).$$

We want to prove that, under the conditions stated in Lemma 4.2.14, we have $\varphi(X,T) \simeq 0$ in D. Let $\pi : T''' \longrightarrow X$ be the map induced by $puq : X''' \longrightarrow X$. If $a : T''' \longrightarrow Y$ denotes the map induced by $g : X''' \longrightarrow Y$, and $j_Y : Y_U \longrightarrow Y$ the pullback of j by f, we have the commutative diagram below.

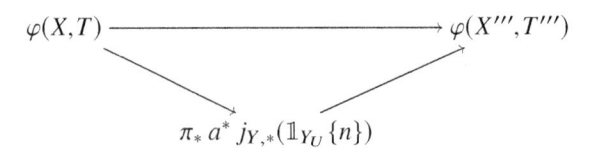

$$\begin{array}{ccc} \varphi(X,T) & \longrightarrow & \varphi(X''',T''') \\ & \searrow \quad \nearrow & \\ & \pi_* a^* j_{Y,*}(\mathbb{1}_{Y_U}\{n\}) & \end{array}$$

By virtue of Lemmas 4.2.17, 4.2.19, and 4.2.18, the horizontal map is a split monomorphism in D. It is thus sufficient to prove that this map vanishes in D, for which it will be sufficient to prove that $\pi_* a^* j_{Y,*}(\mathbb{1}_{Y_U}\{n\})$ is in $\mathscr{T}_0$. The morphism π is finite (by construction, the map $T'' \longrightarrow T'$ is an isomorphism, while the maps $T''' \longrightarrow T''$ and $T' \longrightarrow T$ are finite). Under this condition, $\mathscr{T}_0$ is stable under the operations π_* and a^*. To finish the proof of the theorem, it remains to check that $j_{Y,*}(\mathbb{1}_{Y_U}\{n\})$ is in $\mathscr{T}_0$, which follows from property (c) of Section 4.2.15 (and additivity). □

4 Constructible motives

Definition 4.2.20 We shall say that $\mathscr{T}$ is τ-*compatible* if it satisfies the following two conditions.

(a) For any closed immersion $i : Z \longrightarrow X$ between regular schemes in $\mathscr{S}$, the image of $\mathbb{1}_X\{n\}$, $n \in \tau$, under the exceptional inverse image functor $i^! : \mathscr{T}(X) \longrightarrow \mathscr{T}(Z)$ is constructible.
(b) For any scheme X, any $n \in \tau$, and any constructible object M in $\mathscr{T}(X)$, the object $\mathrm{Hom}_X(\mathbb{1}_X\{n\}, M)$ is constructible.

As usual, when τ is the monoid generated by the Tate twist, instead of τ-*compatible* we say *compatible with Tate twists*.

Remark 4.2.21 Condition (b) of the definition above will come essentially for free if the objects $\mathbb{1}_X\{n\}$ are $\otimes$-invertible with constructible $\otimes$-quasi-inverse (which will hold in practice, essentially by definition).

Example 4.2.22 In practice, condition (a) of the definition above will be a consequence of the *absolute purity theorem*. In particular, the category of Beilinson motives DM_B is compatible with Tate twists as a corollary of the fact the Tate twist is invertible and Theorem 14.4.1.

Lemma 4.2.23 *Assume that $\mathscr{T}$ is τ-compatible. Let $i : Z \longrightarrow X$ be a closed immersion, with X regular, and Z the support of a strict normal crossing divisor. Then $i^!(\mathbb{1}_X\{n\})$ is constructible for any $n \in \tau$. As a consequence, if $j : U \longrightarrow X$ denotes the complement open immersion, then $j_*(\mathbb{1}_U\{n\})$ is constructible for any $n \in \tau$.*

Proof The first assertion follows easily by induction on the number of irreducible components of Z, using Proposition 4.2.6. The second assertion follows from the distinguished triangles

$$i_* i^!(M) \longrightarrow M \longrightarrow j_* j^*(M) \longrightarrow i_* i^!(M)[1]$$

and from Lemma 4.2.9. □

Theorem 4.2.24 *Assume that $\mathscr{T}$ is $\mathbf{Q}$-linear, separated, and τ-compatible. Then, for any morphism of finite type $f : X \longrightarrow Y$ such that Y is quasi-excellent, the functor*

$$f_* : \mathscr{T}(X) \longrightarrow \mathscr{T}(Y)$$

preserves constructible objects.

Proof By virtue of Propositions 4.2.4 and 4.2.11 and of Lemma 4.2.23, if $\mathscr{T}$ is τ-compatible, we can apply Theorem 4.2.16, where $\mathscr{T}$ stands for the subcategory of constructible objects. □

Corollary 4.2.25 *Under the assumptions of the above theorem, for any quasi-excellent scheme X, and for any pair of constructible objects M and N in $\mathscr{T}(X)$, the object $\mathrm{Hom}_X(M, N)$ is constructible.*

Proof It is sufficient to treat the case where $M = f_\sharp(\mathbb{1}_Y\{n\})$, for $n \in \tau$ and $f : Y \longrightarrow X$ a smooth morphism. But then, we have, by transposition of the Sm-projection formula, a natural isomorphism:

$$Hom_X(M, N) \simeq f_* Hom(\mathbb{1}_Y\{n\}, f^*(N)).$$

This corollary then follows immediately from Proposition 4.2.4 and from Theorem 4.2.24. □

Corollary 4.2.26 *Under the assumptions of the above theorem, for any closed immersion $i : Z \longrightarrow X$ such that X is quasi-excellent, the functor*

$$i^! : \mathscr{T}(X) \longrightarrow \mathscr{T}(Z)$$

preserves constructible objects.

Proof Let $j : U \longrightarrow X$ be the complement open immersion. For an object M of $\mathscr{T}(X)$, we have the following distinguished triangle.

$$i_* i^!(M) \longrightarrow M \longrightarrow j_* j^*(M) \longrightarrow i_* i^!(M)[1].$$

By virtue of Proposition 4.2.6 and Theorem 4.2.24, if M is constructible, then $j_* j^*(M)$ have the same property, which completes the proof. □

Lemma 4.2.27 *Let $f : X \longrightarrow Y$ be a separated morphism of finite type. The condition that the functor $f^!$ preserves constructible objects in $\mathscr{T}$ is local over X and over Y for the Zariski topology.*

Proof If $u : X' \longrightarrow X$ is a Zariski cover, then we have, by definition, $u^* = u^!$, so that, by Proposition 4.2.6, the condition that $f^!$ preserves τ-constructibility is equivalent to the condition that the functors $u^* f^! \simeq (fu)^!$ preserve τ-constructibility. Let $v : Y' \longrightarrow Y$ be a Zariski cover, and consider the following pullback square.

$$\begin{array}{ccc} X' & \xrightarrow{u} & X \\ g \downarrow & & \downarrow f \\ Y' & \xrightarrow{v} & Y \end{array}$$

We then have a natural isomorphism $u^* f^! \simeq g^! v^*$, and, as u is still a Zariski cover, we deduce again from Proposition 4.2.6 that, if $g^!$ preserves τ-constructibility, so does $f^!$. □

Corollary 4.2.28 *Under the assumptions of the above theorem, for any separated morphism of finite type $f : X \longrightarrow Y$, the functor*

$$f^! : \mathscr{T}(Y) \longrightarrow \mathscr{T}(X)$$

preserves constructible objects.

4 Constructible motives

Proof By virtue of the preceding lemma, we may assume that f is affine. We can then factor f as an immersion $i : X \longrightarrow \mathbf{A}_Y^n$ followed by the canonical projection $p : \mathbf{A}_Y^n \longrightarrow Y$. The case of an immersion is reduced to the case of an open immersion (Proposition 4.2.4) and to the case of a closed immersion (Corollary 4.2.26). Thus we may assume that $f = p$, in which case $p^! \simeq p^*(-)(n)[2n]$ (according to point (3) of Theorem 2.4.50), so that the result follows by Propositions 4.2.4 and 4.2.9. □

In conclusion, we have proved the following finiteness theorem:

Theorem 4.2.29 *Assume the motivic triangulated category $\mathscr{T}$ is $\mathbf{Q}$-linear, separated and τ-compatible.*[67]
Then constructible objects of $\mathscr{T}$ are closed under the six operations of Grothendieck when restricted to the subcategory $\mathscr{S}'$ of $\mathscr{S}$ comprising quasi-excellent schemes and morphisms of finite type. In particular, $\mathscr{T}_c$ is a τ-generated motivic category over $\mathscr{S}'$.

4.3 Continuity

4.3.1 For the next definition, we consider an admissible class $\mathscr{P}$ of morphisms in $\mathscr{S}$ and an abstract symmetric monoidal $\mathscr{P}$-fibred category $\mathscr{T}$ over $\mathscr{S}$.

Let $(S_\alpha)_{\alpha \in A}$ be a projective system of schemes in $\mathscr{S}$, with affine transition maps, and such that $S = \varprojlim_{\alpha \in A} S_\alpha$ is representable in $\mathscr{S}$ (we assume that A is a partially ordered set to keep the notations simple). For each index α, we denote by $p_\alpha : S \longrightarrow S_\alpha$ the canonical projection. Given an index $\alpha_0 \in A$ and an object E_{α_0} in $\mathscr{T}(S_{\alpha_0})$, we write E_α for the pullback of E_{α_0} along the map $S_\alpha \longrightarrow S_{\alpha_0}$, and put $E = \mathbf{L}p_\alpha^*(E_\alpha)$. We will say that $(S_\alpha)_{\alpha \in A}$ is *dominant* if each transition map is furthermore dominant.

Definition 4.3.2 Consider the assumptions above and let τ be a set of twists of $\mathscr{T}$.

We say that $\mathscr{T}$ is τ-*continuous* (resp. *weakly τ-continuous*), or simply *continuous* (resp. *weakly continuous*) if τ is clearly specified by the context, if it is τ-generated and if, given any projective system (resp. dominant projective system) of schemes (S_α) as above, for any index α_0, any object E_{α_0} in $\mathscr{T}(S_{\alpha_0})$, and any twist $n \in \tau$, the canonical map

$$\varinjlim_{\alpha \geq \alpha_0} \mathrm{Hom}_{\mathscr{T}(S_\alpha)}(\mathbb{1}_{S_\alpha}\{n\}, E_\alpha) \longrightarrow \mathrm{Hom}_{\mathscr{T}(S)}(\mathbb{1}_S\{n\}, E)$$

is bijective.

Example 4.3.3 We will meet the main examples of τ-continuous categories later:
- the $\mathbf{A}^1$-derived category $\mathrm{D}_{\mathbf{A}^1, \Lambda}$ (Example 6.1.13);
- the motivic category DM_{B} of Beilinson motives (Proposition 14.3.1).

---
[67] Also recall that $\mathscr{T}$ is associated with a combinatorial stable premotivic model category.

The triangulated motivic category of motivic complexes DM_Λ, as well as its effective counterpart $\mathrm{DM}_\Lambda^{e\!f\!f}$, is weakly continuous (Theorem 11.1.24). We are only able to prove it is continuous in some special cases (namely when it compares to Beilinson motives, see Theorem 16.1.4).

The interest of the continuity property is to allow a description of constructible objects over S in terms of constructible objects over the S_α's.

Proposition 4.3.4 *Assume, under the hypotheses of Section 4.3.1, that $\mathscr{T}$ is τ-continuous (resp. weakly τ-continuous). Consider a scheme S in $\mathscr{S}$, as well as a projective system of schemes $(S_\alpha)_{\alpha\in A}$ in $\mathscr{S}$ with affine (resp. affine dominant) transition maps and such that $S = \varprojlim_\alpha S_\alpha$.*

Then, for any index α_0, and for any objects C_{α_0} and E_{α_0} in $\mathscr{T}(S_{\alpha_0})$, if C_{α_0} is constructible, then the canonical map

(4.3.4.1) $\quad \varinjlim_{\alpha \geq \alpha_0} \mathrm{Hom}_{\mathscr{T}(S_\alpha)}(C_\alpha, E_\alpha) \longrightarrow \mathrm{Hom}_{\mathscr{T}(S)}(C, E)$

is bijective. Moreover, the canonical functor

(4.3.4.2) $\quad 2\text{-}\varinjlim_\alpha \mathscr{T}_c(S_\alpha) \longrightarrow \mathscr{T}_c(S)$

is an equivalence of monoidal triangulated categories.

Proof To prove the first assertion, we may assume, without loss of generality, that $C_{\alpha_0} = M_{S_{\alpha_0}}(X_{\alpha_0})\{n\}$ for some smooth S_{α_0}-scheme of finite type X_{α_0}, and $n \in \tau$. Consider an object E_{α_0} in $\mathscr{T}(S_{\alpha_0})$. For $\alpha \geq \alpha_0$, write X_α (resp. E_α) for the pullback of X_{α_0} (resp. of E_{α_0}) along the map $S_\alpha \longrightarrow S_{\alpha_0}$. Similarly, write X (resp. E) for the pullback of X_{α_0} (resp. of E_{α_0}) along the map $S \longrightarrow S_{\alpha_0}$. We shall also write E'_α (resp. E') for the pullback of E_α (resp. E) along the smooth map $X_\alpha \longrightarrow S_\alpha$ (resp. $X \longrightarrow S$). Then, (X_α) is a projective system of schemes in $\mathscr{S}$, with affine transition maps, such that $X = \varprojlim_\alpha X_\alpha$. Note that if (S_α) is dominant in the sense of Section 4.3.1, then (X_α) is dominant, as dominant morphisms are stable under smooth base change. Then, by continuity (resp. weak continuity), we have the following natural isomorphism, which proves the first assertion:

$$\varinjlim_\alpha \mathrm{Hom}_{\mathscr{T}(S_\alpha)}(M_{S_\alpha}(X_\alpha)\{n\}, E_\alpha) \simeq \varinjlim_\alpha \mathrm{Hom}_{\mathscr{T}(X_\alpha)}(\mathbb{1}_{X_\alpha}\{n\}, E'_\alpha)$$

$$\simeq \mathrm{Hom}_{\mathscr{T}(X)}(\mathbb{1}_X\{n\}, E')$$

$$\simeq \mathrm{Hom}_{\mathscr{T}(S)}(M_S(X)\{n\}, E).$$

Note that the first assertion implies that the functor (4.3.4.2) is fully faithful. Pseudo-abelian triangulated categories are stable under filtered 2-colimits. In particular, the source of the functor (4.3.4.2) can be seen as a thick subcategory of $\mathscr{T}(S)$. The essential surjectivity of (4.3.4.2) follows from the fact that, for any smooth S-scheme of finite type X, there exists some index α, and some smooth S_α-scheme X_α, such

4 Constructible motives

that $X \simeq S \times_{S_\alpha} X_\alpha$; see [GD67, 8.8.2 and 17.7.8]: this implies that the essential image of the fully faithful functor (4.3.4.2) contains all the objects of shape $M_S(X)\{n\}$ for $n \in \tau$ and X smooth over S, so that it contains $\mathcal{T}_c(S)$, by definition. □

4.3.5 Before showing how the assumption of weak continuity can be used in the case of motivic categories, we state a proposition which later on will allow us to show continuity or weak continuity in concrete cases. Let $\mathcal{M}$ be a symmetric monoidal $\mathscr{P}$-fibred model category $\mathcal{M}$ over $\mathcal{S}$.

We consider again the assumptions and notations of Section 4.3.1, assuming the transition maps of the pro-scheme (S_α) are $\mathscr{P}$-morphisms, with $\mathcal{T} = \mathrm{Ho}(\mathcal{M})$. For each index $\alpha \in A$, we choose a small set I_α (resp. J_α) of generating cofibrations (resp. of generating trivial cofibration) in $\mathrm{Ho}(\mathcal{M})(S_\alpha)$. We also choose a small set I (resp. J) of generating cofibrations (resp. of generating trivial cofibration) in $\mathrm{Ho}(\mathcal{M})(S)$.

Consider the following assumptions:

(a) We have $I \subset \bigcup_{\alpha \in A} p_\alpha^*(I_\alpha)$ and $J \subset \bigcup_{\alpha \in A} p_\alpha^*(J_\alpha)$.
(b) For any index α_0, if C_{α_0} and E_{α_0} are two objects of $\mathcal{M}(S_{\alpha_0})$, with C_{α_0} either a source or a target of a map in $I_{\alpha_0} \cup J_{\alpha_0}$, the natural map

$$\varinjlim_{\alpha \in A} \mathrm{Hom}_{\mathcal{M}(S_\alpha)}(C_\alpha, E_\alpha) \longrightarrow \mathrm{Hom}_{\mathcal{M}(S)}(C, E)$$

is bijective.

Proposition 4.3.6 *Under the assumptions of Section 4.3.5, for any index $\alpha_0 \in A$, the pullback functor $p_{\alpha_0}^* : \mathcal{M}(S_{\alpha_0}) \to \mathcal{M}(S)$ preserves fibrations and trivial fibrations. Moreover, given an index $\alpha_0 \in A$, as well as two objects C_{α_0} and E_{α_0} in $\mathcal{M}(S_{\alpha_0})$, if C_{α_0} belongs to the smallest full subcategory of $\mathcal{T}(S_{\alpha_0})$ which is closed under finite homotopy colimits and which contains the source and targets of I_{α_0}, then, the canonical map*

$$\varinjlim_{\alpha \in A} \mathrm{Hom}_{\mathrm{Ho}(\mathcal{M})(S_\alpha)}(C_\alpha, E_\alpha) \longrightarrow \mathrm{Hom}_{\mathrm{Ho}(\mathcal{M})(S)}(C, E)$$

is bijective.

Proof We shall prove first that, for any index $\alpha_0 \in A$, the pullback functor $p_{\alpha_0}^*$ preserves fibrations and trivial fibrations. By assumption, for any $\alpha \geq \alpha_0$, the pullback functor along the $\mathscr{P}$-morphism $S_\alpha \to S_{\alpha_0}$ is both a left Quillen functor and a right Quillen functor. Let $E_{\alpha_0} \to F_{\alpha_0}$ be a trivial fibration (resp. a fibration) of $\mathcal{M}(S_{\alpha_0})$. Let $i : C \to D$ be a generating cofibration (resp. a generating trivial cofibration) in $\mathcal{M}(S)$. By condition (a) of Section 4.3.5, we may assume that there exists a cofibration (resp. a trivial cofibration) $i_{\alpha_1} : C_{\alpha_1} \to D_{\alpha_1}$, $\alpha_1 \in A$, such that $i = p_{\alpha_1}^*(i_{\alpha_1})$. We want to prove that the map

$$\mathrm{Hom}(D, E) \longrightarrow \mathrm{Hom}(C, E) \times_{\mathrm{Hom}(C, F)} \mathrm{Hom}(D, F)$$

is surjective. By condition (b) of Section 4.3.5, this map is isomorphic to the filtered colimit of the surjective maps

$$\mathrm{Hom}(D_\alpha, E_\alpha) \longrightarrow \mathrm{Hom}(C_\alpha, E_\alpha) \times_{\mathrm{Hom}(C_\alpha, F_\alpha)} \mathrm{Hom}(D_\alpha, F_\alpha)$$

with $\alpha \geq \sup(\alpha_0, \alpha_1)$, which proves the first assertion.

To prove the second assertion, we may assume that C_{α_0} is cofibrant and that E_{α_0} is fibrant. The set of maps from a cofibrant object to a fibrant object in the homotopy category of a model category can be described as homotopy classes of maps. Therefore, using the fact that $p^*_{\alpha_0}$ preserves cofibrations and fibrations, as well as the trivial ones, we see it is sufficient to prove that the map

$$\varinjlim_{\alpha \in A} \mathrm{Hom}_{\mathscr{M}(S_\alpha)}(C_\alpha, E_\alpha) \longrightarrow \mathrm{Hom}_{\mathscr{M}(S)}(C, E)$$

is bijective for some nice cofibrant replacement of C_{α_0}. But the assumptions on C_{α_0} imply that it is weakly equivalent to an object C'_{α_0} such that the map $\varnothing \longrightarrow C'_{\alpha_0}$ belongs to the smallest class of maps in $\mathscr{M}(S_{\alpha_0})$, which contains I_{α_0}, and which is closed under pushouts and (finite) compositions. We may thus assume that $C_{\alpha_0} = C'_{\alpha_0}$. In that case, C_{α_0} is in particular contained in the smallest full subcategory of $\mathscr{M}(S_{\alpha_0})$ which is stable under finite colimits and which contains the source and targets of I_{α_0}. As filtered colimits commute with finite limits in the category of sets, the result follows by using again condition (a) of Section 4.3.5. □

We now go back to the situation of a motivic triangulated category $\mathscr{T}$ satisfying our general assumptions 4.0.1 on page 130

Lemma 4.3.7 *Let $a : X \longrightarrow Y$ be a morphism in $\mathscr{S}$. Assume that $X = \varprojlim_\alpha X_\alpha$, where $(X_\alpha)_{\alpha \in A}$ is a projective system of smooth affine Y-schemes.*

If $\mathscr{T}$ is τ-continuous, then, for any objects E and F in $\mathscr{T}(Y)$, with E constructible, the exchange morphism

$$a^* \, Hom_Y(E, F) \simeq Hom_X(a^*(E), a^*(F)),$$

defined in Section 1.1.33, is an isomorphism.

The same conclusion holds if $\mathscr{T}$ is weakly τ-continuous and the transition maps of (X_α) are dominant.

Proof We have

$$a_* \, Hom_X(a^*(E), a^*(F)) \simeq Hom_Y(E, a_* \, a^*(F)),$$

so that the map $F \longrightarrow a_* \, a^*(F))$ induces a map

$$Hom_Y(E, F) \longrightarrow a_* \, Hom_X(a^*(E), a^*(F)),$$

hence, by adjunction, a map

$$a^* \, Hom_Y(E, F) \longrightarrow Hom_X(a^*(E), a^*(F)).$$

We already know that the latter is an isomorphism whenever a is smooth.

4 Constructible motives

Let us write $a_\alpha : X_\alpha \longrightarrow Y$ for the structural maps. Let C be a constructible object in $\mathscr{T}(X)$. By Proposition 4.3.4, we may assume that there exists an index α_0, and a constructible object C_{α_0} in $\mathscr{T}(X_{\alpha_0})$ such that, if we write C_α for the pullback of C_{α_0} along the map $X_\alpha \longrightarrow X_{\alpha_0}$ for $\alpha \geq \alpha_0$, we have isomorphisms:

$$\begin{aligned}\operatorname{Hom}(C, a^* \operatorname{Hom}_Y(E,F)) &\simeq \varinjlim_\alpha \operatorname{Hom}(C_\alpha, a_\alpha^* \operatorname{Hom}_Y(E,F)) \\ &\simeq \varinjlim_\alpha \operatorname{Hom}(C_\alpha, \operatorname{Hom}_X(a_\alpha^*(E), a_\alpha^*(F))) \\ &\simeq \varinjlim_\alpha \operatorname{Hom}(C_\alpha \otimes_{X_\alpha} a_\alpha^*(E), a_\alpha^*(F)) \\ &\simeq \operatorname{Hom}(C \otimes_X a^*(E), a^*(F)) \\ &\simeq \operatorname{Hom}(C, \operatorname{Hom}_X(a^*(E), a^*(F))) \,.\end{aligned}$$

As constructible objects generate $\mathscr{T}(X)$, this proves the first assertion. The second assertion obviously follows from the same argument. □

4.3.8 Let X be a scheme in $\mathscr{S}$. Assume that, for any point x of X, the corresponding morphism $i_x : \operatorname{Spec}\left(\mathscr{O}_{X,x}^h\right) \longrightarrow X$ is in $\mathscr{S}$ (where $\mathscr{O}_{X,x}^h$ denotes the henselization of $\mathscr{O}_{X,x}$). Lastly, consider a scheme of finite type Y over X, and write

$$a_x : Y_x = \operatorname{Spec}\left(\mathscr{O}_{X,x}^h\right) \times_X Y \longrightarrow Y$$

for the morphism obtained by pullback. Finally, for an object E of $\mathscr{T}(Y)$, let us write

$$E_x = a_x^*(E) \,.$$

Proposition 4.3.9 *If the motivic category $\mathscr{T}$ is weakly τ-continuous, then the family of functors*

$$\mathscr{T}(Y) \longrightarrow \mathscr{T}(Y_x), \quad E \longmapsto E_x, \quad x \in X,$$

is conservative.

Proof Let E be an object of $\mathscr{T}(Y)$ such that $E_x \simeq 0$ for any point x of X. For any constructible object C of $\mathscr{T}(Y)$, we have a presheaf of S^1-spectra on the small Nisnevich site of X:

$$F : U \longmapsto F(U) = \operatorname{Hom}(M_Y(U \times_X Y), \operatorname{Hom}_Y(C, E)) \,.$$

It is sufficient to prove that $F(X)$ is acyclic. As $\mathscr{T}$ satisfies Nisnevich descent (Proposition 3.3.4), it is sufficient to prove that F is acyclic locally for the Nisnevich topology, i.e. that, for any point x of X, the spectrum $F(\operatorname{Spec}\left(\mathscr{O}_{X,x}^h\right))$ is acyclic. Writing $\operatorname{Spec}\left(\mathscr{O}_{X,x}^h\right)$ as the projective limit of the Nisnevich neighborhoods of x in X, we see easily, using Proposition 4.3.4 and Lemma 4.3.7, that, for any integer i, $\pi_i(F(\operatorname{Spec}\left(\mathscr{O}_{X,x}^h\right)) \simeq \operatorname{Hom}(C_x, E_x[i]) \simeq 0$. □

Proposition 4.3.10 *Let S be a quasi-excellent noetherian and henselian scheme. Write $\hat{S}$ for its completion along its closed point, and assume that both S and $\hat{S}$ are in $\mathscr{S}$. Consider an S-scheme of finite type X, and write $i : \hat{S} \times_S X \longrightarrow X$ for the induced map. If $\mathscr{T}$ is τ-continuous, then the pullback functor*

$$i^* : \mathscr{T}(X) \longrightarrow \mathscr{T}(\hat{S} \times_S X)$$

is conservative.

Proof As S is quasi-excellent, the map $\hat{S} \longrightarrow S$ is regular. By Popescu's theorem, we can then write $\hat{S} = \varprojlim_\alpha S_\alpha$, where $\{S_\alpha\}$ is a projective system of schemes with affine transition maps, and such that each scheme S_α is smooth over S. Moreover, as $\hat{S}$ and S have the same residue field, and as S is henselian, each map S_α has a section. Write $X_\alpha = S_\alpha \times_S X$, so that we have $X = \varprojlim_\alpha X_\alpha$. Consider a constructible object C and an object E in $\mathscr{T}(X)$. Then, as the maps $X_\alpha \longrightarrow X$ have sections, it follows from the first assertion of Proposition 4.3.4 that the map

$$\mathrm{Hom}_{\mathscr{T}(X)}(C, E) \longrightarrow \mathrm{Hom}_{\mathscr{T}(\hat{S} \times_S X)}(i^*(C), i^*(E))$$

is a monomorphism (as a filtered colimit of such things). Hence, if $i^*(E) \simeq 0$, for any constructible object C in $\mathscr{T}(X)$, we have $\mathrm{Hom}_{\mathscr{T}(X)}(C, E) \simeq 0$. Therefore, as τ-constructible objects generate $\mathscr{T}(X)$, we get $E \simeq 0$. $\square$

Proposition 4.3.11 *Let $a : X \longrightarrow Y$ be a regular morphism in $\mathscr{S}$. If $\mathscr{T}$ is τ-continuous, then, for any objects E and F in $\mathscr{T}(Y)$, with E constructible, there is a canonical isomorphism*

$$a^* \, \mathrm{Hom}_Y(E, F) \simeq \mathrm{Hom}_X(a^*(E), a^*(F)) \, .$$

Proof We want to prove that the canonical map

$$a^* \, \mathrm{Hom}_Y(E, F) \longrightarrow \mathrm{Hom}_X(a^*(E), a^*(F))$$

is an isomorphism, while we already know it is so whenever a is smooth. Therefore, to prove the general case, we see that the problem is local on X and on Y with respect to the Zariski topology. In particular, we may assume that both X and Y are affine. By Popescu's Theorem 4.1.5, we thus have $X = \varprojlim_\alpha X_\alpha$, where $\{X_\alpha\}$ is a projective system of smooth affine Y-schemes. The result follows from Lemma 4.3.7. $\square$

4.3.12 Consider the following pullback square in $\mathscr{S}$

$$\begin{array}{ccc} X' & \xrightarrow{a} & X \\ {\scriptstyle g}\downarrow & \Delta & \downarrow{\scriptstyle f} \\ Y' & \xrightarrow{b} & Y \end{array}$$

4 Constructible motives

and assume that f is separated of finite type. Then one gets, using the recipe that we have seen several times before, the following exchange transformation:

$$Ex(\Delta^{*!}) : a^*f^! \xrightarrow{ad(b^*,b_*)} a^*f^!b_*b^* \xrightarrow{[Ex(\Delta^!_*)]^{-1}} a^*a_*g^!b^* \xrightarrow{ad'(a^*,a_*)} g^!b^*,$$

where $Ex(\delta^!_*)$ is the exchange isomorphism of Theorem 2.4.50, point (4).

Proposition 4.3.13 *Consider the previous notations and assume that b is regular and $\mathscr{T}$ is τ-continuous. Then the exchange transformation defined above*

$$Ex(\Delta^{*!}) : a^*f^! \longrightarrow g^!b^*$$

is an isomorphism.

Proof The exchange transformation $Ex(\Delta^{*!})$ is invertible whenever b is smooth: this is obvious in the case of an open immersion, so that, by Zariski descent, it is sufficient to treat the case where b is smooth with trivial cotangent bundle of rank d; in this case, by relative purity (Theorem 2.4.50 (3)), this reduces to the canonical isomorphism $a^!f^! \simeq g^!b^!$ evaluated at $E(-d)[-2d]$. To prove the general case, as the condition is local on X and on Y for the Zariski topology, we may assume that f factors as an immersion $X \longrightarrow \mathbf{P}^n_Y$, followed by the canonical projection $\mathbf{P}^n_Y \longrightarrow Y$. We deduce from there that it is sufficient to treat the case where f is either a closed immersion or a smooth morphism of finite type. The case where f (hence also g) is smooth follows by relative purity (Theorem 2.4.50): we can then replace $f^!$ and $g^!$ by f^* and g^* respectively, and the formula follows from the fact that $a^*f^* \simeq g^*b^*$. We may thus assume that f is a closed immersion. As g is a closed immersion as well, the functor $g_!$ is conservative (it is fully faithful). Therefore, it is sufficient to prove that the map

$$b^* f_! f^!(E) \simeq g_! a^* f^!(E) \longrightarrow g_! g^! b^*(E)$$

is invertible. Then, using Proposition 4.3.11 (which makes sense because the functor $f_!$ preserves τ-constructibility by Proposition 4.2.11), and the projection formula, we have

$$b^* f_! f^!(E) \simeq b^* Hom_Y(f_!(\mathbb{1}_X), E)$$
$$\simeq Hom_{Y'}(b^* f_!(\mathbb{1}_X), b^*(E))$$
$$\simeq Hom_{Y'}(g_!(\mathbb{1}_{X'}), b^*(E))$$
$$\simeq g_! g^! b^*(E),$$

which completes the proof. □

Lemma 4.3.14 *Let $f : X \longrightarrow Y$ be a morphism in $\mathscr{S}$. Assume that $X = \varprojlim_\alpha X_\alpha$ and $Y = \varprojlim_\alpha Y_\alpha$, where $\{X_\alpha\}$ and $\{Y_\alpha\}$ are projective systems of schemes with affine (resp. affine and dominant) transition maps, while f is induced by a system of morphisms $f_\alpha : X_\alpha \longrightarrow Y_\alpha$. Let α_0 be some index, C_{α_0} a constructible object* of $\mathscr{T}(Y_{\alpha_0})$, and E_{α_0} an object of $\mathscr{T}(X_{\alpha_0})$. If $\mathscr{T}$ is τ-continuous (resp. weakly τ-continuous), then we have a natural isomorphism of abelian groups

$$\varinjlim_{\alpha \geq \alpha_0} \operatorname{Hom}_{\mathscr{T}(Y_\alpha)}(C_\alpha, f_{\alpha,*}(E_\alpha)) \simeq \operatorname{Hom}_{\mathscr{T}(Y)}(C, f_*(E)).$$

Proof By virtue of Proposition 4.3.4, we have a natural isomorphism

$$\varinjlim_{\alpha \geq \alpha_0} \operatorname{Hom}_{\mathscr{T}(X_\alpha)}(f_\alpha^*(C_\alpha), E_\alpha) \simeq \operatorname{Hom}_{\mathscr{T}(Y)}(f^*(C), E).$$

The expected formula follows by adjunction. □

Proposition 4.3.15 *Consider the following pullback square in $\mathscr{S}$*

$$\begin{array}{ccc} X' & \xrightarrow{a} & X \\ {\scriptstyle g}\downarrow & & \downarrow{\scriptstyle f} \\ Y' & \xrightarrow{b} & Y, \end{array}$$

with b regular. If $\mathscr{T}$ is τ-continuous, then, for any object E in $\mathscr{T}(X)$, there is a canonical isomorphism in $\mathscr{T}(Y')$:

$$b^* f_*(E) \simeq g_* a^*(E).$$

Proof This proposition is true in the case where b is smooth (by definition of Sm-fibred categories), from which we deduce, by Zariski separation, that this property is local on Y and on Y' for the Zariski topology. In particular, we may assume that both Y and Y' are affine. Then, by Popescu's Theorem 4.1.5, we may assume that $Y' = \varprojlim_\alpha Y'_\alpha$, where $\{Y'_\alpha\}$ is a projective system of smooth Y-algebras. Then, using the preceding lemma as well as Proposition 4.3.4, we easily reduce the proposition to the case where b is smooth. □

Proposition 4.3.16 *Assume that $\mathscr{T}$ is weakly τ-continuous, $\mathbf{Q}$-linear and semi-separated, and consider a field k, with inseparable closure k', such that both $\operatorname{Spec}(k)$ and $\operatorname{Spec}(k')$ are in $\mathscr{S}$. Given a k-scheme X write $X' = k' \otimes_k X$, and $f : X' \longrightarrow X$ for the canonical projection. Then the functor*

$$f^* : \mathscr{T}(X) \longrightarrow \mathscr{T}(X')$$

is an equivalence of categories.

Proof Note that X' is a projective limit of k-schemes with affine and dominant (even flat) transition maps. Thus, it follows from weak τ-continuity, Proposition 4.3.4 and Proposition 2.1.9 that the functor

$$f^* : \mathscr{T}_c(X) \longrightarrow \mathscr{T}_c(X')$$

4 Constructible motives

is an equivalence of categories. Similarly, for any objects C and E in $\mathscr{T}(X)$, if C is constructible, the map

$$\operatorname{Hom}_{\mathscr{T}(X)}(C,E) \longrightarrow \operatorname{Hom}_{\mathscr{T}(X)}(f^*(C), f^*(E))$$

is bijective. As constructible objects generate $\mathscr{T}(X)$, this implies that the functor

$$f^* : \mathscr{T}(X) \longrightarrow \mathscr{T}(X')$$

is fully faithful. As the latter is essentially surjective on a set of generators, this implies that it is an equivalence of categories (see Corollary 1.3.20). □

Proposition 4.3.17 *Assume that $\mathscr{T}$ is weakly τ-continuous. Then, for any scheme X in $\mathscr{S}$, the family of functors induced by its points*

$$x^* : \mathscr{T}(X) \longrightarrow \mathscr{T}(\operatorname{Spec}(\kappa(x))), \quad x \in X,$$

is conservative.

Proof We proceed by induction on the dimension d of X. If $d \leq 0$, this is trivial. If $d > 0$, using Proposition 4.3.9, we may assume that X is local. By induction, the proposition is true on the complement of the closed point of x. Therefore, Proposition 2.3.6 completes the proof. □

4.4 Duality

The aim of this section is to prove a local duality theorem in $\mathscr{T}$ (see Theorem 4.4.21 and Corollary 4.4.24).

If we work with rational coefficients, resolution of singularities up to quotient singularities is almost as good as classical resolution of singularities: we have the following replacement of the blow-up formula.

Theorem 4.4.1 *Assume that $\mathscr{T}$ is $\mathbf{Q}$-linear and separated. Let X be a scheme in $\mathscr{S}$. Consider a proper surjective morphism $p : X' \longrightarrow X$ and a finite group G acting on X' over X. Assume that there is a closed subscheme $Z \subset X$ such that $U = X - Z$ is normal, while the induced map $p_U : U' = p^{-1}(U) \longrightarrow U$ is finite, and the map $U'/G \longrightarrow U$ is generically radicial (i.e. is radicial over an open dense subscheme of U) — this situation occurs, for example, when p is a Galois alteration. Then the pullback square*

(4.4.1.1)
$$\begin{array}{ccc} Z' & \xrightarrow{i'} & X' \\ {\scriptstyle q}\downarrow & & \downarrow{\scriptstyle p} \\ Z & \xrightarrow{i} & X \end{array}$$

induces an homotopy pullback square

(4.4.1.2)
$$\begin{array}{ccc} M & \longrightarrow & (\mathbf{R}p_* \, \mathbf{L}p^*(M))^G \\ \downarrow & & \downarrow \\ \mathbf{R}i_* \, \mathbf{L}i^*(M) & \longrightarrow & (\mathbf{R}i_* \mathbf{R}q_* \, \mathbf{L}q^* \, \mathbf{L}i^*(M))^G \end{array}$$

for any object M of $\mathscr{T}(X)$.

Proof We already know that, for any object N of $\mathscr{T}(U)$, the map

$$N \longrightarrow (\mathbf{R}p_{U*} \, \mathbf{L}p_U^*(N))^G$$

is an isomorphism (Corollary 3.3.39). The proof is then similar to the proof of condition (iv) of Theorem 3.3.37. □

Remark 4.4.2 Under the assumptions of the preceding theorem, applying the total derived functor $\mathbf{R}\operatorname{Hom}_X(-, E)$ to the homotopy pullback square (4.4.1.2) for $M = \mathbb{1}_X$, we obtain the homotopy pushout square

(4.4.2.1)
$$\begin{array}{ccc} (i_! \, q_! \, q^! \, i^!(E))_G & \longrightarrow & (p_! \, p^!(E))_G \\ \downarrow & & \downarrow \\ i_! \, i^!(E) & \longrightarrow & E \end{array}$$

for any object E of $\mathscr{T}(X)$.

Corollary 4.4.3 *Assume that $\mathscr{T}$ is $\mathbf{Q}$-linear and separated. Let B be a scheme in $\mathscr{S}$, admitting wide resolution of singularities up to quotient singularities. Consider a separated B-scheme of finite type S, endowed with a closed subscheme $T \subset S$. The category of constructible objects in $\mathscr{T}(S)$ is the smallest thick triangulated subcategory which contains the objects of shape $\mathbf{R}f_*(\mathbb{1}_X\{n\})$ for $n \in \tau$, and for $f : X \longrightarrow S$ a projective morphism, with X regular and connected, such that $f^{-1}(T)_{red}$ is either empty, X itself or the support of a strict normal crossing divisor.*

Proof Let $\mathscr{T}(S)'$ be the smallest thick triangulated subcategory of $\mathscr{T}(S)$ which contains the objects of shape $\mathbf{R}f_*(\mathbb{1}_X\{n\})$ for $n \in \tau$ and $f : X \longrightarrow S$ a projective morphism with X regular and connected, while $f^{-1}(T)_{red}$ is empty, or X itself, or the support of a strict normal crossing divisor. We clearly have $\mathscr{T}(S)' \subset \mathscr{T}_c(S)$ (Proposition 4.2.11). To prove the reverse inclusion, by virtue of Proposition 4.2.13, it is sufficient to prove that, for any $n \in \tau$, and any projective morphism $f : X \longrightarrow S$, the object $\mathbf{R}f_*(\mathbb{1}_X\{n\})$ belongs to $\mathscr{T}(S)'$. We shall proceed by induction on the dimension of X. If X is of dimension ≤ 0, we may replace it by its reduction, which is regular. If X is of dimension > 0, by assumption on B, there exists a Galois alteration $p : X' \longrightarrow X$ of group G, with X' regular and projective over S (and in which T becomes either empty, X' itself, or the support of a strict normal crossing divisor, in each connected component of X'). Choose a closed subscheme $Z \subset X$ such that $U = X - Z$ is a normal dense open subscheme, and such that the induced

4 Constructible motives

map $r : U' = p^{-1}(U) \longrightarrow U$ is a finite morphism, and consider the pullback square (4.4.1.1). As Z and $Z' = p^{-1}(Z)$ are of dimension smaller than the dimension of X, the result follows by considering the homotopy pullback square obtained by applying the functor $\mathbf{R}f_*$ to (4.4.1.2) for $M = \mathbb{1}_X\{n\}$, $n \in \tau$. □

Definition 4.4.4 Let S be a scheme in $\mathscr{S}$. An object R of $\mathscr{T}(S)$ is τ-*dualizing* if it satisfies the following conditions.

(i) The object R is constructible.
(ii) For any constructible object M of $\mathscr{T}(S)$, the natural map

$$M \longrightarrow \mathbf{R}Hom_S(\mathbf{R}Hom_S(M, R), R)$$

is an isomorphism.

Remark 4.4.5 If $\mathscr{T}$ is τ-compatible, $\mathbf{Q}$-linear and separated, then, in particular, the six operations of Grothendieck preserve τ-constructibility in $\mathscr{T}$ (Theorem 4.2.29). Under this assumption, for any scheme X in $\mathscr{S}$, and any $\otimes$-invertible object U in $\mathscr{T}(X)$ which is constructible, its quasi-inverse is constructible: the quasi-inverse of U is simply its dual $U^\wedge = \mathbf{R}Hom(U, \mathbb{1}_X)$, which is constructible by virtue of Corollary 4.2.25.

Proposition 4.4.6 *Assume that $\mathscr{T}$ is τ-compatible, $\mathbf{Q}$-linear and separated, and consider a scheme X in $\mathscr{S}$.*

(i) Let R be a τ-dualizing object, and U be a constructible $\otimes$-invertible object in $\mathscr{T}(X)$. Then $U \otimes_S^{\mathbf{L}} R$ is τ-dualizing.
(ii) Let R and R' be two τ-dualizing objects in $\mathscr{T}(X)$. Then the evaluation map

$$\mathbf{R}Hom_S(R, R') \otimes_S^{\mathbf{L}} R \longrightarrow R'$$

is an isomorphism.

Proof This follows immediately from [Ayo07a, 2.1.139]. □

Proposition 4.4.7 *Consider an open immersion $j : U \longrightarrow X$ in $\mathscr{S}$. If R is a τ-dualizing object in $\mathscr{T}(X)$, then $j^!(R)$ is τ-dualizing in $\mathscr{T}(U)$.*

Proof If M is a constructible object in $\mathscr{T}(U)$, then $j_!(M)$ is constructible, and the map

(4.4.7.1) $\qquad j_!(M) \longrightarrow \mathbf{R}Hom_X(\mathbf{R}Hom_X(j_!(M), R), R)$

is an isomorphism. Using the isomorphisms of type

$$M \simeq j^* j_!(M) = j^! j_!(M) \quad \text{and} \quad j^* \mathbf{R}Hom_X(A, B) \simeq \mathbf{R}Hom_U(j^*(A), j^*(B)),$$

we see that the image of the map (4.4.7.1) under the functor $j^* = j^!$ is isomorphic to the map

(4.4.7.2) $$M \longrightarrow \mathrm{R}Hom_U(\mathrm{R}Hom_U(M, j^!(R)), j^!(R)),$$

which proves the proposition. □

Proposition 4.4.8 *Let X be a scheme in $\mathscr{S}$, and R an object in $\mathscr{T}(X)$. Assume there exists an open cover $X = \bigcup_{i \in I} U_i$ such that the restriction of R on each of the open subschemes U_i is τ-dualizing in $\mathscr{T}(U_i)$. Then R is τ-dualizing.*

Proof We already know that the property of τ-constructibility is local with respect to the Zariski topology (Proposition 4.2.6). Denote by $j_i : U_i \longrightarrow X$ the corresponding open immersions, and put $R_i = j_i^!(R)$. Let M be a constructible object in $\mathscr{T}(X)$. Then, for all $i \in I$, the image under $j_i^* = j_i^!$ of the map

$$M \longrightarrow \mathrm{R}Hom_X(\mathrm{R}Hom_X(M, R), R)$$

is isomorphic to the map

$$j_i^*(M) \longrightarrow \mathrm{R}Hom_{U_i}(\mathrm{R}Hom_{U_i}(j_i^*(M), R_i), R_i).$$

This proposition thus follows from the property of separation with respect to the Zariski topology. □

Corollary 4.4.9 *Let $f : X \longrightarrow Y$ be a separated morphism of finite type in $\mathscr{S}$. Given an object R of $\mathscr{T}(Y)$, the property for $f^!(R)$ of being a τ-dualizing object in $\mathscr{T}(X)$ is local over X and over Y for the Zariski topology.*

Proposition 4.4.10 *Assume that $\mathscr{T}$ is τ-compatible. Let $i : Z \longrightarrow X$ be a closed immersion and R be a τ-dualizing object in $\mathscr{T}(X)$. Then $i^!(R)$ is τ-dualizing in $\mathscr{T}(Z)$.*

Proof As $\mathscr{T}$ is τ-compatible, we already know that $i^!(R)$ is constructible. For any objects M and R of $\mathscr{T}(Z)$ and $\mathscr{T}(X)$, respectively, we have the identification:

$$i_! \, \mathrm{R}Hom_Z(M, i^!(R)) \simeq \mathrm{R}Hom_X(i_!(M), R).$$

Let $j : U \longrightarrow X$ be the complement immersion. Then we have

$$j^! \mathrm{R}Hom_X(i_!(M), R) \simeq \mathrm{R}Hom_U(j^* i_!(M), j^!(R)) \simeq 0,$$

so that

$$\mathrm{R}Hom_X(i_!(M), R) \simeq i_! \, \mathbf{L}i^* \mathrm{R}Hom_X(i_!(M), R).$$

As $i_!$ is fully faithful, this provides a canonical isomorphism

$$\mathbf{L}i^* \mathrm{R}Hom_X(i_!(M), R) \simeq i^! \mathrm{R}Hom_X, (i_!(M), R).$$

Under this identification, we easily see that the map

$$i_!(M) \longrightarrow \mathrm{R}Hom_X(\mathrm{R}Hom_X(i_!(M), R), R)$$

4 Constructible motives

is isomorphic to the image under $i_!$ of the map

$$M \longrightarrow \mathbf{R}Hom_Z(\mathbf{R}Hom_Z(M, i^!(R)), i^!(R)).$$

As $i_!$ is fully faithful, it is conservative, and this ends the proof. $\square$

Proposition 4.4.11 *Assume that $\mathscr{T}$ is τ-compatible, $\mathbf{Q}$-linear and separated, and consider a scheme B in $\mathscr{S}$ which admits wide resolution of singularities up to quotient singularities. Consider a separated B-scheme of finite type S, and a constructible object R in $\mathscr{T}(S)$. The following conditions are equivalent.*

(i) For any separated morphism of finite type $f : X \longrightarrow S$, the object $f^!(R)$ is τ-dualizing.
(ii) For any projective morphism $f : X \longrightarrow S$, the object $f^!(R)$ is τ-dualizing.
(iii) For any projective morphism $f : X \longrightarrow S$, with X regular, the object $f^!(R)$ is τ-dualizing.
(iv) For any projective morphism $f : X \longrightarrow S$, with X regular, and for any $n \in \tau$, the map

(4.4.11.1) $\qquad \mathbb{1}_X\{n\} \longrightarrow \mathbf{R}Hom_X(\mathbf{R}Hom_X(\mathbb{1}_X\{n\}, f^!(R)), f^!(R))$

is an isomorphism in $\mathscr{T}(X)$.

If, furthermore, for any regular separated B-scheme of finite type X, and for any $n \in \tau$, the object $\mathbb{1}_X\{n\}$ is $\otimes$-invertible, then these conditions are equivalent to the following one.

(v) For any projective morphism $f : X \longrightarrow S$, with X regular, the map

(4.4.11.2) $\qquad \mathbb{1}_X \longrightarrow \mathbf{R}Hom_X(f^!(R), f^!(R))$

is an isomorphism in $\mathscr{T}(X)$.

Proof It is clear that (i) implies (ii), which implies (iii), which implies (iv). Let us check that condition (ii) also implies condition (i). Let $f : X \longrightarrow S$ be a morphism of separated B-schemes of finite type, with S regular. We want to prove that $f^!(\mathbb{1}_S)$ is τ-dualizing, while we already know it is true whenever f is projective. In the general case, by virtue of Corollary 4.4.9, we may assume that f is quasi-projective, so that $f = pj$, where p is projective, and j is an open immersion. As $f^! \simeq j^! p^!$, the result follows by Proposition 4.4.7. Under the additional assumption, the equivalence between (iv) and (v) is obvious. It thus remains to prove that (iv) implies (ii). It is in fact sufficient to prove that, under condition (iv), the object R itself is τ-dualizing. To prove that the map

(4.4.11.3) $\qquad M \longrightarrow \mathbf{R}Hom_X(\mathbf{R}Hom_X(M, R), R)$

is an isomorphism for any constructible object M of $\mathscr{T}(S)$, it is sufficient to consider the case where $M = \mathbf{R}f_*(\mathbb{1}_X\{n\}) = f_!(\mathbb{1}_X\{n\})$, where $n \in \tau$ and $f : X \longrightarrow S$ is a projective morphism with X regular (Corollary 4.4.3). For any object A of $\mathscr{T}(X)$, we have canonical isomorphisms

$$\mathbf{R}Hom_S(f_!(A), R) \simeq \mathbf{R}f_* \, \mathbf{R}Hom_X(A, f^!(R))$$
$$= f_! \, \mathbf{R}Hom_X(A, f^!(R)),$$

from which we get a natural isomorphism:

$$\mathbf{R}Hom_S(\mathbf{R}Hom_S(f_!(A), R), R) \simeq f_! \, \mathbf{R}Hom_X(\mathbf{R}Hom_X(A, f^!(R)), f^!(R)) \,.$$

Under these identifications, the map (4.4.11.3) for $M = f_!(\mathbb{1}_X\{n\})$ is the image of the map (4.4.11.1) under the functor $f_!$. As (4.4.11.1) is invertible by assumption, this proves that R is τ-dualizing. □

Lemma 4.4.12 *Let X be a scheme in $\mathscr{S}$, and R be an object of $\mathscr{T}(X)$. The property for R of being $\otimes$-invertible is local over X with respect to the Zariski topology.*

Proof Let $R^\wedge = \mathbf{R}Hom(R, \mathbb{1}_X)$ be the dual of R. The object R is $\otimes$-invertible if and only if the evaluation map

$$R^\wedge \otimes_X^{\mathbf{L}} R \longrightarrow \mathbb{1}_X$$

is invertible. Let $j : U \longrightarrow X$ be an open immersion. Then, for any objects M and N in $\mathscr{T}(X)$, we have the identification

$$j^* \mathbf{R}Hom_X(M, N) \simeq \mathbf{R}Hom_U(j^*(M), j^*(N)) \,.$$

In particular, we have $j^*(R^\wedge) \simeq j^*(R)^\wedge$. As j^* is monoidal, the lemma follows from the fact that $\mathscr{T}$ has the property of separation with respect to the Zariski topology.□

Definition 4.4.13 We shall say that $\mathscr{T}$ is τ-*dualizable* if it satisfies the following conditions:

(i) $\mathscr{T}$ is τ-compatible (Definition 4.2.20);
(ii) for any closed immersion between regular schemes $i : Z \longrightarrow S$ in $\mathscr{S}$, the object $i^!(\mathbb{1}_S)$ is $\otimes$-invertible (i.e. the functor $i^!(\mathbb{1}_S) \otimes_S^{\mathbf{L}} (-)$ is an equivalence of categories);
(ii) for any regular scheme X in $\mathscr{S}$, and for any $n \in \tau$, the map

$$\mathbb{1}_X\{n\} \longrightarrow \mathbf{R}Hom_X(\mathbf{R}Hom_X(\mathbb{1}_X\{n\}, \mathbb{1}_X), \mathbb{1}_X)$$

is an isomorphism.

As in other similar situations, we simply say *dualizable with respect to Tate twists* when the set of twists τ is generated by the Tate twist.

Example 4.4.14 In practice, the property of being dualizable with respect to Tate twists is a consequence of the absolute purity theorem. Our main example is the motivic category DM_B of Beilinson motives over excellent noetherian schemes, as a consequence of Theorem 14.4.1.

4 Constructible motives

Remark 4.4.15 Note that, whenever the set of twists τ consists of rigid objects (which will be the case in practice), conditions (i) and (ii) of the preceding definition are equivalent to the condition that $i^!(\mathbb{1}_X)$ is constructible and $\otimes$-invertible for any closed immersion i between regular separated schemes in $\mathscr{S}$, while condition (iii) is then automatic. This principle easily gives the property of τ-purity when $\mathscr{S}$ comprises schemes of finite type over some perfect field:

Proposition 4.4.16 *Assume that $\mathscr{S}$ consists exactly of schemes of finite type over a field k. If the objects $\mathbb{1}\{n\}$ are rigid with constructible duals in $\mathscr{T}(\operatorname{Spec}(k))$ for all $n \in \tau$, then $\mathscr{T}$ is τ-dualizable.*

Proof For any k-scheme of finite type $f : X \longrightarrow \operatorname{Spec}(k)$, as the functor $\mathbf{L}f^*$ is symmetric monoidal, the objects $\mathbb{1}_X\{n\}$ are rigid in $\mathscr{T}(X)$ for all $n \in \tau$. Therefore, as stated in Remark 4.4.15, we have only to prove that, for any closed immersion $i : Z \longrightarrow X$ between regular k-schemes of finite type, the object $i^!(\mathbb{1}_X)$ is $\otimes$-invertible and constructible. Note that, as k is perfect, X and Z are in fact smooth. Using Lemma 4.4.12 and Proposition 4.2.6, we may also assume that X is quasi-projective and that Z is purely of codimension c in X, while the normal bundle of i is trivial. This proposition is then a consequence of relative purity (Theorem 2.4.50), which gives a canonical isomorphism $i^!(\mathbb{1}_X) \simeq \mathbb{1}_Z(-c)[-2c]$. □

Proposition 4.4.17 *Assume that $\mathscr{S}$ consists of schemes of finite type over a field k and that $\mathscr{T}$ has the following properties:*

(a) it is τ-dualizable;
(b) for any $n \in \tau$, $\mathbb{1}\{n\}$ is rigid;
(c) either k is perfect or $\mathscr{T}$ is continuous.

Then, any constructible object of $\mathscr{T}(k)$ is rigid.

Proof By Proposition 4.3.16, it is sufficient to treat the case where k is perfect. It is well known that rigid objects form a thick subcategory of $\mathscr{T}$. Thus, the result follows easily from Corollary 4.4.3 and Proposition 2.4.31. □

Lemma 4.4.18 *Assume that $\mathscr{T}$ is τ-dualizable. Then, for any projective morphism $f : X \longrightarrow S$ between regular schemes in $\mathscr{S}$, the object $f^!(\mathbb{1}_S)$ is $\otimes$-invertible and constructible.*

Proof Since, for any open immersion $j : U \longrightarrow X$, one has $j^* = j^!$, we easily deduce from Lemma 4.4.12 (resp. Proposition 4.2.6) that the property for $f^!(\mathbb{1}_S)$ of being $\otimes$-invertible (resp. constructible) is local on S for the Zariski topology. Therefore, we may assume that S is separated over B and that f factors as a closed immersion $i : X \longrightarrow \mathbf{P}_S^n$ followed by the canonical projection $p : \mathbf{P}_S^n \longrightarrow S$. Using relative purity for p, we have the following computations:

$$f^!(\mathbb{1}_S) \simeq i^! p^!(\mathbb{1}_S) \simeq i^!(\mathbb{1}_{\mathbf{P}_S^n}(n)[2n]) \simeq i^!(\mathbb{1}_{\mathbf{P}_S^n})(n)[2n].$$

As i is a closed immersion between regular schemes, the object $i^!(\mathbb{1}_{\mathbf{P}_S^n})$ is $\otimes$-invertible and constructible by assumption on $\mathscr{T}$, which implies that $f^!(\mathbb{1}_S)$ is $\otimes$-invertible and constructible as well. □

160   Fibred categories and the six functors formalism

Definition 4.4.19 Let B a scheme in $\mathscr{S}$. We shall say that *local duality holds over B in $\mathscr{T}$* if, for any separated morphism of finite type $f : X \longrightarrow S$, with S regular and of finite type over B, the object $f^!(\mathbb{1}_S)$ is τ-dualizing in $\mathscr{T}(X)$.

Remark 4.4.20 By definition, if $\mathscr{T}$ is τ-compatible, and if local duality holds over B in $\mathscr{T}$, then the restriction of $\mathscr{T}$ to the category of B-schemes of finite type is τ-dualizable. A convenient sufficient condition for local duality to hold in $\mathscr{T}$ is the following (in particular, using the result below as well as Proposition 4.4.16, local duality holds almost systematically over fields).

Theorem 4.4.21 *Assume that $\mathscr{T}$ is τ-dualizable, $\mathbf{Q}$-linear and separated, and consider a scheme B in $\mathscr{S}$ which admits wide resolution of singularities up to quotient singularities (e.g. B might be any scheme which is separated and of finite type over an excellent noetherian scheme of dimension less than or equal to 2 in $\mathscr{S}$; see Corollary 4.1.11). Then local duality holds over B in $\mathscr{T}$.*

Proof Let S be a regular separated B-scheme of finite type. Then, for any separated morphism of finite type $f : X \longrightarrow S$, the object $f^!(\mathbb{1}_S)$ is τ-dualizing: Lemma 4.4.18 immediately implies condition (iv) of Proposition 4.4.11. The general case (without the separation assumption on S) follows easily from Corollary 4.4.8. □

Proposition 4.4.22 *Consider a scheme B in $\mathscr{S}$. Assume that $\mathscr{T}$ is τ-dualizable, and that local duality holds over B in $\mathscr{T}$. Consider a regular B-scheme of finite type S.*

(i) An object of $\mathscr{T}(S)$ is τ-dualizing if and only if it is constructible and $\otimes$-invertible.
(ii) For any separated morphism of S-schemes of finite type $f : X \longrightarrow Y$, and for any τ-dualizing object R in $\mathscr{T}(Y)$, the object $f^!(R)$ is τ-dualizing in $\mathscr{T}(X)$.

Proof As the unit of $\mathscr{T}(S)$ is τ-dualizing by assumption, Proposition 4.4.6 implies that an object of $\mathscr{T}(S)$ is τ-dualizing if and only if it is constructible and $\otimes$-invertible.

Consider a regular B-scheme of finite type S, a separated morphism of S-schemes of finite type $f : X \longrightarrow Y$, and a τ-dualizing object R in $\mathscr{T}(Y)$. To prove that $f^!(R)$ is τ-dualizing, by virtue of Corollary 4.4.8, we may assume that Y is separated over S. Denote by u and v the structural maps from X and Y to S, respectively. As we already know that $v^!(\mathbb{1}_S)$ is τ-dualizing, by virtue of Proposition 4.4.6, there exists a constructible and $\otimes$-invertible object U in $\mathscr{T}(Y)$ such that $U \otimes_Y^\mathbf{L} R \simeq v^!(\mathbb{1}_S)$. As the functor $\mathbf{L}f^*$ is symmetric monoidal, it preserves $\otimes$-invertible objects and their duals, from which we deduce the following isomorphisms:

$$u^!(\mathbb{1}_S) \simeq f^! v^!(\mathbb{1}_S)$$
$$\simeq f^!(U \otimes_Y^\mathbf{L} R)$$
$$\simeq f^! \mathbf{R}Hom_Y(U^\wedge, R)$$
$$\simeq \mathbf{R}Hom_X(\mathbf{L}f^*(U^\wedge), f^!(R))$$
$$\simeq \mathbf{R}Hom_X(\mathbf{L}f^*(U)^\wedge, f^!(R))$$
$$\simeq \mathbf{L}f^*(U) \otimes_X^\mathbf{L} f^!(R).$$

4 Constructible motives

The object $a^!(\mathbb{1}_S)$ being τ-dualizing, while $\mathbf{L}f^*(U)$ is constructible and invertible, we deduce from Proposition 4.4.6 that $f^!(R)$ is τ-dualizing as well. □

4.4.23 Assume that $\mathscr{T}$ is τ-dualizable, **Q**-linear and separated.

Consider a scheme B in $\mathscr{S}$ such that local duality holds over B in $\mathscr{T}$ — this is the case if B admits wide resolution of singularities up to quotient singularities according to the above theorem. Consider a fixed regular B-scheme of finite type S, as well as a constructible and $\otimes$-invertible object R in $\mathscr{T}(S)$ (if S is of pure dimension d, it might be wise to consider $R = \mathbb{1}_S(d)[2d]$, but an arbitrary R as above is eligible by Proposition 4.4.22). Then, for any separated S-scheme of finite type $f : X \longrightarrow S$, we define the *local duality functor*

$$D_X : \mathscr{T}(X)^{op} \longrightarrow \mathscr{T}(X)$$

by the formula
$$D_X(M) = \mathbf{R}Hom_X(M, f^!(R)).$$

This functor D_X is right adjoint to itself.

Corollary 4.4.24 *Under the above assumptions, we have the following properties of the motivic triangulated category $\mathscr{T}$:*

(a) For any separated S-scheme of finite type X, the functor D_X preserves constructible objects.

(b) For any separated S-scheme of finite type X, the natural map

$$M \longrightarrow D_X(D_X(M))$$

is an isomorphism for any constructible object M in $\mathscr{T}(X)$.

(c) For any separated S-scheme of finite type X, and for any objects M and N in $\mathscr{T}(X)$, if N is constructible, then we have a canonical isomorphism

$$D_X(M \otimes_X^{\mathbf{L}} D_X(N)) \simeq \mathbf{R}Hom_X(M, N).$$

(d) For any morphism between separated S-schemes of finite type $f : Y \longrightarrow X$, we have natural isomorphisms

$$D_Y(f^*(M)) \simeq f^!(D_X(M))$$
$$f^*(D_X(M)) \simeq D_Y(f^!(M))$$
$$D_X(f_!(N)) \simeq f_*(D_Y(N))$$
$$f_!(D_Y(N)) \simeq D_X(f_*(N))$$

for any constructible objects M and N in $\mathscr{T}(X)$ and $\mathscr{T}(Y)$, respectively.

This corollary sums up what must be called the *Grothendieck duality* property for the motivic triangulated category $\mathscr{T}$ with respect to the set of twists τ.

Proof Assertions (a) and (b) are only stated for the record;[68] see Corollary 4.2.25. To prove (c), we see that we have an obvious isomorphism

$$D_X(M \otimes_X^{\mathbf{L}} P) \simeq \mathbf{R}Hom_X(M, D_X(P))$$

for any objects M and P. If N is constructible, we may replace P by $D_X(N)$ and get the expected formula using (b). The identification $D_Y f^* \simeq f^! D_X$ is a special case of the formula

$$\mathbf{R}Hom_Y(f^*(A), f^!(B)) \simeq f^! \mathbf{R}Hom_X(A, B).$$

Therefore, we also get:

$$f^* D_X \simeq D_Y^2 f^* D_X \simeq D_Y f^! D_X^2 \simeq D_Y f^!.$$

The two other formulas of (d) follow by adjunction. □

Theorem 4.4.25 *Assume that $\mathscr{S}$ consists of schemes of finite type over a field k. We consider a τ'-generated motivic triangulated category $\mathscr{T}'$ over $\mathscr{S}$ as well as a premotivic morphism*

$$\varphi^* : \mathscr{T} \longrightarrow \mathscr{T}'.$$

We suppose that the following properties hold:

(a) $\mathscr{T}$ is τ-dualizable, $\mathbf{Q}$-linear and separated;
(b) $\mathscr{T}'$ is $\mathbf{Q}$-linear and separated;
(b) the object $\mathbb{1}\{i\}$ is rigid in $\mathscr{T}(k)$ for any $i \in \tau$.

Then, the premotivic morphism

$$\varphi^* : \mathscr{T}_c \longrightarrow \mathscr{T}'$$

commutes with the six operations.

Remark 4.4.26 Note that, as a corollary, we obtain immediately, under the assumptions of the theorem, that $\mathscr{T}'$ is $\varphi^*(\tau)$-dualizable and that the functor φ^* commutes with the duality functors on $\mathscr{T}$ and $\mathscr{T}'$, respectively obtained by applying the above corollary in the case $B = \operatorname{Spec}(k)$.

Proof Given a morphism of finite type $f : X \longrightarrow \operatorname{Spec}(k)$, let us consider the following property.

$(*)_f$ *For any constructible object M in $\mathscr{T}(X)$, the natural exchange map*

$$\varphi^* f_*(M) \longrightarrow f_* \varphi^*(M)$$

is invertible.

---

[68] We have included a lot of assumptions here: in fact, if $\mathscr{T}$ is τ-dualizable and if local duality holds over B in $\mathscr{T}$, the six Grothendieck operations preserve constructible objects on the restriction of $\mathscr{T}$ to B-schemes of finite type; we leave this as a formal exercise for the reader.

4 Constructible motives

We will first prove the theorem assuming that property $(*)_f$ holds for any f. Let $u : X \longrightarrow Y$ be a k-morphism of finite type. We claim that the exchange map

$$\varphi^* u_*(M) \longrightarrow u_* \varphi^*(M)$$

is invertible for any τ-constructible object M of $\mathscr{T}(X)$.

It is sufficient to prove that, for any smooth separated k-morphism of finite type $g : T \longrightarrow X$, any constructible object M in $\mathscr{T}(X)$ and any twist i in τ', the natural map

$$\operatorname{Hom}_{\mathscr{T}'(X)}(g_\sharp(\mathbb{1}_T)\{i\}, \varphi^* u_*(M)) \longrightarrow \operatorname{Hom}_{\mathscr{T}'(X)}(g_\sharp(\mathbb{1}_T)\{i\}, u_* \varphi^*(M))$$

is bijective. Consider the following commutative diagram of morphisms of schemes:

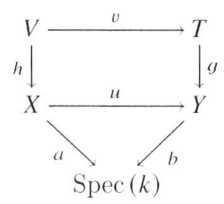

$$\begin{array}{ccc} V & \xrightarrow{v} & T \\ {\scriptstyle h}\downarrow & & \downarrow{\scriptstyle g} \\ X & \xrightarrow{u} & Y \\ & \searrow_{a} \quad \swarrow_{b} & \\ & \operatorname{Spec}(k) & \end{array}$$

in which the square is cartesian. Recall that the functor v_* preserves constructible objects by virtue of Theorem 4.2.16. We have the following computations:

$$\begin{aligned}
\operatorname{Hom}_{\mathscr{T}'(Y)}(g_\sharp(\mathbb{1}_T)\{i\}, \varphi^* u_*(M)) &= \operatorname{Hom}_{\mathscr{T}'(T)}(\mathbb{1}_T\{i\}, g^* \varphi^* u_*(M)) \\
&= \operatorname{Hom}_{\mathscr{T}'(T)}(\mathbb{1}_T\{i\}, \varphi^* g^* u_*(M)) \\
&= \operatorname{Hom}_{\mathscr{T}'(T)}(g^* b^*(\mathbb{1}_k)\{i\}, \varphi^* g^* u_*(M)) \\
&= \operatorname{Hom}_{\mathscr{T}'(T)}(g^* b^*(\mathbb{1}_k)\{i\}, \varphi^* v_* h^*(M)) \\
&= \operatorname{Hom}_{\mathscr{T}'(k)}(\mathbb{1}_k\{i\}, (bg)_* \varphi^* v_* h^*(M)) \\
&= \operatorname{Hom}_{\mathscr{T}'(k)}(\mathbb{1}_k\{i\}, \varphi^* (bg)_* v_* h^*(M)) \quad \text{(by } (*)_{bg}) \\
&= \operatorname{Hom}_{\mathscr{T}'(k)}(\mathbb{1}_k\{i\}, (bgv)_* \varphi^* h^*(M)) \quad \text{(by } (*)_{bgv}) \\
&= \operatorname{Hom}_{\mathscr{T}'(k)}(\mathbb{1}_k\{i\}, (bg)_* g^* u_* \varphi^*(M)) \\
&= \operatorname{Hom}_{\mathscr{T}'(Y)}(g_\sharp(\mathbb{1}_T)\{i\}, u_* \varphi^*(M)).
\end{aligned}$$

From this, we see that, for any k-scheme of finite type X and any τ-constructible objects M and N of $\mathscr{T}(X)$, the natural map

$$\varphi^*(\operatorname{Hom}_X(M, N)) \longrightarrow \operatorname{Hom}_X(\varphi^*(M), \varphi^*(N))$$

is invertible in $\mathscr{T}'(X)$. For this, we may assume that $M = f_\sharp(\mathbb{1}_Y\{i\})$ for a smooth morphism of finite type $f : Y \longrightarrow X$ and a twist i, in which case we have

$$\varphi^*(\operatorname{Hom}_X(M, N)) = \varphi^* f_* f^*(N) \simeq f_* f^* \varphi^*(N) = \operatorname{Hom}_X(\varphi^*(M), \varphi^*(N)).$$

It remains to prove that for any separated k-morphism $f : X \longrightarrow Y$ of finite type and any constructible object N in $\mathscr{T}(X)$, the exchange map:

$$\varphi^* f^!(N) \longrightarrow f^! \varphi^*(N)$$

is an isomorphism. It is easy to see that this property is local for the Zariski topology, both on X and on Y, so that we may assume that the morphism f is affine. Therefore, it is sufficient to consider the situation where f is either a closed immersion or a separated smooth map. In the smooth case, as the functor $f^!$ is of the form $f^*(d)[2d]$, this is obvious. If $f = i$ is a closed immersion with open complement j, as we already know that φ^* commutes with u_* for any morphism u, this property follows straight away from the localization distinguished triangles

$$i_* i^! \longrightarrow 1 \longrightarrow j_* j^* \longrightarrow .$$

It remains to prove property $(*)_f$ for any morphism f of finite type.

We claim it is sufficient to prove that, for any k-scheme of finite type X with structural morphism f, the following property holds:

$(**)_X$ For any twist $i \in \tau$, the natural exchange map

$$\varphi^* f_*(\mathbb{1}_X\{i\}) \longrightarrow f_* \varphi^*(\mathbb{1}_X\{i\})$$

is invertible.

Indeed, by virtue of Theorem 4.2.13, we may assume that $M = w_*(\mathbb{1}_W\{i\})$ for $w : W \longrightarrow X$ a projective k-morphism, and $i \in \tau$. As the exchange map $\varphi^* w_* \longrightarrow w_* \varphi^*$ is invertible (Proposition 2.4.53), we see that we may assume that $M = \mathbb{1}_X\{i\}$ for some twist i.

Let us prove property $(**)_X$ in the case when X is in addition smooth over k. As φ^* is monoidal, for any rigid object M of $\mathscr{T}(k)$, we get the identification:

$$\varphi^*(M^\vee) = \varphi^*(M)^\vee.$$

On the other hand, according to assumption (b), the object $f_\sharp(\mathbb{1}_X)$ is rigid in $\mathscr{T}(k)$ as well as in $\mathscr{T}'(k)$ (because the functor φ^* is symmetric monoidal and commutes with the operations of the form $f_\sharp$ for f smooth). Thus we get:

$$f_*(\mathbb{1}_X\{i\}) = Hom_k(f_\sharp(\mathbb{1}_X), \mathbb{1}_k\{i\}) = f_\sharp(\mathbb{1}_X)^\vee\{i\}.$$

Then property $(**)_X$ readily follows.

We finally prove property $(**)_X$ for any algebraic k-scheme X. We will proceed by induction on the dimension of X.

If $\dim(X) < 0$, the result is obvious. Let us assume $\dim(X) \geq 0$. According to the localization property, we can assume that X is reduced. Let $\bar{k}$ be an inseparable closure of k and $\bar{X} = X \otimes_k \bar{k}$. According to De Jong's theorem applied to $\bar{X}$ (see Theorem 4.1.10 for $S = \mathrm{Spec}\,(\bar{k})$), there exists a Galois alteration $\bar{X}' \longrightarrow \bar{X}$ of group G such that $\bar{X}'$ is smooth over $\bar{k}$.

4 Constructible motives

We can assume that such a smooth alteration exists over a finite inseparable extension field E/k. Because $\mathscr{T}$ (resp. $\mathscr{T}'$) is **Q**-linear and separated, the base change functor π^* associated with the finite morphism $\pi : \mathrm{Spec}\,(E) \longrightarrow \mathrm{Spec}\,(k)$ and relative to the premotivic category $\mathscr{T}$ (resp. $\mathscr{T}'$) is an equivalence of categories (see Proposition 2.1.9). Thus we can replace k by E and assume that there exists a Galois alteration $p : X' \longrightarrow X$ of group G such that X' is a smooth k-scheme. Using the localization property, we can assume X is reduced. Then there exists a nowhere dense closed subscheme $\nu : Z \longrightarrow X$ such that $U = X - Z$ is regular (thus normal) and the induced map $p|_U : p^{-1}(U) \longrightarrow U$ is finite. Thus we can apply Theorem 4.4.1 to the cartesian square:

$$\begin{array}{ccc} Z' & \xrightarrow{\nu'} & X' \\ {\scriptstyle q}\downarrow & & \downarrow{\scriptstyle p} \\ Z & \xrightarrow{\nu} & X \end{array}$$

and we get the distinguished triangle in $\mathscr{T}(X)$ (thus in $\mathscr{T}'(X)$ as well, as the functor φ^* is monoidal and commutes with the operations of the form u_* for any proper morphism u) of the form:

$$\mathbb{1}_X\{i\} \longrightarrow p_*(\mathbb{1}_{X'}\{i\})^G \oplus \nu_*(\mathbb{1}_Z\{i\}) \longrightarrow (\nu q)_*(\mathbb{1}_{Z'}\{i\})^G \xrightarrow{+1}$$

for any twist i. If we consider the triangles in $\mathscr{T}(k)$ and $\mathscr{T}'(k)$ obtained by applying the functor f_*, where f is the structural morphism of X/k, we deduce that property $(**)_X$ follows from properties $(**)_{X'}, (**)_Z, (**)_{Z'}$. Thus the result follows by applying either the case of a smooth k-scheme treated above or the induction hypothesis as $\dim(Z) = \dim(Z') < \dim(X)$. $\square$

Part II
Construction of fibred categories

© Springer Nature Switzerland AG 2019
D.-C. Cisinski and F. Déglise, *Triangulated Categories of Mixed Motives*, Springer Monographs in Mathematics,
https://doi.org/10.1007/978-3-030-33242-6_2

5 Fibred derived categories

5.0.1 In this entire section, we fix a full subcategory $\mathscr{S}$ of the category of noetherian $\mathscr{S}$-schemes satisfying the following properties:

(a) $\mathscr{S}$ is closed under finite sums and pullback along morphisms of finite type.
(b) For any scheme S in $\mathscr{S}$, any quasi-projective S-scheme belongs to $\mathscr{S}$.

We fix an admissible class of morphisms $\mathscr{P}$ of $\mathscr{S}$. All our $\mathscr{P}$-premotivic categories (cf. Definition 1.4.2) are defined over $\mathscr{S}$. Moreover, for any abelian $\mathscr{P}$-premotivic category $\mathscr{A}$ in this section, we assume the following:

(c) $\mathscr{A}$ is a *Grothendieck* abelian $\mathscr{P}$-premotivic category (see Definition 1.3.8 and the recall below).
(d) $\mathscr{A}$ is given with a generating set of twists τ. We sometimes refer to it as *the twists of $\mathscr{A}$*.
(e) We will denote by $M_S(X, \mathscr{A})$, or simply by $M_S(X)$, the geometric section over a $\mathscr{P}$-scheme X/S.

Unless stated otherwise, any scheme will be assumed to be an object of $\mathscr{S}$.

In Section 5.2, except possibly for Section 5.2.a, we assume further:

(f) $\mathscr{P}$ contains the class of smooth morphisms of finite type.

5.0.2 We will sometimes refer to the canonical dg-structure of the category of complexes $C(\mathscr{A})$ over an abelian category $\mathscr{A}$. Recall that to any complexes K and L over $\mathscr{A}$, we associate a complex of abelian groups $\operatorname{Hom}^\bullet_{\mathscr{A}}(K, L)$ whose component in degree $n \in \mathbf{Z}$ is

$$\prod_{p \in \mathbf{Z}} \operatorname{Hom}_{\mathscr{A}}(K^p, L^{p+n})$$

and whose differential in degree $n \in \mathbf{Z}$ is defined by the formula:

$$(f_p)_{p \in \mathbf{Z}} \mapsto \left(d_L \circ f_p - (-1)^n . f_{p+1} \circ d_K\right)_{p \in \mathbf{Z}}.$$

In other words, this is the image of the bicomplex $\operatorname{Hom}_{\mathscr{A}}(K, L)$ under the Tot-product functor, which we denote by Tot^π. Of course, the associated homotopy category is the category $K(\mathscr{A})$ of complexes up to chain homotopy equivalence.

5.1 From abelian premotives to triangulated premotives

5.1.a Abelian premotives: recall and examples

Consider an abelian $\mathscr{P}$-premotivic category $\mathscr{A}$. According to the convention of Section 5.0.1, for any scheme S, $\mathscr{A}_S$ is a Grothendieck abelian closed symmetric monoidal category. Moreover, if τ denotes the twists of $\mathscr{A}$, the essentially small family

$$\bigl(M_S(X)\{i\}\bigr)_{X\in\mathscr{P}/S, i\in\tau}$$

is a family of generators of $\mathscr{A}_S$ in the sense of [Gro57].

Example 5.1.1 Consider a fixed ring Λ. Let $\mathrm{PSh}(\mathscr{P}/S,\Lambda)$ be the category of Λ-presheaves (i.e. presheaves of Λ-modules) on $\mathscr{P}/S$. For any $\mathscr{P}$-scheme X/S, we let $\Lambda_S(X)$ be the free Λ-presheaf on $\mathscr{P}/S$ represented by X. Then $\mathrm{PSh}(\mathscr{P}/S,\Lambda)$ is a Grothendieck abelian category generated by the essentially small family $\bigl(\Lambda_S(X)\bigr)_{X\in\mathscr{P}/S}$.

There is a unique symmetric closed monoidal structure on $\mathrm{PSh}(\mathscr{P}/S,\Lambda)$ such that
$$\Lambda_S(X)\otimes_S \Lambda_S(Y) = \Lambda_S(X\times_S Y).$$

Finally the existence of functors f^*, f_* and, in the case when f is a $\mathscr{P}$-morphism, of $f_\sharp$, follows from general sheaf theory (*cf.* [AGV73]).

Thus, $\mathrm{PSh}(\mathscr{P},\Lambda)$ defines an abelian $\mathscr{P}$-premotivic category.

5.1.2 Consider an abstract abelian $\mathscr{P}$-premotivic category $\mathscr{A}$. To any premotive M of $\mathscr{A}_S$, we can associate a presheaf of abelian groups
$$X \longmapsto \mathrm{Hom}_{\mathscr{A}_S}(M_S(X), M),$$

which we denote by $\gamma_*(M)$.

This defines a functor $\gamma_* : \mathscr{A}_S \longrightarrow \mathrm{PSh}(\mathscr{P}/S,\mathbf{Z})$. It admits the following left adjoint:
$$\gamma^* : \mathrm{PSh}(\mathscr{P}/S,\mathbf{Z}) \longrightarrow \mathscr{A}_S, \quad F \longmapsto \varinjlim_{X/F} M_S(X,\mathscr{A}),$$

where the colimit runs over the category of representable presheaves over F.

It is now easy to check we have defined a morphism of (complete) abelian $\mathscr{P}$-premotivic categories:

(5.1.2.1) $$\gamma^* : \mathrm{PSh}(\mathscr{P},\mathbf{Z}) \rightleftarrows \mathscr{A} : \gamma_*.$$

Moreover, $\mathrm{PSh}(\mathscr{P},\mathbf{Z})$ appears as the initial abelian $\mathscr{P}$-premotivic category.

Note that the functor $\gamma_* : \mathscr{A}_S \longrightarrow \mathrm{PSh}(\mathscr{P}/S,\mathbf{Z})$ is conservative if the set of twists τ of $\mathscr{A}$ is trivial.

Definition 5.1.3 A $\mathscr{P}$-admissible topology t is a Grothendieck pretopology t on the category $\mathscr{S}$ such that any t-covering family consists of $\mathscr{P}$-morphisms.

Note that, for any scheme S in $\mathscr{S}$, such a topology t induces a pretopology on $\mathscr{P}/S$ (which we denote by the same letter). For any morphism (resp. $\mathscr{P}$-morphism) $f : T \longrightarrow S$, the functor f^* (resp. $f_\sharp$) preserves t-covering families.

As $\mathscr{P}$ is fixed throughout this section, we will simply say *admissible* for $\mathscr{P}$-admissible.

Example 5.1.4 Let t be an admissible topology. We denote by $\mathrm{Sh}_t(\mathscr{P}/S,\Lambda)$ the category of t-sheaves of Λ-modules on $\mathscr{P}/S$. Given a $\mathscr{P}$-scheme X/S, we let

5 Fibred derived categories

$\Lambda_S^t(X)$ be the free Λ-linear t-sheaf represented by X. Then, $\mathrm{Sh}_t(\mathscr{P}/S, \Lambda)$ is an abelian Grothendieck category with generators $(\Lambda_S^t(X))_{X \in \mathscr{P}/S}$.

As in the preceding example, the category $\mathrm{Sh}_t(\mathscr{P}/S, \Lambda)$ admits a unique closed symmetric monoidal structure such that $\Lambda_S^t(X) \otimes_S \Lambda_S^t(Y) = \Lambda_S^t(X \times_S Y)$. Finally, for any morphism $f : T \longrightarrow S$ of schemes, the existence of functors f^*, f_* (resp. $f_\sharp$ when f is a $\mathscr{P}$-morphism) follows from the general theory of sheaves (see again [AGV73]: according to our assumption on t and [AGV73, III, 1.6], the functors $f^* : \mathscr{P}/S \longrightarrow \mathscr{P}/T$ and $f_\sharp : \mathscr{P}/T \longrightarrow \mathscr{P}/S$ (for f in $\mathscr{P}$) are continuous).

Thus, $\mathrm{Sh}_t(\mathscr{P}, \Lambda)$ defines an abelian $\mathscr{P}$-premotivic category (with trivial set of twists).

The associated t-sheaf functor induces a morphism

(5.1.4.1) $\qquad a_t^* : \mathrm{PSh}(\mathscr{P}, \Lambda) \rightleftarrows \mathrm{Sh}_t(\mathscr{P}, \Lambda) : a_{t,*}$.

Remark 5.1.5 Recall that the abelian category $\mathrm{Sh}_t(\mathscr{P}/S, \mathbf{Z})$ is a localization of the category $\mathrm{PSh}(S, \mathbf{Z})$ in the sense of Gabriel–Zisman. In particular, given an abstract abelian $\mathscr{P}$-premotivic category $\mathscr{A}$, the canonical morphism

$$\gamma^* : \mathrm{PSh}(\mathscr{P}/S, \mathbf{Z}) \rightleftarrows \mathscr{A}_S : \gamma_*$$

induces a unique morphism

$$\mathrm{Sh}_t(\mathscr{P}/S, \mathbf{Z}) \rightleftarrows \mathscr{A}_S$$

if and only if for any presheaf of abelian groups F on $\mathscr{P}/S$ such that $a_t(F) = F_t = 0$, one has $\gamma^*(F) = 0$.

We leave to the reader the exercise which consists of formulating the universal property of the abelian $\mathscr{P}$-premotivic category $\mathrm{Sh}_t(\mathscr{P}, \mathbf{Z})$.[69]

5.1.b The t-descent model category structure

5.1.6 Consider an abelian $\mathscr{P}$-premotivic category $\mathscr{A}$ with set of twists τ.

We let $\mathrm{C}(\mathscr{A})$ be the $\mathscr{P}$-fibred abelian category over $\mathscr{S}$ whose fibers over a scheme S is the category $\mathrm{C}(\mathscr{A}_S)$ of (unbounded) complexes in $\mathscr{A}_S$. For any scheme S, we let $\iota_S : \mathscr{A}_S \longrightarrow \mathrm{C}(\mathscr{A}_S)$ be the embedding which sends an object of $\mathscr{A}_S$ to the corresponding complex concentrated in degree zero.

If $\mathscr{A}$ is τ-twisted, then the category $\mathrm{C}(\mathscr{A}_S)$ is obviously $(\mathbf{Z} \times \tau)$-twisted. The following lemma is straightforward:

Lemma 5.1.7 *With the notations above, there is a unique structure of an abelian $\mathscr{P}$-premotivic category on $\mathrm{C}(\mathscr{A})$ such that the functor $\iota : \mathscr{A} \longrightarrow \mathrm{C}(\mathscr{A})$ is a morphism of abelian $\mathscr{P}$-premotivic categories.*

---

[69] We will formulate a derived version in the section on descent properties for derived premotives (*cf.* Section 5.2.9).

5.1.8 For a scheme S, let $(\mathscr{P}/S)^{\mathrm{II}}$ be the category introduced in Section 3.2.1. The functor $M_S(-)$ can be extended to $(\mathscr{P}/S)^{\mathrm{II}}$ by associating to a family $(X_i)_{i \in I}$ of $\mathscr{P}$-schemes over S the premotive

$$\bigoplus_{i \in I} M_S(X_i).$$

If $\mathscr{X}$ is a simplicial object of $(\mathscr{P}/S)^{\mathrm{II}}$, we denote by $M_S(\mathscr{X})$ the complex associated with the simplicial object of $\mathscr{A}_S$ obtained by applying degreewise the above extension of $M_S(-)$.

Definition 5.1.9 Let $\mathscr{A}$ be an abelian $\mathscr{P}$-premotivic category and t be an admissible topology.

Let S be a scheme and C be an object of $C(\mathscr{A}_S)$:

1. The complex C is said to be *local* (with respect to the geometric section) if, for any $\mathscr{P}$-scheme X/S and any pair $(n,i) \in \mathbf{Z} \times \tau$, the canonical morphism

$$\mathrm{Hom}_{\mathrm{K}(\mathscr{A}_S)}(M_S(X)\{i\}[n], C) \longrightarrow \mathrm{Hom}_{\mathrm{D}(\mathscr{A}_S)}(M_S(X)\{i\}[n], C)$$

is an isomorphism.
2. The complex C is said to be *t-flasque* if for any t-hypercover $\mathscr{X} \longrightarrow X$ in $\mathscr{P}/S$, for any $(n,i) \in \mathbf{Z} \times \tau$, the canonical morphism

$$\mathrm{Hom}_{\mathrm{K}(\mathscr{A}_S)}(M_S(X)\{i\}[n], C) \longrightarrow \mathrm{Hom}_{\mathrm{K}(\mathscr{A}_S)}(M_S(\mathscr{X})\{i\}[n], C)$$

is an isomorphism.

We say the abelian $\mathscr{P}$-premotivic category $\mathscr{A}$ satisfies *cohomological t-descent* if for any t-hypercover $\mathscr{X} \longrightarrow X$ of a $\mathscr{P}$-scheme X/S, and for any $i \in \tau$, the map

$$M_S(\mathscr{X})\{i\} \longrightarrow M_S(X)\{i\}$$

is a quasi-isomorphism (or equivalently, if any local complex is t-flasque).

We say that $\mathscr{A}$ is *compatible with t* if $\mathscr{A}$ satisfies cohomological t-descent, and if, for any scheme S, any t-flasque complex of $\mathscr{A}_S$ is local.

Example 5.1.10 Consider the notations of Example 5.1.4.

Consider the canonical dg-structure on $C(\mathrm{Sh}_t(\mathscr{P}/S, \Lambda))$ (see Example 5.1.1). By definition, for any complexes D and C of sheaves, we get an equality:

$$\mathrm{Hom}_{\mathrm{K}(\mathrm{Sh}_t(\mathscr{P}/S,\Lambda))}(D, C) = H^0(\mathrm{Hom}^\bullet_{\mathrm{Sh}_t(\mathscr{P}/S,\Lambda)}(D, C))$$
$$= H^0(\mathrm{Tot}^\pi \mathrm{Hom}_{\mathrm{Sh}_t(\mathscr{P}/S,\Lambda)}(D, C)).$$

In the case where $D = \Lambda_S^t(X)$ (resp. $D = \Lambda_S^t(\mathscr{X})$) for a $\mathscr{P}$-scheme X/S (resp. a simplicial $\mathscr{P}$-scheme over S) we obtain the following identification:

$$\mathrm{Hom}_{\mathrm{K}(\mathrm{Sh}_t(\mathscr{P}/S,\Lambda))}(\Lambda_S^t(X), C) = H^0(C(X))$$
$$(\text{resp. } \mathrm{Hom}_{\mathrm{K}(\mathrm{Sh}_t(\mathscr{P}/S,\Lambda))}(\Lambda_S^t(\mathscr{X}), C) = H^0(\mathrm{Tot}^\pi C(\mathscr{X}))).$$

5 Fibred derived categories

Thus, we get the following equivalences:

C is local $\Leftrightarrow$ for any $\mathscr{P}$-scheme X/S, $H_t^n(X,C) \simeq H^n(C(X))$,
C is t-flasque $\Leftrightarrow$ for any t-hypercover $\mathscr{X}^\cdot \longrightarrow X$, $H^n(C(X)) \simeq H^n(\operatorname{Tot}^\pi C(\mathscr{X}^\cdot))$.

According to the computation of cohomology with hypercovers (cf. [Bro74]), if the complex C is t-flasque, it is local. In other words, we have the expected property that the abelian $\mathscr{P}$-premotivic category $\operatorname{Sh}_t(\mathscr{P}, \Lambda)$ is compatible with t.

5.1.11 Consider an abelian $\mathscr{P}$-premotivic category $\mathscr{A}$ and an admissible topology t.

Fix a base scheme S. A morphism $p : C \longrightarrow D$ of complexes on $\mathscr{A}_S$ is called a *t-fibration* if its kernel is a t-flasque complex and if for any $\mathscr{P}$-scheme X/S, any $i \in \tau$ and any integer $n \in \mathbf{Z}$, the map of abelian groups

$$\operatorname{Hom}_{\mathscr{A}_S}(M_S(X)\{i\}, C^n) \longrightarrow \operatorname{Hom}_{\mathscr{A}_S}(M_S(X)\{i\}, D^n)$$

is surjective.

For any object A of $\mathscr{A}_S$, we let $S^n A$ (resp. $D^n A$) be the complex with only one non-trivial term (resp. two non-trivial terms) equal to A in degree n (resp. in degree n and $n+1$, with the identity as the only non-trivial differential). We define the class of *cofibrations* as the smallest class of morphisms of $C(\mathscr{A}_S)$ which:

1. contains the map $S^{n+1}M_S(X)\{i\} \longrightarrow D^n M_S(X)\{i\}$ for any $\mathscr{P}$-scheme X/S, any $i \in \tau$, and any integer n;
2. is stable under pushout, transfinite composition and retract.

A complex C is said to be *cofibrant* if the canonical map $0 \longrightarrow C$ is a cofibration. For instance, for any $\mathscr{P}$-scheme X/S and any $i \in \tau$, the complex $M_S(X)\{i\}[n]$ is cofibrant.

Let $\mathscr{G}_S$ be the essentially small family comprising premotives $M_S(X)\{i\}$ for a $\mathscr{P}$-scheme X/S and a twist $i \in \tau$, and $\mathscr{H}_S$ be the family of complexes of the form $\operatorname{Cone}(M_S(\mathscr{X}^\cdot)\{i\} \longrightarrow M_S(X)\{i\})$ for any t-hypercover $\mathscr{X}^\cdot \longrightarrow X$ and any twist $i \in \tau$. By the very definition, as $\mathscr{A}$ is compatible with t (Definition 5.1.9), $(\mathscr{G}_S, \mathscr{H}_S)$ is a descent structure on $\mathscr{A}_S$ in the sense of [CD09, def. 2.2]. Moreover, it is weakly flat in the sense of [CD09, par. 3.1]. Thus the following proposition is a particular case of [CD09, theorem 2.5, proposition 3.2, and corollary 5.5]:

Proposition 5.1.12 *Let $\mathscr{A}$ be an abelian $\mathscr{P}$-premotivic category, which we assume to be compatible with an admissible topology t. Then for any scheme S, the category $C(\mathscr{A}_S)$ with the preceding definition of fibrations and cofibrations, with quasi-isomorphisms as weak equivalences, is a proper symmetric monoidal model category.*

5.1.13 We will call this model structure on $C(\mathscr{A}_S)$ the *t-descent model category structure* (over S). Note that, for any $\mathscr{P}$-scheme X/S and any twist $i \in \tau$, the complex $M_S(X)\{i\}$ concentrated in degree 0 is cofibrant by definition, as well as any of its suspensions and twists. They form a family of generators for the triangulated category $D(A_S)$.

Observe also that the fibrant objects for the t-descent model category structure are exactly the t-flasque complexes in $\mathscr{A}_S$. Moreover, essentially by definition, a complex of $\mathscr{A}_S$ is local if and only if it is t-flasque (see [CD09, 2.5]).

5.1.14 Consider again the notations and hypotheses of Section 5.1.11.

Consider a morphism of schemes $f : T \longrightarrow S$. Then the functor

$$f^* : C(\mathscr{A}_S) \longrightarrow C(\mathscr{A}_T)$$

sends $\mathscr{G}_S$ to $\mathscr{G}_T$, and $\mathscr{H}_S$ to $\mathscr{H}_T$ because the topology t is admissible. This means it satisfies descent according to the definition of [CD09, 2.4]. Applying theorem 2.14 of *op. cit.*, the functor f^* preserves cofibrations and trivial cofibrations, i.e. the pair of functors (f^*, f_*) is a Quillen adjunction with respect to the t-descent model category structures.

Assume that f is a $\mathscr{P}$-morphism. Then, similarly, the functor

$$f_\sharp : C(\mathscr{A}_T) \longrightarrow C(\mathscr{A}_S)$$

sends $\mathscr{G}_S$ (resp. $\mathscr{H}_S$) to $\mathscr{G}_T$ (resp. $\mathscr{H}_T$) so that $f_\sharp$ also satisfies descent in the sense of *op. cit.* Therefore, it preserves cofibrations and trivial cofibrations, and the pair of adjoint functors $(f_\sharp, f^*)$ is a Quillen adjunction for the t-descent model category structures.

In other words, we have obtained the following result.

Corollary 5.1.15 *Let $\mathscr{A}$ be an abelian $\mathscr{P}$-premotivic category compatible with an admissible topology t. The $\mathscr{P}$-fibred category $C(\mathscr{A})$ with the t-descent model category structure defined in Proposition 5.1.12 is a symmetric monoidal $\mathscr{P}$-fibred model category. Moreover, it is stable, proper and combinatorial.*

5.1.16 Recall the following consequences of this corollary (see also Section 1.3.23 for the general theory). Consider a morphism $f : T \longrightarrow S$ of schemes. Then the pair of adjoint functors (f^*, f_*) admits total left/right derived functors

$$\mathbf{L}f^* : D(\mathscr{A}_S) \rightleftarrows D(\mathscr{A}_T) : \mathbf{R}f_*.$$

More precisely, f_* (resp. f^*) preserves t-local (resp. cofibrant) complexes. For any complex K on $\mathscr{A}_S$, $\mathbf{R}f_*(K) = f_*(K')$ (resp. $\mathbf{L}f^*(K) = f^*(K'')$) where $K' \longrightarrow K$ (resp. $K \longrightarrow K''$) is a t-local (resp. cofibrant) resolution of K.[70]

When f is a $\mathscr{P}$-morphism, the functor f^* is even exact and thus preserves quasi-isomorphisms. This implies that $\mathbf{L}f^* = f^*$. The functor $f_\sharp$ admits a total left derived functor

$$\mathbf{L}f_\sharp : D(\mathscr{A}_T) \rightleftarrows D(\mathscr{A}_S) : \mathbf{R}f^*$$

---

[70] Recall also that fibrant/cofibrant resolutions can be made functorially, because our model categories are cofibrantly generated, so that the left or right derived functors are in fact defined at the level of complexes.

5 Fibred derived categories

defined by the formula $\mathbf{L}f_\sharp(K) = f_\sharp(K'')$ for a complex K on $\mathscr{A}_T$ and a cofibrant resolution $K'' \longrightarrow K$.

Note also that the tensor product (resp. internal Hom) of $C(\mathscr{A}_S)$ admits a total left derived functor (resp. total right derived functor). For any complexes K and L on $\mathscr{A}_S$, this derived functors are defined by the formula:

$$K \otimes_S^{\mathbf{L}} L = K'' \otimes_S L''$$
$$\mathbf{R}Hom_S(K, L) = Hom_S(K'', L')$$

where $K \longrightarrow K''$ and $L \longrightarrow L''$ are cofibrant resolutions and $L' \longrightarrow L$ is a t-local resolution.

It is now easy to check that these functors define a triangulated $\mathscr{P}$-premotivic category $D(\mathscr{A})$, which is τ-generated according to Section 5.1.13.

Definition 5.1.17 Let $\mathscr{A}$ be an abelian $\mathscr{P}$-premotivic category compatible with an admissible topology t.

The triangulated $\mathscr{P}$-premotivic category $D(\mathscr{A})$ defined above is called the *derived $\mathscr{P}$-premotivic category* associated with $\mathscr{A}$.[71]

The geometric section of a $\mathscr{P}$-scheme X/S in the category $D(\mathscr{A})$ is the complex concentrated in degree 0 equal to the object $M_S(X)$. The triangulated $\mathscr{P}$-fibred category is τ-generated and well generated in the sense of Definition 1.3.15. Recall this means that $D(\mathscr{A}_S)$ is equal to the localizing[72] subcategory generated by the family

(5.1.17.1) $\{M_S(X)\{i\}; X/S \; \mathscr{P}\text{-scheme}, i \in \tau\}.$

Example 5.1.18 Given any admissible topology t, the abelian $\mathscr{P}$-premotivic category $Sh_t(\mathscr{P}, \Lambda)$ introduced in Example 5.1.4 is compatible with t (*cf.* Example 5.1.10) and defines the derived $\mathscr{P}$-premotivic category $D(Sh_t(\mathscr{P}, \Lambda))$.

Note also that the abelian $\mathscr{P}$-premotivic category $PSh(\mathscr{P}, \Lambda)$ introduced in Example 5.1.1 is compatible with the coarse topology and gives the derived $\mathscr{P}$-premotivic category $D(PSh(\mathscr{P}, \Lambda))$.

Remark 5.1.19 Recall from Section 5.0.2 that there exists a canonical dg-structure on $C(\mathscr{A}_S)$. Then we can define a derived dg-structure by defining, for any complexes K and L of $\mathscr{A}_S$, the complex of morphisms:

$$\mathbf{R}\operatorname{Hom}_{\mathscr{A}_S}(K, L) = \operatorname{Hom}^\bullet_{\mathscr{A}_S}(Q(K), R(L)),$$

where R and Q are respectively some fibrant and cofibrant (functorial) resolutions for the t-descent model structure. The homotopy category associated with this new dg-structure on $C(\mathscr{A}_S)$ is the derived category $D(\mathscr{A}_S)$. Moreover, for any morphism (resp. $\mathscr{P}$-morphism) of schemes f, the pair $(\mathbf{L}f^*, \mathbf{R}f_*)$ (resp. $(\mathbf{L}f_\sharp, f^*)$) is a dg-adjunction. The same is true for the pair of bifunctors $(\otimes_S^{\mathbf{L}}, \mathbf{R}Hom_S)$.

---
[71] Indeed, note that $D(\mathscr{A})$ does not depend on the topology t.
[72] *i.e.* triangulated and stable under sums.

5.1.20 Consider an abelian $\mathscr{P}$-premotivic category $\mathscr{A}$ compatible with a topology t. According to Section 3.1.b, the 2-functor $D(\mathscr{A})$ can be extended to the category of $\mathscr{S}$-diagrams: to any diagram of schemes $\mathscr{X} : I \longrightarrow \mathscr{S}$ indexed by a small category I, we can associate a closed symmetric monoidal triangulated category $D(\mathscr{A})(\mathscr{X},I)$ which coincides with $D(\mathscr{A})(X)$ when $I = e$, $\mathscr{X} = X$ for a scheme X.

Let us be more specific. The fibred category $\mathscr{A}$ admits an extension to $\mathscr{S}$-diagrams: a section of $\mathscr{A}$ over a diagram of schemes $\mathscr{X} : I \longrightarrow \mathscr{S}$, indexed by a small category I, is the following data:

1. A family $(A_i)_{i \in I}$ such that A_i is an object of $\mathscr{A}_{X_i}$.
2. A family $(a_u)_{u \in Fl(I)}$ such that for any arrow $u : i \longrightarrow j$ in I, $a_u : u^*(A_j) \longrightarrow A_i$ is a morphism in $\mathscr{A}_{X_i}$ and this family of morphisms satisfies a cocyle condition (see Section 3.1.1).

Then, $D(\mathscr{A})(\mathscr{X},I)$ is the derived category of the abelian category $\mathscr{A}(\mathscr{X},I)$. In particular, objects of $D(\mathscr{A})(\mathscr{X},I)$ are complexes of sections of $\mathscr{A}$ over $(\mathscr{X},I)$ (or, what amount to the same thing, families of complexes $(K_i)_{i \in I}$ with transition maps (a_u) as above, relative to the fibred category $C(\mathscr{A})$).

Recall that a morphism of $\mathscr{S}$-diagrams $\varphi : (\mathscr{X},I) \longrightarrow (\mathscr{Y},J)$ is given by a functor $f : I \longrightarrow J$ and a natural transformation $\varphi : \mathscr{X} \longrightarrow \mathscr{Y} \circ f$. We say that φ is a $\mathscr{P}$-morphism if for any $i \in I$, $\varphi_i : \mathscr{X}_i \longrightarrow \mathscr{Y}_{f(i)}$ is a $\mathscr{P}$-morphism. For any morphism (resp. $\mathscr{P}$-morphism) φ, we have defined in Section 3.1.3 adjunctions of (abelian) categories:

$$\varphi^* : \mathscr{A}(\mathscr{Y},J) \rightleftarrows \mathscr{A}(\mathscr{X},I) : \varphi_*$$
$$\text{resp. } \varphi_\sharp : \mathscr{A}(\mathscr{X},I) \rightleftarrows \mathscr{A}(\mathscr{Y},J) : \varphi^*$$

which extend the adjunctions we had on trivial diagrams.

According to Proposition 3.1.11, these respective adjunctions admit left/right derived functors as follows:

(5.1.20.1) $\qquad \mathbf{L}\varphi^* : D(\mathscr{A})(\mathscr{Y},J) \rightleftarrows D(\mathscr{A})(\mathscr{X},I) : \mathbf{R}\varphi_*$

(5.1.20.2) $\qquad \text{resp. } \mathbf{L}\varphi_\sharp : D(\mathscr{A})(\mathscr{X},I) \rightleftarrows D(\mathscr{A})(\mathscr{Y},J) : \mathbf{L}\varphi^* = \varphi^*$.

Again, these adjunctions coincide on trivial diagrams with the map we already had.

Note also that the symmetric closed monoidal structure on $C(\mathscr{A}(\mathscr{X},I))$ can be derived and induces a symmetric monoidal structure on $D(\mathscr{A})(\mathscr{X},I)$ (see Proposition 3.1.24).[73]

Recall from Definition 3.2.5 and Corollary 3.2.7 that, given a topology t' (not necessarily admissible) over $\mathscr{S}$, we say that $D(\mathscr{A})$ satisfies t'-descent if for any t'-hypercover $p : \mathscr{X} \longrightarrow X$ (here $\mathscr{X}$ is considered as a $\mathscr{S}$-diagram), the functor

(5.1.20.3) $\qquad \mathbf{L}p^* : D(\mathscr{A})(X) \longrightarrow D(\mathscr{A})(\mathscr{X})$

---

[73] In fact, $D(\mathscr{A})$ is then a monoidal $\mathscr{P}_{cart}$-fibred category over the category of $\mathscr{S}$-diagrams (Remark 3.1.21).

5 Fibred derived categories

is fully faithful (see Corollary 3.2.7).

Proposition 5.1.21 *Consider the notations and hypotheses introduced above. Let t' be an admissible topology on $\mathscr{S}$. Then the following conditions are equivalent:*

(i) $D(\mathscr{A})$ satisfies t'-descent, in the sense recalled above.
(ii) $\mathscr{A}$ satisfies cohomological t'-descent.

Proof We prove (i) implies (ii). Consider a t'-hypercover $p : \mathscr{X} \longrightarrow X$ in $\mathscr{P}/S$. This is a $\mathscr{P}$-morphism. Thus, by the full faithfulness of (5.1.20.3), the counit map $\mathbf{L}p_\sharp p^* \longrightarrow 1$ is an isomorphism. By applying the latter to the unit object $\mathbb{1}_X$ of $D(\mathscr{A}_X)$, we thus obtain that

$$M_X(\mathscr{X}) \longrightarrow \mathbb{1}_X$$

is an isomorphism in $D(\mathscr{A}_X)$. If $\pi : X \longrightarrow S$ is the structural $\mathscr{P}$-morphism, by applying the functor $\mathbf{L}\pi_\sharp$ to this isomorphism, we obtain that

$$M_S(\mathscr{X}) \longrightarrow M_S(X)$$

is an isomorphism in $D(\mathscr{A}_S)$, from which the result follows.

Conversely, to prove (i), we can restrict to t'-hypercovers $p : \mathscr{X} \longrightarrow X$ which are $\mathscr{P}$-morphisms because t' is admissible. Because $\mathbf{R}p^* = p^*$ admits a left adjoint $\mathbf{L}p_\sharp$, we have to prove that the counit

$$\mathbf{L}p_\sharp p^* \longrightarrow 1$$

is an isomorphism. This is a natural transformation between triangulated functors which commutes with small sums. Thus, according to (5.1.17.1), we have only to check this is an isomorphism when evaluated at a complex of the form $M_X(Y)\{i\}$ for a $\mathscr{P}$-scheme Y/X and a twist $i \in \tau$. The resulting morphism is then $M_X(\mathscr{X} \times_X Y)\{i\} \longrightarrow M_X(Y)\{i\}$ and we can conclude the proof because $\mathscr{X} \times_X Y \longrightarrow Y$ is a t'-hypercover in $\mathscr{P}/S$ (again because t' is admissible). □

5.1.22 Consider the situation of Section 5.1.20. Let S be a scheme. An interesting particular case is given for constant $\mathscr{S}$-diagrams over S; for a small category I, we let I_S be the constant $\mathscr{S}$-diagram $I \longrightarrow \mathscr{S}, i \longmapsto S, u \longmapsto 1_S$. Then the adjunctions (5.1.20.1) for this kind of diagram define a *Grothendieck derivator*

$$I \longmapsto D(\mathscr{A})(I_S).$$

Recall that, if $f : I \longrightarrow e$ is the canonical functor to the terminal category and $\varphi = f_X : I_X \longrightarrow X$ the corresponding morphism of $\mathscr{S}$-diagrams, for any I-diagram $K_\bullet = (K_i)_{i \in I}$ of complexes over $\mathscr{A}_S$, we get right derived limits and left derived colimits:

$$\mathbf{R}\varphi_*(K_\bullet) = \mathbf{R}\varprojlim_{i \in I} K_i,$$

$$\mathbf{L}\varphi_\sharp(K_\bullet) = \mathbf{L}\varinjlim_{i \in I} K_i.$$

5.1.23 The associated derived $\mathscr{P}$-premotivic category is functorial in the following sense.

Consider an adjunction
$$\varphi : \mathscr{A} \rightleftarrows \mathscr{B} : \psi$$

of abelian $\mathscr{P}$-premotivic categories. Let τ (resp. τ') be the set of twists of $\mathscr{A}$ (resp. $\mathscr{B}$), and recall that φ induces a morphism of monoids $\tau \longrightarrow \tau'$ still denoted by φ. Consider two topologies t and t' such that t' is finer than t. Suppose $\mathscr{A}$ (resp. $\mathscr{B}$) is compatible with t (resp. t') and let $(\mathscr{G}_S^{\mathscr{A}}, \mathscr{H}_S^{\mathscr{A}})$ (resp. $(\mathscr{G}_S^{\mathscr{B}}, \mathscr{H}_S^{\mathscr{B}})$) be the descent structure on $\mathscr{A}_S$ (resp. $\mathscr{B}_S$) defined in Section 5.1.11.

For any scheme S, consider the evident extensions
$$\varphi_S : C(\mathscr{A}_S) \rightleftarrows C(\mathscr{B}_S) : \psi_S$$

of the above adjoint functors to complexes. Recall that for any $\mathscr{P}$-scheme X/S and any twist $i \in \tau$, $\varphi_S(M_S(X, \mathscr{A})\{i\}) = M_S(X, \mathscr{B})\{\varphi(i)\}$ by definition. Thus, φ_S sends $\mathscr{G}_S^{\mathscr{A}}$ to $\mathscr{G}_S^{\mathscr{A}}$. Because t' is finer than t, it also sends $\mathscr{H}_S^{\mathscr{A}}$ to $\mathscr{H}_S^{\mathscr{B}}$. In other words, it satisfies descent in the sense of [CD09, par. 2.4] so that the pair (φ_S, ψ_S) is a Quillen adjunction with respect to the respective t-descent and t'-descent model structure on $C(\mathscr{A}_S)$ and $C(\mathscr{B}_S)$.

Considering the derived functors, it is now easy to check we have obtained a $\mathscr{P}$-premotivic adjunction[74]
$$\mathbf{L}\varphi : D(\mathscr{A}) \rightleftarrows D(\mathscr{B}) : \mathbf{R}\psi.$$

Example 5.1.24 Let t be an admissible topology. Consider an abelian $\mathscr{P}$-premotivic category $\mathscr{A}$ compatible with t. Then the morphism of abelian $\mathscr{P}$-premotivic categories (5.1.2.1) induces a morphism of triangulated $\mathscr{P}$-premotivic categories:

(5.1.24.1) $\qquad \mathbf{L}\gamma^* : D(\mathrm{PSh}(\mathscr{P}, \mathbf{Z})) \rightleftarrows D(\mathscr{A}) : \mathbf{R}\gamma_*.$

Similarly, the morphism (5.1.4.1) induces a morphism of triangulated $\mathscr{P}$-premotivic categories

(5.1.24.2) $\qquad a_t^* : D(\mathrm{PSh}(\mathscr{P}, \Lambda)) \rightleftarrows D(\mathrm{Sh}_t(\mathscr{P}, \Lambda)) : \mathbf{R}a_{t,*}.$

Note that $a_t^* = \mathbf{L}a_t^*$ on objects, because the functor a_t^* is exact.

Example 5.1.25 Consider an admissible topology t. Let $\varphi : \Lambda \longrightarrow \Lambda'$ be a morphism of rings. For any scheme S, it induces a pair of adjoint functors:

---

[74] Note also that this adjunction extends on $\mathscr{S}$-diagrams considering the situation described in Section 5.1.20: for any diagram $\mathscr{X} : I \longrightarrow \mathscr{S}$, we get an adjunction
$$\mathbf{L}\varphi_{\mathscr{X}} : D(\mathscr{A})(\mathscr{X}) \rightleftarrows D(\mathscr{B})(\mathscr{X}) : \mathbf{R}\psi_{\mathscr{X}}$$
and this defines a morphism of triangulated monoidal $\mathscr{P}_{cart}$-fibred categories over the $\mathscr{S}$-diagrams (*cf.* Proposition 3.1.32).

5 Fibred derived categories

(5.1.25.1) $\qquad \varphi^* : \mathrm{Sh}_t(\mathscr{P}_S, \Lambda) \rightleftarrows \mathrm{Sh}_t(\mathscr{P}_S, \Lambda') : \varphi_*$

such that φ^* (resp. φ_*) is induced by the obvious extension (resp. restriction) of scalars functor. By definition, for any $\mathscr{P}$-scheme X/S, the functor φ^* sends the representable sheaf of Λ-modules $\Lambda_S^t(X)$ to the representable sheaf of Λ'-modules $\Lambda_S''(X)$. Thus (φ^*, φ_*) defines an adjunction of abelian $\mathscr{P}$-premotivic categories. Applying the results of Section 5.1.23, one deduces a $\mathscr{P}$-premotivic adjunction:

$$\mathbf{L}\varphi^* : \mathrm{D}(\mathrm{Sh}_t(\mathscr{P}, \Lambda)) \rightleftarrows \mathrm{D}(\mathrm{Sh}_t(\mathscr{P}, \Lambda')) : \mathbf{R}\varphi_*.$$

The functor φ_* is exact so that $\mathbf{R}\varphi_* = \varphi_*$. Similarly when Λ'/Λ is flat, $\mathbf{L}\varphi^* = \varphi^*$.

The following result can be used to check the compatibility with a given admissible topology:

Proposition 5.1.26 *Let t be an admissible topology. Consider a morphism of abelian $\mathscr{P}$-premotivic categories*

$$\varphi : \mathscr{A} \rightleftarrows \mathscr{B} : \psi$$

such that:

(a) For any scheme S, ψ_S is exact.
(b) The morphism φ induces an isomorphism of the underlying set of twists of $\mathscr{A}$ and $\mathscr{B}$.

According to the last property, we identify the set of twists of $\mathscr{A}$ and $\mathscr{B}$ with a monoid τ in such a way that φ acts on τ by the identity.

Assume that $\mathscr{A}$ is compatible with t. Then the following conditions are equivalent:

(i) $\mathscr{B}$ is compatible with t.
(ii) $\mathscr{B}$ satisfies cohomological t-descent,

Proof The fact that (i) implies (ii) is clear from the definition, and we prove the converse using the following lemma:

Lemma 5.1.27 *Consider a morphism of $\mathscr{P}$-premotivic abelian categories*

$$\varphi : \mathscr{A} \rightleftarrows \mathscr{B} : \psi$$

satisfying conditions (a) and (b) of the above proposition and a base scheme S.
Given a simplicial scheme $\mathscr{X}$ which is degree-wise a sum of $\mathscr{P}$-schemes over S, a twist $i \in \tau$ and a complex C over $\mathscr{B}_S$, we denote by

$$\epsilon_{\mathscr{X}, i, C} : \mathrm{Hom}_{C(\mathscr{B}_S)}\big(M_S(\mathscr{X}, \mathscr{B})\{i\}, C\big) \longrightarrow \mathrm{Hom}_{C(\mathscr{A}_S)}\big(M_S(\mathscr{X}, \mathscr{A})\{i\}, \psi_S(C)\big)$$

the adjunction isomorphism obtained for the adjoint pair (φ_S, ψ_S).
Then there exists a unique isomorphism $\epsilon'_{\mathscr{X}, i, C}$ making the following diagram commutative:

$$\mathrm{Hom}_{C(\mathscr{B}_S)}\left(M_S(\mathscr{X},\mathscr{B})\{i\},C\right) \xrightarrow{\epsilon_{\mathscr{X},i,C}} \mathrm{Hom}_{C(\mathscr{A}_S)}\left(M_S(\mathscr{X},\mathscr{A})\{i\},\psi_S(C)\right)$$
$$\downarrow \qquad\qquad\qquad\qquad\qquad\qquad \downarrow$$
$$\mathrm{Hom}_{K(\mathscr{B}_S)}\left(M_S(\mathscr{X},\mathscr{B})\{i\},C\right) \xrightarrow{\epsilon'_{\mathscr{X},i,C}} \mathrm{Hom}_{K(\mathscr{A}_S)}\left(M_S(\mathscr{X},\mathscr{A})\{i\},\psi_S(C)\right).$$

Assume moreover that $\mathscr{B}$ satisfies cohomological t-descent. Then there exists an isomorphism $\epsilon''_{\mathscr{X},i,C}$ making the following diagram commutative:

(5.1.27.1)

$$\mathrm{Hom}_{K(\mathscr{B}_S)}\left(M_S(\mathscr{X},\mathscr{B})\{i\},C\right) \xrightarrow{\epsilon'_{\mathscr{X},i,C}} \mathrm{Hom}_{K(\mathscr{A}_S)}\left(M_S(\mathscr{X},\mathscr{A})\{i\},\psi_S(C)\right)$$
$$\pi^{\mathscr{B}}_{\mathscr{X},i,C} \downarrow \qquad\qquad\qquad\qquad\qquad \downarrow \pi^{\mathscr{A}}_{\mathscr{X},i,C}$$
$$\mathrm{Hom}_{D(\mathscr{B}_S)}\left(M_S(\mathscr{X},\mathscr{B})\{i\},C\right) \xrightarrow{\epsilon''_{\mathscr{X},i,C}} \mathrm{Hom}_{D(\mathscr{A}_S)}\left(M_S(\mathscr{X},\mathscr{A})\{i\},\psi_S(C)\right),$$

where $\pi^{\mathscr{A}}_{\mathscr{X},i,C}$ and $\pi^{\mathscr{B}}_{\mathscr{X},i,C}$ are induced by the obvious localization functors. □

The existence and unicity of isomorphism $\epsilon'_{\mathscr{X},i,C}$ follows from the fact that the functors φ_S and ψ_S are additive. Indeed, this implies that the isomorphism $\epsilon_{\mathscr{X},i,C}$ is compatible with chain homotopies.

Consider the injective model structure on $C(\mathscr{A}_S)$ and $C(\mathscr{B}_S)$ (see for example [CD09, 1.2] for the definition). We first treat the case when C is fibrant for this model structure on $C(\mathscr{B}_S)$. Because the premotive $M_S(\mathscr{X},\mathscr{B})\{i\}$ is cofibrant for the injective model structure, we obtain that the canonical map $\pi^{\mathscr{B}}_{\mathscr{X},i,C}$ is an isomorphism. This implies there exists a unique map $\epsilon''_{\mathscr{X},i,C}$ making diagram (5.1.27.1) commutative. On the other hand, the isomorphism $\epsilon'_{\mathscr{X},i,C}$ obtained previously is obviously functorial in $\mathscr{X}$. Thus, because $\mathscr{B}$ satisfies t-descent, we obtain that $\psi_S(C)$ is t-flasque. Because $\mathscr{A}$ is compatible with t, this implies $\psi_S(C)$ is t-local, and because $M_S(\mathscr{X},\mathscr{B})\{i\}$ is cofibrant for the t-descent model structure on $C(\mathscr{A}_S)$, this implies $\pi^{\mathscr{B}}_{\mathscr{X},i,C}$ is an isomorphism. Thus finally, $\epsilon''_{\mathscr{X},i,C}$ is an isomorphism, as required.

To treat the general case, we consider a fibrant resolution $C \longrightarrow D$ for the injective model structure on $C(\mathscr{B}_S)$. Because ψ_S is exact, it preserves isomorphisms. Using the previous case, we define $\epsilon''_{\mathscr{X},i,C}$ by the following commutative diagram:

$$\mathrm{Hom}_{D(\mathscr{B}_S)}\left(M_S(\mathscr{X},\mathscr{B})\{i\},C\right) \xrightarrow{\epsilon''_{\mathscr{X},i,C}} \mathrm{Hom}_{D(\mathscr{A}_S)}\left(M_S(\mathscr{X},\mathscr{A})\{i\},\psi_S(C)\right)$$
$$\sim \downarrow \qquad\qquad\qquad\qquad\qquad \downarrow \sim$$
$$\mathrm{Hom}_{D(\mathscr{B}_S)}\left(M_S(\mathscr{X},\mathscr{B})\{i\},D\right) \xrightarrow{\epsilon''_{\mathscr{X},i,D}} \mathrm{Hom}_{D(\mathscr{A}_S)}\left(M_S(\mathscr{X},\mathscr{A})\{i\},\psi_S(D)\right).$$

The required property for $\epsilon''_{\mathscr{X},i,C}$ then follows easily and the lemma is proved.

To finish the proof that (ii) implies (i), we note the lemma immediately implies, under (ii), that the following two conditions are equivalent:

- C is t-flasque (resp. local) in $C(\mathscr{B}_S)$;

5 Fibred derived categories

- $\psi_S(C)$ is t-flasque (resp. local) in $C(\mathscr{A}_S)$.

This concludes the proof. $\square$

5.1.c Constructible premotivic complexes

Definition 5.1.28 Let $\mathscr{A}$ be an abelian $\mathscr{P}$-premotivic category compatible with an admissible topology t. We will say that t is *bounded in $\mathscr{A}$* if for any scheme S, there exists an essentially small family $\mathscr{N}_S^t$ of bounded complexes which are direct factors of finite sums of objects of type $M_S(X)\{i\}$ in each degree, such that, for any complex C of $\mathscr{A}_S$, the following conditions are equivalent.

(i) C is t-flasque.
(ii) For any H in $\mathscr{N}_S^t$, the abelian group $\mathrm{Hom}_{K(\mathscr{A}_S)}(H,C)$ vanishes.

In this case, we say the family $\mathscr{N}_S^t$ is a *bounded generating family for t-hypercoverings in $\mathscr{A}_S$*.

Example 5.1.29

1. Assume $\mathscr{P}$ contains the open immersions so that the Zariski topology is admissible. Let MV_S be the family of complexes of the form
$$\Lambda_S(U \cap V) \xrightarrow{l_* - k_*} \Lambda_S(U) \oplus \Lambda_S(V) \xrightarrow{i_* + j_*} \Lambda_S(X)$$
for any open cover $X = U \cup V$, where i,j,k,l denotes the obvious open immersions. It follows then from [BG73] that MV_S is a bounded generating family of Zariski hypercovers in $\mathrm{Sh}_{Zar}(\mathscr{P}/S, \Lambda)$.

2. Assume $\mathscr{P}$ contains the étale morphisms so that the Nisnevich topology is admissible. We let BG_S be the family of complexes of the form
$$\Lambda_S(W) \xrightarrow{g_* - l_*} \Lambda_S(U) \oplus \Lambda_S(V) \xrightarrow{j_* + f_*} \Lambda_S(X)$$
for a Nisnevich distinguished square in $\mathscr{S}$ (*cf.* Example 2.1.11)

$$\begin{array}{ccc} W & \xrightarrow{l} & V \\ g \downarrow & & \downarrow f \\ U & \xrightarrow{j} & X. \end{array}$$

Then, by applying Theorem 3.3.2, we see that BG_S is a bounded generating family for Nisnevich hypercovers in $\mathrm{Sh}_{Nis}(\mathscr{P}/S, \Lambda)$.

3. Assume that $\mathscr{P} = \mathscr{S}^{ft}$ is the class of morphisms of finite type in $\mathscr{S}$. We let $PCDH_S$ be the family of complexes of the form
$$\Lambda_S(T) \xrightarrow{g_* - k_*} \Lambda_S(Z) \oplus \Lambda_S(Y) \xrightarrow{i_* + f_*} \Lambda_S(X)$$

for a cdh-distinguished square in $\mathscr{S}$ (cf. Example 2.1.11)

$$\begin{array}{ccc} T & \xrightarrow{k} & Y \\ {\scriptstyle g}\downarrow & & \downarrow{\scriptstyle f} \\ Z & \xrightarrow{i} & X. \end{array}$$

Then, by virtue of Theorem 3.3.8, $CDH_S = BG_S \cup PCDH_S$ is a bounded generating family for cdh-hypercovers in $\mathrm{Sh}_{cdh}\left(\mathscr{S}^{ft}/S,\Lambda\right)$.

4. The étale topology is not bounded in $\mathrm{Sh}_{\acute{e}t}(Sm,\Lambda)$ for an arbitrary ring Λ. However, if $\Lambda = \mathbf{Q}$, it is bounded: by virtue of Theorem 3.3.23, a bounded generating family for étale hypercovers in $\mathrm{Sh}_{\acute{e}t}(Sm,\mathbf{Q})_S$ is the union of the class BG_S and that of complexes of the form $\mathbf{Q}_S(Y)_G \longrightarrow \mathbf{Q}_S(X)$ for any Galois cover $Y \longrightarrow X$ of group G.

5. As in the case of the étale topology, the qfh-topology is not bounded in general, but it is so with rational coefficients. Let $PQFH_S$ be the family of complexes of the form

$$\mathbf{Q}_S(T)_G \xrightarrow{g_*-k_*} \mathbf{Q}_S(Z) \oplus \mathbf{Q}_S(Y)_G \xrightarrow{i_*+f_*} \mathbf{Q}_S(X)$$

for a qfh-distinguished square of group G in $\mathscr{S}$ (cf. Definition 3.3.15)

$$\begin{array}{ccc} T & \xrightarrow{k} & Y \\ {\scriptstyle g}\downarrow & & \downarrow{\scriptstyle f} \\ Z & \xrightarrow{i} & X. \end{array}$$

Then, by virtue of Theorem 3.3.25, $QFH_S = PQFH_S \cup BG_S$ is a bounded generating family for qfh-hypercovers in $\mathrm{Sh}_{qfh}\left(\mathscr{S}^{ft}/S,\mathbf{Q}\right)$.

6. Similarly, by Theorem 3.3.30, $H_S = CDH_S \cup QFH_S$ is a bounded generating family for h-hypercovers in $\mathrm{Sh}_h\left(\mathscr{S}^{ft}/S,\mathbf{Q}\right)$.

Proposition 5.1.30 *Let $\mathscr{A}$ be an abelian $\mathscr{P}$-premotivic category compatible with an admissible topology t. We make the following assumptions:*

(a) t is bounded in $\mathscr{A}$;
(b) for any $\mathscr{P}$-morphism $X \longrightarrow S$ and any $n \in \tau$, the functor $\mathrm{Hom}_{\mathscr{A}_S}(M_S(X)\{n\},-)$ preserves filtered colimits.

Then t-local complexes are stable under filtering colimits.

Proof Let $\mathscr{N}_S^t$ be a bounded generating family for t-hypercovers in $\mathscr{A}_S$. Then a complex C of $\mathscr{A}_S$ is t-flasque if and only if for any $H \in \mathscr{N}_S^t$, the abelian group $\mathrm{Hom}_{K(\mathscr{A}_S)}(H,C)$ is trivial. Hence it is sufficient to prove that the functor

$$C \longmapsto \mathrm{Hom}_{K(\mathscr{A}_S)}(H,C)$$

preserves filtering colimits of complexes. This will follow from the fact that the functor

$$C \longmapsto \mathrm{Hom}_{C(\mathscr{A}_S)}(H,C)$$

5 Fibred derived categories

preserves filtering colimits. As H is a bounded complex that is degreewise compact, this latter property is obvious. □

5.1.31 Consider an abelian $\mathscr{P}$-premotivic category $\mathscr{A}$ compatible with an admissible topology t, with generating set of twists τ. Assume that t is bounded in $\mathscr{A}$ and consider a bounded generating family $\mathscr{N}_S^t$ for t-hypercovers in $\mathscr{A}_S$.

Let $M(\mathscr{P}/S, \mathscr{A})$ be the full subcategory of $\mathscr{A}_S$ spanned by direct factors of finite sums of premotives of shape $M_S(X)\{i\}$ for a $\mathscr{P}$-scheme X/S and a twist $i \in \tau$. This category is additive and we can associate with it its category of complexes up to chain homotopy. We get an obvious triangulated functor

(5.1.31.1) $$K^b(M(\mathscr{P}/S, \mathscr{A})) \longrightarrow D(\mathscr{A}_S).$$

Then the previous functor induces a triangulated functor

$$K^b(M(\mathscr{P}/S, \mathscr{A}))/\mathscr{N}_S^t \longrightarrow D(\mathscr{A}_S),$$

where the left-hand side stands for the Verdier quotient of $K^b(M(\mathscr{P}/S, \mathscr{A}))$ by the thick subcategory generated by $\mathscr{N}_S^t$.

The category $K^b(M(\mathscr{P}/S, \mathscr{A}))/\mathscr{N}_S^t$ may not be pseudo-abelian while the aim of the previous functor is. Thus we can consider its pseudo-abelian envelope and the induced functor

(5.1.31.2) $$\left(K^b(M(\mathscr{P}/S, \mathscr{A}))/\mathscr{N}_S^t\right)^{\natural} \longrightarrow D(\mathscr{A}_S).$$

According to Definition 1.4.9, the image of this functor is the subcategory of τ-constructible premotives of the triangulated $\mathscr{P}$-premotivic category $D(\mathscr{A}_S)$. Then the following proposition is a corollary of [CD09, theorem 6.2]:

Proposition 5.1.32 *Consider the hypotheses and notations above.*

If $\mathscr{A}$ is finitely τ-presented then $D(\mathscr{A})$ is compactly τ-generated. Moreover, the functor (5.1.31.2) is fully faithful.

Let us denote by $D_c(\mathscr{A})$ the subcategory of $D(\mathscr{A})$ comprising τ-constructible premotives in the sense of Definition 1.4.9. Taking into account Proposition 1.4.11, the previous proposition admits the following corollary:

Corollary 5.1.33 *Consider the situation of Section 5.1.31, and assume that $\mathscr{A}$ is finitely τ-presented. For any premotive $\mathscr{M}$ in $D(\mathscr{A}_S)$, the following conditions are equivalent:*

(i) $\mathscr{M}$ is compact.
(ii) $\mathscr{M}$ is τ-constructible.

Moreover, the functor (5.1.31.2) induces an equivalence of categories:

$$\left(K^b(M(\mathscr{P}/S, \mathscr{A}))/\mathscr{N}_S^t\right)^{\natural} \longrightarrow D_c(\mathscr{A}_S).$$

Example 5.1.34 According to Example 5.1.29, we get the following examples:

1. Let $\Lambda(Sm/S) = M(Sm/S, \mathscr{A})$ for $\mathscr{A} = \mathrm{Sh}_{Nis}(Sm/S, \Lambda)$. We obtain a fully faithful functor
$$\left(\mathrm{K}^b\left(\Lambda(Sm/S)\right)/BG_S\right)^{\natural} \longrightarrow \mathrm{D}\left(\mathrm{Sh}_{Nis}(Sm/S, \Lambda)\right)$$
which is essentially surjective on compact objects.

2. Let $\Lambda(\mathscr{S}^{ft}/S) = M(Sm/S, \mathscr{A})$ for $\mathscr{A} = \mathrm{Sh}_{cdh}(\mathscr{S}^{ft}/S, \Lambda)$. We obtain a fully faithful functor
$$\left(\mathrm{K}^b\left(\Lambda(\mathscr{S}^{ft}/S)\right)/BG_S \cup CDH_S\right)^{\natural} \longrightarrow \mathrm{D}\left(\mathrm{Sh}_{cdh}(\mathscr{S}^{ft}/S, \Lambda)\right)$$
which is essentially surjective on compact objects.

3. Let $\mathbf{Q}_{\acute{e}t}(Sm/S) = M(Sm/S, \mathscr{A})$ for $\mathscr{A} = \mathrm{Sh}_{\acute{e}t}(Sm/S, \mathbf{Q})$. We obtain a fully faithful functor
$$\left(\mathrm{K}^b\left(\mathbf{Q}_{\acute{e}t}(Sm/S)\right)/BG_S\right)^{\natural} \longrightarrow \mathrm{D}\left(\mathrm{Sh}_{\acute{e}t}(Sm/S, \mathbf{Q})\right)$$
which is essentially surjective on compact objects.

5.1.35 Consider an abelian $\mathscr{P}$-premotivic category $\mathscr{A}$. We introduce the following property of $\mathscr{A}$:

(C) Consider a projective system $(S_\alpha)_{\alpha \in A}$ of schemes in $\mathscr{S}$ with affine transition maps such that $S = \varprojlim_{\alpha \in A} S_\alpha$ belongs to $\mathscr{S}$. For any index $\alpha_0 \in A$, any object A_{α_0} in $\mathscr{A}_{S_{\alpha_0}}$, and any twist $n \in \tau$, the canonical map
$$\varinjlim_{\alpha \in A/\alpha_0} \mathrm{Hom}_{\mathscr{A}_{S_\alpha}}(\mathbb{1}_{S_\alpha}\{n\}, A_\alpha) \longrightarrow \mathrm{Hom}_{\mathscr{A}_S}(\mathbb{1}_S\{n\}, A)$$
is an isomorphism where A_α (resp. A) is the pullback of A_{α_0} along the canonical map $S_\alpha \longrightarrow S_{\alpha_0}$ (resp. $S \longrightarrow S_{\alpha_0}$).

We will denote by (wC) the analogous property when one restricts pro-objects to those with affine and dominant transition maps.

Proposition 5.1.36 *Consider an abelian $\mathscr{P}$-premotivic category $\mathscr{A}$ compatible with an admissible topology t and satisfying the assumption (C) (resp. (wC)) above.*

Then the derived premotivic category $\mathrm{D}(\mathscr{A})$ is τ-continuous (resp. weakly τ-continuous) — see Definition 4.3.2.

Proof We use Proposition 4.3.6 applied to the t-descent model structure on $C(\mathscr{A}_T)$ for $T = S$ or $T = S_\alpha$ (see Section 5.1.13). Recall from Section 5.1.11 that this model structure is associated with a descent structure. Thus according to [CD09, 2.3], there exists an explicit generating set I (resp. J) for cofibrations (resp. trivial cofibrations). Moreover, the source or target of any map in $I \cup J$ is a complex C satisfying the following assumption:

(rep) for any integer $i \in \mathbf{Z}$, C^i is a sum of premotives of the form $M_T(X)\{n\}$ where X/T is a $\mathscr{P}$-scheme and $n \in \tau$.

Thus, to check the assumption of Proposition 4.3.6 for $C(\mathscr{A})$, we fix a projective system $(S_\alpha)_{\alpha \in A}$ satisfying the assumptions of property (C) (resp. (wC)) above; we have to prove that for any index $\alpha_0 \in A$ and any complexes C_{α_0} and E_{α_0} such that C_{α_0} satisfies (rep), the natural map:

$$\varinjlim_{\alpha \in A/\alpha_0} \mathrm{Hom}_{\mathrm{C}(\mathscr{A}_{S_\alpha})}(C_\alpha, E_\alpha) \longrightarrow \mathrm{Hom}_{\mathrm{C}(\mathscr{A}_S)}(C, E)$$

is bijective.

Given the definition of morphisms in a category of complexes, it is sufficient to check this when the Hom groups are computed as morphisms of $\mathbf{Z}$-graded objects. Thus it is sufficient to treat the case where C_{α_0} and E_{α_0} are concentrated in degree 0. Thus, as C_{α_0} satisfies property (rep), we are exactly reduced to assumption (C) (resp. (wC)) on $\mathscr{A}$. □

Example 5.1.37

1. Assume $\mathscr{P}$ is contained in the class of morphisms of finite type.
   Then the abelian $\mathscr{P}$-premotivic category $\mathrm{PSh}(\mathscr{P}, \Lambda)$ of Example 5.1.1 satisfies assumption (C). Indeed, property (C) when A is a representable presheaf follows from the assumption on $\mathscr{P}$: $\mathscr{P}$-schemes over some base S are always of finite presentation over S – the base S is noetherian according to our general assumption 5.0.1. Then the case of a general presheaf A follows because A is an inductive limit of representable presheaf and the global sections functor commutes with inductive limits of presheaves.
2. Let $\mathscr{S}^{ft}$ be the class of morphisms of finite type and let t be one of the following topologies: $Nis, \acute{e}t, cdh, qfh, h$.
   Then the generalized abelian premotivic category $\mathrm{Sh}_t(\mathscr{S}^{ft}, \Lambda)$ of Example 5.1.4 satisfies assumption (C).
   Indeed, according to the preceding example, we have only to prove that for any morphism $f : X \longrightarrow S$, the functor

$$f^* : \mathrm{PSh}(\mathscr{S}^{ft}_S, \Lambda) \longrightarrow \mathrm{PSh}(\mathscr{S}^{ft}_T, \Lambda)$$

preserves the property of being a t-sheaf.
   If f is a morphism of finite type, the functor f^* admits as a left adjoint the functor $f_\sharp$, which preserves t-covers. Thus the assertion is clear in that case.
   In the general case, we use the fact that X/S is a projective limit of a projective system $(X_\alpha)_{\alpha \in A}$ where X_α is an S-scheme affine and of finite type over S. To check that for a t-sheaf F over S, the presheaf $f^*(F)$ is a t-sheaf, we fix a t-cover $(W_i)_{i \in I}$ of X in $\mathscr{S}^{ft}_X$. As X is noetherian, we can assume I is finite. Moreover, there exists an index $\alpha_0 \in A$ such that for the t-cover $(W_i)_{i \in I}$ can be lifted to X_{α_0}. Then, using property (C) of $\mathrm{PSh}(\mathscr{S}^{ft}, \Lambda)$ applied to F and (X_α), we reduce

to checking that $f_\alpha^*(F)$ is a t-sheaf for $\alpha \geq \alpha_0$. This follows from the first case treated.

3. Let Sm be the class of smooth morphisms and t be one of the topologies: Nis, $\acute{e}t$. As we will see in Example 6.1.1, there exists a canonical enlargement of abelian premotivic categories (see (6.1.1.1)):

$$\rho_\sharp : \mathrm{Sh}_t(Sm, \Lambda) \rightleftarrows \mathrm{Sh}_t\left(\mathscr{S}^{ft}, \Lambda\right) : \rho^*.$$

As the functor $\rho_\sharp$ is fully faithful and commutes with f^* for any morphism of schemes f, we deduce from the preceding point that the abelian premotivic category $\mathrm{Sh}_t(Sm, \Lambda)$ satisfies the above condition (C).

As an application of the previous proposition, we thus obtain that the derived premotivic category $\mathrm{D}(\mathrm{Sh}_t(Sm, \Lambda))$ is τ-continuous.

5.2 The $\mathbf{A}^1$-derived premotivic category

5.2.a Localization of triangulated premotivic categories

5.2.1 Let $\mathscr{A}$ be an abelian $\mathscr{P}$-premotivic category compatible with an admissible topology t and $\mathrm{D}(\mathscr{A})$ be the associated derived $\mathscr{P}$-premotivic category.

Suppose given an essentially small family of morphisms $\mathscr{W}$ in $\mathrm{C}(\mathscr{A})$ which is stable under the operations f^*, $f_\sharp$ (in other words, $\mathscr{W}$ is a sub-$\mathscr{P}$-fibred category of $\mathrm{C}(\mathscr{A})$). Note that the localizing subcategory $\mathscr{T}$ of $\mathrm{D}(\mathscr{A})$ generated by the cones of arrows in $\mathscr{W}$ is again stable under these operations. Moreover, as for any $\mathscr{P}$-morphism $f : X \longrightarrow S$ we have $f_\sharp f^* = M_S(X) \otimes_S (-)$, the category $\mathscr{T}$ is stable under tensor product with a geometric section.

We will say that a complex K over $\mathscr{A}_S$ is $\mathscr{W}$-*local* if for any object T of $\mathscr{T}$ and any integer $n \in \mathbf{Z}$, $\mathrm{Hom}_{\mathrm{D}(\mathscr{A}_S)}(T, K[n]) = 0$. A morphism of complexes $p : C \longrightarrow D$ over $\mathscr{A}_S$ is a $\mathscr{W}$-*equivalence* if for any $\mathscr{W}$-local complex K over $\mathscr{A}_S$, the induced map

$$\mathrm{Hom}_{\mathrm{D}(\mathscr{A}_S)}(D, K) \longrightarrow \mathrm{Hom}_{\mathrm{D}(\mathscr{A}_S)}(C, K)$$

is bijective.

A morphism of complexes over $\mathscr{A}_S$ is called a $\mathscr{W}$-*fibration* if it is a t-fibration with a $\mathscr{W}$-local kernel. A complex over $\mathscr{A}_S$ will be called $\mathscr{W}$-*fibrant* if it is t-local and $\mathscr{W}$-local.

As consequence of [CD09, 4.3, 4.11 and 5.6], we obtain:

Proposition 5.2.2 *Let $\mathscr{A}$ be an abelian $\mathscr{P}$-premotivic category compatible with an admissible topology t and $\mathscr{W}$ be an essentially small family of morphisms in $\mathrm{C}(\mathscr{A})$ stable under f^* and $f_\sharp$.*

Then the category $\mathrm{C}(\mathscr{A}_S)$ is a proper closed symmetric monoidal category with the $\mathscr{W}$-fibrations as fibrations, the cofibrations as defined in Section 5.1.11, and the $\mathscr{W}$-equivalences as weak equivalences.

5 Fibred derived categories

The homotopy category associated with this model category will be denoted by $D(\mathscr{A}_S)[\mathscr{W}_S^{-1}]$. It can be described as the Verdier quotient $D(\mathscr{A}_S)/\mathscr{T}_S$.

In fact, the $\mathscr{W}$-local model category on $C(\mathscr{A}_S)$ is nothing else than the left Bousfield localization of the t-local model category structure. As a consequence, we obtain an adjunction of triangulated categories:

(5.2.2.1) $$\pi_S : D(\mathscr{A}_S) \rightleftarrows D(\mathscr{A}_S)[\mathscr{W}_S^{-1}] : \mathscr{O}_S$$

such that $\mathscr{O}_S$ is fully faithful with essential image the $\mathscr{W}$-local complexes. In fact, the model structure gives a functorial $\mathscr{W}$-fibrant resolution $1 \longrightarrow R_{\mathscr{W}}$

$$R_{\mathscr{W}} : C(\mathscr{A}_S) \longrightarrow C(\mathscr{A}_S),$$

which induces $\mathscr{O}_S$.

Note that the triangulated category $D(\mathscr{A}_S)[\mathscr{W}_S^{-1}]$ is generated by the complexes concentrated in degree 0 of the form $M_S(X)\{i\}$ — or, equivalently, the $\mathscr{W}$-local complexes $R_{\mathscr{W}}(M_S(X)\{i\})$ — for a $\mathscr{P}$-scheme X and a twist $i \in \tau$.

Remark 5.2.3 Another very useful property is that $\mathscr{W}$-equivalences are stable under filtering colimits; see [CD09, prop. 3.8].

5.2.4 Recall from Section 5.1.14 that for any morphism (resp. $\mathscr{P}$-morphism) $f : T \longrightarrow S$, the functor f^* (resp. $f_\sharp$) satisfies descent; as it also preserves $\mathscr{W}$, it follows from [CD09, 4.9] that the adjunction

$$f^* : C(\mathscr{A}_S) \longrightarrow C(\mathscr{A}_T) : f_*$$
$$(\text{resp. } f_\sharp : C(\mathscr{A}_S) \longrightarrow C(\mathscr{A}_T) : f^*)$$

is a Quillen adjunction with respect to the $\mathscr{W}$-local model structures. This gives the following corollary.

Corollary 5.2.5 *The $\mathscr{P}$-fibred category $C(\mathscr{A})$ with the $\mathscr{W}$-local model structure on its fibers defined above is a monoidal $\mathscr{P}$-fibred model category, which is moreover stable, proper and combinatorial.*

We will denote by $D(\mathscr{A})[\mathscr{W}^{-1}]$ the triangulated $\mathscr{P}$-premotivic category whose fiber over a scheme S is the homotopy category of the $\mathscr{W}_S$-local model category $C(\mathscr{A}_S)$. The adjunction (5.2.2.1) readily defines an adjunction of triangulated $\mathscr{P}$-premotivic categories

(5.2.5.1) $$\pi : D(\mathscr{A}) \rightleftarrows D(\mathscr{A})[\mathscr{W}^{-1}] : \mathscr{O}.$$

The $\mathscr{P}$-fibred categories $D(\mathscr{A})$ and $D(\mathscr{A})[\mathscr{W}^{-1}]$ are both τ-generated (and this adjunction is compatible with τ-twists in a strong sense).

Remark 5.2.6 For any scheme S, the category $D(\mathscr{A}_S)[\mathscr{W}_S^{-1}]$ is well generated and has a canonical dg-structure (see also Remark 5.1.19).

5.2.7 With the notations above, let us put $\mathscr{T} = D(\mathscr{A})[\mathscr{W}^{-1}]$ to clarify the following notations. As in Section 5.1.20, the fibred category $\mathscr{T}$ has a canonical extension to $\mathscr{S}$-diagrams $\mathscr{X} : I \longrightarrow \mathscr{S}$.

If we define $\mathscr{W}_{\mathscr{X}}$ as the class of morphisms $(f_i)_{i \in I}$ in $C(\mathscr{A}(\mathscr{X}, I))$ such that for any object i, f_i is a $\mathscr{W}$-equivalence, then $\mathscr{T}(X)$ is the triangulated category $D(\mathscr{A}(\mathscr{X}, I))[\mathscr{W}_{\mathscr{X}}^{-1}]$.

Again, this triangulated category is symmetric monoidal closed and for any morphism (resp. $\mathscr{P}$-morphism) $\varphi : (\mathscr{X}, I) \longrightarrow (\mathscr{Y}, J)$, we get (derived) adjunctions as in Section 5.1.20:

(5.2.7.1) $$\mathbf{L}\varphi^* : \mathscr{T}(\mathscr{Y}, J) \rightleftarrows \mathscr{T}(\mathscr{X}, I) : \mathbf{R}\varphi_*,$$
(5.2.7.2) $$(\text{resp. } \mathbf{L}\varphi_\sharp : \mathscr{T}(\mathscr{X}, I) \rightleftarrows \mathscr{T}(\mathscr{Y}, J) : \mathbf{L}\varphi^* = \varphi^*).$$

In fact, $\mathscr{T}$ is then a complete monoidal $\mathscr{P}_{cart}$-fibred category over the category of diagrams of schemes and the adjunction (5.2.5.1) extends to an adjunction of complete monoidal $\mathscr{P}_{cart}$-fibred categories.

Example 5.2.8 Suppose we are under the hypothesis of Example 5.1.24.2.

Let $\mathscr{W}_{t,S}$ denote the family of maps which are of the form $\Lambda_S(\mathscr{X}) \longrightarrow \Lambda_S(X)$ for a t-hypercover $\mathscr{X} \longrightarrow X$ in $\mathscr{P}/S$. Then $\mathscr{W}_t$ is obviously stable under f^* and $f_\sharp$.

Recall now that a complex of t-sheaves on $\mathscr{P}/S$ is local if and only if its t-hypercohomology and its hypercohomology computed in the coarse topology agree (*cf.* Example 5.1.10).

This readily implies that the adjunction considered in Example 5.1.24.2

$$a_t^* : D(\mathrm{PSh}(\mathscr{P}, \Lambda)) \rightleftarrows D(\mathrm{Sh}_t(\mathscr{P}, \Lambda)) : \mathbf{R}a_{t,*}$$

induces an equivalence of triangulated $\mathscr{P}$-premotivic categories

$$D(\mathrm{PSh}(\mathscr{P}, \Lambda))[\mathscr{W}_t^{-1}] \rightleftarrows D(\mathrm{Sh}_t(\mathscr{P}, \Lambda)).$$

Recall $\mathbf{R}a_{t,*}$ is fully faithful and identifies $D(\mathrm{Sh}_t(S, \Lambda))$ with the full subcategory of $D(\mathrm{PSh}(S, \Lambda))$ made by t-local complexes.

5.2.9 A triangulated $\mathscr{P}$-premotivic category $(\mathscr{T}, M)$ such that there exists:

1. an abelian $\mathscr{P}$-premotivic category $\mathscr{A}$ compatible with an admissible topology t_0 on Sm.
2. an essentially small family $\mathscr{W}$ of morphisms in $C(\mathscr{A})$ stable under f^* and $f_\sharp$
3. an adjunction of triangulated $\mathscr{P}$-premotivic categories $D(\mathscr{A})[\mathscr{W}^{-1}] \simeq \mathscr{T}$

will be called for short a *derived $\mathscr{P}$-premotivic category*. According to Convention 5.0.1(d) and from the above construction, $\mathscr{T}$ is τ-generated for some set of twists τ.[75]

---

[75] In some remarks below, we will formulate universal properties of some derived $\mathscr{P}$-premotivic categories. When doing so, we will restrict to morphisms of derived $\mathscr{P}$-premotivic categories which can be written as

5 Fibred derived categories

Let us denote simply by $M_S(X)$ the geometric sections of $\mathscr{T}$. In this case, using the morphisms (5.1.24.1) and (5.2.5.1), we get a canonical morphism of triangulated $\mathscr{P}$-premotivic categories:

(5.2.9.1) $\qquad \varphi^* : \mathrm{D}(\mathrm{PSh}(\mathscr{P}, \mathbf{Z})) \rightleftarrows \mathscr{T} : \varphi_*.$

By definition, for any premotive $\mathscr{M}$, any scheme X and any integer $n \in \mathbf{Z}$, we get a canonical identification:

(5.2.9.2) $\qquad \mathrm{Hom}_{\mathscr{T}(S)}(M_S(X), \mathscr{M}[n]) = H^n \Gamma(X, \varphi_*(\mathscr{M})).$

Given any simplicial scheme $\mathscr{X}$, we put $M_S(\mathscr{X}) = \varphi^*\big(\mathbf{Z}_S(\mathscr{X})\big)$, so that we also obtain:

(5.2.9.3) $\qquad \mathrm{Hom}_{\mathscr{T}(S)}(M_S(\mathscr{X}), \mathscr{M}[n]) = H^n \big(\mathrm{Tot}^\pi \, \Gamma(\mathscr{X}, \mathbf{R}\gamma_*(\mathscr{M})) \big).$

Proposition 5.2.10 *Consider the above notations and t an admissible topology. The following conditions are equivalent.*

(i) For any t-hypercover $\mathscr{X} \longrightarrow X$ in $\mathscr{P}/S$, the induced map $M_S(\mathscr{X}) \longrightarrow M_S(X)$ is an isomorphism in $\mathscr{T}(S)$.
(i′) For any t-hypercover $p : \mathscr{X} \longrightarrow X$ in $\mathscr{P}/S$, the induced functor $\mathbf{L}p^* : \mathscr{T}(X) \longrightarrow \mathscr{T}(\mathscr{X})$ is fully faithful.
(i″) The triangulated $\mathscr{P}$-premotivic category $\mathscr{T}$ satisfies t-descent (Definition 3.2.5).
(ii) There exists an essentially unique map $\varphi_t^* : \mathrm{D}(\mathrm{Sh}_t(\mathscr{P}/S, \mathbf{Z})) \longrightarrow \mathscr{T}(S)$ making the following diagram essentially commutative:

$$\begin{array}{c} \mathrm{D}(\mathrm{PSh}(\mathscr{P}/S, \mathbf{Z})) \xrightarrow{\varphi^*} \mathscr{T}(S) \\ a_t \downarrow \quad \nearrow \varphi_t^* \\ \mathrm{D}(\mathrm{Sh}_t(\mathscr{P}/S, \mathbf{Z})). \end{array}$$

(ii′) For any complex $C \in \mathrm{C}(\mathrm{PSh}(\mathscr{P}/S, \mathbf{Z}))$ such that $a_t(C) = 0$, $\varphi^*(C) = 0$.
(ii″) For any map $f : C \longrightarrow D$ in $\mathrm{C}(\mathrm{PSh}(\mathscr{P}/S, \mathbf{Z}))$ such that $a_t(f)$ is an isomorphism, $\varphi^*(f)$ is an isomorphism.
(iii) There exists an essentially unique map $\varphi_{t*} : \mathscr{T}(S) \longrightarrow \mathrm{D}(\mathrm{Sh}_t(\mathscr{P}/S, \mathbf{Z}))$ making the following diagram essentially commutative:

$$\begin{array}{c} \mathrm{D}(\mathrm{PSh}(\mathscr{P}/S, \mathbf{Z})) \xleftarrow{\varphi_*} \mathscr{T}(S) \\ \mathbf{R}\mathscr{O}_t \uparrow \quad \nearrow \varphi_{t*} \\ \mathrm{D}(\mathrm{Sh}_t(\mathscr{P}/S, \mathbf{Z})). \end{array}$$

$$\mathbf{L}\varphi : \mathrm{D}(\mathscr{A}_1)[\mathscr{W}_1^{-1}] \longrightarrow \mathrm{D}(\mathscr{A}_2)[\mathscr{W}_2^{-1}]$$

for a morphism $\varphi : \mathscr{A}_1 \longrightarrow \mathscr{A}_2$ of abelian $\mathscr{P}$-premotivic categories compatible with suitable topologies. More natural universal properties could be obtained if one considers the framework of dg-categories or triangulated derivators.

(iii') For any premotive $\mathcal{M}$ in $\mathscr{T}(S)$, the complex $\varphi_*(\mathcal{M})$ is local.
(iii'') For any premotive $\mathcal{M}$ in $\mathscr{T}(S)$, any $\mathscr{P}$-scheme X/S and any integer $n \in \mathbf{Z}$,

$$\mathrm{Hom}_{\mathscr{T}(S)}(M_S(X), \mathcal{M}[n]) = H_t^n(X, \varphi_*(\mathcal{M})).$$

When these conditions are fulfilled for any scheme S, the functors appearing in (ii) and (iii) induce a morphism of triangulated $\mathscr{P}$-premotivic categories:

$$\varphi_t^* : \mathrm{D}(\mathrm{Sh}_t(\mathscr{P}, \mathbf{Z})) \rightleftarrows \mathscr{T} : \varphi_{t*}.$$

Proof The equivalence between conditions (*i*), (*i'*) and (*i''*) is clear (we proceed as in the proof of Proposition 5.1.21). The equivalences (*ii*) ⇔ (*ii'*) ⇔ (*ii''*) and (*iii*) ⇔ (*iii'*) follow from Example 5.2.8 and the definition of a localization. The equivalence (*i*) ⇔ (*ii''*) follows again from *loc. cit.* Finally, the equivalences (*i*) ⇔ (*iii'*) ⇔ (*iii''*) follow from (5.2.9.2), (5.2.9.3), and the characterization of a local complex of sheaves (*cf.* Example 5.1.10). □

Remark 5.2.11 The preceding proposition expresses the fact that the category $\mathrm{D}(\mathrm{Sh}_t(\mathscr{P}, \mathbf{Z}))$ is the universal derived $\mathscr{P}$-premotivic category satisfying *t*-descent.

5.2.12 We end this section by making explicit two particular cases of the descent property for derived $\mathscr{P}$-premotivic categories.

Consider a derived $\mathscr{P}$-premotivic category $\mathscr{T}$ with geometric sections M. Considering any diagram $\mathscr{X} : I \longrightarrow \mathscr{P}/S$ of $\mathscr{P}$-schemes over S, with projection $p : \mathscr{X} \longrightarrow S$, we can associate a premotive in $\mathscr{T}$:

$$M_S(\mathscr{X}) = \mathbf{L}p_\sharp(\mathbb{1}_S) = \mathbf{L}\varinjlim_{i \in I} M_S(\mathscr{X}_i).$$

In particular, when I is the category $\bullet \longrightarrow \bullet$, we associate to every S-morphism $f : Y \longrightarrow X$ of $\mathscr{P}$-schemes over S a canonical[76] *bivariant premotive*

$$M_S(X \xrightarrow{f} Y).$$

When f is an immersion, we will also write $M_S(Y/X)$ for this premotive. Note that in any case, there is a canonical distinguished triangle in $\mathscr{T}(S)$:

$$M_S(X) \xrightarrow{f_*} M_S(Y) \xrightarrow{\pi_f} M_S(X \xrightarrow{f} Y) \xrightarrow{\partial_f} M_S(X)[1].$$

This triangle is functorial in the arrow f – with respect to *commutative* squares. Given a commutative square of $\mathscr{P}$-schemes over S

---

[76] In fact, if $\mathscr{T} = \mathrm{D}(\mathscr{A})[\mathscr{W}^{-1}]$ for an abelian $\mathscr{P}$-premotivic category $\mathscr{A}$, then we can define $M_S(X \longrightarrow Y)$ as the cone of the morphism of complexes (concentrated in degree 0) $M_S(X) \xrightarrow{f_*} M_S(Y)$.

5 Fibred derived categories

(5.2.12.1)
$$\begin{array}{ccc} B & \xrightarrow{e'} & Y \\ g \downarrow & & \downarrow f \\ A & \xrightarrow{e} & X \end{array}$$

we will say that the image square in $\mathscr{T}(S)$

$$\begin{array}{ccc} M_S(B) & \xrightarrow{e'_*} & M_S(Y) \\ g_* \downarrow & & \downarrow f_* \\ M_S(A) & \xrightarrow{e_*} & M_S(X) \end{array}$$

is *homotopy cartesian*[77] if the premotive associated with diagram 5.2.12.1 is zero.

Proposition 5.2.13 *Consider a derived $\mathscr{P}$-premotivic category $\mathscr{T}$. We assume that $\mathscr{P}$ contains the étale morphisms (resp. $\mathscr{P} = \mathscr{S}^{ft}$). Then, with the above definitions, the following conditions are equivalent:*

(i) $\mathscr{T}$ satisfies Nisnevich (resp. proper cdh) descent.
(ii) For any scheme S and any Nisnevich (resp. proper cdh) distinguished square Q of S-schemes, the square $M_S(Q)$ is homotopy cartesian in $\mathscr{T}(S)$.
(iii) For any Nisnevich (resp. proper cdh) distinguished square of shape (5.2.12.1), the canonical map $M_S(Y/B) \xrightarrow{(f/g)_*} M_S(X/A)$ is an isomorphism.

Moreover, under these conditions, to any Nisnevich (resp. proper cdh) distinguished square Q of shape (5.2.12.1), we associate a map

$$\partial_Q : M_S(X) \xrightarrow{\pi_e} M_S(X/A) \xrightarrow{(f/g)_*^{-1}} M_S(Y/B) \xrightarrow{\partial_{e'}} M_S(Y)[1]$$

which defines a distinguished triangle in $\mathscr{T}(S)$:

$$M_S(B) \xrightarrow{\begin{pmatrix} e'_* \\ -g_* \end{pmatrix}} M_Z(Y) \oplus M_S(A) \xrightarrow{(f_*, e_*)} M_S(X) \xrightarrow{\partial_Q} M_S(Y)[1].$$

Proof The equivalence of (i) and (ii) follows from the theorem of Morel–Voevodsky 3.3.2 (resp. the theorem of Voevodsky 3.3.8). To prove the equivalence of (ii) and (iii), we assume $\mathscr{T} = D(\mathscr{A})[\mathscr{W}^{-1}]$. Then, the homotopy colimit of a square of shape 5.2.12.1 is given by the complex

$$\mathrm{Cone}\big(\mathrm{Cone}(M_S(B) \longrightarrow M_S(Y)) \longrightarrow \mathrm{Cone}(M_S(A) \longrightarrow M_S(X))\big).$$

This readily proves the needed equivalence, together with the remaining assertion. □

Remark 5.2.14 In the first of the respective cases of the proposition, condition (ii) is what we usually call the *Brown–Gersten* property (BG) for $\mathscr{T}$, whereas condition

---

[77] If $\mathscr{T} = D(\mathscr{A})[\mathscr{W}^{-1}]$, this amounts to saying that the diagram obtained from complexes by applying the functor $M_S(-)$ is homotopy cartesian in the $\mathscr{W}$-local model category $C(\mathscr{A})$.

(iii) can be called the *excision property*. In the second respective case, condition (ii) will be called the *proper cdh* property for the *generalized premotivic category* $\mathscr{T}$. We say also that $\mathscr{T}$ satisfies the (cdh) property if it satisfies condition (ii) with respect to any *cdh* distinguished square Q.

5.2.b The homotopy relation

5.2.15 Let $\mathscr{A}$ be an abelian $\mathscr{P}$-premotivic category compatible with an admissible topology t.

We consider $\mathscr{W}_{\mathbf{A}^1}$ to be the family of morphisms $M_S(\mathbf{A}^1_X)\{i\} \longrightarrow M_S(X)\{i\}$ for a $\mathscr{P}$-scheme X/S and a twist i in τ. The family $\mathscr{W}_{\mathbf{A}^1}$ is obviously stable under f^* and $f_\sharp$.

Definition 5.2.16 Let $\mathscr{A}$ be an abelian $\mathscr{P}$-premotivic category compatible with an admissible topology t. With the notation above, we define $\mathrm{D}^{\mathit{eff}}_{\mathbf{A}^1}(\mathscr{A}) = \mathrm{D}(\mathscr{A})[\mathscr{W}^{-1}_{\mathbf{A}^1}]$ and refer to it as the (effective) $\mathscr{P}$-premotivic $\mathbf{A}^1$-derived category with coefficients in $\mathscr{A}$.

By definition, the category $\mathrm{D}^{\mathit{eff}}_{\mathbf{A}^1}(\mathscr{A})$ satisfies the homotopy property (Htp) (see Definition 2.1.3). According to the general facts about localization of derived premotivic categories, the triangulated premotivic category $\mathrm{D}^{\mathit{eff}}_{\mathbf{A}^1}(\mathscr{A})$ is τ-generated.

Example 5.2.17 We can divide our examples into two types:
1) Assume $\mathscr{P} = Sm$:

Consider the admissible topology $t = Nis$. Following F. Morel, we define the *(effective)* $\mathbf{A}^1$*-derived category over* S to be $\mathrm{D}^{\mathit{eff}}_{\mathbf{A}^1}(\mathrm{Sh}_{Nis}(Sm/S,\Lambda))$. Indeed, we get a triangulated premotivic category (see also the construction of [Ayo07b]):

$$(5.2.17.1) \qquad \mathrm{D}^{\mathit{eff}}_{\mathbf{A}^1,\Lambda} := \mathrm{D}^{\mathit{eff}}_{\mathbf{A}^1}(\mathrm{Sh}_{Nis}(Sm,\Lambda)).$$

We shall also write its fibers

$$(5.2.17.2) \qquad \mathrm{D}^{\mathit{eff}}_{\mathbf{A}^1}(S,\Lambda) := \mathrm{D}^{\mathit{eff}}_{\mathbf{A}^1,\Lambda}(S) = \mathrm{D}^{\mathit{eff}}_{\mathbf{A}^1}(\mathrm{Sh}_{Nis}(Sm/S,\Lambda))$$

for a scheme S. For $\Lambda = \mathbf{Z}$, we shall often write simply

$$(5.2.17.3) \qquad \mathrm{D}^{\mathit{eff}}_{\mathbf{A}^1} := \mathrm{D}^{\mathit{eff}}_{\mathbf{A}^1}(\mathrm{Sh}_{Nis}(Sm,\mathbf{Z})).$$

Another interesting case is when $t = \acute{e}t$; we get a triangulated premotivic category of *effective étale premotives*:

$$\mathrm{D}^{\mathit{eff}}_{\mathbf{A}^1}(\mathrm{Sh}_{\acute{e}t}(Sm,\Lambda)).$$

In each of these cases, we denote by $\Lambda^t_S(X)$ the premotive associated with a smooth S-scheme X.

5 Fibred derived categories

2) Assume $\mathscr{P} = \mathscr{S}^{ft}$:

Consider the admissible topology $t = h$ (resp. $t = qfh$). In [Voe96], Voevodsky introduced the category of h-motives (resp. qfh-motives). In our formalism, one defines the category of *effective h-motives* (resp. *effective h-motives*) over S with coefficients in Λ as:

$$\underline{DM}_h^{eff}(S,\Lambda) = \mathrm{D}_{\mathbf{A}^1}^{eff}\left(\mathrm{Sh}_h\left(\mathscr{S}^{ft}/S,\Lambda\right)\right)$$

resp. $\underline{DM}_{qfh}^{eff}(S,\Lambda) = \mathrm{D}_{\mathbf{A}^1}^{eff}\left(\mathrm{Sh}_{qfh}\left(\mathscr{S}^{ft}/S,\Lambda\right)\right)$.

In other words, this is the $\mathbf{A}^1$-derived category of h-sheaves (resp. qfh-sheaves) of Λ-modules. Moreover, these categories for various schemes S are the fibers of a generalized premotivic triangulated category. What we have added to the construction of Voevodsky is the functors of the generalized premotivic structure.

We will denote simply by $\underline{\Lambda}_S^t(X)$ the corresponding premotive associated with X in $\underline{DM}_t^{eff}(S,\Lambda)$.

Another interesting case is obtained when $t = cdh$. We get an $\mathbf{A}^1$-derived generalized premotivic category $\mathrm{D}_{\mathbf{A}^1}^{eff}(\mathrm{Sh}_{cdh}\left(\mathscr{S}^{ft},\Lambda\right))$ whose premotives are simply denoted by $\underline{\Lambda}_S^{cdh}(X)$ for any finite type S-scheme X.

5.2.18 Let C be a complex with coefficients in $\mathscr{A}_S$. According to the general case, we say that C is $\mathbf{A}^1$-*local* if for any $\mathscr{P}$-scheme X/S and any $(i,n) \in \tau \times \mathbf{Z}$, the map induced by the canonical projection

$$\mathrm{Hom}_{\mathrm{D}(\mathscr{A}_S)}(M_S(X)\{i\}[n], C) \longrightarrow \mathrm{Hom}_{\mathrm{D}(\mathscr{A}_S)}(M_S(\mathbf{A}_X^1)\{i\}[n], C)$$

is an isomorphism. The adjunction (5.2.2.1) defines a morphism of triangulated $\mathscr{P}$-premotivic categories

$$\mathrm{D}(\mathscr{A}) \rightleftarrows \mathrm{D}_{\mathbf{A}^1}^{eff}(\mathscr{A})$$

such that for any scheme S, $\mathrm{D}_{\mathbf{A}^1}^{eff}(\mathscr{A}_S)$ is identified with the full subcategory of $\mathrm{D}(\mathscr{A}_S)$ comprising $\mathbf{A}^1$-local complexes.

Fibrant objects for the model category structure on $C(\mathscr{A}_S)$ appearing in Proposition 5.2.2 relatively to $\mathscr{W}_{\mathbf{A}^1}$, simply called $\mathbf{A}^1$-*fibrant* objects, are the t-flasque and $\mathbf{A}^1$-local complexes.

We say a morphism $f : C \longrightarrow D$ of complexes of $\mathscr{A}_S$ is an $\mathbf{A}^1$-*equivalence* if it becomes an isomorphism in $\mathrm{D}_{\mathbf{A}^1}^{eff}(\mathscr{A}_S)$. Considering moreover two morphisms $f, g : C \longrightarrow D$ of complexes of $\mathscr{A}_S$, we say they are $\mathbf{A}^1$-*homotopic* if there exists a morphism of complexes

$$H : M_S(\mathbf{A}_S^1) \otimes_S C \longrightarrow D$$

such that $H \circ (s_0 \otimes 1_C) = f$ and $H \circ (s_1 \otimes 1_C) = g$, where s_0 and s_1 are respectively induced by the zero and the unit section of $\mathbf{A}_S^1/S$. When f and g are $\mathbf{A}^1$-homotopic, they are equal as morphisms of $\mathrm{D}_{\mathbf{A}^1}^{eff}(\mathscr{A}_S)$. We say the morphism $p : C \longrightarrow D$ is a *strong* $\mathbf{A}^1$-*equivalence* if there exists a morphism $q : D \longrightarrow C$ such that the morphisms $p \circ q$ and $q \circ p$ are $\mathbf{A}^1$-homotopic to the identity. A complex C is $\mathbf{A}^1$-*contractible* if the map $C \longrightarrow 0$ is a strong $\mathbf{A}^1$-equivalence.

As an example, for any integer $n \in \mathbf{N}$, and any $\mathscr{P}$-scheme X/S, the map

$$p_* : M_S(\mathbf{A}^n_X) \longrightarrow M_S(X)$$

induced by the canonical projection is a strong $\mathbf{A}^1$-equivalence with inverse the zero section $s_{0,*} : M_S(X) \longrightarrow M_S(\mathbf{A}^n_X)$.

5.2.19 The category $\mathrm{D}^{e\!f\!f}_{\mathbf{A}^1}(\mathscr{A})$ is functorial in $\mathscr{A}$.

Let $\varphi : \mathscr{A} \rightleftarrows \mathscr{B} : \psi$ be an adjunction of abelian $\mathscr{P}$-premotivic categories. Consider two topologies t and t' such that t' is finer than t. Suppose $\mathscr{A}$ (resp. $\mathscr{B}$) is compatible with t (resp. t').

For any scheme S, consider the evident extensions $\varphi_S : \mathrm{C}(\mathscr{A}_S) \rightleftarrows \mathrm{C}(\mathscr{B}_S) : \psi_S$ of the above adjoint functors to complexes. We easily check that the functor ψ_S preserves $\mathbf{A}^1$-local complexes. Thus, applying Section 5.1.23, the pair (φ_S, ψ_S) is a Quillen adjunction for the respective $\mathbf{A}^1$-localized model structure on $\mathrm{C}(\mathscr{A}_S)$ and $\mathrm{C}(\mathscr{B}_S)$; see [CD09, 3.11]. Considering the derived functors, it is now easy to check we have obtained an adjunction

$$\mathbf{L}\varphi : \mathrm{D}^{e\!f\!f}_{\mathbf{A}^1}(\mathscr{A}) \rightleftarrows \mathrm{D}^{e\!f\!f}_{\mathbf{A}^1}(\mathscr{B}) : \mathbf{R}\psi$$

of triangulated $\mathscr{P}$-premotivic categories.

Example 5.2.20 Consider the notations of Example 5.2.17. In the case where $\mathscr{P} = Sm$, we get from the adjunction of (5.1.24.2) the following adjunction of triangulated premotivic categories

$$a^*_{\acute{e}t} : \mathrm{D}^{e\!f\!f}_{\mathbf{A}^1, \Lambda} \rightleftarrows \mathrm{D}^{e\!f\!f}_{\mathbf{A}^1}(\mathrm{Sh}_{\acute{e}t}(Sm, \Lambda)) : \mathbf{R}a_{\acute{e}t,*}.$$

Example 5.2.21 Let $\mathscr{T}$ be a derived $\mathscr{P}$-premotivic category as in Section 5.2.9. If $\mathscr{T}$ satisfies the property (Htp), then the canonical morphism (5.2.9.1) induces a morphism

$$\mathrm{D}^{e\!f\!f}_{\mathbf{A}^1}(\mathrm{PSh}(\mathscr{P}, \mathbf{Z})) \rightleftarrows \mathscr{T}.$$

If moreover $\mathscr{T}$ satisfies t-descent for an admissible topology t, we further obtain as in Proposition 5.2.10 a morphism

$$\mathrm{D}^{e\!f\!f}_{\mathbf{A}^1}(\mathrm{Sh}_t(\mathscr{P}, \mathbf{Z})) \rightleftarrows \mathscr{T}.$$

Particularly interesting cases are given by $\mathrm{D}^{e\!f\!f}_{\mathbf{A}^1}$ (resp. $\mathrm{D}^{e\!f\!f}_{\mathbf{A}^1}(\mathrm{Sh}_{cdh}(\mathscr{S}^{ft}, \mathbf{Z}))$), which is the universal derived premotivic category (resp. generalized premotivic category), *i.e.* the initial premotivic category satisfying Nisnevich descent (resp. *cdh* descent) and the homotopy property.

5.2.22 As in Example 5.1.25, let t be an admissible topology and $\varphi : \Lambda \longrightarrow \Lambda'$ be an extension of rings. Then, from the $\mathscr{P}$-premotivic adjunction (5.1.25.1) and according to Section 5.2.19, we get an adjunction of triangulated $\mathscr{P}$-premotivic categories:

5 Fibred derived categories

$$\mathbf{L}\varphi^* : \mathrm{D}^{\mathit{eff}}_{\mathbf{A}^1}\big(\mathrm{Sh}_t(\mathscr{P},\Lambda)\big) \rightleftarrows \mathrm{D}^{\mathit{eff}}_{\mathbf{A}^1}\big(\mathrm{Sh}_t(\mathscr{P},\Lambda')\big) : \mathbf{R}\varphi_*.$$

Consider also complexes C and D of t-sheaves of Λ-modules over $\mathscr{P}_S$. Then there exists a canonical morphism of Λ'-modules:
(5.2.22.1)
$$\mathrm{Hom}_{\mathrm{D}^{\mathit{eff}}_{\mathbf{A}^1}(\mathrm{Sh}_t(\mathscr{P}_S,\Lambda))}(C,D) \otimes_\Lambda \Lambda' \longrightarrow \mathrm{Hom}_{\mathrm{D}^{\mathit{eff}}_{\mathbf{A}^1}(\mathrm{Sh}_t(\mathscr{P}_S,\Lambda'))}\big(\mathbf{L}\varphi^*(C),\mathbf{L}\varphi^*(D)\big).$$

There are two notable cases where this map is an isomorphism:

Proposition 5.2.23 *Consider the above assumptions. Then the map (5.2.22.1) is an isomorphism in the two following cases:*

1. *If Λ' is a free Λ-module and C is compact;*
2. *If Λ' is a free Λ-module of finite rank.*

Proof Note that in any case, the functor φ_* admits a right adjoint $\varphi^!$.[78]
We can assume that $\Lambda' = I.\Lambda$ for a set I. In this case, we get for any sheaf F of Λ-modules:
$$\varphi_*\varphi^*(F) = F \otimes_\Lambda \Lambda' = I.F.$$
Moreover, for any $\mathscr{P}$-scheme X/S, we get:
$$\varphi_*(\Lambda''_S(X)) = \Lambda''_S(X) = I.\Lambda'_S(X).$$
In particular, the functor $\varphi_* : \mathrm{C}(\mathrm{Sh}_t(\mathscr{P}_S,\Lambda')) \longrightarrow \mathrm{C}(\mathrm{Sh}_t(\mathscr{P}_S,\Lambda))$ satisfies descent in the sense of [CD09, 2.4] and preserves the family $\mathscr{W}_{\mathbf{A}^1}$. Thus it is a left Quillen functor with respect to the $\mathbf{A}^1$-local model structures. In particular, because it is also a right Quillen functor, we get: $\mathbf{R}\varphi_* = \varphi_* = \mathbf{L}\varphi_*$. In particular, we get in $\mathrm{D}^{\mathit{eff}}_{\mathbf{A}^1}(\mathrm{Sh}_t(\mathscr{P}_S,\Lambda))$:
$$\mathbf{R}\varphi_*\mathbf{L}\varphi^*(D) = \mathbf{L}\varphi_*\mathbf{L}\varphi^*(D) = \mathbf{L}(\varphi_*\varphi^*)(D) = I.D.$$
Thus the proposition follows as the functor $\mathrm{Hom}(C,-)$ commutes with direct sums if C is compact and with finite direct sums in any case. □

We note the following useful property.

Proposition 5.2.24 *Consider a morphism*
$$\varphi^* : \mathscr{A} \rightleftarrows \mathscr{B} : \varphi_*$$
of abelian $\mathscr{P}$-premotivic categories such that $\mathscr{A}$ (resp. $\mathscr{B}$) is compatible with an admissible topology t (resp. t'). Assume t' is finer than t.

Let S be a base scheme. Assume that $\varphi_* : \mathscr{A}_S \longrightarrow \mathscr{B}_S$ commutes with colimits[79]. *Then $\varphi_* : \mathrm{C}(\mathscr{A}_S) \longrightarrow \mathrm{C}(\mathscr{B}_S)$ respects $\mathbf{A}^1$-equivalences.*

---

[78] It is defined by the formula:
$$\varphi^!(F) = Hom_\Lambda(\Lambda', F)$$
equipped with its canonical structure of sheaf of Λ'-modules.

[79] This amounts to asking that φ_* is exact and commutes with direct sums.

In other words, the right derived functor $\mathbf{R}\varphi_* : D^{\mathit{eff}}_{\mathbf{A}^1}(\mathscr{B}_S) \longrightarrow D^{\mathit{eff}}_{\mathbf{A}^1}(\mathscr{A}_S)$ satisfies the relation $\mathbf{R}\varphi_* = \varphi_*$.

Proof In this proof, we write φ_* for $\varphi_{*,S}$. We first prove that φ_* preserves strong $\mathbf{A}^1$-equivalences (see Section 5.2.18).

Consider two maps $u, v : K \longrightarrow L$ in $C(\mathscr{B}_S)$. To give an $\mathbf{A}^1$-homotopy $H : M_S(\mathbf{A}^1_S, \mathscr{B}) \otimes_S K \longrightarrow L$ between u and v is equivalent by adjunction to giving a map $H' : K \longrightarrow Hom_{\mathscr{B}_S}(M_S(\mathbf{A}^1_S, \mathscr{B}), L)$ which fits into the following commutative diagram:

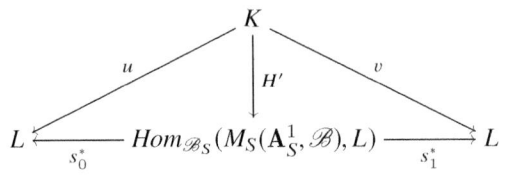

$$\begin{array}{c} K \\ u \swarrow \quad \downarrow H' \quad \searrow v \\ L \xleftarrow{s_0^*} Hom_{\mathscr{B}_S}(M_S(\mathbf{A}^1_S, \mathscr{B}), L) \xrightarrow{s_1^*} L \end{array}$$

where s_0 and s_1 are the respective zero and unit section of $\mathbf{A}^1_S/S$.

Because $M_S(\mathbf{A}^1_S, \mathscr{B}) = \varphi_S^*(M_S(\mathbf{A}^1_S, \mathscr{A}))$, we get a canonical isomorphism (see Section 1.2.9)

$$\varphi_*(Hom_{\mathscr{B}_S}(M_S(\mathbf{A}^1_S, \mathscr{B}), L)) \simeq Hom_{\mathscr{B}_S}(M_S(\mathbf{A}^1_S, \mathscr{A}), \varphi_*(L)).$$

Thus, applying φ_* to the previous commutative diagram and using this identification, we obtain that $\varphi_*(u)$ is $\mathbf{A}^1$-homotopic to $\varphi_*(v)$.

As a consequence, for any $\mathscr{P}$-scheme X over S, and any $\mathscr{B}$-twist i, the map

$$\varphi_*(M_S(\mathbf{A}^1_X, \mathscr{B})\{i\}) \longrightarrow \varphi_*(M_S(X, \mathscr{B})\{i\})$$

induced by the canonical projection is a strong $\mathbf{A}^1$-equivalence, thus an $\mathbf{A}^1$-equivalence.

The functor $\varphi_* : \mathscr{B}_S \longrightarrow \mathscr{A}_S$ commutes with colimits. Thus it admits a right adjoint that we will denote by $\varphi^!$. Consider the injective model structure on $C(A_S)$ and $C(\mathscr{B}_S)$ (see [CD09, 2.1]). Because φ_* is exact, it is a left Quillen functor for these model structures. Thus, the right derived functor $\mathbf{R}\varphi^!$ is well defined. From the above result, we see that $\mathbf{R}\varphi^!$ preserves $\mathbf{A}^1$-local objects, and this readily implies that $\mathbf{L}\varphi_* = \varphi_*$ preserves $\mathbf{A}^1$-equivalences. □

5.2.25 To relate the category $D^{\mathit{eff}}_{\mathbf{A}^1}(S)$ with the homotopy category of schemes of Morel and Voevodsky [MV99], we have to consider the category of simplicial Nisnevich sheaves of sets denoted by $\Delta^{op}\,\mathrm{Sh}(Sm/S)$. Considering the free abelian sheaf functor, we obtain an adjunction of categories

$$\Delta^{op}\,\mathrm{Sh}(Sm/S) \rightleftarrows C(\mathrm{Sh}(Sm/S, \mathbf{Z})).$$

If we consider Blander's projective $\mathbf{A}^1$-model structure [Bla03] on the category $\Delta^{op}\,\mathrm{Sh}(Sm/S)$, we can easily see that this is a Quillen pair, so that we obtain a $\mathscr{P}$-premotivic adjunction of simple $\mathscr{P}$-premotivic categories

5 Fibred derived categories

$$N : \mathcal{H} \rightleftarrows D_{\mathbf{A}^1}^{\mathit{eff}} : K.$$

Note that the functor N sends cofiber sequences in $\mathcal{H}(S)$ to distinguished triangles in $D_{\mathbf{A}^1}^{\mathit{eff}}(S)$.

5.2.c Explicit $\mathbf{A}^1$-resolution

5.2.26 Consider an abelian $\mathcal{P}$-premotivic category $\mathcal{A}$ compatible with an admissible topology t.

Consider the canonically split exact sequence

$$0 \longrightarrow \mathbb{1}_S \xrightarrow{s_0} M_S(\mathbf{A}_S^1) \longrightarrow U \longrightarrow 0$$

where the map $s_0 : \mathbb{1}_S \longrightarrow M_S(\mathbf{A}_S^1)$ is induced by the zero section of $\mathbf{A}^1$. The section corresponding to 1 in $\mathbf{A}^1$ defines another map

$$s_1 : \mathbb{1}_S \longrightarrow M_S(\mathbf{A}_S^1)$$

which does not factor through s_0, so that we get canonically a non-trivial map $u : \mathbb{1}_S \longrightarrow U$. This defines for any complex C of $\mathcal{A}_S$ a map, called the *evaluation at* 1,

$$\mathit{Hom}(U,C) = \mathbb{1}_S \otimes_S \mathit{Hom}(U,C) \xrightarrow{u \otimes 1} U \otimes_S \mathit{Hom}(U,C) \xrightarrow{ev} C.$$

We define the complex $R_{\mathbf{A}^1}^{(1)}(C)$ to be

$$R_{\mathbf{A}^1}^{(1)}(C) = \mathrm{Cone}\bigl(\mathit{Hom}(U,C) \longrightarrow C\bigr).$$

We have by construction a map

$$r_C : C \longrightarrow R_{\mathbf{A}^1}^{(1)}(C).$$

This defines a morphism of functors from the identity functor to $R_{\mathbf{A}^1}^{(1)}$. For an integer $n \geq 1$, we define by induction a complex

$$R_{\mathbf{A}^1}^{(n+1)}(C) = R_{\mathbf{A}^1}^{(1)}(R_{\mathbf{A}^1}^{(n)}(C)),$$

and a map

$$r_{R_{\mathbf{A}^1}^{(n)}(C)} : R_{\mathbf{A}^1}^{(n)}(C) \longrightarrow R_{\mathbf{A}^1}^{(n+1)}.$$

We finally define a complex $R_{\mathbf{A}^1}(C)$ by the formula

$$R_{\mathbf{A}^1}(C) = \varinjlim_n R_{\mathbf{A}^1}^{(n)}(C).$$

We have a functorial map
$$C \longrightarrow R_{\mathbf{A}^1}(C).$$

Lemma 5.2.27 *With the above hypotheses and notations, the map $C \longrightarrow R_{\mathbf{A}^1}(C)$ is an $\mathbf{A}^1$-equivalence.*

Proof For any closed symmetric monoidal category $\mathscr{C}$ and any objects A, B, C and I in $\mathscr{C}$, we have
$$\mathrm{Hom}(I \otimes \mathit{Hom}(B,C), \mathit{Hom}(A,C)) = \mathrm{Hom}(\mathit{Hom}(B,C), \mathit{Hom}(I, \mathit{Hom}(A,C)))$$
$$= \mathrm{Hom}(\mathit{Hom}(B,C), \mathit{Hom}(I \otimes A, C)).$$

Hence any map $I \otimes A \longrightarrow B$ induces a map $I \otimes \mathit{Hom}(B,C) \longrightarrow \mathit{Hom}(A,C)$ for any object C. If we apply this to $\mathscr{C} = C(\mathscr{A}_S)$ and $I = M_S(\mathbf{A}^1)$, we see immediately that the functor $\mathit{Hom}(-,C)$ preserves strong $\mathbf{A}^1$-homotopy equivalences. In particular, for any complex C, the map $C \longrightarrow \mathit{Hom}(M_S(\mathbf{A}^1_X), C)$ is a strong $\mathbf{A}^1$-homotopy equivalence. This implies that $\mathit{Hom}(U,C) \longrightarrow 0$ is an $\mathbf{A}^1$-equivalence, so that the map r_C is an $\mathbf{A}^1$-equivalence as well. As $\mathbf{A}^1$-equivalences are stable under filtering colimits, this implies our result. □

Proposition 5.2.28 *Consider the above notations and hypotheses, and assume that t is bounded in $\mathscr{A}$.*

For any t-flasque complex C of $\mathscr{A}_S$, the complex $R_{\mathbf{A}^1}(C)$ is t-flasque and $\mathbf{A}^1$-local. Moreover, the morphism $C \longrightarrow R_{\mathbf{A}^1}(C)$ is an $\mathbf{A}^1$-equivalence. If furthermore C is t-flasque, so is $R_{\mathbf{A}^1}(C)$.

Proof The last assertion is a particular case of Lemma 5.2.27. The functor $R^{(1)}_{\mathbf{A}^1}$ preserves t-flasque complexes. By virtue of Proposition 5.1.30, the functor $R_{\mathbf{A}^1}$ has the same gentle property. It thus remains to prove that the functor $R_{\mathbf{A}^1}$ sends t-flasque complexes to $\mathbf{A}^1$-local ones. We shall use that the derived category $\mathrm{D}(\mathscr{A}_S)$ is compactly generated; see Proposition 5.1.30.

Let C be a t-flasque complex of $\mathscr{A}_S$. To prove $R_{\mathbf{A}^1}(C)$ is $\mathbf{A}^1$-local, we are reduced to proving that the map
$$R_{\mathbf{A}^1}(C) \longrightarrow \mathit{Hom}(M_S(\mathbf{A}^1_X), R_{\mathbf{A}^1}(C))$$
is a quasi-isomorphism, or, equivalently, that the complex $\mathit{Hom}(U, R_{\mathbf{A}^1}(C))$ is acyclic. As U is a direct factor of $M_S(\mathbf{A}^1_X, \mathscr{A})$, for any $\mathscr{P}$-scheme X over S and any i in I, the object $\mathbf{Z}_S(X;\mathscr{A})\{i\} \otimes_S U$ is compact. This implies that the canonical map
$$\varinjlim_n \mathit{Hom}(U, R^{(n)}_{\mathbf{A}^1}(C)) \longrightarrow \mathit{Hom}(U, R_{\mathbf{A}^1}(C))$$
is an isomorphism of complexes. As filtering colimits preserve quasi-isomorphisms, the complex $\mathit{Hom}(U, R_{\mathbf{A}^1}(C))$ (resp. $R_{\mathbf{A}^1}(C)$) can be considered as the homotopy colimit of the complexes $\mathit{Hom}(U, R^{(n)}_{\mathbf{A}^1}(C))$ (resp. $R^{(n)}_{\mathbf{A}^1}(C)$). In particular, for any compact object K of $\mathrm{D}(\mathscr{A}_S)$, the canonical morphisms

5 Fibred derived categories

$$\varinjlim_{n} \mathrm{Hom}(K, Hom(U, R^{(n)}_{\mathbf{A}^1}(C))) \longrightarrow \mathrm{Hom}(K, Hom(U, R_{\mathbf{A}^1}(C)))$$

$$\varinjlim_{n} \mathrm{Hom}(K, R^{(n)}_{\mathbf{A}^1}(C)) \longrightarrow \mathrm{Hom}(K, R_{\mathbf{A}^1}(C))$$

are bijective.

By construction, we have distinguished triangles

$$Hom(U, R^{(n)}_{\mathbf{A}^1}(C)) \longrightarrow R^{(n)}_{\mathbf{A}^1}(C) \longrightarrow R^{(n+1)}_{\mathbf{A}^1}(C) \longrightarrow Hom(U, R^{(n)}_{\mathbf{A}^1}(C))[1].$$

This implies that the evaluation at 1 morphism

$$ev_1 : Hom(U, R_{\mathbf{A}^1}(C)) \longrightarrow R_{\mathbf{A}^1}(C)$$

induces the zero map

$$\mathrm{Hom}_{D(\mathscr{A}_S)}(K, Hom(U, R_{\mathbf{A}^1}(C))) \longrightarrow \mathrm{Hom}_{D(\mathscr{A}_S)}(K, R_{\mathbf{A}^1}(C))$$

for any compact object K of $D(\mathscr{A}_S)$. Hence the induced map

$$a = Hom(U, ev_1) : Hom(U, Hom(U, R_{\mathbf{A}^1}(C))) \longrightarrow Hom(U, R_{\mathbf{A}^1}(C))$$

has the same property: for any compact object K, the map

$$\mathrm{Hom}_{D(\mathscr{A}_S)}(K, Hom(U, Hom(U, R_{\mathbf{A}^1}(C)))) \longrightarrow \mathrm{Hom}_{D(\mathscr{A}_S)}(K, Hom(U, R_{\mathbf{A}^1}(C)))$$

is zero.

The multiplication map $\mathbf{A}^1 \times \mathbf{A}^1 \longrightarrow \mathbf{A}^1$ induces a map

$$\mu : U \otimes_S U \longrightarrow U$$

such that the composition of

$$\mu^* : Hom(U, R_{\mathbf{A}^1}(C)) \longrightarrow Hom(U \otimes_S U, R_{\mathbf{A}^1}(C)) = Hom(U, Hom(U, R_{\mathbf{A}^1}(C)))$$

with a is the identity of $Hom(U, R_{\mathbf{A}^1}(C))$. As $D(\mathscr{A}_S)$ is compactly generated, this implies that $Hom(U, R_{\mathbf{A}^1}(C)) = 0$ in the derived category $D(\mathscr{A}_S)$. □

Remark 5.2.29 Consider a t-flasque resolution functor (*i.e.* a fibrant resolution for the t-local model structure) $R_t : C(\mathscr{A}_S) \longrightarrow C(\mathscr{A}_S)$, $1 \longrightarrow R_t$. As a corollary of the proposition, the composite functor $R_{\mathbf{A}^1} \circ R_t$ is a resolution functor by t-local and $\mathbf{A}^1$-local complexes.

Example 5.2.30 Consider an admissible topology t and the $\mathscr{P}$-premotivic $\mathbf{A}^1$-derived category $D = D^{\mathit{eff}}_{\mathbf{A}^1}(\mathrm{Sh}_t(\mathscr{P}, \Lambda))$. Suppose that t is bounded for abelian t-sheaves (for example, this is the case for the Zariski and the Nisnevich topologies, see Example 5.1.29).

Let C be a complex of abelian t-sheaves on $\mathscr{P}/S$. If C is $\mathbf{A}^1$-local, then

$$\mathrm{Hom}_{D(S)}(\Lambda_S^t(X), C) = H_t^n(X; C)$$

(this is true without any condition on t).

Consider a t-local resolution C_t of C in $C\big(\mathrm{Sh}_t(\mathscr{P}/S, \Lambda)\big)$. Then we get the following formula:

$$\mathrm{Hom}_{D(S)}\big(\Lambda_S^t(X), C[n]\big) = H^n\big(\Gamma\big(X, R_{\mathbf{A}^1}(C_t)\big)\big).$$

Corollary 5.2.31 *Consider a morphism of abelian $\mathscr{P}$-premotivic categories*

$$\varphi : \mathscr{A} \rightleftarrows \mathscr{B} : \psi.$$

Suppose there are admissible topologies t and t', with t' finer than t, such that the following conditions are verified.

(i) $\mathscr{A}$ is compatible with t and $\mathscr{B}$ is compatible with t'.
(ii) $\mathscr{B}$ and $D(\mathscr{B})$ are compactly τ-generated.
(iii) For any scheme S, the functor $\psi_S : \mathscr{B}_S \longrightarrow \mathscr{A}_S$ preserves filtering colimits.

Then, $\psi_S : C(\mathscr{B}_S) \longrightarrow C(\mathscr{A}_S)$ preserves $\mathbf{A}^1$-equivalences between t'-flasque objects. If moreover ψ_S is exact, the functor ψ_S preserves $\mathbf{A}^1$-equivalences.

Proof We already know that ψ_S is a right Quillen functor, so that it preserves local objects and $\mathbf{A}^1$-fibrant objects. This also implies that ψ_S preserves $\mathbf{A}^1$-equivalences between $\mathbf{A}^1$-fibrant objects (this is Ken Brown's lemma [Hov99, 1.1.12]). Let D be a t'-flasque complex of $\mathscr{B}_S$. Then $\psi_S(D)$ is a t-flasque complex of $\mathscr{A}_S$. It follows from Proposition 5.2.28 that $R_{\mathbf{A}^1}(D)$ is $\mathbf{A}^1$-local and that $D \longrightarrow R_{\mathbf{A}^1}(D)$ is an $\mathbf{A}^1$-equivalence. Lemma 5.2.27 implies the map

$$\psi_S(D) \longrightarrow R_{\mathbf{A}^1}(\psi_S(D)) = \psi_S(R_{\mathbf{A}^1}(D))$$

is an $\mathbf{A}^1$-equivalence. This implies the first assertion.

The last assertion is a direct consequence of the first one. □

5.2.32 Consider the usual cosimplicial scheme $\Delta^\bullet$ defined by

$$\Delta^n = \mathrm{Spec}\,(\mathbf{Z}[t_0, \ldots, t_n]/(t_1 + \cdots + t_n - 1)) \simeq \mathbf{A}^n$$

(see [MV99]). For any scheme S, we get a cosimplicial object of $\mathscr{A}_S$, namely $M_S(\Delta_S^\bullet)$. Given any complex C of $\mathscr{A}_S$, we define its associated *Suslin singular complex* as

(5.2.32.1) $$\underline{C}^*(C) = \mathrm{Tot}^\oplus Hom(M_S(\Delta_S^\bullet), C),$$

where $Hom(M_S(\Delta_S^\bullet), C)$ is considered as a bicomplex by the Dold–Kan correspondence. The canonical map $M_S(\Delta_S^\bullet) \longrightarrow \mathbb{1}_S$ induces a map

$$C \longrightarrow \underline{C}^*(C).$$

5 Fibred derived categories

Lemma 5.2.33 *For any complex C of $\mathscr{A}_S$, the map*

$$\underline{C}^*(C) \longrightarrow Hom(M_S(\mathbf{A}_S^1), \underline{C}^*(C)) = \underline{C}^*(Hom(M_S(\mathbf{A}_S^1), C))$$

is a chain homotopy equivalence.

Proof The composite morphism

$$(s_0 p \times \mathrm{Id})_* : M_S(\mathbf{A}^1 \times \Delta_S^\bullet) \longrightarrow M_S(\mathbf{A}^1 \times \Delta_S^\bullet),$$

where s_0 is the map induced by the zero section, and p is the map induced by the obvious projection of $\mathbf{A}^1$ on its base, is chain homotopic to the identity. Indeed, the homotopy relation is given by the formula

$$s_n = \sum_{i=0}^{n} (-1)^i . (1 \otimes_S \psi_i) : M_S(\mathbf{A}^1 \times \Delta_S^{n+1}) \longrightarrow M_S(\mathbf{A}^1 \times \Delta_S^n)$$

where 1 is the identity of $M_S(\mathbf{A}_S^1)$, and ψ_i is induced by the map $\Delta_S^{n+1} \longrightarrow \mathbf{A}^1 \times \Delta_S^n$ which sends the j-th vertex $v_{j,n+1}$ to either $0 \times v_{j,n}$, if $j \leq i$, or to $1 \times v_{j-1,n}$ otherwise. This implies the lemma. □

Lemma 5.2.34 *For any t-flasque complex C of $\mathscr{A}_S$, we have a canonical isomorphism*

$$\underline{C}^*(C) \simeq \mathbf{L}\varinjlim_n \mathbf{R}Hom(M_S(\Delta_S^n), C)$$

in $\mathrm{D}(\mathscr{A}_S)$.

This is a variation on the Dold–Kan correspondence. As a direct consequence, we get:

Lemma 5.2.35 *For any complex C of $\mathscr{A}_S$, the map $C \longrightarrow \underline{C}^*(C)$ is an $\mathbf{A}^1$-equivalence.*

Proposition 5.2.36 *If t is bounded in $\mathscr{A}$, then, for any t-flasque complex C of $\mathscr{A}_S$, $\underline{C}^*(C)$ is $\mathbf{A}^1$-local.*

Proof Using the first premotivic adjunction of Example 5.2.21 and the fact that $\mathrm{D}(\mathscr{A})$ is compactly generated (Proposition 5.1.30), we can reduce the proposition to the case where $\mathscr{A}_S$ is the category of presheaves of abelian groups over $\mathscr{P}/S$, in which case this is well known. □

5.2.d Constructible $\mathbf{A}^1$-local premotives

5.2.37 Consider an abelian $\mathscr{P}$-premotivic category $\mathscr{A}$ compatible with an admissible topology t. Assume that t is bounded in $\mathscr{A}$ (see Definition 5.1.28) and consider a bounded generating family $\mathscr{N}_S^t$ for t-hypercovers in $\mathscr{A}_S$.

Let $\mathscr{T}_{\mathbf{A}_S^1}$ be the family of complexes of $C(\mathscr{A}_S)$ of shape

$$M_S(\mathbf{A}_X^1)\{i\} \longrightarrow M_S(X)\{i\}$$

for a $\mathscr{P}$-scheme X over S and a twist $i \in I$. Then the functor (5.1.31.1) obviously induces the following functor

(5.2.37.1) $\qquad \left(K^b\left(M(\mathscr{P}/S,\mathscr{A})\right)/\mathscr{N}_S^t \cup \mathscr{T}_{\mathbf{A}_S^1}\right)^{\natural} \longrightarrow \mathrm{D}_{\mathbf{A}^1}^{\mathit{eff}}(\mathscr{A}_S),$

where the category on the left is the pseudo-abelian category associated to the Verdier quotient of $K^b\left(M(\mathscr{P}/S,\mathscr{A})\right)$ by the thick subcategory generated by $\mathscr{N}_S^t \cup \mathscr{T}_{\mathbf{A}_S^1}$. Applying Thomason's localization theorem [Nee01], we get from Proposition 5.1.32 the following result:

Proposition 5.2.38 *Consider the previous hypotheses and notations and assume that $\mathscr{A}$ is finitely τ-presented.*

Then $\mathrm{D}_{\mathbf{A}^1}^{\mathit{eff}}(\mathscr{A})$ is compactly τ-generated. Moreover, the functor (5.2.37.1) is fully faithful.

Let us denote by $\mathrm{D}_{\mathbf{A}^1,c}^{\mathit{eff}}(\mathscr{A})$ the subcategory of $\mathrm{D}_{\mathbf{A}^1}^{\mathit{eff}}(\mathscr{A})$ comprising τ-constructible premotives in the sense of Definition 1.4.9. Taking into account Proposition 1.4.11, we deduce from the above proposition the following corollary:

Corollary 5.2.39 *Under the assumptions of Proposition 5.2.38, for any premotive $\mathscr{M}$ in $\mathrm{D}_{\mathbf{A}^1}^{\mathit{eff}}(\mathscr{A}_S)$, the following conditions are equivalent:*

(i) $\mathscr{M}$ is compact;
(ii) $\mathscr{M}$ is τ-constructible.

Moreover, the functor (5.2.37.1) induces an equivalence of categories:

$$\left(K^b\left(M(\mathscr{P}/S,\mathscr{A})\right)/\mathscr{N}_S^t \cup \mathscr{T}_{\mathbf{A}_S^1}\right)^{\natural} \longrightarrow \mathrm{D}_{\mathbf{A}^1,c}^{\mathit{eff}}(\mathscr{A}_S).$$

Example 5.2.40 With the notations of Example 5.1.34, we get the following equivalences of categories:

$$\left(\mathrm{K}^b\left(\Lambda(Sm/S)\right)/(BG_S \cup \mathscr{T}_{\mathbf{A}_S^1})\right)^{\natural} \longrightarrow \mathrm{D}_{\mathbf{A}^1,c}^{\mathit{eff}}(S,\Lambda),$$

$$\left(\mathrm{K}^b\left(\Lambda(\mathscr{S}^{ft}/S)\right)/CDH_S \cup \mathscr{T}_{\mathbf{A}_S^1}\right)^{\natural} \longrightarrow \mathrm{D}_{\mathbf{A}^1,c}^{\mathit{eff}}\left(\mathrm{Sh}_{cdh}(\mathscr{S}^{ft}/S,\Lambda)\right).$$

This statement is the analog of the embedding theorem [VSF00, chap. 5, 3.2.6].

Proposition 5.2.41 *Assume $\mathscr{P} = \mathscr{S}^{ft}$ is the class of finite type (resp. separated and of finite type) morphisms.*

Let $\mathscr{A}$ be an abelian generalized premotivic category compatible with an admissible topology t and satisfying the property (C) (resp. (wC)) of Section 5.1.35.

Then the triangulated generalized premotivic category $D^{eff}_{\mathbf{A}^1}(\mathscr{A})$ is τ-continuous (resp. weakly τ-continuous) — see Definition 4.3.2.

Proof The proof relies on the following lemma:

Lemma 5.2.42 *Under the assumptions of the preceding proposition, for any morphism of schemes $f : T \longrightarrow S$, the functor*

$$\mathbf{L}f^* : D(\mathscr{A}_S) \longrightarrow D(\mathscr{A}_T)$$

preserves $\mathbf{A}^1$-local complexes. □

When f is a morphism of finite type (resp. separated of finite type), the functor $\mathbf{L}f^*$ admits $\mathbf{L}f_\sharp$ as a left adjoint and the lemma is clear. In the general case, one can write f as a projective limit of a projective system of morphisms of scheme $(f_\alpha : T_\alpha \longrightarrow S)_{\alpha \in A}$ such that f_α is affine of finite type. Recall from Proposition 5.1.36 that $D(\mathscr{A})$ is τ-continuous. Thus, to check that for an $\mathbf{A}^1$-local complex C in $D(\mathscr{A}_S)$, the complex $\mathbf{L}f^*(C)$ is $\mathbf{A}^1$-local, we thus are reduced to proving that $\mathbf{L}f^*_\alpha(C)$ is $\mathbf{A}^1$-local, which follows from the first treated case. The lemma is proven.

Given the full embedding $D^{eff}_{\mathbf{A}^1}(\mathscr{A}) \longrightarrow D(\mathscr{A})$ whose image is comprised of $\mathbf{A}^1$-local complexes, the proposition now directly follows from the previous lemma and the fact that $D(\mathscr{A})$ is τ-continuous. □

Example 5.2.43 Taking into account the second point of Example 5.1.37, the previous proposition can be applied to the category $\mathrm{Sh}_t(\mathscr{S}^{ft}, \mathbf{Z})$, where $t = Nis, \acute{e}t, cdh, qfh, h$.

Remark 5.2.44 The previous proposition will be extended to the (non-generalized) premotivic case in Corollary 6.1.12.

5.3 The stable $\mathbf{A}^1$-derived premotivic category

5.3.a Modules

Let $\mathscr{A}$ be an abelian $\mathscr{P}$-premotivic category with generating set of twists τ.

A *cartesian commutative monoid* R of $\mathscr{A}$ is a cartesian section of the fibred category $\mathscr{A}$ over $\mathscr{S}$ such that for any scheme S, R_S has a commutative monoid structure in $\mathscr{A}_S$ and for any morphism of schemes $f : T \longrightarrow S$, the structural transition maps $\phi_f : f^*(R_S) \longrightarrow R_T$ are isomorphisms of monoids.

Let us fix a cartesian commutative monoid R of $\mathscr{A}$.

Consider a base scheme S. We denote by R_S-mod the category of modules in the monoidal category $\mathscr{A}_S$ over the monoid R_S. For any $\mathscr{P}$-scheme X/S and any twist $i \in \tau$, we put

$$R_S(X)\{i\} = R_S \otimes_S M_S(X)\{i\}$$

endowed with its canonical R_S-module structure. The category R_S-mod is a Grothendieck abelian category such that the forgetful functor $U_S : R_S\text{-mod} \longrightarrow \mathscr{A}_S$

is exact and conservative. A family of generators for R_S-mod is given by the modules $R_S(X)\{i\}$ for a $\mathscr{P}$-scheme X/S and a twist $i \in \tau$. As A_S is commutative, R_S-mod has a unique symmetric monoidal structure such that the free R_S-module functor is symmetric monoidal. We denote this tensor product by $\otimes_R$. Note that $R_S(X) \otimes_R R_S(Y) = R_S(X \times_S Y)$. Finally, the categories of modules R_S-mod form a symmetric monoidal $\mathscr{P}$-fibred category, such that the following proposition holds (see Section 7.2.10).

Proposition 5.3.1 *Let $\mathscr{A}$ be a τ-generated abelian $\mathscr{P}$-premotivic category and R be a cartesian commutative monoid of $\mathscr{A}$.*

Then the category R-mod equipped with the structures introduced above is a τ-generated abelian $\mathscr{P}$-premotivic category.

Moreover, we have an adjunction of abelian $\mathscr{P}$-premotivic categories:

(5.3.1.1) $$R \otimes (-) : \mathscr{A} \rightleftarrows R\text{-mod} : U.$$

Remark 5.3.2 With the hypotheses of the preceding proposition, for any morphism of schemes $f : T \longrightarrow S$, the exchange transformation $f^* U_S \longrightarrow U_T f^*$ is an isomorphism by construction of R-mod (Section 7.2.10).

Proposition 5.3.3 *Let $\mathscr{A}$ be a τ-generated abelian $\mathscr{P}$-premotivic category compatible with an admissible topology t. Consider a cartesian commutative monoid R of $\mathscr{A}$ such that for any scheme S, tensoring quasi-isomorphisms between cofibrant complexes by R_S gives quasi-isomorphisms (e.g. R_S might be cofibrant (as a complex concentrated in degree zero), or flat). Then the abelian $\mathscr{P}$-premotivic category R-mod is compatible with t.*

Proof In view of Proposition 5.1.26, we have only to show that R-mod satisfies cohomological t-descent. Consider a t-hypercover $p : \mathscr{X} \longrightarrow X$ in $\mathscr{P}/S$. We prove that the map $p_* : R_S(\mathscr{X}) \longrightarrow R_S(X)$ is a quasi-isomorphism in $C(R_S\text{-mod})$. The functor U_S is conservative, and $U_S(p_*)$ is equal to the map:

$$R_S \otimes_S M_S(\mathscr{X}) \longrightarrow R_S \otimes_S M_S(X).$$

But this is a quasi-isomorphism in $C(\mathscr{A}_S)$ by assumption on R_S. □

Remark 5.3.4 According to Lemma 5.1.27, for any simplicial $\mathscr{P}$-scheme $\mathscr{X}$ over S, any twist $i \in \tau$ and any R_S-module C, we get canonical isomorphisms:

(5.3.4.1) $\operatorname{Hom}_{K(R_S\text{-mod})}\left(R_S(\mathscr{X})\{i\}, C\right) \simeq \operatorname{Hom}_{K(\mathscr{A}_S)}(M_S(\mathscr{X})\{i\}, C),$

(5.3.4.2) $\operatorname{Hom}_{D(R_S\text{-mod})}(R_S(\mathscr{X})\{i\}, C) \simeq \operatorname{Hom}_{D(\mathscr{A}_S)}(M_S(\mathscr{X})\{i\}, C).$

5.3.b Symmetric sequences

Let $\mathscr{A}$ be an abelian category.

Let G be a group. An action of G on an object $A \in \mathscr{A}_S$ is a morphism of groups $G \longrightarrow \mathrm{Aut}_{\mathscr{A}}(A), g \mapsto \gamma_g^A$. We say that A is a G-object of $\mathscr{A}$. A G-equivariant morphism $A \xrightarrow{f} B$ of G-objects of $\mathscr{A}$ is a morphism f in $\mathscr{A}$ such that $\gamma_g^B \circ f = f \circ \gamma_g^A$.

If E is any object of $\mathscr{A}$, we put $G \times E = \bigoplus_{g \in G} E$, considered as a G-object via the permutation isomorphisms of the summands.

If H is a subgroup of G, and E is an H-object, $G \times E$ has two actions of H: the first one, say γ, is obtained via the inclusion $H \subset G$, and the second one, denoted by γ', is obtained using the structural action of H on E. We define $G \times_H E$ as the coequalizer of the family of morphisms $(\gamma_\sigma - \gamma'_\sigma)_{\sigma \in H}$, and consider it equipped with its induced action of G.

Definition 5.3.5 Let $\mathscr{A}$ be an abelian category.

A *symmetric sequence* of $\mathscr{A}$ is a sequence $(A_n)_{n \in \mathbf{N}}$ such that for each $n \in \mathbf{N}$, A_n is a $\mathfrak{S}_n$-object of $\mathscr{A}$. A morphism of symmetric sequences of $\mathscr{A}$ is a collection of $\mathfrak{S}_n$-equivariant morphisms $(f_n : A_n \longrightarrow B_n)_{n \in \mathbf{N}}$.

We let $\mathscr{A}^{\mathfrak{S}}$ be the category of symmetric sequences of $\mathscr{A}$.

It is straightforward to check $\mathscr{A}^{\mathfrak{S}}$ is abelian. For any integer $n \in \mathbf{N}$, we define the *n-th evaluation functor* as follows:

$$ev_n : \mathscr{A}^{\mathfrak{S}} \longrightarrow \mathscr{A}, A_* \mapsto A_n.$$

Any object A of $\mathscr{A}$ can be considered as the trivial symmetric sequence $(A, 0, \ldots)$. The functor $i_0 : A \mapsto (A, 0, \ldots)$ is obviously left adjoint to ev_0 and we obtain an adjunction

$$(5.3.5.1) \qquad i_0 : \mathscr{A} \rightleftarrows \mathscr{A}^{\mathfrak{S}} : ev_0.$$

Note that i_0 is also right adjoint to ev_0. Thus, i_0 preserves every limit and colimit.

For any integer $n \in \mathbf{N}$ and any symmetric sequence A_* of $\mathscr{A}$, we put

$$(5.3.5.2) \qquad (A_*\{-n\})_m = \begin{cases} \mathfrak{S}_m \times_{\mathfrak{S}_{m-n}} A_{m-n} & \text{if } m \geq n \\ 0 & \text{otherwise.} \end{cases}$$

This defines an endofunctor on $\mathscr{A}^{\mathfrak{S}}$, and we have $A_*\{-n\}\{-m\} = A_*\{-n-m\}$ (through a canonical isomorphism). Observe finally that for any integer $n \in \mathbf{N}$, the functor

$$i_n : \mathscr{A} \longrightarrow \mathscr{A}^{\mathfrak{S}}, A \mapsto (i_0(A))\{-n\}$$

is left adjoint to ev_n.

Remark 5.3.6 Let $\mathfrak{S}$ be the category of finite sets with bijective maps as morphisms. Then the category of symmetric sequences is canonically equivalent to the category of functors $\mathfrak{S} \longrightarrow \mathscr{A}$. This presentation is useful when defining a tensor product on $\mathscr{A}^{\mathfrak{S}}$.

Definition 5.3.7 Let $\mathscr{A}$ be a symmetric closed monoidal abelian category. Given two functors $A_*, B_* : \mathfrak{S} \longrightarrow \mathscr{A}$, we put:

$$E \otimes^{\mathfrak{S}} F : \mathfrak{S} \longmapsto \mathscr{A}$$
$$N \longmapsto \bigoplus_{N=P \sqcup Q} E(P) \otimes F(Q).$$

If $\mathbb{1}_{\mathscr{A}}$ is the unit object of the monoidal category $\mathscr{A}$, the category $\mathscr{A}^{\mathfrak{S}}$ is then a symmetric closed monoidal category with unit object $i_0(\mathbb{1}_{\mathscr{A}})$.

5.3.8 Let A be an object of $\mathscr{A}$. Then the n-th tensor power $A^{\otimes n}$ of A is endowed with a canonical action of the group $\mathfrak{S}_n$ through the structural permutation isomorphism of the symmetric structure on $\mathscr{A}$. Thus the sequence $\mathrm{Sym}(A) = (A^{\otimes n})_{n \in \mathbb{N}}$ is a symmetric sequence.

Moreover, the isomorphism $A^{\otimes n} \otimes A^{\otimes m} \longrightarrow A^{\otimes n+m}$ is $\mathfrak{S}_n \times \mathfrak{S}_m$-equivariant. Thus it induces a morphism $\mu : \mathrm{Sym}(A) \otimes^{\mathfrak{S}} \mathrm{Sym}(A) \longrightarrow \mathrm{Sym}(A)$ of symmetric sequences. We also consider the obvious morphism $\eta : i_0(\mathbb{1}_{\mathscr{A}}) = i_0(A^{\otimes 0}) \longrightarrow \mathrm{Sym}(A)$. One can easily check that $\mathrm{Sym}(A)$ equipped with the multiplication μ and the unit η is a commutative monoid in the monoidal category $\mathscr{A}^{\mathfrak{S}}$.

Definition 5.3.9 Let $\mathscr{A}$ be an abelian symmetric monoidal category. The commutative monoid $\mathrm{Sym}(A)$ of $\mathscr{A}^{\mathfrak{S}}$ defined above will be called the symmetric monoid generated by A.

Remark 5.3.10 One can describe $\mathrm{Sym}(A)$ by a universal property: given a commutative monoid R in $\mathscr{A}^{\mathfrak{S}}$, to give a morphism of commutative monoids $\mathrm{Sym}(A) \longrightarrow R$ is equivalent to giving a morphism $A \longrightarrow R_1$ in $\mathscr{A}$.

5.3.11 Consider an abelian $\mathscr{P}$-premotivic category $\mathscr{A}$.

Consider a base scheme S. According to the previous paragraph, the category $\mathscr{A}_S^{\mathfrak{S}}$ is an abelian category, endowed with a symmetric tensor product $\otimes_S^{\mathfrak{S}}$. For any $\mathscr{P}$-scheme X/S and any integer $n \in \mathbb{N}$, using (5.3.5.2), we put

$$M_S(X, \mathscr{A}^{\mathfrak{S}})\{-n\} = i_0(M_S(X, \mathscr{A}))\{-n\}.$$

It is immediate that the class of symmetric sequences of the form $M_S(X, \mathscr{A}^{\mathfrak{S}})\{-n\}$ for a smooth S-scheme X and an integer $n \geq 0$ is a generating family for the abelian category $\mathscr{A}_S^{\mathfrak{S}}$, which is therefore a Grothendieck abelian category. It is clear that for any $\mathscr{P}$-schemes X and Y over S,

$$M_S(X, \mathscr{A}^{\mathfrak{S}})\{-n\} \otimes_S^{\mathfrak{S}} M_S(Y, \mathscr{A}^{\mathfrak{S}})\{-n\} = M_S(X \times_S Y, \mathscr{A}^{\mathfrak{S}})\{-n\}.$$

Given a morphism (resp. $\mathscr{P}$-morphism) of schemes $f : T \longrightarrow S$ and a symmetric sequence A_* of $\mathscr{A}_S$, we put $f_{\mathfrak{S}}^*(A_*) = (f^* A_n)_{n \in \mathbb{N}}$ (resp. $f_{\sharp}^{\mathfrak{S}}(A_*) = (f_{\sharp} A_n)_{n \in \mathbb{N}}$). This defines a functor $f_{\bar{z}}^* : \mathcal{A}_S^{\bar{z}} \longrightarrow \mathcal{A}_T^{\bar{z}}$ (resp. $f_\sharp^{\bar{z}} : A_T^{\bar{z}} \longrightarrow \mathcal{A}_S^{\bar{z}}$) which is obviously right exact. Thus, the functor $f_{\bar{z}}^*$ admits a right adjoint, which we denote by $f_*^{\bar{z}}$. When f is in $\mathscr{P}$, we easily check that the functor $f_\sharp^{\bar{z}}$ is left adjoint to $f_{\bar{z}}^*$.

From Proposition 1.1.42 and Lemma 1.2.13, we easily check the following proposition:

Proposition 5.3.12 *Consider the previous hypotheses and notations.*

The association $S \longmapsto \mathcal{A}_S^{\bar{z}}$ *together with the structures introduced above defines an* $\mathbf{N} \times \tau$-*generated abelian* $\mathscr{P}$-*premotivic category.*

Moreover, the different adjunctions of the form (5.3.5.1) *over each fiber over a scheme S define an adjunction of* $\mathscr{P}$-*premotivic categories:*

(5.3.12.1) $$i_0 : \mathcal{A} \rightleftarrows \mathcal{A}^{\bar{z}} : ev_0.$$

Indeed, i_0 is trivially compatible with twists.

Proposition 5.3.13 *Let* $\mathcal{A}$ *be an abelian* $\mathscr{P}$-*premotivic category, and t be an admissible topology. If* $\mathcal{A}$ *is compatible with t then* $\mathcal{A}^{\bar{z}}$ *is compatible with t.*

Proof This is based on the following lemma (see [CD09, 7.5, 7.6]):

Lemma 5.3.14 *For any complex C of* $\mathcal{A}_S$, *any complex E of* $\mathcal{A}_S^{\bar{z}}$ *and any integer* $n \geq 0$, *there are canonical isomorphisms:*

(5.3.14.1) $\quad\quad\quad\quad \mathrm{Hom}_{\mathrm{K}(\mathcal{A}_S^{\bar{z}})}(i_0(C)\{-n\}, E) \simeq \mathrm{Hom}_{\mathrm{K}(\mathcal{A}_S)}(C, E_n),$

(5.3.14.2) $\quad\quad\quad\quad \mathrm{Hom}_{\mathrm{D}(\mathcal{A}_S^{\bar{z}})}(i_0(C)\{-n\}, E) \simeq \mathrm{Hom}_{\mathrm{D}(\mathcal{A}_S)}(C, E_n).$

If $\mathcal{A}$ is compatible with t, this implies that E is local (resp. t-flasque) if and only if for any $n \geq 0$, E_n is local (resp. t-flasque). This concludes the proof. □

5.3.c Symmetric Tate spectra

5.3.15 Consider an abelian $\mathscr{P}$-premotivic category $\mathcal{A}$.

For any scheme S, the unit point of $\mathbf{G}_{m,S}$ defines a split monomorphism of $\mathcal{A}$-premotives $\mathbb{1}_S \longrightarrow M_S(\mathbf{G}_{m,S})$. We denote by $\mathbb{1}_S\{1\}$ the cokernel of this monomorphism and call it the *suspended Tate S-premotive* with coefficients in $\mathcal{A}$. The collection of these objects for any scheme S is a cartesian section of $\mathcal{A}$ denoted by $\mathbb{1}\{1\}$. For any integer $n \geq 0$, we denote by $\mathbb{1}\{n\}$ its n-th tensor power.

With the notations of Definition 5.3.9, we define the *symmetric Tate spectrum* over S as the symmetric sequence $\mathbb{1}_S\{*\} = Sym(\mathbb{1}_S\{1\})$ in $\mathcal{A}_S^{\bar{z}}$. The corresponding collection defines a cartesian commutative monoid of the fibred category $\mathcal{A}^{\bar{z}}$, called the *absolute Tate spectrum*.

Definition 5.3.16 Consider an abelian $\mathscr{P}$-premotivic category $\mathscr{A}$.

We denote by $\mathrm{Sp}(\mathscr{A})$ the abelian $\mathscr{P}$-premotivic category of modules over $\mathbb{1}\{*\}$ in the category $\mathscr{A}^{\tilde{\mathfrak{S}}}$. The objects of $\mathrm{Sp}(\mathscr{A})$ are called the abelian (symmetric) Tate spectra.[80]

The category $\mathrm{Sp}(\mathscr{A})$ is $(\mathbf{N} \times \tau)$-generated. Composing the adjunctions (5.3.1.1) and (5.3.12.1), we get an adjunction

(5.3.16.1) $$\Sigma^\infty : \mathscr{A} \rightleftarrows \mathrm{Sp}(\mathscr{A}) : \Omega^\infty$$

of abelian $\mathscr{P}$-premotivic categories.

Let us now give the definition. An abelian Tate spectrum (E, σ) is the data of:

1. for any $n \in \mathbf{N}$, an object E_n of $\mathscr{A}_S$ endowed with an action of $\mathfrak{S}_n$,
2. for any $n \in \mathbf{N}$, a morphism $\sigma_n : E_n\{1\} \longrightarrow E_{n+1}$ in $\mathscr{A}_S$

such that the composite map

$$E_m\{n\} \xrightarrow{\sigma_m\{n-1\}} E_{m+1}\{n-1\} \longrightarrow \ldots \xrightarrow{\sigma_{m+n-1}} E_{m+n}$$

is $\mathfrak{S}_n \times \mathfrak{S}_m$-equivariant with respect to the canonical action of $\mathfrak{S}_n$ on $\mathbb{1}_S\{n\}$ and the structural action of $\mathfrak{S}_m$ on E_m. By definition, $\Omega^\infty(E) = E_0$. Thus, the functor Ω^∞ is exact.

Given an object A of $\mathscr{A}_S$, the abelian Tate spectrum $\Sigma^\infty A$ is defined such that $(\Sigma^\infty A)_n = A\{n\}$ with the action of $\mathfrak{S}_n$ given by its action on $\mathbb{1}_S\{n\}$ by permutations of the factors.

Be careful: we consider the category $\mathrm{Sp}(\mathscr{A}_S)$ as $\mathbf{N}$-twisted by negative twists. For any abelian Tate spectrum E_*, $(E_*\{-n\})_m = \mathfrak{S}_m \times_{\mathfrak{S}_{m-n}} E_{m-n}$ for $n \geq m$.

5.3.17 Consider a morphism

$$\varphi : \mathscr{A} \longrightarrow \mathscr{B}$$

of abelian $\mathscr{P}$-premotivic categories. Then as $\varphi(\mathbb{1}^{\mathscr{A}}\{1\}) = \mathbb{1}^{\mathscr{B}}\{1\}$, φ can be extended to abelian Tate spectra in such a way that the following diagram commutes:

$$\begin{array}{ccc} \mathscr{A} & \xrightarrow{\varphi} & \mathscr{B} \\ \Sigma^\infty_{\mathscr{A}} \downarrow & & \downarrow \Sigma^\infty_{\mathscr{B}} \\ \mathrm{Sp}(\mathscr{A}) & \xrightarrow{\mathrm{Sp}(\varphi)} & \mathrm{Sp}(\mathscr{B}). \end{array}$$

(Of course the obvious diagram for the corresponding right adjoints also commutes.)

---

[80] As we will almost never consider non-symmetric spectra, we will omit the word "symmetric" in our terminology.

5 Fibred derived categories

Definition 5.3.18 For any scheme S, a complex of abelian Tate spectra over S will be called simply a *Tate spectrum* over S.

A Tate spectrum E is a bigraded object. In the notation E_n^m, the index m corresponds to the (cochain) complex structure and the index n to the symmetric sequence structure.

From Propositions 5.3.3 and 5.3.13, we get the following:

Proposition 5.3.19 *Let $\mathscr{A}$ be an abelian $\mathscr{P}$-premotivic category compatible with an admissible topology t. Then $\mathrm{Sp}(\mathscr{A})$ is compatible with t.*

Note also that Remark 5.3.4 and Lemma 5.3.14 imply that for any simplicial $\mathscr{P}$-scheme $\mathscr{X}$ over S, any integer $n \in \mathbf{N}$, and any Tate spectrum E, we have canonical isomorphisms:

(5.3.19.1)
$$\mathrm{Hom}_{\mathrm{K}(\mathrm{Sp}(\mathscr{A}_S))}(\Sigma^\infty M_S(\mathscr{X},\mathscr{A})\{-n\}, E) \simeq \mathrm{Hom}_{\mathrm{K}(\mathscr{A}_S)}(\Sigma^\infty M_S(\mathscr{X},\mathscr{A}), E_n),$$
(5.3.19.2)
$$\mathrm{Hom}_{\mathrm{D}(\mathrm{Sp}(\mathscr{A}_S))}(\Sigma^\infty M_S(\mathscr{X},\mathscr{A})\{-n\}, E) \simeq \mathrm{Hom}_{\mathrm{D}(\mathscr{A}_S)}(\Sigma^\infty M_S(\mathscr{X},\mathscr{A}), E_n).$$

According to the proposition, the category $\mathrm{C}(\mathrm{Sp}(\mathscr{A}_S))$ of Tate spectra over S has a t-descent model structure. The previous isomorphisms allow us to describe this structure as follows:

1. For any simplicial $\mathscr{P}$-scheme $\mathscr{X}$ over S, and any integer $n \geq 0$, the Tate spectrum $\Sigma^\infty M_S(\mathscr{X},\mathscr{A})\{-n\}$ is cofibrant.
2. A Tate spectrum E over S is fibrant if and only if for any integer $n \geq 0$, the complex E_n over $\mathscr{A}_S$ is local (*i.e.* t-flasque).
3. Let $f : E \longrightarrow F$ be a morphism of Tate spectra over S. Then f is a fibration (resp. quasi-isomorphism) if and only if for any integer $n \geq 0$, the morphism $f_n : E_n \longrightarrow F_n$ of complexes over $\mathscr{A}_S$ is a fibration (resp. quasi-isomorphism).

Note that properties (2) and (3) follow from (5.3.4.1) and (5.3.14.1).

5.3.20 We can also introduce the $\mathbf{A}^1$-localization of this model structure. The corresponding homotopy category is the $\mathbf{A}^1$-derived $\mathscr{P}$-premotivic category $\mathrm{D}_{\mathbf{A}^1}^{\mathit{eff}}(\mathrm{Sp}(\mathscr{A}))$ introduced in Definition 5.2.16. The isomorphism (5.3.19.2) gives the following assertion: From the above, a Tate spectrum E is $\mathbf{A}^1$-local if and only if for any integer $n \geq 0$, E_n is $\mathbf{A}^1$-local.

1. A Tate spectrum E over S is $\mathbf{A}^1$-local if and only if for any integer $n \geq 0$, the complex E_n over $\mathscr{A}_S$ is $\mathbf{A}^1$-local.
2. Let $f : E \longrightarrow F$ be a morphism of Tate spectra over S. Then f is an $\mathbf{A}^1$-local fibration (resp. weak $\mathbf{A}^1$-equivalence) if and only if for any integer $n \geq 0$, the morphism $f_n : E_n \longrightarrow F_n$ of complexes over $\mathscr{A}_S$ is an $\mathbf{A}^1$-local fibration (resp. weak $\mathbf{A}^1$-equivalence).

As a consequence, the isomorphism (5.3.19.2) induces an isomorphism

(5.3.20.1)
$$\operatorname{Hom}_{D^{\mathit{eff}}_{\mathbf{A}^1}(\operatorname{Sp}(\mathscr{A}_S))}(\Sigma^\infty M_S(\mathscr{X},\mathscr{A})\{-n\},E) \simeq \operatorname{Hom}_{D^{\mathit{eff}}_{\mathbf{A}^1}(\mathscr{A}_S)}(\Sigma^\infty M_S(\mathscr{X},\mathscr{A}),E_n).$$

Similarly, the adjunction (5.3.16.1) induces an adjunction of triangulated $\mathscr{P}$-premotivic categories

(5.3.20.2) $\quad \mathbf{L}\Sigma^\infty : D^{\mathit{eff}}_{\mathbf{A}^1}(\mathscr{A}) \rightleftarrows D^{\mathit{eff}}_{\mathbf{A}^1}(\operatorname{Sp}(\mathscr{A})) : \mathbf{R}\Omega^\infty.$

5.3.d Symmetric Tate Ω-spectra

5.3.21 The final step is to localize further the category $D^{\mathit{eff}}_{\mathbf{A}^1}(\operatorname{Sp}(\mathscr{A}))$. The aim is to relate the positive twists on $D^{\mathit{eff}}_{\mathbf{A}^1}(\mathscr{A})$ obtained by tensoring with $\mathbb{1}_S\{1\}$ and the negative twists on $D^{\mathit{eff}}_{\mathbf{A}^1}(\operatorname{Sp}(\mathscr{A}))$ induced by the consideration of symmetric sequences.

Let X be a $\mathscr{P}$-scheme over S. From the definition of Σ^∞, there is a canonical morphism of abelian Tate spectra:
$$\left[\Sigma^\infty(\mathbb{1}_S\{1\})\right]\{-1\} \longrightarrow \Sigma^\infty \mathbb{1}_S.$$

Tensoring this map by $\Sigma^\infty M_S(X,\mathscr{A})\{-n\}$ for any $\mathscr{P}$-scheme X over S and any integer $n \in \mathbf{N}$, we obtain a family of morphisms of Tate spectra concentrated in cohomological degree 0:
$$\left[\Sigma^\infty(M_S(X,\mathscr{A})\{1\})\right]\{-n-1\} \longrightarrow \Sigma^\infty M_S(X,\mathscr{A})\{-n\}.$$

We denote this family by $\mathscr{W}_\Omega$ and put $\mathscr{W}_{\Omega,\mathbf{A}^1} = \mathscr{W}_\Omega \cup \mathscr{W}_{\mathbf{A}^1}$. Obviously, $\mathscr{W}_{\Omega,\mathbf{A}^1}$ is stable under the operations f^* and $f_\sharp$.

Definition 5.3.22 Let $\mathscr{A}$ be an abelian $\mathscr{P}$-premotivic category compatible with an admissible topology t. With the notations introduced above, we define the *stable* $\mathbf{A}^1$-*derived $\mathscr{P}$-premotivic category with coefficients in* $\mathscr{A}$ as the derived $\mathscr{P}$-premotivic category
$$D_{\mathbf{A}^1}(\mathscr{A}) := D(\operatorname{Sp}(\mathscr{A}))[\mathscr{W}_{\Omega,\mathbf{A}^1}^{-1}]$$
defined in Corollary 5.2.5.

5.3.23 According to this definition, we get the following identification:
$$D_{\mathbf{A}^1}(\mathscr{A}) = D^{\mathit{eff}}_{\mathbf{A}^1}(\operatorname{Sp}(\mathscr{A}))[\mathscr{W}_\Omega^{-1}].$$

Using the left Bousfield localization of the $\mathbf{A}^1$-local model structure on $C(\operatorname{Sp}(\mathscr{A}))$, we thus obtain a canonical adjunction of triangulated $\mathscr{P}$-fibred premotivic categories
$$D^{\mathit{eff}}_{\mathbf{A}^1}(\operatorname{Sp}(\mathscr{A})) \rightleftarrows D^{\mathit{eff}}_{\mathbf{A}^1}(\operatorname{Sp}(\mathscr{A}))[\mathscr{W}_\Omega^{-1}],$$

5 Fibred derived categories

which allows us to describe $D_{\mathbf{A}^1}(\mathscr{A}_S)$ as the full subcategory of $D^{\mathit{eff}}_{\mathbf{A}^1}(\mathrm{Sp}(\mathscr{A}_S))$ comprising Tate spectra which are $\mathscr{W}_\Omega$-local in $D^{\mathit{eff}}_{\mathbf{A}^1}(\mathrm{Sp}(\mathscr{A}_S))$. Recall a Tate spectrum E is a sequence of complexes $(E_n)_{n \in \mathbf{N}}$ over $\mathscr{A}_S$ together with suspension maps in $C(\mathscr{A}_S)$

$$\sigma_n : \mathbb{1}_S\{1\} \otimes E_n \longrightarrow E_{n+1}.$$

From this, we deduce a canonical morphism $\mathbb{1}_S\{1\} \otimes^{\mathbf{L}} E_n \longrightarrow E_{n+1}$ in $D^{\mathit{eff}}_{\mathbf{A}^1}(\mathscr{A})$ whose adjoint morphism we denote by

(5.3.23.1) $\qquad u_n : E_n \longrightarrow \mathbf{R}Hom_{D^{\mathit{eff}}_{\mathbf{A}^1}(\mathscr{A}_S)}(\mathbb{1}_S\{1\}, E_{n+1}).$

According to (5.3.20.1), the condition that E is $\mathscr{W}_\Omega$-local in $D^{\mathit{eff}}_{\mathbf{A}^1}(\mathrm{Sp}(\mathscr{A}))$ is equivalent to asking that for any integer $n \geq 0$, the map (5.3.23.1) is an isomorphism in $D^{\mathit{eff}}_{\mathbf{A}^1}(\mathrm{Sp}(\mathscr{A}))$.

Considering the adjunction (5.3.20.2), we finally obtain an adjunction of triangulated $\mathscr{P}$-fibred categories:

(5.3.23.2) $\qquad \Sigma^\infty : D^{\mathit{eff}}_{\mathbf{A}^1}(\mathscr{A}) \rightleftarrows D^{\mathit{eff}}_{\mathbf{A}^1}(\mathrm{Sp}(\mathscr{A})) \rightleftarrows D_{\mathbf{A}^1}(\mathscr{A}) : \Omega^\infty.$

Note that tautologically, the Tate spectrum $\Sigma^\infty(\mathbb{1}_S\{1\})$ has a tensor inverse given by the spectrum $(\Sigma^\infty \mathbb{1}_S)\{-1\}$ in $D_{\mathbf{A}^1}(\mathscr{A}_S)$. Thus, we have obtained from the abelian premotivic category $\mathscr{A}$ a triangulated premotivic category $D_{\mathbf{A}^1}(\mathscr{A}_S)$ which satisfies the properties:

- the homotopy property (Htp);
- the stability property (Stab);
- the t-descent property.

As we will see in the following, the construction satisfies a universality property that the reader can already guess.

Definition 5.3.24 Consider the assumptions of Definition 5.3.22.

For any scheme S, we say that a Tate spectrum E over S is a *Tate Ω-spectrum* if the following conditions are fulfilled:

(a) For any integer $n \geq 0$, E_n is t-flasque and $\mathbf{A}^1$-local.
(b) For any integer $n \geq 0$, the adjoint of the structural suspension map

$$E_n \longrightarrow Hom_{C(\mathscr{A}_S)}(\mathbb{1}_S\{1\}, E_{n+1})$$

is a quasi-isomorphism.

In particular, a Tate Ω-spectrum is $\mathscr{W}_\Omega$-local in $D^{\mathit{eff}}_{\mathbf{A}^1}(\mathrm{Sp}(\mathscr{A}_S))$. In fact, it is also $\mathscr{W}_{\Omega,\mathbf{A}^1}$-local in the category $D(\mathrm{Sp}(\mathscr{A}_S))$ so that the category $D_{\mathbf{A}^1}(\mathscr{A})$ is also equivalent to the full subcategory of $D(\mathrm{Sp}(\mathscr{A}_S))$ spanned by Tate Ω-spectra.

The fibrant objects of the $\mathscr{W}_{\Omega,\mathbf{A}^1}$-local model category on $C(\mathrm{Sp}(\mathscr{A}))$ obtained in Definition 5.3.22 are precisely the Tate Ω-spectra.

Proposition 5.3.25 *Consider the above notations. Let S be a base scheme.*

1. *If the endofunctor*

$$D^{\mathit{eff}}_{\mathbf{A}^1}(\mathscr{A}_S) \longrightarrow D^{\mathit{eff}}_{\mathbf{A}^1}(\mathscr{A}_S), C \longmapsto \mathbf{R}\mathit{Hom}_{D^{\mathit{eff}}_{\mathbf{A}^1}(\mathscr{A}_S)}(\mathbb{1}_S\{1\}, C)$$

is conservative, then the functor Ω_S^∞ is conservative.
2. *If the Tate twist $E \longmapsto E(1)$ is fully faithful in $D^{\mathit{eff}}_{\mathbf{A}^1}(\mathscr{A}_S)$, then Σ_S^∞ is fully faithful.*
3. *If the Tate twist $E \longmapsto E(1)$ induces an auto-equivalence of $D^{\mathit{eff}}_{\mathbf{A}^1}(\mathscr{A}_S)$, then $(\Sigma_S^\infty, \Omega_S^\infty)$ are adjoint equivalences of categories.*

Remark 5.3.26 Similar statements can be obtained for the derived categories rather than the $\mathbf{A}^1$-derived categories. We leave their formulation to the reader.

Proof Consider point (1). We have to prove that for any $\mathscr{W}_\Omega$-local Tate spectrum E in $D^{\mathit{eff}}_{\mathbf{A}^1}(\mathrm{Sp}(\mathscr{A}_S))$, if $\mathbf{R}\Omega^\infty(E) = 0$, then $E = 0$. But $\mathbf{R}\Omega^\infty(E) = \Omega^\infty(E) = E_0$ (see Section 5.3.20). Because for any integer $n \geq 0$, the map (5.3.23.1) is an $\mathbf{A}^1$-equivalence, we deduce that for any integer $n \in \mathbf{Z}$, the complex E_n is (weakly) $\mathbf{A}^1$-acyclic. According to (5.3.20.1), this implies $E = 0$ — because $D_{\mathbf{A}^1}(\mathscr{A}_S)$ is N-generated.

Consider point (2). We want to prove that for any complex C over $\mathscr{A}_S$, the counit map $C \longrightarrow \mathbf{R}\Omega^\infty \mathbf{L}\Sigma^\infty(C)$ is an isomorphism. It is enough to treat the case where C is cofibrant.

Considering the left adjoint $\mathbf{L}\Sigma^\infty$ of (5.3.20.2), we first prove that $\mathbf{L}\Sigma^\infty(C)$ is $\mathscr{W}_\Omega$-local. Because C is cofibrant, this Tate spectrum is equal in degree n to the complex $C\{n\}$ (with its natural action of $\mathfrak{S}_n$). Moreover, the suspension map is given by the isomorphism (in the monoidal category $C(\mathscr{A}_S)$)

$$\sigma_n : \mathbb{1}_S\{1\} \otimes_S C\{n\} \longrightarrow C\{n+1\}.$$

In particular, the corresponding map in $D^{\mathit{eff}}_{\mathbf{A}^1}(\mathscr{A}_S)$

$$\sigma'_n : \mathbb{1}_S\{1\} \otimes^L_S C\{n\} \longrightarrow C\{n+1\}$$

is canonically isomorphic to

$$\mathbb{1}_S\{1\} \otimes^L_S C\{n\} \xrightarrow{1 \otimes 1} \mathbb{1}_S\{1\} \otimes^L_S C\{n\}.$$

Thus, because the Tate twist is fully faithful in $D^{\mathit{eff}}_{\mathbf{A}^1}(\mathscr{A}_S)$, the adjoint map to σ'_n is an $\mathbf{A}^1$-equivalence. In other words, $\mathbf{L}\Sigma^\infty(C)$ is $\mathscr{W}_\Omega$-local. But then, as C is cofibrant, $C = \Omega^\infty \Sigma^\infty(C) = \mathbf{R}\Omega^\infty \mathbf{L}\Sigma^\infty(C)$.

Point (3) is then a consequence of (1) and (2). □

5 Fibred derived categories

Remark 5.3.27

1. The construction of the triangulated category $D_{\mathbb{A}^1}(\mathscr{A})$ can also be obtained using the more general construction of [CD09, §7] — see also [Hov01, 7.11] and [Ayo07b, chap. 4] for even more general accounts. Here, we exploit the simplification arising from the fact that we invert a complex concentrated in degree 0: this allowed us to describe $D_{\mathbb{A}^1}(\mathscr{A})$ simply as a Verdier quotient of the derived category of an abelian category. However, we can also consider the category of symmetric spectra in $C(\mathscr{A}_S)$ with respect to one of the complexes $\mathbb{1}_S(1)[2]$ or $\mathbb{1}_S(1)$ and this leads to the equivalent categories; see [Hov01, 8.3].
2. Point (3) of Proposition 5.3.25 is a particular case of [Hov01, 8.1].

5.3.28 Consider a morphism of abelian $\mathscr{P}$-premotivic categories

$$\varphi : \mathscr{A} \rightleftarrows \mathscr{B} : \psi$$

such that $\mathscr{A}$ (resp. $\mathscr{B}$) is compatible with a system of topology t (resp. t'). Suppose t' is finer than t. According to Section 5.3.17, we obtain an adjunction of abelian $\mathscr{P}$-premotivic categories

$$\varphi : C(\mathrm{Sp}(\mathscr{A})) \rightleftarrows C(\mathrm{Sp}(\mathscr{B})) : \psi.$$

The pair (φ_S, ψ_S) is a Quillen adjunction for the stable model structures (apply again [CD09, prop. 3.11]). Thus we obtain a morphism of triangulated $\mathscr{P}$-premotivic categories:

$$\mathbf{L}\varphi : D_{\mathbb{A}^1}(\mathscr{A}) \rightleftarrows D_{\mathbb{A}^1}(\mathscr{B}) : \mathbf{R}\psi.$$

Remark 5.3.29 Under the light of Proposition 5.3.25, the category $D_{\mathbb{A}^1}(\mathscr{A})$ might be considered as the universal derived $\mathscr{P}$-premotivic category $\mathscr{T}$ with a morphism $D(\mathscr{A}) \longrightarrow \mathscr{T}$, and such that $\mathscr{T}$ satisfies the homotopy and the stability property. This can be made precise in the setting of algebraic derivators or of dg-categories (or any other kind of stable ∞-categories).

Proposition 5.3.30 *Let t and t' be two admissible topologies, with t' finer than t. Then $D_{\mathbb{A}^1}(\mathrm{Sh}_{t'}(\mathscr{P}, \Lambda))$ is canonically equivalent to the full subcategory of $D_{\mathbb{A}^1}(\mathrm{Sh}_t(\mathscr{P}, \Lambda))$ spanned by the objects which satisfy t'-descent.*

Proof It is sufficient to prove this proposition in the case where t is the coarse topology. We deduce from [Ayo07b, 4.4.42] that, for any scheme S in $\mathscr{S}$, we have

$$D_{\mathbb{A}^1}(\mathrm{Sh}_{t'}(\mathscr{P}/S, \Lambda)) = D(\mathrm{PSh}(\mathscr{P}/S, \Lambda))[\mathscr{W}^{-1}],$$

with $\mathscr{W} = \mathscr{W}_{t'} \cup \mathscr{W}_{\mathbb{A}^1} \cup \mathscr{W}_\Omega$, where $\mathscr{W}_{t'}$ is the set of maps of shape

$$\Sigma^\infty M_S(\mathscr{X})\{n\}[i] \longrightarrow \Sigma^\infty M_S(X)\{n\}[i],$$

for any t'-hypercover $\mathscr{X} \longrightarrow X$ and any integers $n \leq 0$ and i. The assertion is then a particular case of the description of the homotopy category of a left Bousfield localization. □

Example 5.3.31 We have the stable versions of the $\mathscr{P}$-premotivic categories introduced in Example 5.2.17:
1) Consider the admissible topology $t = Nis$. Following F. Morel, we define the *stable $\mathbf{A}^1$-derived premotivic category* as (see also the construction of [Ayo07b]):

$$D_{\mathbf{A}^1, \Lambda} := D_{\mathbf{A}^1}(\mathrm{Sh}_{Nis}(Sm, \Lambda)) \quad \text{and} \quad \underline{D}_{\mathbf{A}^1, \Lambda} := D_{\mathbf{A}^1}\left(\mathrm{Sh}_{Nis}\left(\mathscr{S}^{ft}, \Lambda\right)\right),$$

as well as the *generalized stable $\mathbf{A}^1$-derived premotivic category*[81]

(5.3.31.1) $$\underline{D}_{\mathbf{A}^1, \Lambda} := D_{\mathbf{A}^1}\left(\mathrm{Sh}_{Nis}\left(\mathscr{S}^{ft}, \Lambda\right)\right).$$

Given a scheme S, we shall also write:

(5.3.31.2) $$D_{\mathbf{A}^1}(S, \Lambda) := D_{\mathbf{A}^1, \Lambda}(S) \quad \text{and} \quad \underline{D}_{\mathbf{A}^1}(S, \Lambda) := \underline{D}_{\mathbf{A}^1, \Lambda}(S).$$

In the case when $t = \acute{e}t$, we get the triangulated premotivic categories of *étale premotives*:

$$D_{\mathbf{A}^1}(\mathrm{Sh}_{\acute{e}t}(Sm, \Lambda)) \quad \text{and} \quad D_{\mathbf{A}^1}\left(\mathrm{Sh}_{\acute{e}t}\left(\mathscr{S}^{ft}, \Lambda\right)\right).$$

In each of these cases, we denote by $\Sigma^\infty \Lambda_S^t(X)$ the premotive associated with a smooth S-scheme X.

From the adjunction (5.1.24.2), we get an adjunction of triangulated premotivic categories:

$$a_{\acute{e}t} : D_{\mathbf{A}^1, \Lambda} \rightleftarrows D_{\mathbf{A}^1}(\mathrm{Sh}_{\acute{e}t}(Sm, \Lambda)) : RO_{\acute{e}t}.$$

2) Assume $\mathscr{P} = \mathscr{S}^{ft}$:
Consider the $\mathscr{S}^{ft}$-admissible topology $t = h$ (resp. $t = qfh$). In [Voe96], Voevodsky introduced the category of effective h-motives (resp. qfh-motives). According to the theory presented above, one can extend this definition to the stable setting: one defines the category of stable h-*motives* (resp. qfh-*motives*) over S with coefficients in Λ as:

$$\underline{DM}_h(S, \Lambda) := D_{\mathbf{A}^1}\left(\mathrm{Sh}_h\left(\mathscr{S}^{ft}/S, \Lambda\right)\right),$$

resp. $$\underline{DM}_{qfh}(S, \Lambda) := D_{\mathbf{A}^1}\left(\mathrm{Sh}_{qfh}\left(\mathscr{S}^{ft}/S, \Lambda\right)\right).$$

In other words, this is the stable $\mathbf{A}^1$-derived category of h-sheaves (resp. qfh-sheaves) of Λ-modules. Moreover, we get the *generalized triangulated premotivic category of h-motives (resp. qfh-motives)* with coefficients in Λ over $\mathscr{S}$:

---

[81] We will see in Example 6.1.10 that the generalized version contains the usual one as a full subcategory.

5 Fibred derived categories

$$\underline{DM}_{h,\Lambda} := D_{\mathbf{A}^1}\left(\mathrm{Sh}_h\left(\mathscr{S}^{ft},\Lambda\right)\right),$$

$$\text{resp. } \underline{DM}_{qfh,\Lambda} := D_{\mathbf{A}^1}\left(\mathrm{Sh}_{qfh}\left(\mathscr{S}^{ft},\Lambda\right)\right).$$

For an S-scheme of finite type X, we will denote by $\Sigma^\infty \underline{\Lambda}_S^h(X)$ (resp $\Sigma^\infty \underline{\Lambda}_S^{qfh}(X)$) the corresponding premotive associated with X in $\underline{DM}_t(S,\Lambda)$. Note that the h-sheafification functor induces a premotivic adjunction (see Section 5.3.28):

(5.3.31.3) $$\underline{DM}_{qfh,\Lambda} \rightleftarrows \underline{DM}_{h,\Lambda}.$$

These generalized premotivic categories are too big to be reasonable (in particular for the localization property — see Remark 2.3.4). Therefore, we introduce the triangulated category $DM_t(S,\Lambda)$ as the localizing subcategory of $\underline{DM}_t(S,\Lambda)$ generated by objects of shape $\Sigma^\infty \Lambda_S^t(X)(p)[q]$ for any smooth S-scheme of finite type X and any integers p and q. The fibred category $DM_{h,\Lambda}$ (resp. $DM_{qfh,\Lambda}$) defined above is premotivic. We call it the *premotivic category of h-motives (resp. qfh-motives)*. The family of inclusions

(5.3.31.4) $$DM_t(S,\Lambda) \longrightarrow \underline{DM}_t(S,\Lambda)$$

indexed by a scheme S defines a premotivic morphism (the existence of right adjoints is ensured by the Brown representability theorem).

Remark 5.3.32 When $\Lambda = \mathbf{Q}$, we will show that the categories $DM_{h,\mathbf{Q}}$ and $DM_{qfh,\mathbf{Q}}$ are equivalent and satisfy the axioms of a motivic category. In fact, they are equivalent to the category of Beilinson motives. See Theorem 16.1.2 for all these results.

Proposition 5.3.33 *Consider the notations of the second point in the above example. Then the premotivic category $DM_{t,\Lambda}$ satisfies t-descent.*

Proof This is true for $\underline{DM}_{t,\Lambda}$ by construction, which implies formally the assertion for $DM_{t,\Lambda}$. □

Remark 5.3.34 According to Proposition 5.2.10 and Remark 5.3.29, for any admissible topology t, $D_{\mathbf{A}^1}(\mathrm{Sh}_t(\mathscr{P},\mathbf{Z}))$ is the universal derived $\mathscr{P}$-premotivic category satisfying t-descent as well as the homotopy and stability properties.

A crucial example for us: the stable $\mathbf{A}^1$-derived premotivic category $D_{\mathbf{A}^1}$ is the universal derived premotivic category satisfying the properties of homotopy, of stability and of Nisnevich descent.

5.3.35 We assume $\mathscr{P} = Sm$. Let $\mathrm{Sh}_\bullet(Sm)$ be the category of pointed Nisnevich sheaves of sets. Consider the pointed version of the adjunction of $\mathscr{P}$-premotivic categories

$$N : \Delta^{op}\mathrm{Sh}_\bullet(Sm) \rightleftarrows C(\mathrm{Sh}_{Nis}(Sm,\mathbf{Z})) : K$$

constructed in Section 5.2.25.

If we consider on the left-hand side the $\mathbf{A}^1$-model category defined by Blander [Bla03], (N_S, K_S) is a Quillen adjunction for any scheme S.

We consider $(\mathbf{G}_m, 1)$ as a constant pointed simplicial sheaf. The construction of symmetric $\mathbf{G}_m$-spectra respectively to the model category $\Delta^{op}\,\mathrm{Sh}_\bullet(Sm)$ can now be carried out following [Jar00] or [Ayo07b] and yields a symmetric monoidal model category whose homotopy category is the stable homotopy category of Morel and Voevodsky $\mathrm{SH}(S)$.

Using the functoriality statements [Hov01, th. 8.3 and 8.4], we finally obtain a $\mathscr{P}$-premotivic adjunction

(5.3.35.1) $$N : \mathrm{SH} \rightleftarrows \mathrm{D}_{\mathbf{A}^1} : K.$$

The functor K is the analog of the Eilenberg–MacLane functor in algebraic topology; in fact, this adjunction is actually induced by the Eilenberg–MacLane functor (see [Ayo07b, chap. 4]). In particular, as the rational model category of topological (symmetric) S^1-spectra is Quillen equivalent to the model category of complexes of $\mathbf{Q}$-vector spaces, we have a natural equivalence of premotivic categories

(5.3.35.2) $$\mathrm{SH}_\mathbf{Q} \rightleftarrows \mathrm{D}_{\mathbf{A}^1,\mathbf{Q}},$$

(where $\mathrm{SH}_\mathbf{Q}(S)$ denotes the Verdier quotient of $\mathrm{SH}(S)$ by the localizing subcategory generated by compact torsion objects).

5.3.36 We can extend the considerations of Example 5.1.25 and Section 5.2.22 on changing coefficients in categories of sheaves.

Let t be an admissible topology and $\varphi : \Lambda \longrightarrow \Lambda'$ be an extension of rings. Using the $\mathscr{P}$-premotivic adjunction (5.1.25.1) and according to Section 5.3.28, we get an adjunction of triangulated $\mathscr{P}$-premotivic categories:

$$\mathbf{L}\varphi^* : \mathrm{D}_{\mathbf{A}^1}\big(\mathrm{Sh}_t(\mathscr{P},\Lambda)\big) \rightleftarrows \mathrm{D}_{\mathbf{A}^1}\big(\mathrm{Sh}_t(\mathscr{P},\Lambda')\big) : \mathbf{R}\varphi_*.$$

Given two Tate spectra C and D of t-sheaves of Λ-modules over $\mathscr{P}_S$, we get a canonical morphism of Λ'-modules:
(5.3.36.1)
$$\mathrm{Hom}_{\mathrm{D}_{\mathbf{A}^1}(\mathrm{Sh}_t(\mathscr{P}_S,\Lambda))}(C,D) \otimes_\Lambda \Lambda' \longrightarrow \mathrm{Hom}_{\mathrm{D}_{\mathbf{A}^1}(\mathrm{Sh}_t(\mathscr{P}_S,\Lambda'))}\big(\mathbf{L}\varphi^*(C), \mathbf{L}\varphi^*(D)\big).$$

Then the stable version of Proposition 5.2.23 holds (the proof is the same):

Proposition 5.3.37 *Consider the above assumptions. Then the map* (5.3.36.1) *is an isomorphism in the two following cases:*

1. If Λ' is a free Λ-module and C is compact;
2. If Λ' is a free Λ-module of finite rank.

5.3.e Constructible premotivic spectra

Lemma 5.3.38 *Let $\mathscr{A}$ be an abelian $\mathscr{P}$-premotivic category compatible with a topology t and such that the category $\mathbf{A}^1$-derived category $\mathrm{D}_{\mathbf{A}^1}^{\mathit{eff}}(\mathscr{A})$ satisfies Nisnevich descent.*

5 Fibred derived categories

Then, for any scheme S, the non-trivial cyclic permutation (123) *of order 3 acts as the identity on the premotive* $\mathbb{1}_S\{1\}^{\otimes 3}$ *in* $\mathrm{D}^{\mathit{eff}}_{\mathbf{A}^1}(\mathscr{A}_S)$.

Proof Using Example 5.2.21, it is sufficient to prove this in $\mathrm{D}^{\mathit{eff}}_{\mathbf{A}^1,\Lambda}(S)$, which is well known; see for example [Ayo07b, 4.5.65]. □

Proposition 5.3.39 *Consider the hypotheses of the previous lemma and assume that the triangulated premotivic category* $\mathrm{D}^{\mathit{eff}}_{\mathbf{A}^1}(\mathscr{A})$ *is compactly τ-generated.*

Then, for any scheme S, any pair of integers (i,a), *any compact object C of* $\mathrm{D}^{\mathit{eff}}_{\mathbf{A}^1}(\mathscr{A}_S)$ *and any Tate spectrum E in* $\mathscr{A}_S$, *we have a canonical isomorphism*

$$\mathrm{Hom}_{\mathrm{D}_{\mathbf{A}^1}(\mathscr{A}_S)}(\mathbf{L}\Sigma^\infty(C)\{a\}, E[i]) \simeq \varinjlim_{r \gg 0} \mathrm{Hom}_{\mathrm{D}^{\mathit{eff}}_{\mathbf{A}^1}(\mathscr{A}_S)}(C\{a+r\}, E_r[i]).$$

Proof Given the previous lemma, this is a direct consequence of [Ayo07b, theorems 4.3.61 and 4.3.79]. □

Corollary 5.3.40 *Under the assumptions of the preceding proposition, the triangulated category* $\mathrm{D}_{\mathbf{A}^1}(\mathscr{A}_S)$ *is compactly* $(\mathbf{Z} \times \tau)$-*generated where the factor* $\mathbf{Z}$ *corresponds to the Tate twist.*

More precisely, if $\mathrm{D}^{\mathit{eff}}_{\mathbf{A}^1,c}(\mathscr{A}_S)$ *denotes the category of compact objects in* $\mathrm{D}^{\mathit{eff}}_{\mathbf{A}^1}(\mathscr{A}_S)$, *then the category of compact objects in* $\mathrm{D}_{\mathbf{A}^1}(\mathscr{A}_S)$ *is canonically equivalent to the pseudo-abelian completion of the category obtained as the 2-colimit of the following diagram:*

$$\mathrm{D}^{\mathit{eff}}_{\mathbf{A}^1,c}(\mathscr{A}_S) \xrightarrow{\otimes \mathbb{1}_S\{1\}} \mathrm{D}^{\mathit{eff}}_{\mathbf{A}^1,c}(\mathscr{A}_S) \longrightarrow \cdots \longrightarrow \mathrm{D}^{\mathit{eff}}_{\mathbf{A}^1,c}(\mathscr{A}_S) \xrightarrow{\otimes \mathbb{1}_S\{1\}} \mathrm{D}^{\mathit{eff}}_{\mathbf{A}^1,c}(\mathscr{A}_S) \longrightarrow \cdots$$

5.3.41 Let $\mathscr{A}$ be an abelian $\mathscr{P}$-premotivic category compatible with an admissible topology t. Assume that:

- The topology t is bounded in $\mathscr{A}$ (Definition 5.1.28).
- The abelian $\mathscr{P}$-premotivic category $\mathscr{A}$ is finitely τ-presented.

We will denote by $\mathscr{N}_S^t$ a bounded generating family for t-hypercovers in $\mathscr{A}_S$.

Recall from Proposition 5.2.38 that the category of compact objects of the triangulated category $\mathrm{D}^{\mathit{eff}}_{\mathbf{A}^1}(\mathscr{A}_S)$ is canonically equivalent to the triangulated monoidal category:

$$\left(K^b \left(\mathbf{Z}_S(Sm/S; \mathscr{A}) \right) / (\mathscr{N}_S^t \cup \mathscr{T}_{\mathbf{A}_S^1}) \right)^{\natural}.$$

Let us denote by $\mathrm{D}_{\mathbf{A}^1,gm}(\mathscr{A}_S)$ the category obtained from the monoidal category on the left-hand side of the above functor by formally inverting the Tate twist $\mathbf{Z}_S^{\mathscr{A}}(1)$. Because $\mathrm{D}_{\mathbf{A}^1}(\mathscr{A})$ satisfies the stability property by construction, we readily obtain a canonical monoidal functor

(5.3.41.1) $$\mathrm{D}_{\mathbf{A}^1,gm}(\mathscr{A}_S) \longrightarrow \mathrm{D}_{\mathbf{A}^1}(\mathscr{A}_S).$$

Then applying Proposition 5.2.38, the above corollary and Proposition 1.4.11, we deduce:

Corollary 5.3.42 *Consider the above hypotheses and notations.*
Then the triangulated premotivic category $D_{A^1}(\mathscr{A})$ *is compactly* $(\mathbf{Z} \times \tau)$-*generated. For any premotive $\mathscr{M}$ in* $D_{A^1}(\mathscr{A}_S)$ *the following conditions are equivalent:*

(i) $\mathscr{M}$ is compact;
(ii) $\mathscr{M}$ is $(\mathbf{Z} \times \tau)$-constructible.

Moreover, the functor (5.3.41.1) is fully faithful and has for essential image the compact (i.e. τ-constructible) objects of $D_{A^1}(\mathscr{A}_S)$.

Example 5.3.43 From the considerations of Example 5.2.40, we get that, for any scheme S, the category of compact objects of $D_{A^1}(S, \Lambda)$ (resp., of its cdh-local counterpart $D_{A^1}\left(Sh_{cdh}(\mathscr{S}^{ft}/S, \Lambda)\right)$) is obtained from the monoidal triangulated category

$$K^b\left(\Lambda(Sm/S)\right) \text{ (resp. } K^b\left(\Lambda(\mathscr{S}^{ft}/S)\right)\text{)}$$

by the following steps:

- one mods out by the triangulated subcategories $\mathscr{T}_{A^1_S}$ and BG_S (resp. CDH_S) corresponding to the A^1-homotopy property and the Brown–Gersten triangles (resp. cdh-triangles),
- one takes the pseudo-abelian envelope,
- one formally inverts the Tate twist.

Proposition 5.3.44 *Assume $\mathscr{P} = \mathscr{S}^{ft}$ is the class of finite type (resp. separated and of finite type) morphisms.*
Let $\mathscr{A}$ be an abelian generalized premotivic category compatible with an admissible topology t such that:

- *$\mathscr{A}$ satisfies property (C) (resp. (wC)) of Section 5.1.35.*
- *The A^1-derived category $D^{eff}_{A^1}(\mathscr{A})$ is compactly τ-generated and satisfies Nisnevich descent.*

Then the stable A^1-derived premotivic category $D_{A^1}(\mathscr{A})$ is $(\mathbf{Z} \times \tau)$-continuous (resp. weakly $(\mathbf{Z} \times \tau)$-continuous) — see Definition 4.3.2.

Proof This is an immediate corollary of Proposition 5.2.41 combined with Proposition 5.3.39. □

Example 5.3.45 According to the previous proposition and the second point of Example 5.1.37, the generalized triangulated premotivic category $\underline{D}_{A^1, \Lambda}$ is continuous. We also refer the reader to Corollary 6.1.12 for an extension of this result to the non-generalized case.

6 Localization and the universal derived example

6.0.1 In this section, $\mathscr{S}$ is an adequate category of $\mathscr{S}$-schemes as in Section 2.0.1. In Sections 6.2 and 6.3, we assume in addition that the schemes in $\mathscr{S}$ are finite-dimensional.

We will apply the definitions of the preceding section to the admissible class comprising morphisms of finite type (resp. smooth morphisms of finite type) in $\mathscr{S}$, denoted by $\mathscr{S}^{ft}$ (resp. Sm).

Recall the general convention of Section 1.4:

- *premotivic* means Sm-premotivic;
- *generalized premotivic* means $\mathscr{S}^{ft}$-premotivic.

6.1 Generalized derived premotivic categories

Example 6.1.1 Let t be an $\mathscr{S}^{ft}$-admissible topology. For a scheme S, we denote by $\mathrm{Sh}_t(\mathscr{S}^{ft}/S, \Lambda)$ the category of sheaves of abelian groups on $\mathscr{S}^{ft}/S$ for the topology t_S. For an S-scheme of finite type X, we let $\underline{\Lambda}_S(X)$ be the free t-sheaf of Λ-modules represented by X. Recall $\mathrm{Sh}_t(\mathscr{S}^{ft}, \Lambda)$ is a generalized abelian premotivic category (see Example 5.1.4).

Let $\rho : Sm/S \longrightarrow \mathscr{S}^{ft}/S$ be the obvious inclusion functor and let us denote by t_S the initial topology on Sm/S such that ρ is continuous. Then it induces (cf. [AGV73, IV, 4.10]) a sequence of adjoint functors

$$\mathrm{Sh}_t(Sm/S, \Lambda) \xleftarrow[\rho_*]{\overset{\rho_\sharp}{\underset{\rho^*}{\longrightarrow}}} \mathrm{Sh}_t(\mathscr{S}^{ft}/S, \Lambda)$$

and we easily check that this induces an enlargement of abelian premotivic categories:

(6.1.1.1) $\qquad \rho_\sharp : \mathrm{Sh}_t(Sm, \Lambda) \rightleftarrows \mathrm{Sh}_t(\mathscr{S}^{ft}, \Lambda) : \rho^*$.

Remark 6.1.2 Note that for any scheme S, the abelian category $\mathrm{Sh}_t(Sm/S, \Lambda)$ can be described as the Gabriel quotient of the abelian category $\mathrm{Sh}_t(\mathscr{S}^{ft}/S, \Lambda)$ with respect to the sheaves $\underline{F}$ over $\mathscr{S}^{ft}/S$ such that $\rho^*(\underline{F}) = 0$.

An example of such a sheaf in the case where $t = Nis$ and $\dim(S) > 0$ is the Nisnevich sheaf $\underline{\Lambda}_S(Z)$ on $\mathscr{S}^{ft}/S$ represented by a nowhere dense closed subscheme Z of S which is zero when restricted to Sm/S.

6.1.3 Consider an abelian premotivic category $\mathscr{A}$ compatible with an admissible topology t on Sm and a generalized abelian premotivic category $\underline{\mathscr{A}}$ compatible with an admissible topology t' on $\mathscr{S}$. We denote by M (resp. $\underline{M}$) the geometric sections of $\mathscr{A}$ (resp. $\underline{\mathscr{A}}$). We assume that t' restricted to Sm is finer than t, and

consider an adjunction of abelian premotivic categories:
$$\rho_\sharp : \mathscr{A} \rightleftarrows \underline{\mathscr{A}} : \rho^*.$$

Let S be a scheme in $\mathscr{S}$. The functors $\rho_\sharp$ and ρ^* induce a derived adjunction (see Section 5.2.19):
$$\mathbf{L}\rho_\sharp : \mathrm{D}^{\mathit{eff}}_{\mathbf{A}^1}(\mathscr{A}_S) \rightleftarrows \mathrm{D}^{\mathit{eff}}_{\mathbf{A}^1}(\underline{\mathscr{A}}_S) : \mathbf{R}\rho^*$$
(where $\underline{\mathscr{A}}$ is considered as an Sm-fibred category).

Proposition 6.1.4 *Consider the previous hypotheses, and fix a scheme S. Assume furthermore that we have the following properties.*

(i) The functor $\rho_\sharp : \mathscr{A}_S \longrightarrow \underline{\mathscr{A}}_S$ is fully faithful.
(ii) The functor $\rho^* : \underline{\mathscr{A}}_S \longrightarrow \mathscr{A}_S$ commutes with small colimits.

Then, the following conditions hold:

(a) The induced functor
$$\rho^* : \mathrm{C}(\underline{\mathscr{A}}_S) \longrightarrow \mathrm{C}(\mathscr{A}_S)$$
preserves $\mathbf{A}^1$-equivalences.
(b) The $\mathbf{A}^1$-derived functor $\mathbf{L}\rho_\sharp : \mathrm{D}^{\mathit{eff}}_{\mathbf{A}^1}(\mathscr{A}_S) \longrightarrow \mathrm{D}^{\mathit{eff}}_{\mathbf{A}^1}(\underline{\mathscr{A}}_S)$ is fully faithful.

Proof Point (a) follows from Proposition 5.2.24. To prove (b), we have to prove that the unit map
$$M \longrightarrow \rho^* \mathbf{L}\rho_\sharp(M)$$
is an isomorphism for any object M of $\mathrm{D}^{\mathit{eff}}_{\mathbf{A}^1}(\mathscr{A}_S)$. For this purpose, we may assume that M is cofibrant, so that we have
$$M \simeq \rho^* \rho_\sharp(M) \simeq \rho^* \mathbf{L}\rho_\sharp(M)$$
(where the first isomorphism holds already in $\mathrm{C}(\mathscr{A}_S)$). □

Corollary 6.1.5 *Consider the hypotheses of the previous proposition. Then the family of adjunctions $\mathbf{L}\rho_\sharp : \mathrm{D}^{\mathit{eff}}_{\mathbf{A}^1}(\mathscr{A}_S) \longrightarrow \mathrm{D}^{\mathit{eff}}_{\mathbf{A}^1}(\underline{\mathscr{A}}_S) : \mathbf{R}\rho^*$ indexed by a scheme S induces an enlargement of triangulated premotivic categories*
$$\mathbf{L}\rho_\sharp : \mathrm{D}^{\mathit{eff}}_{\mathbf{A}^1}(\mathscr{A}) \rightleftarrows \mathrm{D}^{\mathit{eff}}_{\mathbf{A}^1}(\underline{\mathscr{A}}) : \mathbf{R}\rho^*.$$

Example 6.1.6 Considering the situation of Example 6.1.1, we will be particularly interested in the case of the Nisnevich topology. We denote by $\underline{\mathrm{D}}^{\mathit{eff}}_{\mathbf{A}^1,\Lambda}$ the generalized $\mathbf{A}^1$-derived premotivic category associated with $\mathrm{Sh}(\mathscr{S}^{ft}, \Lambda)$ (see also Example 5.3.31). The preceding corollary gives a canonical enlargement:

(6.1.6.1) $$\mathrm{D}^{\mathit{eff}}_{\mathbf{A}^1,\Lambda} \rightleftarrows \underline{\mathrm{D}}^{\mathit{eff}}_{\mathbf{A}^1,\Lambda}.$$

6 Localization and the universal derived example

6.1.7 Consider again the hypotheses of Section 6.1.3. We denote simply by M (resp. $\underline{M}$) the geometric sections of the premotivic triangulated category $D_{\mathbf{A}^1}(\mathscr{A})$ (resp. $D_{\mathbf{A}^1}(\underline{\mathscr{A}})$).

Recall from Section 5.3.15 that we have defined $\mathbb{1}_S\{1\}$ (resp. $\underline{\mathbb{1}}_S\{1\}$) as the cokernel of the canonical map $\mathbb{1}_S \longrightarrow M_S(\mathbf{G}_{m,S})$ (resp. $\underline{\mathbb{1}}_S \longrightarrow \underline{M}_S(\mathbf{G}_{m,S})$). Thus, it is obvious that we get a canonical identification $\rho_\sharp(\mathbb{1}_S\{1\}) = \underline{\mathbb{1}}_S\{1\}$. Therefore, the enlargement $\rho_\sharp$ can be extended canonically to an enlargement

$$\rho_\sharp : \mathrm{Sp}(\mathscr{A}) \rightleftarrows \mathrm{Sp}(\underline{\mathscr{A}}) : \rho^*$$

of abelian premotivic categories in such a way that for any scheme S, the following diagram commutes:

$$\begin{array}{ccc} \mathscr{A}_S & \xrightarrow{\rho_\sharp} & \underline{\mathscr{A}}_S \\ \Sigma^\infty_{\mathscr{A}} \downarrow & & \downarrow \Sigma^\infty_{\underline{\mathscr{A}}} \\ \mathrm{Sp}(\mathscr{A}_S) & \xrightarrow{\rho_\sharp} & \mathrm{Sp}(\underline{\mathscr{A}}_S). \end{array}$$

According to Proposition 5.3.13, $\mathrm{Sp}(\mathscr{A})$ (resp. $\mathrm{Sp}(\underline{\mathscr{A}})$) is compatible with t (resp. t'), and we obtain an adjoint pair of functors (Section 5.3.28):

$$\mathbf{L}\rho_\sharp : D_{\mathbf{A}^1}(\mathscr{A}_S) \rightleftarrows D_{\mathbf{A}^1}(\underline{\mathscr{A}}_S) : \mathbf{R}\rho^*.$$

From the preceding commutative square, we get the identification:

$$(6.1.7.1) \qquad \mathbf{L}\rho_\sharp \circ \Sigma^\infty_{\mathscr{A}} = \Sigma^\infty_{\underline{\mathscr{A}}} \circ \mathbf{L}\rho_\sharp.$$

As in the non-effective case, we get the following result:

Proposition 6.1.8 *Keep the assumptions of Proposition 6.1.4, and suppose furthermore that both $D^{\mathrm{eff}}_{\mathbf{A}^1}(\mathscr{A})$ and $D^{\mathrm{eff}}_{\mathbf{A}^1}(\underline{\mathscr{A}})$ are compactly τ-generated. Then the derived functor $\mathbf{L}\rho_\sharp : D_{\mathbf{A}^1}(\mathscr{A}_S) \longrightarrow D_{\mathbf{A}^1}(\underline{\mathscr{A}}_S)$ is fully faithful.*

Proof We have to prove that for any Tate spectrum E of $D_{\mathbf{A}^1}(\mathscr{A}_S)$, the adjunction morphism

$$E \longrightarrow \mathbf{L}\rho^* \mathbf{R}\rho_\sharp(E)$$

is an isomorphism. According to Proposition 1.3.19, the functor $\mathbf{L}\rho^*$ admits a right adjoint. Thus, applying Lemma 1.1.43, it is sufficient to consider the case where $E = M_S(X)\{i\}[n]$ for a smooth S-scheme X, and a pair $(n,i) \in \mathbf{Z} \times \tau$.

Moreover, it is sufficient to prove that for another smooth S-scheme Y and an integer $j \in \mathbf{Z}$, the induced morphism

$$\mathrm{Hom}(\Sigma^\infty M_S(Y)\{j\}, \Sigma^\infty M_S(X)\{i\}[n]) \longrightarrow \mathrm{Hom}(\Sigma^\infty \underline{M}_S(Y)\{j\}, \Sigma^\infty \underline{M}_S(X)\{i\}[n])$$

is an isomorphism. Using the identification (6.1.7.1), the result follows by Propositions 5.3.39 and 6.1.4. □

Corollary 6.1.9 *If the assumptions of Proposition 6.1.8 hold for any scheme S in $\mathscr{S}$, then we obtain an enlargement of triangulated premotivic categories*

$$\mathbf{L}\rho_\sharp : D_{\mathbf{A}^1}(\mathscr{A}) \rightleftarrows \underline{D}_{\mathbf{A}^1}(\underline{\mathscr{A}}) : \mathbf{R}\rho^*.$$

Example 6.1.10 Consider again the situation of Example 6.1.1, in the case of the Nisnevich topology. We denote by $\underline{D}_{\mathbf{A}^1,\Lambda}$ the generalized stable $\mathbf{A}^1$-derived premotivic category associated with $\mathrm{Sh}(\mathscr{S}^{ft}, \Lambda)$. The preceding corollary gives a canonical enlargement:

(6.1.10.1) $$\mathbf{L}\rho_\sharp : D_{\mathbf{A}^1,\Lambda} \rightleftarrows \underline{D}_{\mathbf{A}^1,\Lambda} : \mathbf{R}\rho^*$$

which is compatible with the enlargement (6.1.6.1) in the sense that the following diagram is essentially commutative:

$$\begin{array}{ccc} D^{e\!f\!f}_{\mathbf{A}^1,\Lambda} & \longrightarrow & \underline{D}^{e\!f\!f}_{\mathbf{A}^1,\Lambda} \\ {\scriptstyle \Sigma^\infty}\downarrow & & \downarrow{\scriptstyle \underline{\Sigma}^\infty} \\ D_{\mathbf{A}^1,\Lambda} & \longrightarrow & \underline{D}_{\mathbf{A}^1,\Lambda}. \end{array}$$

Corollary 6.1.11 *Consider a Grothendieck topology t on our category of schemes $\mathscr{S}$. Let S be a scheme in $\mathscr{S}$, and M an object of $D_{\mathbf{A}^1,\Lambda}(S)$. Then M satisfies t-descent in $D_{\mathbf{A}^1,\Lambda}(S)$ if and only if $\mathbf{L}\rho_\sharp(M)$ satisfies t-descent in $\underline{D}_{\mathbf{A}^1,\Lambda}(S)$.*

Proof Let $f : \mathscr{X} \longrightarrow S$ be a diagram of S-schemes of finite type. Define

$$H^q(\mathscr{X}, M(p)) = \mathrm{Hom}_{D_{\mathbf{A}^1,\Lambda}(S)}(\Lambda_{\mathscr{X}}, \mathbf{L}f^*(M)(p)[q]),$$

$$\underline{H}^q(\mathscr{X}, M(p)) = \mathrm{Hom}_{\underline{D}_{\mathbf{A}^1,\Lambda}(S)}(\underline{\Lambda}_{\mathscr{X}}, \mathbf{L}f^* \mathbf{L}\rho_\sharp(M)(p)[q])$$

for any integers p and q. The full faithfulness of $\mathbf{L}\rho_\sharp$ ensures that the comparison map

$$H^q(\mathscr{X}, M(p)) \longrightarrow \underline{H}^q(\mathscr{X}, M(p))$$

is always bijective. This proposition follows then from the fact that M (resp. $\mathbf{L}\rho_\sharp(M)$) satisfies t-descent if and only if, for any integers p and q, for any S-scheme of finite type X, and any t-hypercover $\mathscr{X} \longrightarrow X$, the induced map

$$H^q(X, M(p)) \longrightarrow H^q(\mathscr{X}, M(p)) \text{ (resp. } \underline{H}^q(X, M(p)) \longrightarrow \underline{H}^q(\mathscr{X}, M(p)) \text{)}$$

is bijective. □

We conclude this section with another interesting application of the preceding results.

Corollary 6.1.12 *Consider the hypotheses and assumptions of Proposition 6.1.4. We suppose furthermore that the generalized abelian premotivic category $\mathscr{A}$ satisfies condition (C) of Section 5.1.35.*

1. Then the triangulated premotivic category $D^{eff}_{\mathbf{A}^1}(\mathscr{A})$ is τ-continuous.
2. Assume furthermore that $D^{eff}_{\mathbf{A}^1}(\mathscr{A})$ and $D^{eff}_{\mathbf{A}^1}(\underline{\mathscr{A}})$ are compactly τ-generated. Then the triangulated premotivic category $D_{\mathbf{A}^1}(\mathscr{A})$ is τ-continuous.

Proof According to Proposition 5.2.41, the category $D^{eff}_{\mathbf{A}^1}(\mathscr{A})$ is τ-continuous. According to Corollary 6.1.5, the functor $\mathbf{L}\rho_\sharp : D^{eff}_{\mathbf{A}^1}(\mathscr{A}) \longrightarrow D^{eff}_{\mathbf{A}^1}(\underline{\mathscr{A}}) : \mathbf{R}\rho^*$ is fully faithful and commutes with $\mathbf{L}f^*$. Thus Point (1) follows.

In the assumption of Point (2), we deduce from Proposition 5.3.44 that $D_{\mathbf{A}^1}(\underline{\mathscr{A}})$ is $(\mathbf{Z} \times \tau)$-continuous. Thus it is sufficient to apply Corollary 6.1.9 as in the effective case to get the assertion of Point (2). □

Example 6.1.13 According to the second point of Example 5.1.37, we can apply this corollary to the enlargement

$$\mathrm{Sh}_{Nis}(Sm, \Lambda) \longrightarrow \mathrm{Sh}_{Nis}\left(\mathscr{S}^{ft}, \Lambda\right).$$

Thus, we deduce that the triangulated premotivic categories $D^{eff}_{\mathbf{A}^1, \Lambda}$ and $D_{\mathbf{A}^1, \Lambda}$ are both continuous.

6.2 The fundamental example

Recall the following theorem of Ayoub [Ayo07b]:

Theorem 6.2.1 *The triangulated premotivic categories $D^{eff}_{\mathbf{A}^1, \Lambda}$ and $D_{\mathbf{A}^1, \Lambda}$ satisfy the localization property.*

Corollary 6.2.2

1. *The premotivic category $D_{\mathbf{A}^1, \Lambda}$ is a motivic category.*
2. *It is compactly generated by the Tate twist.*
3. *Suppose that $\mathscr{T}$ is a derived premotivic category (see Section 5.2.9) which is a motivic category. Then there exists a canonical morphism of derived premotivic categories:*

$$D_{\mathbf{A}^1, \mathbf{Z}} \longrightarrow \mathscr{T}.$$

Proof The first assertion follows from the previous theorem and Remark 2.4.47. The second one follows from Corollary 5.3.42. The last one follows from Proposition 3.3.5 and Example 5.3.34. □

Remark 6.2.3 Thus, Theorem 2.4.50 can be applied to $D_{\mathbf{A}^1, \Lambda}$. In particular, for any separated morphism of finite type $f : T \longrightarrow S$, there exists a pair of adjoint functors

$$f_! : D_{\mathbf{A}^1, \Lambda}(T) \rightleftarrows D_{\mathbf{A}^1, \Lambda}(S) : f^!$$

as in the theorem *loc. cit.* so that we have removed the quasi-projective assumption in [Ayo07a].

6.2.4 Because the cdh topology is finer than the Nisnevich topology, we get an adjunction of generalized premotivic categories:

$$a^*_{cdh} : \underline{D}_{\mathbf{A}^1, \Lambda} \rightleftarrows D_{\mathbf{A}^1}\left(\mathrm{Sh}_{cdh}\left(\mathscr{S}^{ft}, \Lambda\right)\right) : \mathbf{R}a_{cdh,*}.$$

Corollary 6.2.5 *For any scheme S, the composite functor*

$$D_{\mathbf{A}^1}(S, \Lambda) \longrightarrow \underline{D}_{\mathbf{A}^1}(S, \Lambda) \xrightarrow{a_{cdh}} D_{\mathbf{A}^1}\left(\mathrm{Sh}_{cdh}\left(\mathscr{S}^{ft}/S, \Lambda\right)\right)$$

is fully faithful.

Moreover, it induces an enlargement of premotivic categories:

(6.2.5.1) $$\underline{D}_{\mathbf{A}^1, \Lambda} \rightleftarrows D_{\mathbf{A}^1}\left(\mathrm{Sh}_{cdh}\left(\mathscr{S}^{ft}, \Lambda\right)\right).$$

Remark 6.2.6 This corollary is a generalization in our derived setting of the main theorem of [Voe10c]. Note that if $\dim(S) > 0$, there is no hope that the above composite functor is essentially surjective because as soon as Z is a nowhere dense closed subscheme of S, the premotive $\underline{M}_S^{cdh}(Z, \Lambda)$ does not belong to its image (*cf.* Remark 6.1.2).

Proof According to Corollary 6.2.2 and Proposition 3.3.10, any Tate spectrum E of $D_{\mathbf{A}^1}(S, \Lambda)$ satisfies *cdh*-descent in the derived premotivic category $\underline{D}_{\mathbf{A}^1, \Lambda}$, and this implies the first assertion by Proposition 5.3.30 and Corollary 6.1.11. The second assertion then follows from the fact the forgetful functor

$$D_{\mathbf{A}^1}\left(\mathrm{Sh}_{cdh}\left(\mathscr{S}^{ft}/S, \Lambda\right)\right) \longrightarrow \underline{D}_{\mathbf{A}^1}(S, \Lambda)$$

commutes with direct sums (its left adjoint preserves compact objects). □

6.3 Nearly Nisnevich sheaves

6.3.1 Throughout this section, we fix an abelian premotivic category $\mathscr{A}$ and we consider the canonical premotivic adjunction (5.1.2.1) associated with $\mathscr{A}$.

We assume $\mathscr{A}$ satisfies the following properties.

(i) $\mathscr{A}$ is compatible with Nisnevich topology, so that we have from (5.1.2.1) a premotivic adjunction:

(6.3.1.1) $$\gamma^* : \mathrm{Sh}_{Nis}(Sm, \mathbf{Z}) \rightleftarrows \mathscr{A} : \gamma_*.$$

(ii) $\mathscr{A}$ is finitely presented (*i.e.* the functors $\mathrm{Hom}_{\mathscr{A}_S}(M_S(X), -)$ preserve filtered colimits and form a conservative family, Definition 1.3.11).

(iii) For any scheme S, and for any open immersion $U \longrightarrow X$ of smooth S-schemes, the map $M_S(U) \longrightarrow M_S(X)$ is a monomorphism.

6 Localization and the universal derived example

(iv) For any scheme S, the functor $\gamma_* : \mathscr{A}_S \longrightarrow \mathrm{Sh}_{Nis}(Sm/S, \mathbf{Z})$ is exact.

Note that the functor $\gamma_* : \mathscr{A}_S \longrightarrow \mathrm{Sh}_{Nis}(Sm/S, \mathbf{Z})$ is exact and conservative. As it also preserves filtered colimits, this functor preserves in fact small colimits.

Observe also that, according to assumptions (i)–(iv), the abelian premotivic category of Tate spectra $\mathrm{Sp}(\mathscr{A})$ is compatible with the Nisnevich topology and **N**-generated. Moreover, we get a canonical premotivic adjunction

(6.3.1.2) $$\gamma^* : \mathrm{Sp}(\mathrm{Sh}_{Nis}(Sm, \mathbf{Z})) \rightleftarrows \mathrm{Sp}(\mathscr{A}) : \gamma_*$$

such that γ_* is conservative and preserves small colimits.

In the following, we show how one can deduce properties of the premotivic triangulated categories $\mathrm{D}^{\mathit{eff}}_{\mathbf{A}^1}(\mathscr{A})$ and $\mathrm{D}_{\mathbf{A}^1}(\mathscr{A})$ from the good properties of $\mathrm{D}^{\mathit{eff}}_{\mathbf{A}^1, \mathbf{Z}}$ and $\mathrm{D}_{\mathbf{A}^1, \mathbf{Z}}$.

6.3.a Support property (effective case)

Proposition 6.3.2 *For any scheme S, the functor $\gamma_* : \mathrm{C}(\mathscr{A}_S) \longrightarrow \mathrm{C}(\mathrm{Sh}_{Nis}(Sm/S, \mathbf{Z}))$ preserves and detects $\mathbf{A}^1$-equivalences.*

Proof It follows immediately from Corollary 5.2.31 that γ_* preserves $\mathbf{A}^1$-equivalences. The fact that it detects them can be rephrased by saying that the induced functor

$$\gamma_* : \mathrm{D}^{\mathit{eff}}_{\mathbf{A}^1}(\mathscr{A}_S) \longrightarrow \mathrm{D}^{\mathit{eff}}_{\mathbf{A}^1, \mathbf{Z}}(S)$$

is conservative. This is obviously true once we observe that its left adjoint is essentially surjective on generators. □

Corollary 6.3.3 *The right derived functor*

$$\mathbf{R}\gamma_* = \gamma_* : \mathrm{D}^{\mathit{eff}}_{\mathbf{A}^1}(\mathscr{A}_S) \longrightarrow \mathrm{D}^{\mathit{eff}}_{\mathbf{A}^1, \mathbf{Z}}(S)$$

is conservative.

Proposition 6.3.4 *Let $f : S' \longrightarrow S$ be a finite morphism of schemes. Then the induced functor*

$$f_* : \mathrm{C}(\mathscr{A}_{S'}) \longrightarrow \mathrm{C}(\mathscr{A}_S)$$

preserves colimits and $\mathbf{A}^1$-equivalences.

Proof We first prove that f_* preserves colimits. We know the functors γ_* preserve colimits and are conservative. As we have the identification $\gamma_* f_* = f_* \gamma_*$, it is sufficient to prove the property for $\mathscr{A} = \mathrm{Sh}_{Nis}(Sm, \mathbf{Z})$. Let X be a smooth S-scheme. It is sufficient to prove that, for any point x of X, if X_x^h denotes the henselization of X at x, the functor

$$\mathrm{Sh}_{Nis}(Sm/S', \mathbf{Z}) \longrightarrow \mathscr{A}b \quad , \quad F \longmapsto f_*(F)(X_x^h) = F(S' \times_S X_x^h)$$

commutes with colimits. Moreover, the scheme $S' \times_S X_x^h$ is finite over X_x^h, so that we have $S' \times_S X_x^h = \amalg_i Y_i$, where the Y_i's are a finite family of henselian local schemes over $S' \times_S X_x^h$. Hence, we have to check that the functor $F \mapsto \bigoplus_i F(Y_i)$ preserves colimits. As colimits commute with sums, it is thus sufficient to prove that the functors $F \mapsto F(Y_i)$ commute with colimits. This follows from the fact that the local henselian schemes Y_i are points of the topos of sheaves over the small Nisnevich site of X.

We are left to prove that the functor $f_* : C(\mathscr{A}_{S'}) \longrightarrow C(\mathscr{A}_S)$ respects $\mathbf{A}^1$-equivalences. For this, we shall study the behavior of f_* with respect to the $\mathbf{A}^1$-resolution functor constructed in Section 5.2.26. Note that f_* commutes with limits because it has a left adjoint. In particular, we know that f_* is exact. Moreover, one checks easily that $f_* R_{\mathbf{A}^1}^{(n)} = f_* R_{\mathbf{A}^1}^{(n)}$. As f_* commutes with colimits, this gives the formula $f_* R_{\mathbf{A}^1} = R_{\mathbf{A}^1} f_*$. Let C be a complex of Nisnevich sheaves of abelian groups on Sm/S'. Choose a quasi-isomorphism $C \longrightarrow C'$ with C' a Nis-flasque complex. Applying Proposition 5.2.28, we know that $R_{\mathbf{A}^1}(C')$ is $\mathbf{A}^1$-fibrant and that we get a canonical $\mathbf{A}^1$-equivalence

$$f_*(C) \longrightarrow f_*(C') \longrightarrow f_*(R_{\mathbf{A}^1}(C')) = R_{\mathbf{A}^1}(f_*(C')).$$

Hence, we are reduced to proving that f_* preserves $\mathbf{A}^1$-equivalences between $\mathbf{A}^1$-fibrant objects. But such $\mathbf{A}^1$-equivalences are quasi-isomorphisms, so the result follows from the exactness of f_*. $\square$

Proposition 6.3.5 *For any open immersion of schemes $j : U \longrightarrow S$, the exchange transformation $j_\sharp \gamma_* \longrightarrow \gamma_* j_\sharp$ is an isomorphism of functors.*

Proof Let X be a scheme, and F a Nisnevich sheaf of abelian groups on Sm/X. Define the category $\mathscr{C}_F$ as follows. The objects are the pairs (Y, s), where Y is a smooth scheme over X, and s is a section of F over Y. The arrows $(Y, s) \longrightarrow (Y', s')$ are the morphisms $f \in \mathrm{Hom}_{\mathrm{Sh}_{Nis}(Sm/X,\mathbf{Z})}(\mathbf{Z}_X(Y), \mathbf{Z}_X(Y'))$ such that $f^*(s') = s$. We have a canonical functor

$$\varphi_F : \mathscr{C}_F \longrightarrow \mathrm{Sh}_{Nis}(Sm/X, \mathbf{Z})$$

defined by $\varphi_F(Y, s) = \mathbf{Z}_X(Y)$, and one easily checks that the canonical map

$$\varinjlim_{\mathscr{C}_F} \varphi_F = \varinjlim_{(Y,s) \in \mathscr{C}_F} \mathbf{Z}_X(Y) \longrightarrow F$$

is an isomorphism in $\mathrm{Sh}_{Nis}(Sm/X, \mathbf{Z})$ (this is essentially a reformulation of the Yoneda lemma).

Consider now an object F in the category $\mathscr{A}_U$. We get two categories $\mathscr{C}_{\gamma_*(F)}$ and $\mathscr{C}_{\gamma_*(j_\sharp(F))}$. There is a functor

$$i : \mathscr{C}_{\gamma_*(F)} \longrightarrow \mathscr{C}_{\gamma_*(j_\sharp(F))}$$

6 Localization and the universal derived example

which is defined by the formula $i(Y, s) = (Y, j_\sharp(s))$. To explain our notation, let us say that we view s as a morphism from $M_S(U, \mathscr{A})$ to F, so that $j_\sharp(s)$ is a morphism from $M_S(Y, \mathscr{A}) = j_\sharp M_S(U, \mathscr{A})$ to $j_\sharp(F)$. This functor i has right adjoint

$$i' : \mathscr{C}_{\gamma_*(j_\sharp(F))} \longrightarrow \mathscr{C}_{\gamma_*(F)}$$

defined by $i'(Y, s) = (Y_U, s_U)$, where $Y_U = Y \times_S U$, and s_U is the section of $\gamma_*(F)$ over Y_U that corresponds to the section $j^*(s)$ of $j^* j_\sharp \gamma_*(F)$ over Y_U under the canonical isomorphism $\gamma_*(F) \simeq j^* j_\sharp \gamma_*(F)$ (here, we use strongly the fact that the functor $j_\sharp$ is fully faithful). The existence of a right adjoint implies i is cofinal. This latter property is sufficient for the canonical morphism

$$\varinjlim_{\mathscr{C}_{\gamma_*(F)}} \varphi_{\gamma_*(j_\sharp(F))} \circ i \longrightarrow \varinjlim_{\mathscr{C}_{\gamma_*(j_\sharp(F))}} \varphi_{\gamma_*(j_\sharp(F))} = \gamma_*(j_\sharp(F))$$

to be an isomorphism. But the functor $\varphi_{\gamma_*(j_\sharp(F))} \circ i$ is exactly the composition of the functor $j_\sharp$ with $\varphi_{\gamma_*(F)}$. As the functor $j_\sharp$ commutes with colimits, we have

$$\varinjlim_{\mathscr{C}_{\gamma_*(F)}} \varphi_{\gamma_*(j_\sharp(F))} \circ i = \varinjlim_{\mathscr{C}_{\gamma_*(F)}} j_\sharp \varphi_{\gamma_*(F)} \simeq j_\sharp \varinjlim_{\mathscr{C}_{\gamma_*(F)}} \varphi_{\gamma_*(F)} \simeq j_\sharp(\gamma_*(F)).$$

Hence we obtain a canonical isomorphism $j_\sharp(\gamma_*(F)) \simeq \gamma_*(j_\sharp(F))$. It is easily seen that the corresponding map $\gamma_*(F) \longrightarrow j^*(\gamma_*(j_\sharp(F))) = \gamma_*(j^* j_\sharp(F))$ is the image under γ_* of the unit map $F \longrightarrow j^* j_\sharp(F)$. This shows the isomorphism we have constructed is the exchange morphism. □

Corollary 6.3.6 *For any open immersion of schemes $j : U \longrightarrow S$, the functor $j_\sharp : \mathscr{A}_U \longrightarrow \mathscr{A}_S$ is exact. Moreover, the induced functor*

$$j_\sharp : C(\mathscr{A}_U) \longrightarrow C(\mathscr{A}_S)$$

preserves $\mathbf{A}^1$-equivalences.

Proof Using the fact that γ_* is exact and conservative, and Propositions 6.3.2 and 6.3.5, it is sufficient to prove this corollary when $\mathscr{A} = \mathrm{Sh}_{Nis}(Sm, \mathbf{Z})$. It is straightforward to prove exactness using Nisnevich points. The fact that $j_\sharp$ preserves $\mathbf{A}^1$-equivalences follows from the exactness property and from the obvious fact it preserves strong $\mathbf{A}^1$-equivalences. □

Corollary 6.3.7 *Let $j : U \longrightarrow S$ be an open immersion of schemes. For any object M of $\mathrm{D}^{e\!f\!f}_{\mathbf{A}^1}(\mathscr{A}_U)$ the exchange morphism*

(6.3.7.1) $$\mathbf{L}j_\sharp(\mathbf{R}\gamma_*(M)) \longrightarrow \mathbf{R}\gamma_*(\mathbf{L}j_\sharp(M))$$

is an isomorphism in $\mathrm{D}^{e\!f\!f}_{\mathbf{A}^1}(S, \mathbf{Z})$.

6.3.b Support property (stable case)

6.3.8 Recall from Section 5.3.17 that the premotivic adjunction (γ^*, γ_*) induces a canonical adjunction of abelian premotivic categories that we denote by:

$$\tilde{\gamma}^* : \mathrm{Sp}(\mathrm{Sh}_{Nis}(Sm, \mathbf{Z})) \rightleftarrows \mathrm{Sp}(\mathscr{A}_S) : \tilde{\gamma}_*.$$

Proposition 6.3.9 *For any scheme S, the induced functor*

$$\tilde{\gamma}_* : \mathrm{C}\left(\mathrm{Sp}(\mathscr{A}_S)\right) \rightleftarrows \mathrm{C}\left(\mathrm{Sp}(\mathrm{Sh}_{Nis}(Sm/S, \mathbf{Z}))\right)$$

preserves and detects stable $\mathbf{A}^1$*-equivalences.*

Proof Using the equivalence between symmetric Tate spectra and non-symmetric Tate spectra, we are reduced to proving this for complexes of non-symmetric Tate spectra. Consider a non-symmetric Tate spectrum $(E_n)_{n \in \mathbf{N}}$ with suspension maps $\sigma_n : E_n\{1\} \longrightarrow E_{n+1}$. The non-symmetric Tate spectrum $\tilde{\gamma}_*(E)$ is equal to $\gamma_*(E_n)$ in degree $n \in \mathbf{Z}$, and the suspension map is given by the composite:

$$\mathbb{1}_S\{1\} \otimes_S \gamma_*(E_n) \longrightarrow \gamma_*(\gamma^*(\mathbb{1}_S\{1\}) \otimes_S E_n) = \gamma_*(E_n\{1\}) \xrightarrow{\gamma_*(\sigma_n)} E_{n+1}.$$

Thus, the result follows from Propositions 6.3.2 and 5.3.40. □

Corollary 6.3.10 *The right derived functor*

$$\mathbf{R}\gamma_* = \gamma_* : \mathrm{D}_{\mathbf{A}^1}(\mathscr{A}_S) \longrightarrow \mathrm{D}_{\mathbf{A}^1, \mathbf{Z}}(S)$$

is conservative.

Proposition 6.3.11 *Let* $j : U \longrightarrow X$ *be an open immersion of schemes. For any object M of* $\mathrm{D}_{\mathbf{A}^1}(\mathscr{A}_U)$*, the exchange morphism*

$$\mathbf{L}j_\sharp(\mathbf{R}\gamma_*(M)) \longrightarrow \mathbf{R}\gamma_*(\mathbf{L}j_\sharp(M))$$

is an isomorphism in $\mathrm{D}_{\mathbf{A}^1, \mathbf{Z}}(X)$.

Proof From Corollary 6.3.6 and the $\mathscr{P}$-base change formula for the open immersion j, one easily deduces that $j_\sharp$ preserves stable $\mathbf{A}^1$-equivalences of (non-symmetric) Tate spectra. Moreover, Proposition 6.3.5 shows that $j_\sharp \gamma_* = \gamma_* j_\sharp$ at the level of Tate spectra. This concludes the proof. □

Corollary 6.3.12 *The triangulated premotivic category* $\mathrm{D}_{\mathbf{A}^1}(\mathscr{A})$ *satisfies the support property.*

Proof According to Corollary 6.3.10, the functor $\mathbf{R}\gamma_*$ is conservative. Thus, by virtue of the preceding proposition, to prove the support property in the case of $\mathrm{D}_{\mathbf{A}^1}(\mathscr{A})$ it is sufficient to prove it in the case where $\mathscr{A} = \mathrm{Sh}_{Nis}(Sm, \mathbf{Z})$. This follows from Theorems 6.2.1 and 2.4.50. □

6.3.c Localization for smooth schemes

Lemma 6.3.13 *Let $i : Z \longrightarrow S$ be a closed immersion which admits a smooth retraction $p : S \longrightarrow Z$. Then the exchange transformation*

$$L\gamma^* Ri_* \longrightarrow Ri_* L\gamma^*$$

is an isomorphism in $D^{eff}_{\mathbf{A}^1}(\mathscr{A}_S)$ (resp. $D_{\mathbf{A}^1}(\mathscr{A}_S)$).

Proof We first remark that for any object C of $C(\mathscr{A}_Z)$ (resp. $C(Sp(\mathscr{A}_Z))$) the canonical sequence

$$j_\sharp(pj)^*(C) \longrightarrow p^*(C) \longrightarrow i_*(C)$$

is a cofiber sequence in $D^{eff}_{\mathbf{A}^1}(\mathscr{A}_S)$ (resp. $D_{\mathbf{A}^1}(\mathscr{A})_S)$). Indeed, we can check this after applying the exact conservative functor γ_*. The sequence we obtain is canonically isomorphic through exchange transformations to

$$j_\sharp j^* p^*(\gamma_* C) \longrightarrow p^*(\gamma_* C) \longrightarrow i_* i^* p^*(\gamma_* C)$$

using Corollary 6.3.7, the commutation of γ_* with j^*, p^* and i_* (recall it is the right adjoint of a premotivic adjunction) and the relation $pi = 1$. But this last sequence is a cofiber sequence in $D^{eff}_{\mathbf{A}^1, Z}(S)$ (resp. $D_{\mathbf{A}^1, Z}(S)$) because it satisfies the localization property (see Theorem 6.2.1).

Using exchange transformations, we obtain a morphism of distinguished triangles in $DM^{eff}_Z(S)$

$$\begin{array}{ccccccc} \gamma^* j_\sharp j^* p^*(C) & \longrightarrow & \gamma^* p^*(C) & \longrightarrow & \gamma^* i_*(C) & \longrightarrow & \gamma^* j_\sharp j^* p^*(C)[1] \\ \parallel & & \parallel & & \downarrow Ex(\gamma^*, i_*) & & \parallel \\ j_\sharp j^* p^*(\gamma^* C) & \longrightarrow & p^*(\gamma^* C) & \longrightarrow & i_*(\gamma^* C) & \longrightarrow & j_\sharp j^* p^*(\gamma^* C)[1]. \end{array}$$

The first two vertical arrows are isomorphisms as γ^* is the left adjoint of a premotivic adjunction; thus the morphism $Ex(\gamma^*, i_*)$ is also an isomorphism. $\square$

Proposition 6.3.14 *Let $i : Z \longrightarrow S$ be a closed immersion. If i admits a smooth retraction, then $D^{eff}_{\mathbf{A}^1}(\mathscr{A})$ satisfies (Loc_i).*

Proof This follows from Proposition 2.3.19 and the preceding lemma. $\square$

Corollary 6.3.15 *Let S be a scheme. Then the premotivic category $D^{eff}_{\mathbf{A}^1}(\mathscr{A})$ (resp. $D_{\mathbf{A}^1}(\mathscr{A})$) satisfies localization with respect to any closed immersion between smooth S-schemes.*

Proof Let $i : Z \longrightarrow X$ be closed immersion between smooth S-schemes. We want to prove that $D^{eff}_{\mathbf{A}^1}(\mathscr{A})$ (resp. $D_{\mathbf{A}^1}(\mathscr{A})$) satisfies localization with respect to i. According to Corollary 2.3.18, it is sufficient to prove that for any smooth S-scheme S, the canonical map

$$M_S(X/X - X_Z) \longrightarrow i_* M_Z(X_Z)$$

is an isomorphism, where we use the notation of *loc. cit.* and $M(\cdot, \mathscr{A})$ denotes the geometric sections of $D_{\mathbf{A}^1}^{eff}(\mathscr{A})$ (resp. $D_{\mathbf{A}^1}(\mathscr{A})$). But the premotivic triangulated category $D_{\mathbf{A}^1}(\mathscr{A})$ (resp. $D_{\mathbf{A}^1}^{eff}(\mathscr{A})$) satisfies the Nisnevich separation property and the Sm-base change property. Thus, we can argue locally in S for the Nisnevich topology. Thus, the statement is reduced to the preceding proposition as i admits locally for the Nisnevich topology a smooth retraction (see for example [Dég07, 4.5.11]). □

7 Basic homotopy commutative algebra

7.1 Rings

Definition 7.1.1 A symmetric monoidal model category $\mathscr{V}$ satisfies the *monoid axiom* if, for any trivial cofibration $A \longrightarrow B$ and any object X, the smallest class of maps of $\mathscr{V}$ which contains the map $X \otimes A \longrightarrow X \otimes B$ and is stable under pushouts and transfinite compositions is contained in the class of weak equivalences.

7.1.2 Let $\mathscr{V}$ be a symmetric monoidal category. We denote by $Mon(\mathscr{V})$ the category of monoids in $\mathscr{V}$. If $\mathscr{V}$ has small colimits, the forgetful functor

$$U : Mon(\mathscr{V}) \longrightarrow \mathscr{V}$$

has a left adjoint

$$F : \mathscr{V} \longrightarrow Mon(\mathscr{V}).$$

Theorem 7.1.3 *Let $\mathscr{V}$ be a symmetric monoidal combinatorial model category which satisfies the monoid axiom. The category of monoids $Mon(\mathscr{V})$ is endowed with the structure of a combinatorial model category whose weak equivalences (resp. fibrations) are the morphisms of commutative monoids which are weak equivalences (resp. fibrations) in $\mathscr{V}$. In particular, the forgetful functor $U : Mon(\mathscr{V}) \longrightarrow \mathscr{V}$ is a right Quillen functor. Moreover, if the unit object of $\mathscr{V}$ is cofibrant, then any cofibrant object of $Mon(\mathscr{V})$ is cofibrant as an object of $\mathscr{V}$.*

Proof This is a very particular case of the third assertion of [SS00, Theorem 4.1] (the fact that $Mon(\mathscr{V})$ is combinatorial whenever $\mathscr{V}$ is follows for instance from [Bek00, Proposition 2.3]). □

Definition 7.1.4 A symmetric monoidal model category $\mathscr{V}$ is *strongly* Q-*linear* if the underlying category of $\mathscr{V}$ is additive and Q-linear (i.e. all the objects of $\mathscr{V}$ are uniquely divisible).

Remark 7.1.5 If $\mathscr{V}$ is a strongly Q-linear stable model category, then it is Q-linear in the sense of Section 3.2.14.

7 Basic homotopy commutative algebra

Lemma 7.1.6 *Let $\mathcal{V}$ be a strongly $\mathbf{Q}$-linear model category, G a finite group, and $u : E \longrightarrow F$ an equivariant morphism of representations of G in $\mathcal{V}$. Then, if u is a cofibration in $\mathcal{V}$, so is the induced map $E_G \longrightarrow F_G$ (where the subscript G denotes the coinvariants under the action of the group G).*

Proof The map $E_G \longrightarrow F_G$ is easily seen to be a direct factor (retract) of the cofibration $E \longrightarrow F$. □

7.1.7 If $\mathcal{V}$ is a symmetric monoidal category, we denote by $Comm(\mathcal{V})$ the category of commutative monoids in $\mathcal{V}$. If $\mathcal{V}$ has small colimits, the forgetful functor

$$U : Comm(\mathcal{V}) \longrightarrow \mathcal{V}$$

has a left adjoint

$$F : \mathcal{V} \longrightarrow Comm(\mathcal{V}).$$

Theorem 7.1.8 *Let $\mathcal{V}$ be a symmetric monoidal combinatorial model category. Assume that $\mathcal{V}$ is left proper and tractable, satisfies the monoid axiom, and is strongly $\mathbf{Q}$-linear. Then the category of commutative monoids $Comm(\mathcal{V})$ is endowed with the structure of a combinatorial model category whose weak equivalences (resp. fibrations) are the morphisms of commutative monoids which are weak equivalences (resp. fibrations) in $\mathcal{V}$. In particular, the forgetful functor $U : Comm(\mathcal{V}) \longrightarrow \mathcal{V}$ is a right Quillen functor.*

If moreover the unit object of $\mathcal{V}$ is cofibrant, then any cofibrant object of $Comm(\mathcal{V})$ is cofibrant as an object of $\mathcal{V}$.

Proof We will observe first that $\mathcal{V}$ is freely powered in the sense of [Lur17, Definition 4.5.4.2]. Therefore, the existence of this model category structure will follow from a general result of Lurie [Lur17, Proposition 4.5.4.6]. For this, it is sufficient to check that a G-equivariant map $f : A \longrightarrow B$ in $\mathcal{V}$ which is a trivial cofibration when we forget the G-action has the left lifting property with respect to any G-equivariant map $p : X \longrightarrow Y$ which is a fibration in $\mathcal{V}$ (after forgetting the G-action). In other words, we have to check that the map induced by f and p in $\mathcal{V}$

$$\mathrm{Hom}_V(B, X) \longrightarrow \mathrm{Hom}_\mathcal{V}(A, X) \times_{\mathrm{Hom}_\mathcal{V}(A,X)} \mathrm{Hom}_\mathcal{V}(Y, B)$$

will induce a surjective map after we apply the G-invariants functor (we let the reader construct a natural G-action on $\mathrm{Hom}_V(B, X)$, the G-invariants of which gives the $\mathbf{Q}$-vector space of G-equivariant maps from B to X). Since G is a finite group, the G-invariant subspace functor is exact, hence this is obvious. This proves the first assertion. The second assertion of the theorem is true by definition.

The last assertion is proved by a careful analysis of pushouts by free maps in $Comm(\mathcal{V})$ as follows. For two cofibrations $u : A \longrightarrow B$ and $v : C \longrightarrow D$ in $\mathcal{V}$, write $u \wedge v$ for the map

$$u \wedge v : A \otimes D \amalg_{A \otimes C} B \otimes C \longrightarrow B \otimes D$$

(which is a cofibration by definition of monoidal model categories). By iterating this construction, we get, for a cofibration $u : A \longrightarrow B$ in $\mathcal{V}$, a cofibration

$$\wedge^n(u) = \underbrace{u \wedge \cdots \wedge u}_{n \text{ times}} : \square^n(u) \longrightarrow B^{\otimes n} .$$

Note that the symmetric group $\mathfrak{S}_n$ acts naturally on $B^{\otimes n}$ and $\square^n(u)$. We define

$$Sym^n(B) = (B^{\otimes n})_{\mathfrak{S}_n} \quad \text{and} \quad Sym^n(B,A) = \square^n(u)_{\mathfrak{S}_n} .$$

By virtue of Lemma 7.1.6, we get a cofibration of $\mathscr{V}$:

$$\sigma^n(u) : Sym^n(B,A) \longrightarrow Sym^n(B).$$

The free map $F(u) : F(A) \longrightarrow F(B)$ can be filtered by $F(A)$-modules as follows. Define $D_0 = F(A)$. As $A = Sym^1(B,A)$, we have a natural morphism $F(A) \otimes Sym^1(B,A) \longrightarrow F(A)$. The objects D_n are then defined by induction with the pushouts below:

$$\begin{array}{ccc} F(A) \otimes Sym^n(B,A) & \xrightarrow{1_{F(A)} \otimes \sigma^n(u)} & F(A) \otimes Sym^n(B) \\ \downarrow & & \downarrow \\ D_{n-1} & \longrightarrow & D_n . \end{array}$$

We get natural maps $D_n \longrightarrow F(B)$ which induce an isomorphism

$$\varinjlim_{n \geq 0} D_n \simeq F(B)$$

in such a way that the morphism $F(u)$ correspond to the canonical map

$$F(A) = D_0 \longrightarrow \varinjlim_{n \geq 0} D_n .$$

Hence, if $F(A)$ is cofibrant, all the maps $D_{n-1} \longrightarrow D_n$ are cofibrations, so that the map $F(A) \longrightarrow F(B)$ is a cofibration in $\mathscr{V}$. In the particular case where A is the initial object of $\mathscr{V}$, we see that for any cofibrant object B of $\mathscr{V}$, the free commutative monoid $F(B)$ is cofibrant as an object of $\mathscr{V}$ (because the initial object of $Comm(\mathscr{V})$ is the unit object of $\mathscr{V}$). This also implies that, if u is a cofibration between cofibrant objects, the map $F(u)$ is a cofibration in $\mathscr{V}$.

This description of $F(u)$ also allows us to compute the pushouts of $F(u)$ in $Comm(\mathscr{V})$ in $\mathscr{V}$ as follows. Consider a pushout

$$\begin{array}{ccc} F(A) & \xrightarrow{F(u)} & F(B) \\ \downarrow & & \downarrow \\ R & \xrightarrow{v} & S \end{array}$$

7 Basic homotopy commutative algebra 233

in $Comm(\mathscr{V})$. For $n \geq 0$, define R_n by the pushouts of $\mathscr{V}$:

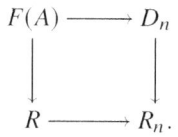

$$\begin{array}{ccc} F(A) & \longrightarrow & D_n \\ \downarrow & & \downarrow \\ R & \longrightarrow & R_n. \end{array}$$

We then have an isomorphism
$$\varinjlim_{n \geq 0} R_n \simeq S.$$

In particular, if u is a cofibration between cofibrant objects, the morphism of commutative monoids $v : R \longrightarrow S$ is then a cofibration in $\mathscr{V}$. As the forgetful functor U preserves filtered colimits, we conclude easily from there (by the small object argument [Hov99, Theorem 2.1.14]) that any cofibration of $Comm(\mathscr{V})$ is a cofibration of $\mathscr{V}$. Using again that the unit object of $\mathscr{V}$ is cofibrant in $\mathscr{V}$ (i.e. that the initial object of $Comm(\mathscr{V})$ is cofibrant in $\mathscr{V}$) this proves the last assertion of the theorem. □

Corollary 7.1.9 *Let $\mathscr{V}$ be a symmetric monoidal combinatorial model category. Assume that $\mathscr{V}$ is left proper and tractable, satisfies the monoid axiom, and is strongly $\mathbf{Q}$-linear. Consider a small set H of maps of $\mathscr{V}$, and denote by $L_H \mathscr{V}$ the left Bousfield localization of $\mathscr{V}$ by H; see [Bar10, Theorem 4.7]. Define the class of H-equivalences in $\mathrm{Ho}(\mathscr{V})$ to be the class of maps which become invertible in $\mathrm{Ho}(L_H \mathscr{V})$. If H-equivalences are stable under (derived) tensor products in $\mathrm{Ho}(\mathscr{V})$, then $L_H \mathscr{V}$ is a symmetric monoidal combinatorial model category (which is again left proper and tractable, satisfies the monoid axiom, and is strongly $\mathbf{Q}$-linear).*

In particular, under these assumptions, there exists a morphism of commutative monoids $\mathbb{1} \longrightarrow R$ in $\mathscr{V}$ which is a weak equivalence of $L_H \mathscr{V}$, with R a cofibrant and fibrant object of $L_H \mathscr{V}$.

Proof The first assertion is a triviality. The last assertion follows immediately: the map $\mathbb{1} \longrightarrow R$ is simply obtained as a fibrant replacement of $\mathbb{1}$ in the model category $Comm(L_H \mathscr{V})$ obtained from Theorem 7.1.8 applied to $L_H \mathscr{V}$. □

7.1.10 Consider now a category $\mathscr{S}$, as well as a closed symmetric monoidal bifibred category $\mathscr{M}$ over $\mathscr{S}$. We shall also assume that the fibers of $\mathscr{M}$ admit limits and colimits.

Then the categories $Mon(\mathscr{M}(X))$ (resp. $Comm(\mathscr{M}(X))$) define a bifibred category over $\mathscr{S}$ as follows. Given a morphism $f : X \longrightarrow Y$, the functor

$$f^* : \mathscr{M}(Y) \longrightarrow \mathscr{M}(X)$$

is symmetric monoidal, so that it preserves monoids (resp. commutative monoids) as well as morphisms between them. It thus induces a functor

(7.1.10.1) $$\begin{array}{c} f^* : Mon(\mathscr{M}(Y)) \longrightarrow Mon(\mathscr{M}(X)) \\ (\text{resp. } f^* : Comm(\mathscr{M}(Y)) \longrightarrow Comm(\mathscr{M}(X))\,). \end{array}$$

As $f^* : \mathcal{M}(Y) \longrightarrow \mathcal{M}(X)$ is symmetric monoidal, its right adjoint f_* is lax monoidal: there is a natural morphism

(7.1.10.2) $$\mathbb{1}_Y \longrightarrow f_*(\mathbb{1}_X) = f_* f^*(\mathbb{1}_Y),$$

and, for any objects A and B of $\mathcal{M}(X)$, there is a natural morphism

(7.1.10.3) $$f_*(A) \otimes_Y f_*(B) \longrightarrow f_*(A \otimes_X B)$$

which corresponds by adjunction to the map

$$f^*(f_*(A) \otimes_Y f_*(B)) \simeq f^* f_*(A) \otimes f^* f_*(B) \longrightarrow A \otimes B.$$

Hence the functor f_* also preserves monoids (resp. commutative monoids) as well as morphisms between them, so that we get a functor

(7.1.10.4) $$f_* : Mon(\mathcal{M}(X)) \longrightarrow Mon(\mathcal{M}(Y))$$
$$(\text{resp. } f_* : Comm(\mathcal{M}(X)) \longrightarrow Comm(\mathcal{M}(Y))).$$

By construction, the functor f^* of (7.1.10.1) is a left adjoint to the functor f_* of (7.1.10.4). These constructions extend to morphisms of $\mathscr{S}$-diagrams in a similar way.

Proposition 7.1.11 *Let $\mathcal{M}$ be a symmetric monoidal combinatorial fibred model category over $\mathscr{S}$. Assume that, for any object X of $\mathscr{S}$, the model category $\mathcal{M}(X)$ satisfies the monoid axiom (resp. is left proper and tractable, satisfies the monoid axiom, and is strongly **Q**-linear).*

(a) For any object X of $\mathscr{S}$, the category $Mon(\mathcal{M})(X)$ (resp. $Comm(\mathcal{M})(X)$) of monoids (resp. of commutative monoids) in $\mathcal{M}(X)$ is a combinatorial model category structure whose weak equivalences (resp. fibrations) are the morphisms of commutative monoids which are weak equivalences (resp. fibrations) in $\mathcal{M}(X)$. This turns $Mon(\mathcal{M})$ (resp. $Comm(\mathcal{M})$) into a combinatorial fibred model category over $\mathscr{S}$.

(b) For any morphism of $\mathscr{S}$-diagrams $\varphi : (\mathscr{X}, I) \longrightarrow (Y, J)$, the adjunction

$$\varphi^* : Mon(\mathcal{M})(\mathscr{Y}, J) \rightleftarrows Mon(\mathcal{M})(\mathscr{X}, I) : \varphi_*$$

(resp. $\varphi^* : Comm(\mathcal{M})(\mathscr{Y}, J) \rightleftarrows Comm(\mathcal{M})(\mathscr{X}, I) : \varphi_*$)

is a Quillen adjunction (where the categories of monoids $Mon(\mathcal{M})(\mathscr{X}, I)$ (resp. of commutative monoids $Comm(\mathcal{M})(\mathscr{X}, I)$) are endowed with the injective model category structure obtained from Proposition 3.1.7 applied to $Mon(\mathcal{M})$ (resp. to $Comm(\mathcal{M})$).

(d) If moreover, for any object X of $\mathscr{S}$, the unit $\mathbb{1}_X$ is cofibrant in $\mathcal{M}(X)$, then, for any morphism of $\mathscr{S}$-diagrams $\varphi : (\mathscr{X}, I) \longrightarrow (Y, J)$, the square

(7.1.11.1)
$$\begin{array}{ccc} \operatorname{Ho}(Mon(\mathcal{M}))(\mathcal{Y},J) & \xrightarrow{\mathbf{L}\varphi^*} & \operatorname{Ho}(Mon(\mathcal{M}))(\mathcal{X},I) \\ U\downarrow & & \downarrow U \\ \operatorname{Ho}(\mathcal{M})(\mathcal{Y},J) & \xrightarrow{\mathbf{L}\varphi^*} & \operatorname{Ho}(\mathcal{M})(\mathcal{X},I) \end{array}$$

is essentially commutative. Similarly, in the respective case, the square

(7.1.11.2)
$$\begin{array}{ccc} \operatorname{Ho}(Comm(\mathcal{M}))(\mathcal{Y},J) & \xrightarrow{\mathbf{L}\varphi^*} & \operatorname{Ho}(Comm(\mathcal{M}))(\mathcal{X},I) \\ U\downarrow & & \downarrow U \\ \operatorname{Ho}(\mathcal{M})(\mathcal{Y},J) & \xrightarrow{\mathbf{L}\varphi^*} & \operatorname{Ho}(\mathcal{M})(\mathcal{X},I) \end{array}$$

is essentially commutative.

Proof Assertion (a) is an immediate consequence of Theorem 7.1.3 (resp. of Theorem 7.1.8), and assertion (b) is a particular case of Proposition 3.1.11 (beware that the injective model category structure on $Comm(\mathcal{M})(\mathcal{X},I)$ does not necessarily coincide with the model category structure given by Theorem 7.1.3 (resp. of Theorem 7.1.8) applied to the injective model structure on $\mathcal{M}(\mathcal{X},I)$). For assertion (d), we see by the second assertion of Proposition 3.1.6 that it is sufficient to prove it when $\varphi : X \longrightarrow Y$ is simply a morphism of $\mathcal{S}$. In this case, by construction of the total left derived functor of a left Quillen functor, this follows from the fact that φ^* commutes with the forgetful functor and from the fact that, by virtue of the last assertion of Theorem 7.1.3 (resp. of Theorem 7.1.8), the forgetful functor U preserves weak equivalences and cofibrant objects. $\square$

Remark 7.1.12 The main application of the preceding corollary will come from assertion (d): it says that, given a monoid (resp. a commutative monoid) R in $\mathcal{M}(Y)$ and a morphism $f : X \longrightarrow Y$, the image of R under the functor

$$\mathbf{L}f^* : \operatorname{Ho}(\mathcal{M})(Y) \longrightarrow \operatorname{Ho}(\mathcal{M})(X)$$

is canonically endowed with the structure of a monoid (resp. of a commutative monoid) in the strongest sense possible. Under the assumptions of assertion (c) of Proposition 7.1.11, we shall often abuse the terminology by saying that $\mathbf{L}f^*(R)$ is a monoid (resp. a commutative monoid) in $\mathcal{M}(X)$ without explicit reference to the model category structure on $Mon(\mathcal{M})(X)$ (resp. on $Comm(\mathcal{M})(X)$). Similarly, for any monoid (resp. commutative monoid) R in $\mathcal{M}(X)$, $\mathbf{R}f_*(R)$ will be canonically endowed with the structure of a monoid (resp. a commutative monoid) in $\mathcal{M}(Y)$. In particular, for any monoid (resp. commutative monoid) R in $\mathcal{M}(Y)$, the adjunction map

$$R \longrightarrow \mathbf{R}f_* \mathbf{L}f^*(R)$$

is a morphism of monoids (i.e. is a map in the homotopy category $\operatorname{Ho}(Mon(\mathcal{M}))(X)$ (resp. $\operatorname{Ho}(Comm(\mathcal{M}))(X)$)), and, for any monoid (resp. commutative monoid) R in $\mathcal{M}(X)$, the adjunction map

$$Lf^* \mathbf{R} f_*(R) \longrightarrow R$$

is a morphism of monoids (i.e. is a map in the homotopy category $\operatorname{Ho}(Mon(\mathcal{M}))(Y)$ (resp. $\operatorname{Ho}(Comm(\mathcal{M}))(Y)$)).

Remark 7.1.13 In order to get a good homotopy theory of commutative monoids without the strongly **Q**-linear assumption, we should replace commutative monoids by E_∞-algebras (i.e. objects endowed with the structure of a commutative monoid up to a bunch of coherent homotopies). More generally, we should prove the analog of Theorem 7.1.3 and of Theorem 7.1.8 by replacing $Mon(\mathcal{V})$ by the category of algebras of some 'well-behaved' operad, and then get as a consequence the analog of Proposition 7.1.11. All this is a consequence of the general constructions and results of [Spi01, BM03, BM09].

However, in the case we are interested in, the homotopy theory of commutative monoids in some category of spectra $\mathcal{V}$, it seems that some version of Shipley's *positive stable model structure* (*cf.* [Shi04, Proposition 3.1]) would provide a good model category for commutative monoids, which, by Lurie's strictification theorem [Lur17, Theorem 4.5.4.7], would be equivalent to the homotopy theory of E_∞-algebras in $\mathcal{V}$. This kind of technique is now available in the context of the stable homotopy theory of schemes, which provides a good setting to speak of motivic commutative ring spectra; see [Hor13, GG16, GG18, PS18]. Therefore, Theorem 7.1.8 and Proposition 7.1.11 are in fact true in SH for genuine commutative monoids without any **Q**-linearity assumption.

7.2 Modules

7.2.1 Given a monoid R in a symmetric monoidal category $\mathcal{V}$, we shall write $R\text{-mod}(\mathcal{V})$ for the category of (left) R-modules. The forgetful functor

$$U : R\text{-mod}(\mathcal{V}) \longrightarrow \mathcal{V}$$

is a left adjoint to the free R-module functor

$$R \otimes (-) : \mathcal{V} \longrightarrow R\text{-mod}(\mathcal{V}).$$

If $\mathcal{V}$ has enough small colimits, and if R is a commutative monoid, the category $R\text{-mod}(\mathcal{V})$ is endowed with a unique symmetric monoidal structure such that the functor $R \otimes (-)$ is naturally symmetric monoidal. We shall denote by $\otimes_R$ the tensor product of $R\text{-mod}(\mathcal{V})$.

Theorem 7.2.2 *Let $\mathcal{V}$ be a combinatorial symmetric model category which satisfies the monoid axiom.*

(i) *For any monoid R in $\mathcal{V}$, the category of right (resp. left) R-modules is a combinatorial model category with weak equivalences (resp. fibrations) the morphisms of R-modules which are weak equivalences (resp. fibrations) in $\mathcal{V}$.*

(ii) *For any commutative monoid R in $\mathscr{V}$, the model category of R-modules given by (i) is a combinatorial symmetric monoidal model category which satisfies the monoid axiom.*

Proof Assertions (i) and (ii) are particular cases of the first two assertions of [SS00, Theorem 4.1]. □

Definition 7.2.3 A symmetric monoidal model category $\mathscr{V}$ is *perfect* if it has the following properties:

(a) $\mathscr{V}$ is combinatorial and tractable (Definition 3.1.27);
(b) $\mathscr{V}$ satisfies the monoid axiom;
(c) for any weak equivalence of monoids $R \longrightarrow S$, the functor $M \longmapsto S \otimes_R M$ is a left Quillen equivalence from the category of left R-modules to the category of left S-modules;
(d) weak equivalences are stable under small sums in $\mathscr{V}$.

Remark 7.2.4 If $\mathscr{V}$ is a perfect symmetric monoidal model category, then, for any commutative monoid R, the symmetric monoidal model category of R-modules in $\mathscr{V}$ given by Theorem 7.2.2 (ii) is also perfect: condition (c) is quite obvious, and condition (d) comes from the fact that the forgetful functor $U : R\text{-mod} \longrightarrow \mathscr{V}$ commutes with small sums, while it preserves and detects weak equivalences. Note that condition (d) implies that the functor $U : \text{Ho}(R\text{-mod}) \longrightarrow \text{Ho}(\mathscr{V})$ preserves small sums.

Remark 7.2.5 If $\mathscr{V}$ is a stable symmetric monoidal model category which satisfies the monoid axiom, then for any monoid R of $\mathscr{V}$, the model category of (left) R-modules given by Theorem 7.2.2 is stable as well: the suspension functor of $\text{Ho}(R\text{-mod})$ is given by the derived tensor product with the R-bimodule $R[1]$, which is clearly invertible with inverse $R[-1]$.

In this work, a basic example of perfect model categories is provided by those coming from stable $\mathbf{A}^1$-derived premotivic categories (cf. Definition 5.3.22):

Proposition 7.2.6 *Let t be an admissible topology. Then, for any scheme S in $\mathscr{S}$, the symmetric monoidal model structure on $C(\text{Sp}(\text{Sh}_t(\mathscr{P}/S,\mathbf{Z})))$ underlying the triangulated category $D_{\mathbf{A}^1}(\text{Sh}_t(\mathscr{P}/S,\mathbf{Z}))$ is perfect.*

Proof The generating family of $\text{Sh}_t(\mathscr{P}/S,\mathbf{Z})$ is flat in the sense of [CD09, 3.1], so that, by virtue of [CD09, prop. 7.22 and cor. 7.24], the assumptions of Proposition 7.2.9 are fulfilled. □

Proposition 7.2.7 *Let $\mathscr{V}$ be a stable perfect symmetric monoidal model category. Assume furthermore that $\text{Ho}(\mathscr{V})$ admits a small family $\mathscr{G}$ of compact generators (as a triangulated category). For any monoid R in $\mathscr{V}$, the triangulated category $\text{Ho}(R\text{-mod}(\mathscr{V}))$ admits the set $\{R \otimes^{\mathbf{L}} E \mid E \in \mathscr{G}\}$ as a family of compact generators.*

Proof We have a derived adjunction

$$R \otimes^{\mathbf{L}} (-) : \mathrm{Ho}(\mathscr{V}) \rightleftarrows \mathrm{Ho}(R\text{-}\mathrm{mod}(\mathscr{V})) : U.$$

As the functor U preserves small sums the functor $R \otimes^{\mathbf{L}} (-)$ preserves compact objects. But U is also conservative, so that $\{R \otimes^{\mathbf{L}} E \mid E \in \mathscr{G}\}$ is a family of compact generators of $\mathrm{Ho}(R\text{-}\mathrm{mod}(\mathscr{V}))$. □

Remark 7.2.8 If $\mathscr{V}$ is a combinatorial symmetric model category which satisfies the monoid axiom, then there are two ways to derive the tensor product. The first one consists in deriving the left Quillen bifunctor $(-)\otimes(-)$, which gives the usual derived tensor product

$$(-) \otimes^{\mathbf{L}} (-) : \mathrm{Ho}(\mathscr{V}) \times \mathrm{Ho}(\mathscr{V}) \longrightarrow \mathrm{Ho}(\mathscr{V}).$$

Recall that, by construction, $A \otimes^{\mathbf{L}} B = A' \otimes B'$, where A' and B' are cofibrant replacements of A and B, respectively. On the other hand, the monoid axiom gives that, for any object A of $\mathscr{V}$, the functor $A \otimes (-)$ preserves weak equivalences between cofibrant objects, which implies that it has a total left derived functor

$$A \otimes^{\mathbf{L}} (-) : \mathrm{Ho}(\mathscr{V}) \longrightarrow \mathrm{Ho}(\mathscr{V}).$$

Despite the fact we have adopted very similar (not to say identical) notations for these two derived functors, there is no reason why they would coincide in general: by construction, the second one is defined by $A \otimes^{\mathbf{L}} B = A \otimes B'$, where B' is some cofibrant replacement of B. However, they coincide quite often in practice (e.g. for simplicial sets, for the good reason that all of them are cofibrant, or for symmetric S^1-spectra, or for complexes of quasi-coherent $\mathscr{O}_X$-modules over a quasi-compact and quasi-separated scheme X).

Proposition 7.2.9 *Let $\mathscr{V}$ be a stable combinatorial symmetric monoidal model category which satisfies the monoid axiom. Assume furthermore that, for any cofibrant object A of $\mathscr{V}$, the functor $A \otimes (-)$ preserves weak equivalences (in other words, that the two ways to derive the tensor product explained in Remark 7.2.8 coincide), and that weak equivalences are stable under small sums in $\mathscr{V}$. Then the symmetric monoidal model category $\mathscr{V}$ is perfect.*

Proof We just have to check condition (c) of Definition 7.2.3. Consider a weak equivalence of monoids $R \longrightarrow S$. We then get a derived adjunction

$$S \otimes^{\mathbf{L}}_R (-) : \mathrm{Ho}(R\text{-}\mathrm{mod}(\mathscr{V})) \rightleftarrows \mathrm{Ho}(S\text{-}\mathrm{mod}(\mathscr{V})) : U,$$

where $S \otimes^{\mathbf{L}}_R (-)$ is the left derived functor of the functor $M \longmapsto S \otimes_R M$. We have to prove that, for any left R-module M, the map

$$M \longrightarrow S \otimes^{\mathbf{L}}_R M$$

is an isomorphism in $\mathrm{Ho}(\mathscr{V})$. As this is a morphism of triangulated functors which commutes with sums, and as $\mathrm{Ho}(R\text{-}\mathrm{mod}(\mathscr{V}))$ is well generated in the sense of

7 Basic homotopy commutative algebra

Neeman [Nee01] (as the localization of a stable combinatorial model category), it is sufficient to check this when M runs over a small family of generators of $\text{Ho}(R\text{-}\mathrm{mod}(\mathscr{V}))$. Let us choose a small family of generators $\mathscr{G}$ of $\text{Ho}(\mathscr{V})$. As the forgetful functor from $\text{Ho}(R\text{-}\mathrm{mod}(\mathscr{V}))$ to $\text{Ho}(\mathscr{V})$ is conservative, we see that $\{R \otimes^{\mathbf{L}} E \mid E \in \mathscr{G}\}$ is a small generating family of $\text{Ho}(R\text{-}\mathrm{mod}(\mathscr{V}))$. We are thus reduced to proving that the map

$$R \otimes^{\mathbf{L}} E \longrightarrow S \otimes^{\mathbf{L}}_R (R \otimes^{\mathbf{L}} E) \simeq S \otimes^{\mathbf{L}} E$$

is an isomorphism for any object E in $\mathscr{G}$. For this, we can assume that E is cofibrant, and this follows then from the fact that the functor $(-) \otimes E$ preserves weak equivalences by assumption. □

7.2.10 Let $\mathscr{S}$ be a category endowed with an admissible class of morphisms $\mathscr{P}$, and $\mathscr{M}$ a cocomplete symmetric monoidal $\mathscr{P}$-fibred category. Consider a monoid R in the symmetric monoidal category $\mathscr{M}(1_{\mathscr{S}}, \mathscr{S})$ (i.e. a section of the fibred category $Mon(\mathscr{M})$ over $\mathscr{S}$). In other words, R consists of the data of a monoid R_X for each object X of $\mathscr{S}$, and of a morphism of monoids $a_f : f^*(R_Y) \longrightarrow R_X$ for each map $f : X \longrightarrow Y$ in $\mathscr{S}$, subject to coherence relations; see Section 3.1.2.

For an object X of $\mathscr{S}$, we shall write $R\text{-}\mathrm{mod}(X)$ for the category of (left) R_X-modules in $\mathscr{M}(X)$, i.e.

$$R\text{-}\mathrm{mod}(X) = R_X\text{-}\mathrm{mod}(\mathscr{M}(X)).$$

This defines a fibred category $R\text{-}\mathrm{mod}$ over $\mathscr{S}$ as follows.

For a morphism $f : X \longrightarrow Y$, the inverse image functor

(7.2.10.1) $$f^* : R\text{-}\mathrm{mod}(Y) \longrightarrow R\text{-}\mathrm{mod}(X)$$

is defined by

(7.2.10.2) $$M \longmapsto R_X \otimes_{f^*(R_Y)} f^*(M)$$

(where, on the right-hand side, f^* stands for the inverse image functor in $\mathscr{M}$). The functor (7.2.10.1) has a right adjoint

(7.2.10.3) $$f_* : R\text{-}\mathrm{mod}(X) \longrightarrow R\text{-}\mathrm{mod}(Y)$$

which is simply the functor induced by $f_* : \mathscr{M}(X) \longrightarrow \mathscr{M}(Y)$ (as the latter sends R_X-modules to $f_*(R_X)$-modules, which are themselves R_Y-modules via the map a_f).

If the map f is a $\mathscr{P}$-morphism, then, for any R_X-module M, the object $f_\sharp(M)$ has the natural structure of an R_Y-module: using the map a_f, M has the natural structure of an $f^*(R_Y)$-module

$$f^*(R_Y) \otimes_X M \longrightarrow M,$$

and applying $f_\sharp$, we get by the $\mathscr{P}$-projection formula (1.1.26) a morphism

$$R_Y \otimes f_\sharp(M) \simeq f_\sharp(f^*(R_Y) \otimes M) \longrightarrow f_\sharp(M)$$

which defines a natural R_Y-module structure on $f_\sharp(M)$. For a $\mathscr{P}$-morphism $f : X \longrightarrow Y$, we define a functor

(7.2.10.4) $$f_\sharp : R\text{-}\mathrm{mod}(X) \longrightarrow R\text{-}\mathrm{mod}(Y)$$

as the functor induced by $f_\sharp : \mathscr{M}(X) \longrightarrow \mathscr{M}(Y)$. Note that the functor (7.2.10.4) is a left adjoint to the functor (7.2.10.1) whenever the map $a_f : f^*(R_Y) \longrightarrow R_X$ is an isomorphism in $\mathscr{M}(X)$.

We shall say that R is a *cartesian monoid in $\mathscr{M}$ over $\mathscr{S}$* if R is a monoid of $\mathscr{M}(1_{\mathscr{C}}, \mathscr{C})$ such that all the structural maps $f^*(R_Y) \longrightarrow R_X$ are isomorphisms (i.e. if R is a cartesian section of the fibred category $\mathrm{Mon}(\mathscr{M})$ over $\mathscr{S}$).

If R is a cartesian monoid in $\mathscr{M}$ over $\mathscr{S}$, then R-mod is a $\mathscr{P}$-fibred category over $\mathscr{S}$: to see this, it remains to prove that, for any pullback square of $\mathscr{S}$

$$\begin{array}{ccc} X' & \xrightarrow{g} & X \\ {\scriptstyle f'}\downarrow & & \downarrow{\scriptstyle f} \\ Y' & \xrightarrow{h} & Y \end{array}$$

in which f is a $\mathscr{P}$-morphism, and for any R_X-module M, the base change map

$$f'_\sharp g^*(M) \longrightarrow h^* f_\sharp(M)$$

is an isomorphism, which follows immediately from the analogous formula for $\mathscr{M}$.

Similarly, we see that if R is a commutative monoid of $\mathscr{M}(1_{\mathscr{S}}, \mathscr{S})$ (i.e. R_X is a commutative monoid in $\mathscr{M}(X)$ for all X in $\mathscr{S}$), then R-mod is a symmetric monoidal $\mathscr{P}$-fibred category.

Proposition 7.2.11 *Let $\mathscr{M}$ be a combinatorial symmetric monoidal $\mathscr{P}$-fibred model category over $\mathscr{S}$ which satisfies the monoid axiom, and R a monoid in $\mathscr{M}(1_{\mathscr{S}}, \mathscr{S})$ (resp. a cartesian monoid in $\mathscr{M}$ over $\mathscr{S}$). Then Theorem 7.2.2 (i) applied termwise turns R-mod into a combinatorial fibred model category (resp. a combinatorial $\mathscr{P}$-fibred model category).*

If moreover R is commutative, then R-mod is a combinatorial symmetric monoidal fibred model category (resp. a combinatorial symmetric monoidal $\mathscr{P}$-fibred model category).

Proof Choose, for each object X of $\mathscr{S}$, two small sets of maps I_X and J_X which generate the class of cofibrations and the class of trivial cofibrations in $\mathscr{M}(X)$ respectively. Then $R_X \otimes_X I_X$ and $R_X \otimes_X J_X$ generate the class of cofibrations and the class of trivial cofibrations in $R\text{-}\mathrm{mod}(X)$, respectively. For a map $f : X \longrightarrow Y$ in $\mathscr{S}$, we see from formula (7.2.10.2) that the functor (7.2.10.1) sends these generating cofibrations and trivial cofibrations to cofibrations and trivial cofibrations respectively, from which we deduce that the functor (7.2.10.1) is a left Quillen functor. In the respective case, if f is a $\mathscr{P}$-morphism, then we deduce similarly from the projection formula (1.1.26) in $\mathscr{M}$ that the functor (7.2.10.4) sends generating cofibrations

7 Basic homotopy commutative algebra

and trivial cofibrations to cofibrations and trivial cofibrations, respectively. The last assertion follows easily by applying Theorem 7.2.2 (ii) termwise. □

Definition 7.2.12 Let $\mathcal{M}$ be a symmetric monoidal $\mathcal{P}$-fibred model category over $\mathcal{S}$. A *homotopy cartesian monoid* R in $\mathcal{M}$ will be a homotopy cartesian section of $\mathrm{Mon}(\mathcal{M})$.

Proposition 7.2.13 *Let $\mathcal{M}$ be a perfect symmetric monoidal $\mathcal{P}$-fibred model category over $\mathcal{S}$, and consider a homotopy cartesian monoid R in $\mathcal{M}$ over $\mathcal{S}$. Then $\mathrm{Ho}(R\text{-mod})$ is a $\mathcal{P}$-fibred category over $\mathcal{S}$, and*

$$R \otimes^{\mathbf{L}} (-) : \mathrm{Ho}(\mathcal{M}) \longrightarrow \mathrm{Ho}(R\text{-mod})$$

is a morphism of $\mathcal{P}$-fibred categories. In the case where R is commutative, $\mathrm{Ho}(R\text{-mod})$ is even a symmetric monoidal $\mathcal{P}$-fibred category.

Moreover, for any weak equivalence between homotopy cartesian monoids $R \longrightarrow S$ over $\mathcal{S}$, the Quillen morphism

$$S \otimes_R (-) : R\text{-mod} \longrightarrow S\text{-mod}$$

induces an equivalence of $\mathcal{P}$-fibred categories over $\mathcal{S}$

$$S \otimes_R^{\mathbf{L}} (-) : \mathrm{Ho}(R\text{-mod}) \longrightarrow \mathrm{Ho}(S\text{-mod}).$$

Proof It is sufficient to prove these assertions by restricting everything over $\mathcal{S}/S$, where S runs over all the objects of $\mathcal{S}$. In particular, we may (and shall) assume that $\mathcal{S}$ has a terminal object S. As $\mathcal{M}$ is perfect, it follows from condition (c) of Definition 7.2.3 that we can replace R by any of its cofibrant resolutions. In particular, we may assume that R_S is a cofibrant object of $\mathrm{Mon}(\mathcal{M})(S)$. We can thus define a termwise cofibrant cartesian monoid R' as the family of monoids $f^*(R_S)$, where $f : X \longrightarrow S$ runs over all the objects of $\mathcal{S} \simeq \mathcal{S}/S$. There is a canonical morphism of homotopy cartesian monoids $R' \longrightarrow R$ which is a termwise weak equivalence. We thus get, by condition (c) of Definition 7.2.3, an equivalence of fibred categories

$$R \otimes_{R'}^{\mathbf{L}} (-) : \mathrm{Ho}(R'\text{-mod}) \longrightarrow \mathrm{Ho}(R\text{-mod}).$$

We can thus replace R by R', which just means that we can assume that R is cartesian and termwise cofibrant. The first assertion then follows easily from Proposition 7.2.11. In the case where R is commutative, we prove that $\mathrm{Ho}(R\text{-mod})$ is a $\mathcal{P}$-fibred symmetric monoidal category as follows. Let $f : X \longrightarrow Y$ a morphism of $\mathcal{S}$. We would like to prove that, for any object M in $\mathrm{Ho}(R\text{-mod})(X)$ and any object N in $\mathrm{Ho}(R\text{-mod})(Y)$, the canonical map

(7.2.13.1) $$\mathbf{L}f_\sharp(M \otimes_R^{\mathbf{L}} f^*(N)) \longrightarrow \mathbf{L}f_\sharp(M) \otimes_R^{\mathbf{L}} N$$

is an isomorphism. By adjunction, this is equivalent to proving that, for any objects N and E in $\mathrm{Ho}(R\text{-mod})(Y)$, the map

(7.2.13.2) $$f^*\mathbf{R}Hom_R(N, E) \longrightarrow \mathbf{R}Hom_R(f^*(N), f^*(E))$$

is an isomorphism in $\mathrm{Ho}(R\text{-}\mathrm{mod})(X)$ (where $\mathbf{R}Hom_R$ stands for the internal Hom of $\mathrm{Ho}(R\text{-}\mathrm{mod})$). But the forgetful functors

$$U : \mathrm{Ho}(R\text{-}\mathrm{mod})(X) \longrightarrow \mathrm{Ho}(\mathscr{M})(X)$$

are conservative, commute with f^* for any $\mathscr{P}$-morphism f, and commute with internal Hom: by adjunction, this follows immediately from the fact that the functors

$$R \otimes^{\mathbf{L}} (-) : \mathrm{Ho}(\mathscr{M})(X) \longrightarrow \mathrm{Ho}(R\text{-}\mathrm{mod})(X) \simeq \mathrm{Ho}(R'\text{-}\mathrm{mod})(X)$$

are symmetric monoidal and define a morphism of $\mathscr{P}$-fibred categories (and thus, in particular, commute with $f_\sharp$ for any $\mathscr{P}$-morphism f). Hence, to prove that (7.2.13.2) is an isomorphism, it is sufficient to prove that its analog in $\mathrm{Ho}(\mathscr{M})$ is so, which follows immediately from the fact that the analog of (7.2.13.1) is an isomorphism in $\mathrm{Ho}(\mathscr{M})$ by assumption.

For the last assertion, we are also reduced to the case where R and S are cartesian and termwise cofibrant, in which case this follows easily again from condition (c) of Definition 7.2.3. □

Proposition 7.2.14 *Let $\mathscr{M}$ be a combinatorial symmetric monoidal model category over $\mathscr{S}$ which satisfies the monoid axiom. Then, for any cartesian monoid R in $\mathscr{M}$ over $\mathscr{S}$ we have a Quillen morphism*

$$R \otimes (-) : \mathscr{M} \longrightarrow R\text{-}\mathrm{mod} \ .$$

If, for any object X of $\mathscr{S}$, the unit object $\mathbb{1}_X$ is cofibrant in $\mathscr{M}(X)$ and the monoid R_X is cofibrant in $\mathrm{Mon}(\mathscr{M})(X)$, then the forgetful functors also define a Quillen morphism

$$U : R\text{-}\mathrm{mod} \longrightarrow \mathscr{M} \ .$$

Proof The first assertion is obvious. For the second one, note that, for any object X of $\mathscr{S}$, the monoid R_X is also cofibrant as an object of $\mathscr{M}(X)$; see Theorem 7.1.3. This implies that the forgetful functor

$$U : R_X\text{-}\mathrm{mod} \longrightarrow \mathscr{M}(X)$$

is a left Quillen functor: by the small object argument and by definition of the model category structure of Theorem 7.2.2 (i), this follows from the trivial fact that the endofunctor

$$R_X \otimes (-) : \mathscr{M}(X) \longrightarrow \mathscr{M}(X)$$

is a left Quillen functor itself whenever R_X is cofibrant in $\mathscr{M}(X)$. □

Remark 7.2.15 The results of the preceding proposition (as well as their proofs) are also true in terms of $\mathscr{P}_{cart}$-fibred categories (Remark 3.1.21) over the category of $\mathscr{S}/S$-diagrams for any object S of $\mathscr{S}$ (whence over all $\mathscr{S}$-diagrams whenever $\mathscr{S}$ has a terminal object).

7.2.16 Consider now a noetherian scheme S of finite dimension. We choose a full subcategory of the category of separated noetherian S-schemes of finite dimension which is stable under finite limits, contains separated S-schemes of finite type, and such that, for any étale S-morphism $Y \longrightarrow X$, if X is in $\mathscr{S}/S$, so is Y. We denote by $\mathscr{S}/S$ this chosen category of S-schemes.

We also fix an admissible class $\mathscr{P}$ of morphisms of $\mathscr{S}/S$ which contains the class of étale morphisms.

Definition 7.2.17 A property P of $\mathrm{Ho}(\mathscr{M})$, for $\mathscr{M}$ a stable combinatorial $\mathscr{P}$-fibred model category over $\mathscr{S}/S$, is *homotopy linear* if the following implications are true.

(a) If $\gamma : \mathscr{M} \longrightarrow \mathscr{M}'$ is a Quillen equivalence (i.e. a Quillen morphism which is termwise a Quillen equivalence) between stable combinatorial $\mathscr{P}$-fibred model categories over $\mathscr{S}/S$, then $\mathscr{M}$ has property P if and only if $\mathscr{M}'$ has property P.
(b) If $\mathscr{M}$ is a stable combinatorial symmetric monoidal $\mathscr{P}$-model category which satisfies the monoid axiom, and such that the unit $\mathbb{1}_X$ of $\mathscr{M}(X)$ is cofibrant, then, for any cartesian and termwise cofibrant monoid R in $\mathscr{M}$ over $\mathscr{S}/S$, R-mod has property P.

Proposition 7.2.18 *The following properties are homotopy linear:* $\mathbf{A}^1$-*homotopy invariance,* $\mathbf{P}^1$-*stability, the localization property, the property of proper transversality, separability, semi-separability, and* t-*descent (for a given Grothendieck topology t on $\mathscr{S}/S$).*

Proof Property (a) of the definition above is obvious. Property (b) comes from the fact that the forgetful functors

$$U : \mathrm{Ho}(R\text{-}\mathrm{mod}) \longrightarrow \mathrm{Ho}(\mathscr{M})$$

are conservative and commute with all the operations: $\mathbf{L}f^*$ and $\mathbf{R}f_*$ for any morphism f, as well as $\mathbf{L}f_\sharp$ for any $\mathscr{P}$-morphism (by Proposition 7.2.14). Hence any property formulated in terms of equations involving only these operations is homotopy linear. $\square$

Part III
Motivic complexes and relative cycles

Throughout this part, we adopt the special convention that smooth means smooth separated of finite type. This also concerns the framework of premotivic categories: we assume the admissible class Sm comprises smooth separated morphisms of finite type.

This assumption is required when using the theory of finite correspondences (see more precisely Example 9.1.4).

8 Relative cycles

8.0.1 In this entire section, $\mathscr{S}$ is the category of noetherian schemes; any scheme is assumed to be noetherian. We fix a subring $\Lambda \subset \mathbf{Q}$ which will be the ring of coefficients of the algebraic cycles considered in the following section. When we want to be precise, we say Λ-*cycle* for "algebraic cycle with coefficients in Λ". Otherwise, we simply say *cycle* and the reader must assume that all algebraic cycles have their coefficients in the ring Λ.

8.1 Definitions

8.1.a The category of cycles

8.1.1 Let X be a scheme. As usual, an element of the underlying set of X will be called a *point* and a morphism $\mathrm{Spec}\,(k) \longrightarrow X$ where k is a field will be called a *geometric point*. We often identify a point $x \in X$ with the corresponding geometric point $\mathrm{Spec}\,(\kappa_x) \longrightarrow X$. However, the explicit expression "the point $\mathrm{Spec}\,(k) \longrightarrow X$" always refers to a geometric point.

As our schemes are assumed to be noetherian, any immersion $f : X \longrightarrow Y$ is quasi-compact. Thus, according to [GD60, 9.5.10], the *schematic closure* $\bar{X}$ of X in Y exists, which gives a unique factorization of f

$$X \xrightarrow{j} \bar{X} \xrightarrow{i} Y$$

such that i is a closed immersion and j is an open immersion with dense image.[82] Note that when Y is reduced, $\bar{X}$ coincides with the topological closure of X in Y with its induced reduced subscheme structure. In this case, we simply call $\bar{Y}$ the closure of Y in X.

Definition 8.1.2 A Λ-*cycle* is a pair (X, α) such that X is a scheme and α is a Λ-linear combination of points of X. A generic point of (X, α) is a point which appears in the

---
[82] Recall the scheme $\bar{X}$ is characterized by the property of being the smallest sub-scheme of Y for which such a factorization exists.

Λ-linear combination α with a non-zero coefficient. The support $\mathrm{Supp}(\alpha)$ of α is the closure of the generic points of α, regarded as a reduced closed subscheme of X.

A morphism of Λ-cycles $(Y,\beta) \longrightarrow (X,\alpha)$ is a morphism of schemes $f : Y \longrightarrow X$ such that $f(\mathrm{Supp}(\beta)) \subset \mathrm{Supp}(\alpha)$. We say this morphism is *pseudo-dominant* if for any generic point y of (Y,β), $f(y)$ is a generic point of (X,α).

When considering such a pair (X,α), we will denote it simply by α and refer to X as the *domain* of α. We also use the notation $\alpha \subset X$ to mean the domain of the cycle α is the scheme X.

The category of Λ-cycles is functorial in Λ with respect to morphisms of integral rings. In what follows, cycles are assumed to have coefficients in Λ unless explicitly stated (following our conventions for this section, see Section 8.0.1).

8.1.3 Given a property ($\mathscr{P}$) of morphisms of schemes, we will say that a morphism $f : \beta \longrightarrow \alpha$ of cycles satisfies property ($\mathscr{P}$) if the induced morphism $f|_{\mathrm{Supp}(\beta)}^{\mathrm{Supp}(\alpha)}$ satisfies property ($\mathscr{P}$).

Definition 8.1.4 Let X be a scheme. We denote by $X^{(0)}$ the set of generic points of X. We define as usual the *cycle associated with* X as the cycle with domain X:

$$\langle X \rangle = \sum_{x \in X^{(0)}} \mathrm{lg}(\mathscr{O}_{X,x}).x.$$

The integer $\mathrm{lg}(\mathscr{O}_{X,x})$, the length of an artinian local ring, is called the *geometric multiplicity* of x in X.

When no confusion is possible, we usually omit the delimiters in the notation $\langle X \rangle$. As an example, we say that α *is a cycle over* X to mean the existence of a structural morphism of cycles $\alpha \longrightarrow \langle X \rangle$.

8.1.5 When Z is a closed subscheme of a scheme X, we denote by $\langle Z \rangle_X$ the cycle $\langle Z \rangle$ considered as a cycle with domain X.

Consider a cycle α with domain X. Let $(Z_i)_{i \in I}$ be the family of the reduced closure of generic points of α. Then we can write α uniquely as $\alpha = \sum_{i \in I} n_i . \langle Z_i \rangle_X$. We call this the *standard form* of α for short.

Definition 8.1.6 Let $\alpha = \sum_{i \in I} n_i . x_i$ be a cycle with domain X and $f : X \longrightarrow Y$ be any morphism.

For any $i \in I$, put $y_i = f(x_i)$. Then f induces an extension field $\kappa(x_i)/\kappa(y_i)$ between the residue fields. We let d_i be the degree of this extension field if it is finite and 0 otherwise.

We define the *pushforward* of α by f as the cycle with domain Y

$$f_*(\alpha) = \sum_{i \in I} n_i d_i . f(x_i).$$

Thus, when f is an immersion, $f_*(\alpha)$ is the same cycle as α but viewed as a cycle with domain X. Note also that we obtain the following equality

(8.1.6.1) $$f_*(\langle X \rangle) = \langle \bar{X} \rangle_Y$$

where $\bar{X}$ is the schematic closure of X in Y (indeed X is a dense open subscheme in $\bar{X}$). When f is clear, we sometimes abusively put: $\langle X \rangle_Y := f_*(\langle X \rangle)$.

By transitivity of degrees, we obviously have $f_* g_* = (fg)_*$ for a composable pair of morphisms (f, g).

Definition 8.1.7 Let $\alpha = \sum_{i \in I} n_i . x_i$ be a cycle over a scheme S with domain $f : X \longrightarrow S$ and $U \subset S$ be an open subscheme. Let $I' = \{ i \in I \mid f(x_i) \in U \}$. We define the *restriction* of α over U as the cycle $\alpha|_U = \sum_{i \in I'} n_i . x_i$ with domain $X \times_S U$ considered as a cycle over U.

If $\alpha = \sum_{i \in I} n_i . \langle Z_i \rangle_X$, then obviously $\alpha|_U = \sum_{i \in I} n_i . \langle Z_i \times_S U \rangle_{X_U}$. We state the following obvious lemma for convenience:

Lemma 8.1.8 *Let S be a scheme, $U \subset S$ an open subscheme and X be an S-scheme. Let $j : X_U \longrightarrow X$ be the obvious open immersion.*

(i) For any cycle (X_U, α'), $(j_*(\alpha'))|_U = \alpha'$.
(ii) Assume $U = S$. For any cycle (X, α) pseudo-dominant over S, $j_*(\alpha|_U) = \alpha$.

8.1.b Hilbert cycles

8.1.9 Recall that a finite-dimensional scheme X is equidimensional – we will say *absolutely equidimensional* – if its irreducible components all have the same dimension.

We will say that a flat morphism $f : X \longrightarrow S$ is equidimensional if it is of finite type and for any connected component X' of X, there exists an integer $e \in \mathbf{N}$ such that for any generic point η in X', the fiber $f^{-1}[f(\eta)]$ is absolutely equidimensional of dimension e.

Definition 8.1.10 Let S be a scheme.

Let α be a cycle over S with domain X. We say that α is a *Hilbert cycle* over S if there exists a finite family $(Z_i)_{i \in I}$ of closed subschemes of X which are flat equidimensional over S and a finite family $(n_i)_{i \in I} \in \Lambda^I$ such that

$$\alpha = \sum_{i \in I} n_i . \langle Z_i \rangle_X.$$

Example 8.1.11 Any cycle over a field k is a Hilbert cycle over $\mathrm{Spec}\,(k)$. Let S be the spectrum of a discrete valuation ring. A cycle $\alpha = \sum_{i \in I} n_i . x_i$ over S is a Hilbert cycle if and only if each point x_i lies over the generic points of S. Indeed, an integral S-scheme is flat if and only if it is dominant.

The following lemma follows almost directly from a result of [SVoob]:

Lemma 8.1.12 *Let $f : S' \longrightarrow S$ be a morphism of schemes and X be an S-scheme of finite type. Put $X' = X \times_S S'$.*
Let $(Z_i)_{i \in I}$ be a finite family of closed subschemes of X such that each Z_i is flat equidimensional over S. We assume the following relation:

(8.1.12.1) $$\sum_{i \in I} n_i.\langle Z_i \rangle_X = 0.$$

Then the following equality holds:

$$\sum_{i \in I} n_i.\langle Z_i \times_S S' \rangle_{X'} = 0.$$

Proof When we assume that for any index $i \in I$, Z_i/S is equidimensional of dimension e, this lemma is exactly [SVoob, Prop. 3.2.2]. We show how to reduce to this case.

Up to adding more members to the family (Z_i), we can always assume that Z_i is connected. Then, because Z_i/S is equidimensional by assumption, there exists an integer e_i such that for any point $x \in Z_i^{(0)}$, the fiber $f^{-1}[f(x)]$ is absolutely equidimensional of dimension e_i. In particular, the transcendence degree d_x of the residual extension $\kappa_x/\kappa_{f(x)}$ satisfies the relation: $d_x = e_i$.

For any integer $e \in \mathbf{N}$, we define the following subset of I:

$$I_e = \{i \in I \mid \forall x \in Z_i^{(0)}, d_x = e\}.$$

Thus $(I_e)_{e \in \mathbf{N}}$ is a partition of I.

One can rewrite the assumption (8.1.12.1) as follows: for any point $x \in X$,

$$\sum_{i \in I \mid x \in Z_i^{(0)}} n_i.\lg(\mathcal{O}_{Z_i,x}) = 0.$$

In particular, given any integer $e \in \mathbf{N}$, we deduce that the family $(Z_i)_{i \in I_e}$ still satisfies the relation (8.1.12.1). As any member of this family is equidimensional of dimension e, we can apply [SVoob, Prop. 3.2.2] to $(Z_i)_{i \in I_e}$. This concludes the proof. □

8.1.13 Consider a Hilbert S-cycle $\alpha \subset X$ and a morphism of schemes $f : S' \longrightarrow S$. Put $X' = X \times_S S'$. We choose a finite family $(Z_i)_{i \in I}$ of flat equidimensional S-schemes and a finite family $(n_i)_{i \in I} \in \Lambda^I$ such that $\alpha = \sum_{i \in I} n_i.\langle Z_i \rangle_X$. The previous lemma says precisely that the cycle

$$\sum_{i \in I} n_i.\langle Z_i \times_S S' \rangle_{X'}$$

depends only on α and not on the chosen families.

8 Relative cycles

Definition 8.1.14 Adopting the preceding notations and hypotheses, we define the *pullback cycle* of α along the morphism $f : S' \longrightarrow S$ as the cycle with domain X'

$$\alpha \otimes_S^\flat S' = \sum_{i \in I} n_i . \langle Z_i \times_S S' \rangle_{X'}.$$

In this setting the following lemma is obvious:

Lemma 8.1.15 *Let α be a Hilbert cycle over S, and $S'' \longrightarrow S' \longrightarrow S$ be morphisms of schemes.*
Then $(\alpha \otimes_S^\flat S') \otimes_{S'}^\flat S'' = \alpha \otimes_S^\flat S''$.

We will use another important computation from [SVoob] (it is a particular case of *loc. cit.*, 3.6.1).

Proposition 8.1.16 *Let R be a discrete valuation ring with residue field k. Let $\alpha \subset X$ be a Hilbert cycle over $\mathrm{Spec}\,(R)$ and $f : X \longrightarrow Y$ a morphism over $\mathrm{Spec}\,(R)$. We denote by $f' : X' \longrightarrow Y'$ the pullback of f over $\mathrm{Spec}\,(k)$.*
Suppose that the support of α is proper with respect to f.
Then $f_*(\alpha)$ is a Hilbert cycle over R and the following equality of cycles holds in X':

$$f'_*(\alpha \otimes_S^\flat k) = f_*(\alpha) \otimes_S^\flat k.$$

Definition 8.1.17 Let $p : \tilde{S} \longrightarrow S$ be a birational morphism. Let C be the minimal closed subset of S such that p induces an isomorphism $(\tilde{S} - \tilde{S} \times_S C) \longrightarrow (S - C)$.
Consider $\alpha = \sum_{i \in I} n_i . \langle Z_i \rangle_X$ a cycle over S written in standard form.
We define the *strict transform* $\tilde{Z}_i$ of the closed subscheme Z_i in X along p as the schematic closure of $(Z_i - Z_i \times_S C) \times_S \tilde{S}$ in $X \times_S \tilde{S}$. We define the *strict transform* of α along p as the cycle over $\tilde{S}$

$$\tilde{\alpha} = \sum_{i \in I} n_i . \langle \tilde{Z}_i \rangle_{X \times_S \tilde{S}}.$$

As in [SVoob], we remark that a corollary of the platification theorem of Gruson–Raynaud is the following:

Lemma 8.1.18 *Let S be a reduced scheme and α be a pseudo-dominant cycle over S.*
Then there exists a dominant blow-up $p : \tilde{S} \longrightarrow S$ such that the strict transform $\tilde{\alpha}$ of α along p is a Hilbert cycle over $\tilde{S}$.

We conclude this part by recalling an elementary lemma about cycles and Galois descent which will be used extensively in the next sections:

Lemma 8.1.19 *Let L/K be an extension of fields and X be a K-scheme. We put $X_L = X \times_K \mathrm{Spec}\,(L)$ and consider the faithfully flat morphism $f : X_L \longrightarrow X$.*
Denote by $\mathrm{Cycl}(X)$ (resp. $\mathrm{Cycl}(X_L)$) the cycles with domain X (resp. X_L).

1. *The morphism* $f^* : \mathrm{Cycl}(X) \longrightarrow \mathrm{Cycl}(X_L), \beta \mapsto \beta \otimes_K^\flat L$ *is a monomorphism.*
2. *Suppose* L/K *is finite. Then for any* K-*cycle* $\beta \in \mathrm{Cycl}(X)$,

$$f_*(\beta \otimes_K^\flat L) = [L:K].\beta.$$

3. *Suppose* L/K *is finite normal with Galois group* G. *Then the cycles in the image of* f^* *are invariant under the action of* G *and for any cycle* $\beta \in \mathrm{Cycl}(X_L)^G$, *there exists a unique cycle* $\beta_K \in \mathrm{Cycl}(X)$ *such that*

$$\beta_K \otimes_K^\flat L = [L:K]_i.\beta,$$

where $[L:K]_i$ *is the* inseparable degree *of* L/K.

8.1.c Specialization

The aim of this section is to give conditions on cycles so that one can define a *relative tensor product* on them.

Definition 8.1.20 Consider two cycles $\alpha = \sum_{i \in I} n_i.s_i$ and $\beta = \sum_{j \in J} m_j.x_j$. Let S be the support of α.

A morphism $\beta \xrightarrow{f} \alpha$ of cycles is said to be *pre-special* if it is of finite type and for any $j \in J$, there exists an $i \in I$ such that $f(x_j) = s_i$ and $n_i | m_j$ in Λ. We define the reduction of β/α as the cycle over S

$$\beta_0 = \sum_{j \in J, f(x_j) = s_i} \frac{m_j}{n_i}.x_j.$$

Example 8.1.21 Let S be a scheme and α a Hilbert S-cycle. Then the canonical morphism of cycles $\alpha \longrightarrow \langle S \rangle$ is pre-special. If S is the spectrum of a discrete valuation ring, an S-cycle α is pre-special if and only if it is a Hilbert S-cycle.

Definition 8.1.22 Let α be a cycle.

A *point* (resp. *trait*) of α will be a morphism of the form $\mathrm{Spec}\,(k) \xrightarrow{x} \alpha$ (resp. $\mathrm{Spec}\,(R) \xrightarrow{\tau} \alpha$) such that k is a field (resp. R is a discrete valuation ring). We simply say that x (resp. τ) is *dominant* if the image of the generic point in the domain of α is a generic point of α.

Let $x : \mathrm{Spec}\,(k_0) \longrightarrow \alpha$ be a point. An extension of x will be a point y on α of the form $\mathrm{Spec}\,(k) \longrightarrow \mathrm{Spec}\,(k_0) \xrightarrow{x} \alpha$.

A *fat point* of α will be a pair of morphisms

$$\mathrm{Spec}\,(k) \xrightarrow{s} \mathrm{Spec}\,(R) \xrightarrow{\tau} \alpha$$

such that τ is a dominant trait and the image of s is the closed point of $\mathrm{Spec}\,(R)$.

Given a point $x : \mathrm{Spec}\,(k) \longrightarrow \alpha$, a fat point over x is a factorization of x through a dominant trait as above.

8 Relative cycles

In the situation of the last definition, we denote simply by (R,k) a fat point over x, without indicating in the notation the morphisms s and τ.

Remark 8.1.23 With our choice of terminology, a point of α is in general an extension of a specialization of a generic point of α. As a further example, a dominant point of α is an extension of a generic point of α.

Lemma 8.1.24 *For any cycle α and any non-dominant point $x : \mathrm{Spec}\,(k_0) \longrightarrow \alpha$, there exists an extension $y : \mathrm{Spec}\,(k) \longrightarrow \alpha$ of x and a fat point (R, k) over y.*

Proof Replacing α by its support S, we can assume $\alpha = \langle S \rangle$. Let s be the image of x in S, κ its residue field. We can assume S is reduced, irreducible by taking one irreducible component containing s, and local with closed point s. Let $S = \mathrm{Spec}\,(A)$, $K = \mathrm{Frac}(A)$. According to [GD61, 7.1.7], there exists a discrete valuation ring R such that $A \subset R \subset K$, and R/A is an extension of local rings. Then any composite extension k/κ of k_0 and the residue field of R over κ gives the desired fat point (R, k). □

Definition 8.1.25 Let $\beta \longrightarrow \alpha$ be a pre-special morphism of cycles. Consider S the support of α and X the domain of β. Let $\beta_0 = \sum_{j \in J} m_j.\langle Z_j \rangle_X$ be the reduction of β/α written in standard form.

1. Let $\mathrm{Spec}\,(K) \longrightarrow \alpha$ be a dominant point. We define the following cycle over $\mathrm{Spec}\,(K)$ with domain $X_K = X \times_S \mathrm{Spec}\,(K)$:

$$\beta_K = \sum_{j \in J} m_j.\langle Z_j \times_S \mathrm{Spec}\,(K) \rangle_{X_K}.$$

2. Let $\mathrm{Spec}\,(R) \xrightarrow{\tau} S$ be a dominant trait, K be the fraction field of R and $j : X_K \longrightarrow X_R$ be the canonical open immersion. We define the following cycle over R with domain X_R:

$$\beta_R = j_*(\beta_K).$$

According to Example 8.1.11, β_R is a Hilbert cycle over R.

3. Let $x : \mathrm{Spec}\,(k) \longrightarrow \alpha$ be a point on α and (R, k) be a fat point over x. We define the *specialization of β along the fat point (R, k)* as the cycle

$$\beta_{R,k} := \beta_R \otimes_R^b k$$

using the above notation and Definition 8.1.14. It is a cycle over $\mathrm{Spec}\,(k)$ with domain $X_k = X \times_S \mathrm{Spec}\,(k)$.

Remark 8.1.26 Let β be an S-cycle, $x : \mathrm{Spec}\,(K) \longrightarrow S$ be a dominant point and U be an open neighborhood of x in S. Then if β is pre-special over S, $\beta|_U$ is pre-special over U and $\beta_K = (\beta|_U)_K$.
If $\tau : \mathrm{Spec}\,(R) \longrightarrow S$ (resp. (R, k)) is a trait (resp. fat point) with generic point x, we also get $\beta_R = (\beta|_U)_R$ (resp. $\beta_{R,k} = (\beta|_U)_{R,k}$).

8.1.27 Let S be a reduced scheme, and $\beta = \sum_{i \in I} n_i.x_i$ be an S-cycle with domain X. For any index $i \in I$, let κ_i be the residue field of x_i.

Consider a dominant point $x : \operatorname{Spec}(K) \longrightarrow S$. Let η be its image in S and F be the residue field of η. We put $I' = \{i \in I \mid f(x_i) = \eta\}$, where $f : X \longrightarrow S$ is the structural morphism. With these notations, we get

$$\beta_K = \sum_{i \in I'} n_i.\langle \operatorname{Spec}(\kappa_i \otimes_F K)\rangle_{X_K},$$

and for a dominant trait $\operatorname{Spec}(R) \longrightarrow S$ with generic point x,

(8.1.27.1) $$\beta_R = \sum_{i \in I'} n_i.\langle \operatorname{Spec}(\kappa_i \otimes_F K)\rangle_{X_R},$$

where $\operatorname{Spec}(\kappa_i \otimes_F K)$ is regarded as a subscheme of X_K (resp. X_R).

Consider a fat point (R, k) with generic point x and write $\beta = \sum_{i \in I} n_i.\langle Z_i\rangle_X$ in standard form (*i.e.* Z_i is the closure of $\{x_i\}$ in X). Then according to (8.1.6.1), we obtain[83]

$$\beta_{R,k} = \sum_{i \in I'} n_i.\left\langle \overline{Z_{i,K} \times_R \operatorname{Spec}(k)}\right\rangle_{X_k},$$

where $Z_{i,K} = Z_i \times_S \operatorname{Spec}(K)$ is considered as a subscheme of X_K and the schematic closure is taken in X_R.

Considering the description of the schematic closure for the generic fiber of an R-scheme (*cf.* [GD67, 2.8.5]), we obtain the following way to compute $\beta_{R,k}$. By definition, R is an F-algebra. For $i \in I'$, let A_i be the image of the canonical morphism

$$\kappa_i \otimes_F R \longrightarrow \kappa_i \otimes_F K.$$

It is an R-algebra without R-torsion. Moreover, the factorization

$$\operatorname{Spec}(\kappa_i \otimes_F K) \longrightarrow \operatorname{Spec}(A_i) \longrightarrow \operatorname{Spec}(\kappa_i \otimes_F R)$$

defines $\operatorname{Spec}(A_i)$ as the schematic closure of the left-hand side in the right-hand side (*cf.* [GD67, 2.8.5]). In particular, we get an immersion $\operatorname{Spec}(A_i \otimes_R k) \longrightarrow X_k$ and the nice formula:

$$\beta_{R,k} = \sum_{i \in I'} n_i.\langle \operatorname{Spec}(A_i \otimes_R k)\rangle_{X_k}.$$

Definition 8.1.28 Consider a morphism of cycles $f : \beta \longrightarrow \alpha$ and a point $x : \operatorname{Spec}(k_0) \longrightarrow \alpha$. We say that f is special at x if it is pre-special and for any extension $y : \operatorname{Spec}(k) \longrightarrow \alpha$ of x, for any fat points (R, k) and (R', k) over y, the equality $\beta_{R,k} = \beta_{R',k}$ holds in X_k. Equivalently, we say that β/α is special at x.

We say that f is special (or that β is special over α) if it is special at every point of α.

---

[83] This shows that our definition coincides with the one given in [SV00b] (p. 23, paragraph preceding 3.1.3) in the case where $\alpha = \langle S\rangle$, S reduced.

8 Relative cycles

8.1.29 Here is a dictionary to compare the above definition with that of Suslin and Voevodsky in [SV00b, 3.1.3].

Consider a pre-special morphism β/α. Let X be the domain of β, S be the support of α and β_0 be the reduction (see Definition 8.1.20) of β/α, regarded as a pre-special S-cycle.

Then the following conditions are equivalent:

(i) β/α is special;
(ii) β_0/S is special.

This follows from the very definition of the specialization of β/α along fat points (Definition 8.1.25).

Moreover, condition (ii) says exactly that β_0 is a relative cycle on X over S in the sense of Definition 3.1.3 of [SV00b].

Remark 8.1.30

1. Trivially, f is special at every dominant point of α.
2. Given an extension y of x, it is equivalent for f to be special at x or at y (use Lemma 8.1.19(1)). Thus, in the case where $\alpha = \langle S \rangle$, we can restrict our attention to the points $s \in S$.
3. According to Remark 8.1.26, the property that β/S is special at $s \in S$ depends only on an open neighborhood U of s in S. More precisely, the following conditions are equivalent:

   (i) β is special at s over S.
   (ii) $\beta|_U$ is special at s over U.

Example 8.1.31 Let S be a scheme and β be a Hilbert cycle over S. We have already seen that $\beta \longrightarrow \langle S \rangle$ is pre-special. The next lemma shows this morphism is in fact special.

Lemma 8.1.32 *Let S be a scheme and β be a Hilbert cycle over S. Consider a point $x : \mathrm{Spec}(k) \longrightarrow S$ and a fat point (R, k) over x.*

Then $\beta_{R,k} = \beta \otimes_S^b k$.

Proof According to the preceding definition and Lemma 8.1.15 it is sufficient to prove $\beta_R = \beta \otimes_S^b R$. As the two sides of this equation are unchanged when replacing β by the reduction β_0 of β/S, we can assume that S is reduced. By additivity, we are reduced to the case where $\beta = \langle X \rangle$ is the fundamental cycle associated with a flat S-scheme X. According to (8.1.6.1), $\beta_R = \langle \overline{X_K} \rangle_{X_R}$. Applying now [GD67, 2.8.5], $\overline{X_K}$ is the unique closed subscheme Z of X_R such that Z is flat over $\mathrm{Spec}(R)$ and $Z \times_R \mathrm{Spec}(K) = X_K$. Thus, as X_R is flat over $\mathrm{Spec}(R)$, we get $\overline{X_K} = X_R$ and this concludes the proof. □

Lemma 8.1.33 *Let $p : \tilde{S} \longrightarrow S$ be a birational morphism and consider a commutative diagram*

$$\operatorname{Spec}(k) \longrightarrow \operatorname{Spec}(R) \begin{array}{c} \longrightarrow \tilde{S} \\ \downarrow p \\ \longrightarrow S \end{array}$$

such that (R, k) is a fat point of $\tilde{S}$ and S.

Consider a pre-special cycle β over S and $\tilde{\beta}$ its strict transform along p. Then, $\tilde{\beta}$ is pre-special and $\tilde{\beta}_{R,k} = \beta_{R,k}$.

Proof Using Remark 8.1.26, we reduce to the case where p is an isomorphism, which is trivial. □

Lemma 8.1.34 *Let S be a reduced scheme, $x : \operatorname{Spec}(k_0) \longrightarrow S$ be a point and α be a pre-special cycle over S. Let $p : \tilde{S} \longrightarrow S$ be a dominant blow-up such that the strict transform $\tilde{\alpha}$ of α along p is a Hilbert cycle over $\tilde{S}$. Then the following conditions are equivalent:*

(i) α is special at x;
(ii) for every pair of points $x_1, x_2 : \operatorname{Spec}(k) \longrightarrow \tilde{S}$ such that $p \circ x_1 = p \circ x_2$ and $p \circ x_1$ is an extension of x, $\tilde{\alpha} \otimes_{\tilde{S}}^{\flat} x_1 = \tilde{\alpha} \otimes_{\tilde{S}}^{\flat} x_2$.

Proof The case where x is a dominant point follows from the definitions and the fact that p is an isomorphism at the generic point. We thus assume x is non-dominant.
$(i) \Rightarrow (ii)$: Applying Lemma 8.1.24 to x_i, $i = 1, 2$, we can find an extension x_i' : $\operatorname{Spec}(k_i) \longrightarrow \tilde{S}$ of x_i and a fat point (R_i, k_i) over x_i'. Taking a composite extension L of k_1 and k_2 over k, we can further assume $L = k_1 = k_2$ and $p \circ x_1' = p \circ x_2'$. Then for $i = 1, 2$, we get

$$\left(\tilde{\alpha} \otimes_{\tilde{S}}^{\flat} x_i\right) \otimes_k^{\flat} L \stackrel{8.1.15}{=\!=\!=} \tilde{\alpha} \otimes_{\tilde{S}}^{\flat} x_i' \stackrel{8.1.32}{=\!=\!=} \tilde{\alpha}_{R_i, L} \stackrel{8.1.33}{=\!=\!=} \alpha_{R_i, L},$$

and the result follows by Lemma 8.1.19(1).
$(ii) \Rightarrow (i)$: Consider an extension $y : \operatorname{Spec}(k) \longrightarrow \alpha$ over x and two fat points (R_1, k), (R_2, k) over y. Fix $i \in \{1, 2\}$. As p is proper birational, the trait $\operatorname{Spec}(R_i)$ on S can be extended (uniquely) to $\tilde{S}$. Let $x_i : \operatorname{Spec}(k) \longrightarrow \operatorname{Spec}(R_i) \longrightarrow \tilde{S}$ be the induced point. Then the result follows from the computation:

$$\alpha_{R_i, k} \stackrel{8.1.33}{=\!=\!=} \tilde{\alpha}_{R_i, k} \stackrel{8.1.32}{=\!=\!=} \tilde{\alpha} \otimes^{\flat} x_i \qquad \square$$

8.1.d Pullback

8.1.35 In this part, we construct a *pullback* which extends the pullback defined by Suslin and Voevodsky in [SV00b, 3.3.1] to the case of morphisms of cycles. Consider the situation of a diagram of cycles

8 Relative cycles

$$\begin{array}{ccc} & \beta & \\ & \downarrow f & \\ \alpha' \longrightarrow & \alpha & \end{array} \quad \subset \quad \begin{array}{ccc} & X & \\ & \downarrow & \\ S' \longrightarrow & S & \end{array}$$

where the diagram on the right is the domain of the one on the left. Let n be the exponential characteristic of $\mathrm{Supp}(\alpha')$.

The pullback of β, considered as an α-cycle, over α' will be a $\Lambda[1/n]$-cycle denoted by $\beta \otimes_\alpha \alpha'$. It will fit into the following commutative diagram of cycles

$$\begin{array}{ccc} \beta \otimes_\alpha \alpha' \longrightarrow & \beta \\ \downarrow & \downarrow \\ \alpha' \longrightarrow & \alpha \end{array} \quad \subset \quad \begin{array}{ccc} X \times_S S' \longrightarrow & X \\ \downarrow & \downarrow \\ S' \longrightarrow & S \end{array}$$

where the right commutative square is again the support of the left one.

It will be defined under an assumption on β/α and is therefore non symmetric[84]. This assumption will imply that β/α is pre-special, and the first property of $\beta \otimes_\alpha \alpha'$ is that it is pre-special over α'.

We define this product in three steps in which the following properties[85] will be a guideline:

(P1) Let S_0 be the support of α and β_0 be the reduction (see Definition 8.1.20) of β/α, as an S_0-cycle. Consider the canonical factorization $\alpha' \longrightarrow S_0 \longrightarrow \alpha$. Then, $\beta \otimes_\alpha \alpha' = \beta_0 \otimes_{S_0} \alpha'$.

(P2) Consider a commutative diagram

$$\begin{array}{ccc} \mathrm{Spec}(E) \longrightarrow \mathrm{Spec}(R') \longrightarrow \mathrm{Spec}(R) \\ \downarrow \quad (*) \quad \downarrow \\ \alpha' \longrightarrow \alpha \end{array}$$

such that (R, E) (resp. (R', E)) is a fat point on α (resp. α'). Then, $(\beta \otimes_\alpha \alpha')_{R',E} = \beta_{R,E}$.

Assume $\alpha' \longrightarrow \alpha = \langle S' \longrightarrow S \rangle$.

(P3) If β is a Hilbert cycle over S, then $\beta \otimes_S S' = \beta \otimes_S^\flat S'$.

(P4) Consider a factorization $S' \longrightarrow U \xrightarrow{j} S$ such that j is an open immersion. Then $\beta \otimes_S S' = \beta|_U \otimes_U S'$.

(P5) Consider a factorization $S' \longrightarrow \tilde{S} \xrightarrow{p} S$ such that p is a birational morphism. Then $\beta \otimes_S S' = \tilde{\beta} \otimes_{\tilde{S}} S'$.

---

[84] See further Corollary 8.2.3 for this question.
[85] All these properties except (P3) will be particular cases of the associativity of the pullback.

Lemma 8.1.36 *Consider the hypotheses of Section 8.1.35 in the case where $\alpha' = \mathrm{Spec}\,(k)$ is a point x of α.*

We suppose that f is special at x.

Then the pre-special $\Lambda[1/n]$-cycle $\beta \otimes_\alpha k$ exists and is uniquely determined by property (P2) above. We also put $\beta_k := \beta \otimes_\alpha k$.

The properties (P1) to (P5) are fulfilled and in addition:
(P6) For any extension fields L/k, $\beta_L = \beta_k \otimes_k^b L$.

Proof According to Lemma 8.1.24 there always exists a fat point (R, E) over an extension of x. Thus the unicity statement follows from Lemma 8.1.19(1).

For the existence, we first consider the case where $\alpha = \langle S \rangle$ is a reduced scheme. Applying Lemma 8.1.18, there exists a blow-up $p : \tilde{S} \longrightarrow S$ such that the strict transform $\tilde{\beta}$ of β along p is a Hilbert cycle over $\tilde{S}$.

As p is surjective, the fiber $\tilde{S}_k$ is a non-empty algebraic k-scheme. Thus, it admits a closed point given by a finite extension k_0' of k. Let k'/k be a normal closure of k_0'/k and G be its Galois group. As β/S is special at x by hypothesis, Lemma 8.1.34 implies that $\tilde{\beta} \otimes_{\tilde{S}}^b k'$ is G-invariant. Thus, applying Lemma 8.1.19, there exists a unique cycle $\beta_k \subset X_k$ with coefficients in $\Lambda[1/n]$ such that $\beta_k \otimes_k^b k' = \tilde{\beta} \otimes_{\tilde{S}}^b k'$.

We prove (P2). Given a diagram (∗) with $\alpha' = \mathrm{Spec}\,(k)$, we first note that $(\beta_k)_{R',E} = \beta_k \otimes_k^b E$. As p is proper birational, the dominant trait $\mathrm{Spec}\,(R) \longrightarrow S$ lifts to a dominant trait $\mathrm{Spec}\,(R) \longrightarrow \tilde{S}$. Let E'/k be a composite extension of k'/k and E/k. With these notations, we get the following computation:

$$\beta_{R,E} \otimes_E^b E' \stackrel{8.1.33}{=\!=\!=} \tilde{\beta}_{R,E} \otimes_E^b E' \stackrel{8.1.32}{=\!=\!=} \tilde{\beta} \otimes_{\tilde{S}}^b E' \stackrel{8.1.15}{=\!=\!=} (\tilde{\beta} \otimes_{\tilde{S}}^b k') \otimes_E^b E' =\!=\!= \beta_k \otimes_k^b E',$$

so that we can conclude (P2) by applying Lemma 8.1.19(1).

In the general case, we consider the support S of α and β_0/S the reduction of β/α. According to (P1), we are led to put $\beta_k := (\beta_0)_k$ with the help of the preceding case. Considering the definition of specialization along fat points, we easily check this cycle satisfies property (P2).

Finally, property (P6) (resp. (P3), (P5)) follows from the unicity statement applying Lemmas 8.1.24 and 8.1.19(1) (resp. and moreover Lemmas 8.1.32 and 8.1.33). □

Remark 8.1.37 In the case where x is a dominant point, the cycle β_k defined in the previous proposition agrees with the one defined in Definition 8.1.25(1).

Lemma 8.1.38 *Consider the hypotheses of Section 8.1.35 in the case where $\alpha' = \mathrm{Spec}\,(O)$ is a trait of α. Let K be the fraction field of O and x the corresponding point on α.*

We suppose that f is special at x.

Then the pre-special $\Lambda[1/n]$-cycle $\beta \otimes_\alpha O$ exists and is uniquely defined by the property $(\beta \otimes_\alpha O) \otimes_O^b K = \beta_K$ with the notations of the preceding lemma. We also put $\beta_O := \beta \otimes_\alpha O$.

The properties (P1) to (P5) are fulfilled and in addition:
(P6') For any extension O'/O of discrete valuation rings, $\beta_{O'} = \beta_O \otimes_O^b O'$.

8 Relative cycles

Proof Observe that, with the notation of Definition 8.1.7, $\beta_O \otimes_O^b K = \beta_O|_{\mathrm{Spec}(K)}$. For the first statement, we simply apply Lemma 8.1.8 and put $\beta_O = j_*(\beta_K)$, where $j : X_K \longrightarrow X_O$ is the canonical open immersion.

Then properties (P1), (P3), (P4), (P5) and (P6') of the case considered in this lemma follows easily from the uniqueness statement and the corresponding properties in the preceding lemma (applying again Lemma 8.1.8).

It remains to prove (P2). According to (P1), we reduce to the case $\alpha = \langle S \rangle$ for a reduced scheme S. We choose a birational morphism $p : \tilde{S} \longrightarrow S$ such that the proper transform $\tilde{\beta}$ is a Hilbert $\tilde{S}$-cycle. Consider a diagram of the form (∗) in this case. According to property (P3), we can assume $R' = O$.

Note that the trait $\mathrm{Spec}(R) \longrightarrow S$ admits an extension $\mathrm{Spec}(R) \longrightarrow \tilde{S}$ as p is proper. The point x admits an extension K'/K which lifts to a point x' : $\mathrm{Spec}(K') \longrightarrow \tilde{S}$ – again $\tilde{S}_K$ is a non-empty algebraic scheme. The discrete valuation corresponding to $O \subset K$ extends to a discrete valuation on K' as K'/K is finite. Let $O' \subset K'$ be the corresponding valuation ring. The corresponding trait $\mathrm{Spec}(O') \longrightarrow S$ thus admits a lifting to $\tilde{S}$ corresponding to the point x' as p is proper. Considering a composite extension E'/K of K'/K and E/K, we have obtained a commutative diagram

$$\begin{array}{ccc} \mathrm{Spec}(E') \longrightarrow \mathrm{Spec}(O') \longrightarrow \mathrm{Spec}(R) \\ \| & & \downarrow \\ \mathrm{Spec}(O') \longrightarrow \tilde{S} \end{array}$$

which lifts our original diagram (∗). Let x_1 (resp. x_2) be the point $\mathrm{Spec}(E)' \longrightarrow \tilde{S}$ corresponding to the composite through the upper way (resp. lower way) in the preceding diagram.

Then, $\beta_{R,E} \otimes_E^b E' = \tilde{\beta}_{x_1}$. Moreover, we get

$$(\beta \otimes_S O)_{O,E} \otimes_E^b E' \stackrel{8.1.32}{=\!=} (\beta \otimes_S O) \otimes_O^b E' \stackrel{(P5)+(P6')}{=\!=\!=\!=} (\tilde{\beta} \otimes_{\tilde{S}} O') \otimes_{O'}^b E' \stackrel{(P3)}{=\!=} \tilde{\beta}_{x_2}.$$

By hypothesis, β/α is special at $\mathrm{Spec}(K') \longrightarrow S$. Thus Lemma 8.1.34 concludes the proof. □

Theorem 8.1.39 *Consider the hypotheses of Section 8.1.35.*
Assume f is special at the generic points of α'.
Then the pre-special $\Lambda[1/n]$-cycle $\beta \otimes_\alpha \alpha'$ exists and is uniquely determined by property (P2).
It satisfies all the properties (P1) to (P5).

Proof According to Lemma 8.1.24, for any point s of S' with residue field κ, there exists an extension E/κ and a fat point (R, E) (resp. (R', E)) of α (resp. α') over $\mathrm{Spec}(E) \longrightarrow \alpha$ (resp. $\mathrm{Spec}(E) \longrightarrow \alpha'$). The uniqueness statement follows by applying Lemma 8.1.19(1).

For the existence, we write $\alpha' = \sum_{i \in I} n_i . \langle Z_i \rangle_{S'}$ in standard form.

For any $i \in I$, let K_i be the function field of Z_i and consider the canonical morphism $\mathrm{Spec}(K_i) \longrightarrow \alpha$. Let $\beta_{K_i} \subset X_{K_i}$ be the $\Lambda[1/n]$-cycle defined in Lemma

8.1.36. Let $j_i : X_{K_i} \longrightarrow X'$ be the canonical immersion and put:

$$(8.1.39.1) \qquad \beta \otimes_\alpha \alpha' = \sum_{i \in I} n_i \cdot j_{i*}(\beta_{K_i}).$$

Then properties (P1), (P3), (P4) and (P5) are direct consequences of this definition and of the corresponding properties of Lemma 8.1.36.

We check property (P2). Given a diagram of the form (∗), there exists a unique $i \in I$ such that $\mathrm{Spec}(R')$ dominates Z_i. Thus we get for this choice of $i \in I$ that $(\beta \otimes_\alpha \alpha')_{R',E} = (j_{i*}(\beta_{K_i}))_{R',E}$. Let K' be the fraction field of R' and consider the open immersion $j' : X_{K'} \longrightarrow X_{R'}$. The following computation then concludes the proof:

$$(j_{i*}(\beta_{K_i}))_{R',E} = j'_*(j_{i*}(\beta_{K_i})_{K'}) \otimes^\flat_{R'} E \stackrel{8.1.26}{=\!=\!=} j'_*(\beta_{K'}) \otimes^\flat_{R'} E \stackrel{8.1.38}{=\!=\!=} \beta_{R'} \otimes^\flat_{R'} E \stackrel{8.1.38(P2)}{=\!=\!=\!=} \beta_{R,E}.$$

Definition 8.1.40 In the situation of the previous theorem, we call the $\Lambda[1/n]$-cycle $\beta \otimes_\alpha \alpha'$ the pullback of β/α by α'.

8.1.41 By construction, the cycle $\beta \otimes_\alpha \alpha'$ is bilinear with respect to addition of cycles in the following sense:

(P7) Consider the hypotheses of Section 8.1.35. Let α'_1, α'_2 be cycles with domain S' such that $\alpha = \alpha'_1 + \alpha'_2$. If β/α is special at the generic points of α_1 and α_2, then the following cycles are equal in $X \times_S S'$:

$$\beta \otimes_\alpha (\alpha'_1 + \alpha'_2) = \beta \otimes_\alpha \alpha'_1 + \beta \otimes_\alpha \alpha'_2.$$

(P7') Consider the hypotheses of Section 8.1.35. Let β_1, β_2 be cycles with domain X such that $\beta = \beta_1 + \beta_2$. If β_1 and β_2 are special over α at the generic points of α', then β/α is special at the generic points of α' and the following cycles are equal in $X \times_S S'$:

$$(\beta_1 + \beta_2) \otimes_\alpha \alpha' = \beta_1 \otimes_\alpha \alpha' + \beta_2 \otimes_\alpha \alpha'.$$

In the theorem above, we can assume that X (resp. S, S') is the support of β (resp. α, α'). Thus the support of $\beta \otimes_\alpha \alpha'$ is included in $X \times_S S'$. More precisely:

Lemma 8.1.42 *Consider the hypotheses of Section 8.1.35 and assume that X (resp. S, S') is the support of β (resp. α, α'). Then, if β/α is special at the generic points of α', we obtain:*

(i) *Let $(X \times_S S')^{(0)}$ be the generic points of $X \times_S S'$. Then, we can write*

$$\beta \otimes_\alpha \alpha' = \sum_{x \in (X \times_S S')^{(0)}} m_x \cdot x.$$

(ii) *For any generic point x of $X \times_S S'$, if $m_x \neq 0$, the image of x in S' is a generic point s' and the multiplicity of s' in α' divides m_x in $\Lambda[1/n]$.*

8 Relative cycles

Proof Point (ii) is just a translation of the fact that $\beta \otimes_\alpha \alpha'$ is pre-special over α'. For point (i), we reduce easily to the case where α is the scheme S and S is reduced. We can also assume that α' is the spectrum of a field k. It is sufficient to check point (i) after an extension of k. Thus we can apply Lemma 8.1.18 to reduce to that case where β is a Hilbert cycle over S. This case is obvious. □

Definition 8.1.43 In the situation of the previous lemma, we put

$$m^{SV}(x; \beta \otimes_\alpha \alpha') := m_x \in \Lambda[1/n]$$

and we call them the Suslin–Voevodsky multiplicities (in the operation of pullback).

Remark 8.1.44 Consider the notations of the previous lemma:

1. Assume that α is the spectrum of a field k. Then the product $\beta \otimes_k \alpha'$ is always defined and agrees with the classical *exterior product* (according to (P3)).
2. According to the previous lemma, the irreducible components of $X \times_S S'$ which do not dominate an irreducible component of S' have multiplicity 0: they correspond to the "non-proper components" with respect to the operation $\beta \otimes_\alpha \alpha'$.
3. Assume $\alpha' \longrightarrow \alpha = \langle S' \xrightarrow{p} S \rangle$, $\beta = \sum_{i \in I} n_i . x_i$. Let y be a generic point of $X \times_S S'$ lying over a generic point s' of S'. Let S'_0 be the irreducible component of S' corresponding to s'. Consider *any* irreducible component S_0 of S which contains $p(s')$ and let $\beta_0 = \sum_i n_i . x_i$, where the sum runs over the indexes i such that x_i lies over S_0. Then, according to (8.1.39.1),

$$m^{SV}(y; \beta \otimes_S \langle S' \rangle) = m^{SV}(y; \beta_0 \otimes_{S_0} \langle S'_0 \rangle).$$

This is a key property of the Suslin–Voevodsky multiplicities which explains why we have to consider the property that β/α is special at s' (see Corollary 8.3.25 for a refined statement).

Lemma 8.1.45 *Consider a morphism of cycles $\alpha' \longrightarrow \alpha$ and a pre-special morphism $f : \beta \longrightarrow \alpha$ which is special at the generic points of α. Consider a commutative square*

$$\begin{array}{ccc} \operatorname{Spec}(k') & \xrightarrow{x'} & \alpha' \\ \downarrow & & \downarrow \\ \operatorname{Spec}(k) & \xrightarrow{x} & \alpha \end{array}$$

such that k and k' are fields. Then the following conditions are equivalent:

(i) f is special at x.
(ii) $\beta \otimes_\alpha \alpha' \longrightarrow \alpha'$ is special at x'.

Proof This follows easily from Lemma 8.1.24 and property (P2). □

Corollary 8.1.46 *Let $f : \beta \longrightarrow \alpha$ be a special morphism. Then for any morphism $\alpha' \longrightarrow \alpha$, $\beta \otimes_\alpha \alpha' \longrightarrow \alpha'$ is special.*

Definition 8.1.47 Let $f : \beta \longrightarrow \alpha$ be a morphism of cycles and $x : \mathrm{Spec}\,(k) \longrightarrow \alpha$ be a point. We say that f is Λ-universal at x if it is special at x and the cycle $\beta \otimes_\alpha k$ has coefficients in Λ.

In the situation of this definition, let s be the image of x in the support of α, and κ_s be its residue field. Then according to (P6), $\beta_k = \beta_{\kappa_s} \otimes^\flat_{\kappa_s} k$. Thus f is Λ-universal at x if and only if it is Λ-universal at s. Furthermore, the following lemma follows easily:

Lemma 8.1.48 *Let $f : \beta \longrightarrow \alpha$ be a morphism of cycles. The following conditions are equivalent:*

(i) For any point $s \in \overline{\alpha}$, f is Λ-universal at s.
(ii) For any point $x : \mathrm{Spec}\,(k) \longrightarrow \alpha$, f is Λ-universal at x.
(iii) For any morphism of cycles $\alpha' \longrightarrow \alpha$, $\beta \otimes_\alpha \alpha'$ has coefficients in Λ.

Definition 8.1.49 We say that a morphism of cycles f is Λ-universal if it satisfies the equivalent properties of the preceding lemma.

Of course, Λ-universal morphisms are stable by base change. These definitions will be applied similarly to morphisms of schemes by considering the associated morphism of cycles.

Example 8.1.50 According to property (P3) of the pullback, a flat equidimensional morphism of schemes is Λ-universal.

8.1.51 Let β/α be a morphism of Λ-cycles.

Let S be the support of α and consider the obvious morphism of cycles $S \longrightarrow \alpha$. Recall from property (P1) of Section 8.1.35 that the cycle

$$\beta_0 := \beta \otimes_\alpha S$$

is the reduction of β/α (Definition 8.1.20). This is a special Λ-cycle over S (see Section 8.1.29)

Moreover, it follows from the definition of the product that the following conditions are equivalent:

(i) β/α is Λ-universal;
(ii) β_0/S is Λ-universal.

In particular, condition (ii) appears in Lemma 3.3.9 of [SV00b] (with a restriction on the relative dimension that in fact is not needed).

Remark 8.1.52 Though Lemma 3.3.9 of [SV00b] does not give rise to any definition in *loc. cit.*, it is central in the theory of Suslin and Voevodsky. In particular, it appears in the definition of the groups $z(X/S,r)$, $c(X/S,r)$,... that is given right after Lemma 3.3.9.

8 Relative cycles

Our definition has the advantage of:

- working properly over non-reduced schemes;
- having a local formulation (this is essential for the theorems of constructibility in Section 8.3.a);
- being free of unnecessary assumptions such as the relative dimension of fibers (the integer r that appear in $z(X/S,r)$).

Besides, the categorical language introduced, obviously inspired by E.G.A., is very natural and will prove to be useful in the treatment of finite correspondences (see for example the definition of the composition product, Definition 9.1.5, and the short proof of the properties of this composition product, Proposition 9.1.7).

The following proposition shows that one can bound the denominators that can appear after an arbitrary number of base changes.

Proposition 8.1.53 *Let β/α be a special morphism of Λ-cycles. Then there exists an integer $N > 0$ such that $N.\beta$ is Λ-universal.*

Proof According to Section 8.1.51, one can reduce to the case where α is a reduced scheme S. We then prove by noetherian induction on S the following assertion: for any closed subscheme $Z \subset S$, and any special Λ-cycle α on S, there exists an integer $N > 0$ such that $N.\alpha$ is Λ-special

Take a special Λ-cycle α on S. According to Lemma 8.1.18, there exists a birational morphism $p : \tilde{S} \longrightarrow S$ such that the strict transform $\tilde{\alpha}$ of α along p is a Hilbert cycle, thus Λ-universal. Let U be a dense open subscheme of S above which p is an isomorphism. Thus for any point $s \in U$, with inverse image t in $p^{-1}(U)$, we obtain that the cycle $\alpha_s = \tilde{\alpha}_t$ has Λ-coefficients.

Let Z be the complement of U in S, with its reduced schematic structure. Then, by construction, the pullback $\alpha \otimes_S Z$ is an $\Lambda[1/N]$-cycle. In particular, $\alpha_0 = N.\alpha \otimes_S Z$ is a special Λ-cycle over Z. As Z is a proper closed subscheme of S, we can apply the Noetherian induction hypothesis to Z and α_0. We find an integer $N' > 0$ such that $N'.\alpha_0$ is Λ-universal. By transitivity of pullbacks (which follows easily from the uniqueness statement of Theorem 8.1.39; see Proposition 8.2.4), we thus obtain that $(NN').\alpha$ is Λ-universal over S. □

Recall that Λ is a sub-ring of the ring of rationals. One easily deduces from the preceding proposition the following result.

Corollary 8.1.54 *For any Λ-cycle α special over a (noetherian) scheme S, there exists an integer $N > 0$ such that $N.\alpha$ is $\mathbf{Z}$-universal over S.*

8.2 Intersection-theoretic properties

8.2.a Commutativity

Lemma 8.2.1 *Consider morphisms of cycles with support in the left diagram*

$$\begin{array}{ccc} \beta & & X \\ \downarrow & \subset & \downarrow f \\ \gamma \longrightarrow \alpha & & T \xrightarrow{g} S \end{array}$$

such that β/α is pre-special and γ/α is pseudo-dominant.
Assume

$$\alpha = \sum_{i \in I} n_i.s_i, \quad \beta = \sum_{j \in J} m_j.x_j, \quad \gamma = \sum_{l \in H} p_l.t_l$$

and denote by κ_{s_i} (resp. κ_{x_j}, κ_{t_l}) the residue field of s_i (resp. x_j, t_l) in S (resp. X, T). Considering $(i,j,l) \in I \times J \times H$ such that $f(x_j) = g(t_l) = s_i$, we denote by $\nu_{j,l} : \mathrm{Spec}\left(\kappa_{x_j} \otimes_{\kappa_{s_i}} \kappa_{t_l}\right) \longrightarrow X \times_S T$ the canonical immersion.
Then the following assertions hold:

(i) β is special at the generic points of γ.
(ii) The cycle $\beta \otimes_\alpha \gamma$ has coefficients in Λ.
(iii) The following equality of cycles holds

$$\beta \otimes_\alpha \gamma = \sum_{i,j,l} \frac{m_j}{n_i} p_l.\nu_{j,l*}(\langle \mathrm{Spec}\left(\kappa_{y_j} \otimes_{\kappa_{x_i}} \kappa_{z_l}\right)\rangle)$$

where the sum runs over $(i,j,l) \in I \times J \times H$ such that $f(x_j) = g(t_j) = s_i$.

Proof Assertion (i) is in fact the first point of Remark 8.1.30. Assertion (ii) follows from assertion (iii), which is a consequence of the defining formula (8.1.39.1) and Remark 8.1.37. □

Corollary 8.2.2 *Let $g : T \longrightarrow S$ be a flat morphism and $\beta = \sum_{j \in J} m_j.\langle Z_j \rangle_X$ be a pre-special S-cycle written in standard form.*
Then β/S is pre-special at the generic points of T and

$$\beta \otimes_S \langle T \rangle = \sum_{j \in J} m_j.\langle Z_j \times_S T \rangle.$$

The pullback $\beta \otimes_\alpha \gamma$, as it is defined only when β/α is special, is in general non-symmetric in β and γ. However the previous lemma implies it is symmetric whenever it makes sense:

Corollary 8.2.3 *Consider pre-special morphisms of cycles $\beta \longrightarrow \alpha$ and $\gamma \longrightarrow \alpha$.*
Then β (resp. γ) is special at the generic points of γ (resp. β) and the following equality holds: $\beta \otimes_\alpha \gamma = \gamma \otimes_\alpha \beta$.

8.2.b Associativity

Proposition 8.2.4 *Consider morphisms of cycles* $\beta \xrightarrow{f} \alpha$, $\alpha'' \to \alpha' \to \alpha$ *such that f is special at the generic points of α' and of α''. Let n be the exponential characteristic of α''.*
Then the following assertions hold:

(i) The relative cycle $(\beta \otimes_\alpha \alpha')/\alpha'$ is special at the generic points of α''.
(ii) The cycle $(\beta \otimes_\alpha \alpha') \otimes_{\alpha'} \alpha''$ has coefficients in $\Lambda[1/n]$.
(iii) $(\beta \otimes_\alpha \alpha') \otimes_{\alpha'} \alpha'' = \beta \otimes_\alpha \alpha''$.

Proof Assertion (i) is a corollary of Lemma 8.1.45. Assertion (ii) is in fact a corollary of assertion (iii), which in turn follows easily from the uniqueness statement in Theorem 8.1.39. □

Lemma 8.2.5 *Let* $\gamma \xrightarrow{g} \beta \xrightarrow{f} \alpha$ *be two pre-special morphisms of cycles with domains $Y \to X \to S$. Consider a fat point (R, k) over α such that γ/β is special at the generic points of $\beta_{R,k}$.*
Then γ/α is pre-special and the following equality of cycles holds in Y_k:

$$\gamma_{R,k} = \gamma \otimes_\beta (\beta_{R,k}).$$

Proof The first statement is obvious.
We first prove: $\gamma_R = \gamma \otimes_\beta \beta_R$.
Note that $\beta_R \to \beta$ is pseudo-dominant. Thus γ/β is special at the generic points of β_R and the right-hand side of the preceding equality is well defined. Moreover, according to Lemma 8.2.1, we can restrict to the case where $\alpha = s, \beta = x$ and $\gamma = y$, with multiplicity 1. Let $\kappa_s, \kappa_x, \kappa_y$ be the corresponding respective residue fields, and K be the fraction field of R.
Then, according to (8.1.27.1), $\gamma_R = \langle \kappa_y \otimes_{\kappa_s} K \rangle_{Y_R}$ and $\beta_R = \langle \kappa_x \otimes_{\kappa_s} K \rangle_{X_R}$. But Lemma 8.2.1 implies that $\gamma \otimes_\beta \beta_R = \langle \kappa_y \otimes_{\kappa_x} (\kappa_x \otimes_{\kappa_s} K) \rangle_{X_R}$. Thus the result follows from the associativity of the tensor product of fields.
From this equality and Proposition 8.2.4, we deduce that:

$$\gamma_R \otimes_{\beta_R} \beta_{R,k} = (\gamma \otimes_\beta \beta_R) \otimes_{\beta_R} \beta_{R,k} = \gamma \otimes_\beta \beta_{R,k}.$$

Thus, the equality we have to prove can be written $\gamma_R \otimes_R^\flat k = \gamma_R \otimes_{\beta_R} (\beta_R \otimes_R^\flat k)$ and we are reduced to the case $\alpha = \operatorname{Spec}(R)$. In this case, we can assume $\beta = \langle X \rangle$ with X integral. Let us consider a blow-up $\tilde{X} \xrightarrow{p} X$ such that the proper transform $\tilde{\gamma}$ of γ along p is a Hilbert cycle over $\tilde{X}$ (Lemma 8.1.18). We easily get (from (P3) and Lemma 8.1.15) that

$$\tilde{\gamma}_k = \tilde{\gamma} \otimes_{\tilde{X}} \langle \tilde{X}_k \rangle.$$

Let Y (resp. $\tilde{Y}$) be the support of γ (resp. $\tilde{\gamma}$) and $q : \tilde{Y} \to Y$ the canonical projection. We consider the cartesian square obtained by pullback along $\operatorname{Spec}(k) \to \operatorname{Spec}(R)$:

$$\tilde{Y}_k \xrightarrow{q_k} Y_k$$
$$\downarrow \qquad \downarrow$$
$$\tilde{X}_k \xrightarrow{p_k} X_k.$$

As $X_k \subset X$ (resp. $Y_k \subset Y$) is purely of codimension 1, the proper morphism p_k (resp. q_k) is still birational. As a consequence, $q_{k*}(\tilde{\gamma}) = \gamma$. Let y be a point in $\tilde{Y}_k^{(0)} \simeq Y_k^{(0)}$ which lies above a point x in $\tilde{X}_k^{(0)} \simeq X_k^{(0)}$. Then, according to (P5) and using the notations of Definition 8.1.43, we get

$$m^{SV}(y; \tilde{\gamma} \otimes_{\tilde{X}} \langle \tilde{X}_k \rangle) = m^{SV}(y; \gamma \otimes_X \langle X_k \rangle).$$

This readily implies $q_{k*}(\tilde{\gamma} \otimes_{\tilde{X}} \langle \tilde{X}_k \rangle) = \gamma \otimes_X \langle X_k \rangle$ and completes the proof. $\square$

As a corollary of this lemma, using the uniqueness statement in Theorem 8.1.39, we obtain:

Corollary 8.2.6 *Let* $\gamma \xrightarrow{g} \beta \xrightarrow{f} \alpha$ *be pre-special morphisms of cycles.*

Let $x : \mathrm{Spec}(k) \longrightarrow \alpha$ *be a point. If* β/α *is special (resp. Λ-universal) at x and γ/β is special (resp. Λ-universal) at the generic points of β_k, then γ/α is special at x.*

Let $\alpha' \longrightarrow \alpha$ *be any morphism of cycles with domain* $S' \longrightarrow S$ *and n be the exponential characteristic of α'. Then, whenever it is well defined, the following equality of $\Lambda[1/n]$-cycles holds:*

$$\gamma \otimes_\beta (\beta \otimes_\alpha \alpha') = \gamma \otimes_\alpha \alpha'.$$

A consequence of the transitivity formulas is the associativity of the pullback:

Corollary 8.2.7 *Suppose given the following morphisms of cycles*

$$\alpha \searrow \quad \beta \swarrow_f \searrow \quad \gamma \swarrow_g$$
$$\delta \qquad \sigma$$

such that f and g are pre-specials.

Then, whenever it is well defined, the following equality of cycles holds:

$$\gamma \otimes_\sigma (\beta \otimes_\delta \alpha) = (\gamma \otimes_\sigma \beta) \otimes_\delta \alpha$$

Proof Indeed, by the transitivity formulas of Proposition 8.2.4 and Corollary 8.2.6, both members of the equation are equal to $(\gamma \otimes_\sigma \beta) \otimes_\beta (\beta \otimes_\delta \alpha)$. $\square$

8.2.c Projection formulas

Proposition 8.2.8 *Consider morphisms of cycles with support in the left diagram*

$$\begin{array}{ccc} \beta & & X \\ \downarrow & \subset & \downarrow \\ \alpha' \longrightarrow \alpha & & S' \xrightarrow{q} S \end{array}$$

such that β/α is special at the generic points of α'.
Consider a factorization $S' \xrightarrow{g} T \longrightarrow S$.
Then β/α is special at the generic points of $g_*(\alpha)$ and the following equality of cycles holds in $X \times_S T$:

$$\beta \otimes_\alpha g_*(\alpha') = (1_X \times_S g)_*(\beta \otimes_\alpha \alpha').$$

Proof The first assumption is obvious. By linearity, we can assume S' is integral and α' is the generic point s of S' with multiplicity 1. Let L (resp. E) be the residue field of s (resp. $g(s)$).

Consider the pullback square $\begin{array}{ccc} X_L & \xrightarrow{g_0} & X_E \\ j \downarrow & & \downarrow i \\ X \times_S S' & \xrightarrow{g_X} & X \times_S T \end{array}$ where i and j are the natural immersions.

Let d be the degree of L/E if it is finite and 0 otherwise. We are reduced to proving the equality $g_{X*}(j_*(\beta_L)) = d.i_*(\beta_E)$. Using the functoriality of pushforward and property (P6), it is sufficient to prove the equality $g_{0*}(\beta_E \otimes_E^b L) = d.\beta_E$. If $d = 0$, the morphism g_0 induces an infinite extension of fields on any point of X_L from which the result follows. If L/E is finite, g_0 is finite flat and $\beta_E \otimes_E^b L$ is the usual pullback by g_0. Then the needed equality follows easily (see [Ful98, 1.7.4]). □

Lemma 8.2.9 *Let $\beta \longrightarrow \alpha$ be a pre-special morphism of cycles with domain $X \xrightarrow{p} S$. Let (R, k) be a fat point over α and $X \xrightarrow{f} Y \longrightarrow S$ be a factorization of p. Let f_k be the pullback of f over $\mathrm{Spec}(k)$.*

Suppose that the support of β is proper with respect to f. Then $f_*(\beta)$ is pre-special over α and the equality of cycles $\left(f_*(\beta)\right)_{R,k} = f_{k*}(\beta_{R,k})$ holds in Y_k.

Proof As usual, considering the support S of α, we reduce to the case where $\alpha = \langle S \rangle$. Let K be the fraction field of R. As $\mathrm{Spec}(K)$ maps to a generic point of S, we can assume S is integral. Let F be its function field. We can assume by linearity that β is a point x in X with multiplicity 1.

Let L (resp. E) be the residue field of x (resp. $y = f(x)$). Let d be the degree of L/E if it is finite and 0 otherwise. Consider the following pullback square

$$\begin{array}{ccc} \mathrm{Spec}(L \otimes_F K) & \xrightarrow{j} & X \times_S \mathrm{Spec}(R) = X_R \\ f_0 \downarrow & & \downarrow f_R \\ \mathrm{Spec}(E \otimes_F K) & \xrightarrow{i} & Y \times_S \mathrm{Spec}(R) = Y_R. \end{array}$$

According to the formula (8.1.27.1), we obtain:

$$f_{R*}(\beta_R) = f_{R*}j_*(\langle L \otimes_F K \rangle) = i_* f_{0*}(\langle L \otimes_F K \rangle)$$
$$= i_* f_{0*}(f_0^*(\langle E \otimes_F K \rangle)) = i_*(d.\langle E \otimes_F K \rangle) = \langle f_*(\beta) \rangle_R.$$

We are finally reduced to the case when $S = \mathrm{Spec}\,(R)$ and β is a Hilbert cycle over $\mathrm{Spec}\,(R)$. Note that $f_*(\beta)$ is still a Hilbert cycle over $\mathrm{Spec}\,(R)$. As $\beta_{R,k} = \beta \otimes_R^\flat k$, the result follows now from Proposition 8.1.16. □

Corollary 8.2.10 *Consider morphisms of cycles with support in the left diagram*

$$\begin{array}{ccc} \beta & & X \\ \downarrow & \subset & \downarrow p \\ \alpha' \longrightarrow \alpha & & S' \longrightarrow S \end{array}$$

such that β/α is special at the generic points of α' (resp. Λ-universal).

Consider a factorization $X \xrightarrow{f} Y \longrightarrow S$ of p.

Suppose that the support of β is proper with respect to f. Then $f_*(\beta)/\alpha$ is special at the generic points of α' (resp. Λ-universal) and the following equality of cycles holds in $X \times_S S'$:

$$(f \times_S 1_{S'})_*(\beta \otimes_\alpha \alpha') = (f_*(\beta)) \otimes_\alpha \alpha'.$$

8.3 Geometric properties

8.3.1 We introduce a notation which will often be used in the next section. Let S be a scheme and $\alpha = \sum_{i \in I} n_i.\langle Z_i \rangle_X$ an S-cycle written in standard form.

Let s be a point of S and $\mathrm{Spec}\,(k) \xrightarrow{\bar{s}} S$ be a geometric point of S with k separably closed. Let S' be one of the following local schemes: the localization of S at s, the Hensel localization of S at s, the strict localization of S at $\bar{s}$.

We then define the cycle with coefficients in Λ and domain $X \times_S S'$ as:

$$\alpha|_{S'} = \sum_{i \in I} n_i \langle Z_i \times_S S' \rangle_{X \times_S S'}.$$

Remark 8.3.2 The canonical morphism $S' \longrightarrow S$ is flat. In particular, α/S is special at the generic points of S' and we easily get: $\alpha|_{S'} = \alpha \otimes_S S'$.

8.3.a Constructibility

Definition 8.3.3 Let S be a scheme and $s \in S$ a point. We say that a pre-special S-cycle α is *trivial* at s if it is special at s and $\alpha \otimes_S s = 0$.

8 Relative cycles

Naturally, we say that α is trivial if it is zero. Thus α is trivial if and only if it is trivial at the generic points of S.

Recall from [GD67, 1.9.6] that an ind-constructible subset of a noetherian scheme X is a union of locally closed subsets of X.

Lemma 8.3.4 *Let S be a noetherian scheme, and α/S be a pre-special cycle. Then the set*
$$T = \{s \in S \mid \alpha/S \text{ is special (resp. trivial, } \Lambda\text{-universal) at } s\}$$
is ind-constructible in S.

Proof Let s be a point of T, and Z be its closure in S with its reduced subscheme structure. Put $\alpha_Z = \alpha \otimes_S Z$, defined because α is special at the generic point of Z. Given any point t of Z, we know that α/S is special at t if and only if α_Z/Z is special at t (cf. Lemma 8.1.45). But there exists a dense open subset U_s of Z such that $\alpha_Z|_{U_Z}$ is a Hilbert cycle over U_Z. Thus, α/S is special at each point of U_s and $U_s \subset T$. The same argument proves the respective statements, and this concludes the proof. □

8.3.5 Let I be a left filtering category and $(S_i)_{i \in I}$ be a projective system of noetherian schemes with affine transition morphisms. We let S be the projective limit of (S_i) and we assume the following:

1. S is noetherian.
2. There exists an index $i \in I$ such that for all $j \geq i$, the canonical projection $S \xrightarrow{p_i} S_j$ is dominant.

In this case, there exists an index j/i such that for any k/j, the map p_k induces an isomorphism $S^{(0)} \longrightarrow S_k^{(0)}$ on the generic points (cf. [GD67, 8.4.2, 8.4.2.1]). Thus, replacing I by I/j, we can assume that this property is satisfied for all indexes $i \in I$. As a consequence, the following properties are consequences of the previous ones:

(3) For any $i \in I$, $p_i : S \longrightarrow S_i$ is pseudo-dominant and p_i induces an isomorphism $S^{(0)} \longrightarrow S_i^{(0)}$.
(4) For any arrow $j \longrightarrow i$ of I, $p_{ji} : S_j \longrightarrow S_i$ is pseudo-dominant and p_{ji} induces an isomorphism $S_j^{(0)} \longrightarrow S_i^{(0)}$.

Proposition 8.3.6 *Consider the notations and hypotheses above. Assume we are given a projective system of cycles $(\alpha_i)_{i \in I}$ such that α_i is a pre-special cycle over S_i and for any $j \longrightarrow i$, $\alpha_j = \alpha_i \otimes_{S_i} S_j$. Put $\alpha = \alpha_i \otimes_{S_i} S$ for an index $i \in I$.*[86]
The following conditions are equivalent:

(i) α/S is special (resp. Λ-universal).
(ii) There exists an $i \in I$ such that α_i/S_i is special (resp. Λ-universal).
(iii) There exists an $i \in I$ such that for all j/i, α_j/S_j is special (resp. Λ-universal).

Let s be point of S and s_i its image in S_i. Then the following conditions are equivalent:

---
[86] The pullback is well defined because of point (3) and (4) of the hypotheses above.

(i) α/S is special (resp. Λ-universal) at s.
(ii) There exists an $i \in I$ such that α_i/S_i is special (resp. Λ-universal) at s_i.
(iii) There exists an $i \in I$ such that for all j/i, α_j/S_j is special (resp. Λ-universal) at s_j.

Proof Let P be one of the respective properties: "special", "trivial", "Λ-universal". Using the fact that being P at s is an ind-constructible property (from Lemma 8.3.4), it is sufficient to apply [GD67, th. 8.3.2] to the following family of sets:

$$F_i = \{s_i \in S_i \mid \alpha_i \text{ satisfies P at } s_i\}, \quad F = \{s \in S \mid \alpha \text{ satisfies P at } s\}.$$

To get the two sets of equivalent conditions of the statement from *op. cit.* we have to prove the following relations:

$$(1) : \forall (j \longrightarrow i) \in \mathrm{Fl}(I), p_{ji}^{-1}(F_i) \subset F_j,$$

$$(2) : F = \cup_{i \in I} p_i^{-1}(F_i).$$

We consider the case where P is the property "special". For relation (1), we apply Lemma 8.1.45, which implies the stronger relation $p_{ji}^{-1}(F_i) = F_j$. For relation (2), another application of Lemma 8.1.45 gives in fact the stronger relation $F = p_i^{-1}(F_i)$ for any $i \in I$.

Consider a point $s_j \in S$ and put $s_i = p_{ji}(s_j)$. Assume α_i is special at s_i. Then, applying Proposition 8.2.4 and (P3), we get:

(8.3.6.1) $\qquad \alpha_j \otimes_{S_j} s_j = (\alpha_i \otimes_{S_i} s_i) \otimes_{\kappa(s_i)}^\flat \kappa(s_j).$

Similarly, given $s \in S_j$, $s_i = p_i(s)$, and assuming α_i is special at s_i, we get:

(8.3.6.2) $\qquad \alpha \otimes_S s = (\alpha_i \otimes_{S_i} s_i) \otimes_{\kappa(s_i)}^\flat \kappa(s).$

We consider now the case where P is the property "trivial". Then relation (1) follows from (8.3.6.1). Relation (2) follows from (8.3.6.1) and Lemma 8.1.19(1).

We finally consider the case when P is the property "Λ-universal". Relation (1) in this case is again a consequence of (8.3.6.1). According to (8.3.6.2), we get the inclusion $\cup_{i \in I} f_i^{-1}(F_i) \subset F$. We have to prove the reverse inclusion.

Consider a point $s \in S$ with residue field k such that α/S is Λ-universal at s. For any $i \in I$, we put $s_i = p_i(s)$ and denote by k_i its residue field. It is sufficient to find an index $i \in I$ such that $\alpha_i \otimes_{S_i} s_i$ has coefficients in Λ. Thus we are reduced to the following lemma:

Lemma 8.3.7 Let $(k_i)_{i \in I^{op}}$ be an ind-field and put: $k = \varinjlim_{i \in I^{op}} k_i$.

Consider a family $(\beta_i)_{i \in I}$ such that β_i is a k_i-cycle of finite type with coefficients in $\mathbf{Q}$ and for any j/i, $\beta_j = \beta_i \otimes_{k_i}^\flat k_j$. We put $\beta = \beta_i \otimes_{k_i}^\flat k$.

If for an index $i \in I$, $\beta_i \otimes_{k_i}^\flat k$ has coefficients in Λ, then there exists j/i such that β_j has coefficients in Λ. $\square$

8 Relative cycles

We can assume that for any j/i, β_j has positive coefficients. Let X_j (resp. X) be the support of β_j (resp. β). We obtain a pro-scheme $(X_j)_{j/i}$ such that $X = \varprojlim_{i \in I} X_i$. The transition maps of $(X_j)_{j/i}$ are dominant. Thus, by enlarging i, we can assume that for any j/i, the induced map $\pi_0(X_i) \longrightarrow \pi_0(X_j)$ is a bijection. Thus we can consider each element of $\pi_0(X)$ separately and assume that all the X_i are integrals: for any j/i, $\beta_j = n_j.\langle X_j \rangle$ for a positive element $n_j \in \mathbf{Q}$. Arguing generically, we can further assume $X_j = \mathrm{Spec}\,(L_j)$ for a field extension of finite type L_j of k_j. By assumption now, for any j/i, $L_i \otimes_{k_i} k_j$ is an Artinian ring whose reduction is the field L_j. Moreover, $n_j = n_i.\lg(L_i \otimes_{k_i} k_j)$ and we know that $n := n_i.\lg(L_i \otimes_{k_i} k)$ belongs to Λ.

Let p be a prime not invertible in Λ such that $v_p(n_i) < 0$ where v_p denotes the p-adic valuation on $\mathbf{Q}$. It is sufficient to find an index j/i such that $v_p(n_j) \geq 0$. Let $L = (L_i \otimes_{k_i} k)_{red}$. Note that $L = \varinjlim_{i \in I^{op}} L_i$. It is a field extension of finite type of k. Consider elements $a_1, ..., a_n$ algebraically independent over k such that L is a finite extension of $k(a_1, ..., a_n)$. By enlarging i, we can assume that $a_1, ..., a_n$ belongs to L_i. Thus L_i is a finite extension of $k_i(a_1, ..., a_n)$: replacing k_i by $k_i(a_1, ..., a_n)$, we can assume that L_i/k_i is finite.

Let L' be the subextension of L over k generated by the p-th roots of elements of k. As L/k is finite, L'/k is finite, generated by elements $b_1, ..., b_r \in L$. Consider an index j/i such that $b_1, ..., b_r$ belongs to L_j. It follows that $v_p(\lg(L_i \otimes_{k_i} k_j)) = v_p(\lg(L_i \otimes_{k_i} k))$. Thus $v_p(n_j) = v_p(n) \geq 0$ and we are done. $\square$

Corollary 8.3.8 *Let S be a scheme and α be a pre-special S-cycle. Let $\bar{s}$ be a geometric point of S, with image s in S, and S' be the strict localization of S at $\bar{s}$. Then the following conditions are equivalent:*

(i) α/S is special at s.
(i') α/S is special at $\bar{s}$.
(ii) $(\alpha|_{S'})/S'$ is special at $\bar{s}$ (notation of Section 8.3.1).
(iii) There exists an étale neighborhood V of $\bar{s}$ in S such that $(\alpha \otimes_S V)/V$ is special at $\bar{s}$.

Proof The equivalence of (i) and (i') follows trivially from the definition (cf. Remark 8.1.30). Recall from Section 8.3.1 that $\alpha|_{S'} = \alpha \otimes_S S'$. Thus (i') $\Rightarrow$ (ii) is easy (see Lemma 8.1.45). Moreover, (ii) $\Rightarrow$ (iii) is a consequence of the previous proposition applied to the pro-scheme of étale neighborhood of $\bar{s}$. Finally, (iii) $\Rightarrow$ (i) follows from Lemma 8.1.45. $\square$

Proposition 8.3.9 *Consider the notations and hypotheses of Section 8.3.5. Assume that S and S_i are reduced for any $i \in I$.*

Suppose given a projective system $(X_i)_{i \in I}$ of S_i-schemes of finite type such that for any j/i, $X_j = X_i \times_{S_i} S_j$. We let X be the projective limit of (X_i).

Then for any pre-special (resp. special, Λ-universal) S-cycle $\alpha \subset X$, there exists an $i \in I$ and a pre-special (resp. special, Λ-universal) S_i-cycle $\alpha_i \subset X_i$ such that $\alpha = \alpha_i \otimes_{S_i} S$.[87]

---

[87] This pullback is defined in any case because of point (3) of the hypotheses above.

Proof Using Proposition 8.3.6, we are reduced to the first of the respective cases of the proposition. Write $\alpha = \sum_{r \in \Theta} n_r.\langle Z_r \rangle_X$ in standard form.

Consider $r \in \Theta$. As X is noetherian, there exists an index $i \in I$ and a closed subscheme $Z_{r,i} \subset X_i$ such that $Z_r = Z_{r,i} \times_{S_i} S$. Moreover, replacing $Z_{r,i}$ by the reduced closure of the image of the canonical map $Z_r \xrightarrow{(*)} Z_{r,i}$, we can assume that the map $(*)$ is dominant. For any $j \in I/i$, we put $Z_{r,j} = Z_{r,i} \times_{S_i} S_j$. The limit of the pro-scheme $(Z_{r,j})_{j \in I/i^{op}}$ is the integral scheme Z_r. Thus, applying [GD67, 8.2.2], we see that by enlarging i, we can assume that for any $j \in I/i$, $Z_{r,j}$ is irreducible (but not necessarily reduced).

We repeat this construction for every $r \in \Theta$, enlarging i at each step. Fix now an element $j \in I/i$. The scheme $Z_{r,j}$ may not be reduced. However, its reduction $Z'_{r,j}$ is an integral scheme such that $Z'_{r,j} \times_{S_j} S = Z_r$. We put

$$\alpha_j = \sum_{r \in \Theta} n_r \langle Z'_{r,j} \rangle_{X_j}.$$

Let $z_{r,j}$ be the generic point of $Z'_{r,j}$, and $s_{r,j}$ be its image in S_j. It is a generic point and corresponds uniquely to a generic point s_r of S according to the point (3) of the hypotheses 8.3.5. Thus α_j/S_j is pre-special. Moreover, we get from the above that $\kappa(z_{r,j}) \otimes_{\kappa(s_{r,j})} \kappa(s_r) = \kappa(z_r)$, where z_r is the generic point of Z_r. Thus the relation $\alpha_j \otimes_{S_j} S = \alpha$ follows from Lemma 8.2.1. □

8.3.b Samuel multiplicities

8.3.10 We give a review of Samuel multiplicities, following as a general reference [Bou93, VIII.§7].

Let A be a noetherian local ring with maximal ideal $\mathfrak{m}$. Let $M \neq 0$ be an A-module of finite type and $\mathfrak{q} \subset \mathfrak{m}$ an ideal of A such that $M/\mathfrak{q}M$ has finite length. Let d be the dimension of the support of M. Recall from *loc. cit.* that the *Samuel multiplicity* of M at $\mathfrak{q}$ is defined as the integer:

$$e_\mathfrak{q}^A(M) := \lim_{n \to \infty} \left(\frac{d!}{n^d} \lg_A(M/\mathfrak{q}^n M) \right).$$

In the case when $M = A$, we simply put $e_\mathfrak{q}(A) := e_\mathfrak{q}^A(A)$ and $e(A) := e_\mathfrak{m}^A(A)$.

We will use the following properties of these multiplicities that we recall for the convenience of the reader; let A be a local noetherian ring with maximal ideal $\mathfrak{m}$:

Let Φ be the generic points $\mathfrak{p}$ of $\operatorname{Spec}(A)$ such that $\dim(A/\mathfrak{p}A) = \dim A$. Then according to proposition 3 of *loc. cit.*:

($\mathscr{S}$1) $\qquad e_\mathfrak{q}(A) = \sum_{\mathfrak{p} \in \Phi} \lg(A_\mathfrak{p}).e_\mathfrak{q}(A/\mathfrak{p}).$

8 Relative cycles

Let B be a local flat A-algebra such that $B/\mathfrak{m}B$ has finite length over B. Then according to proposition 4 of *loc. cit.*:

$$(\mathscr{S}2) \qquad \frac{e_{\mathfrak{m}B}(B)}{e(A)} = \lg_B(B/\mathfrak{m}B).$$

Let B be a local flat A-algebra such that $\mathfrak{m}B$ is the maximal ideal of B. Let $\mathfrak{q} \subset A$ be an ideal such that $A/\mathfrak{q}A$ has finite length. Then according to the corollary of proposition 4 in *loc. cit.*:

$$(\mathscr{S}3) \qquad e_{\mathfrak{q}B}(B) = e_\mathfrak{q}(A).$$

Assume A is integral with fraction field K. Let B be a finite local A-algebra such that $B \supset A$. Let k_B/k_A be the extension of the residue fields of B/A. Then, according to proposition 5 and point b) of the corollary of proposition 4 in *loc. cit.*,

$$(\mathscr{S}4) \qquad \frac{e_{\mathfrak{m}B}(B)}{e(A)} = \frac{\dim_K(B \otimes_A K)}{[k_B : k_A]}.$$

Definition 8.3.11 (i) Let $S = \mathrm{Spec}\,(A)$ be a local scheme, $s = \mathfrak{m}$ the closed point of S.
Let Z be an S-scheme of finite type with special fiber Z_s. For any generic point z of Z_s, denoting by B the local ring of Z at z, we define the *Samuel multiplicity of Z at z over S* as the rational number:

$$m^\mathscr{S}(z, Z/S) = \frac{e_{\mathfrak{m}B}(B)}{e(A)}.$$

In the case where Z is integral, we define the *Samuel specialization of the S-cycle* $\langle Z \rangle$ *at s* as the cycle with rational coefficients and domain Z_s:

$$\langle Z \rangle \otimes_S^\mathscr{S} s = \sum_{z \in Z_s^{(0)}} m^\mathscr{S}(z, Z/S).z.$$

Consider an S-cycle of finite type $\alpha = \sum_{i \in I} n_i.\langle Z_i \rangle_X$ written in standard form. We define the *Samuel specialization of the S-cycle α at s* as the cycle with domain X_s:

$$\alpha \otimes_S^\mathscr{S} s = \sum_{i \in I} n_i.\langle Z_i \rangle \otimes_S^\mathscr{S} s.$$

(ii) Let S be a scheme. For any point s of S, we let $S_{(s)}$ be the localized scheme of S at s.
Let $f : Z \longrightarrow S$ be an S-scheme of finite type, and z a point of Z which is generic in its fiber. Put $s = f(z)$. We define the *Samuel multiplicity of Z/S at z* as the integer

$$m^\mathscr{S}(z, Z/S) := m^\mathscr{S}(z, Z \times_S S_{(s)}/S_{(s)}).$$

Consider an S-cycle of finite type α with domain X and a point s of S. We define the *Samuel specialization of the S-cycle α at s* as the cycle with rational coefficients:
$$\alpha \otimes_S^{\mathscr{S}} s = \left(\alpha|_{S_{(s)}}\right) \otimes_{S_{(s)}}^{\mathscr{S}} s.$$

Lemma 8.3.12 *Let S be a scheme, and $p : Z' \longrightarrow Z$ an S-morphism which is a birational universal homeomorphism. Then for any point $s \in S$,*
$$\langle Z' \rangle \otimes_S^{\mathscr{S}} s = \langle Z \rangle \otimes_S^{\mathscr{S}} s$$
in $(Z'_s)_{red} = (Z_s)_{red}$.

Proof By hypothesis, p induces an isomorphism $Z'^{(0)} \simeq Z^{(0)}$ between the generic points. Given any irreducible component T' of Z' corresponding to the irreducible component T of Z, we get by hypothesis:
$$T'_{red} \simeq T_{red} \text{ (as schemes)}, \ \lg\left(\mathcal{O}_{Z',T'}\right) = \lg\left(\mathcal{O}_{Z,T}\right).$$
Thus, the result follows easily from the definition. $\square$

8.3.13 Let $Z \xrightarrow{f} S$ be a morphism of finite type and z a point of Z, $s = f(z)$. Assume z is a generic point of Z_s. We introduce the following condition:
$$\mathscr{D}(z, Z/S) : \begin{cases} \text{For any irreducible component } T \text{ of } Z_{(z)}, \\ T_s = \varnothing \text{ or } \dim(T) = \dim(Z_{(z)}). \end{cases}$$

Remark 8.3.14 This condition is in particular satisfied if $Z_{(z)}$ is absolutely equidimensional (and *a fortiori* if Z is absolutely equidimensional).

An immediate translation of $(\mathscr{S}1)$ gives:

Lemma 8.3.15 *Let S be a local scheme with closed point s and Z be an S-scheme of finite type such that Z_s is irreducible with generic point z.*
If the condition $\mathscr{D}(z, Z/S)$ is satisfied, then $\langle Z \rangle \otimes_S^{\mathscr{S}} s = m^{\mathscr{S}}(z, Z/S).z$.

We get directly from $(\mathscr{S}2)$ the following lemma:

Lemma 8.3.16 *Let S be a scheme, s be a point of S, and $\alpha = \sum_{i \in I} n_i . \langle Z_i \rangle_X$ be an S-cycle in standard form such that Z_i is a flat S-scheme of finite type.*
Then α is a Hilbert S-cycle and $\alpha \otimes_S^{\mathscr{S}} s = \alpha \otimes_S^{\flat} s$.

With the notations of Section 8.3.1, we get from $(\mathscr{S}3)$:

Lemma 8.3.17 *Let S be a scheme, s a point of S with residue field k and α an S-cycle of finite type.*
(i) Let S' be the Hensel localization of S at s. Then, $\alpha \otimes_S^{\mathscr{S}} s = \left(\alpha|_{S'}\right) \otimes_{S'}^{\mathscr{S}} s$.
(ii) Let $\bar{k}$ be a separable closure and $\bar{s}$ the corresponding geometric point of S. Let $S_{(\bar{s})}$ be the strict localization of S at $\bar{s}$. Then,
$$\left(\alpha \otimes_S^{\mathscr{S}} s\right) \otimes_k^{\flat} \bar{k} = \left(\alpha|_{S_{(\bar{s})}}\right) \otimes_{S_{(\bar{s})}}^{\mathscr{S}} \bar{s}.$$

8 Relative cycles

Let us recall from [GD67, 13.3.2] the following definition:

Definition 8.3.18 Let $f : X \longrightarrow S$ be a morphism of finite type between noetherian schemes, and x a point of X.

We say f is *equidimensional* at x if there exists an open neighborhood U of x in X and a quasi-finite pseudo-dominant S-morphism $U \longrightarrow \mathbf{A}_S^d$ for $d \in \mathbf{N}$. The integer d is independent of the choice of U: it is called the relative dimension of f at x.

We say f is *equidimensional* if it is equidimensional at every point of X.

Remark 8.3.19 A quasi-finite morphism is equidimensional if and only if it is pseudo-dominant. According to [GD67, 12.1.1.5], this definition agrees with the convention stated in Section 8.1.9 in the case of flat morphisms.

Note that a direct translation of $(\mathscr{S}4)$ gives:

Lemma 8.3.20 *Let $S = \mathrm{Spec}\,(A)$ be an integral local scheme with closed point s and fraction field K. Let Z be a finite equidimensional S-scheme and z a generic point of Z_s. Let B be the local ring of Z at z.*
Then,
$$m^{\mathscr{S}}(z, Z/S) = \frac{\dim_K(B \otimes_A K)}{[\kappa(z) : \kappa(s)]}.$$

8.3.21 Recall that a scheme S is said to be *unibranch* (resp. *geometrically unibranch*) *at a point* $s \in S$ if the henselisation (resp. strict henselisation) of the local ring $\mathscr{O}_{S,s}$ is irreducible (see [GD67, 6.15.1, 18.8.16]). The scheme S is said to be *unibranch* (resp. *geometrically unibranch*) if it is so at any point $s \in S$.

The following result is the key point of this subsection.

Proposition 8.3.22 *Consider a cartesian square*

$$\begin{array}{ccc} Z' & \xrightarrow{g'} & Z \\ f' \downarrow & & \downarrow f \\ S' & \xrightarrow{g} & S \end{array}$$

and a point s' of S', $s = g(s')$. Let k (resp. k') be the residue field of s (resp. s'). We assume the following conditions:

1. *S (resp. S') is geometrically unibranch at s (resp. s').*
2. *f and f' are equidimensional of dimension n.*
3. *For any generic point z of Z_s (resp. z' of $Z_{s'}$) the condition $\mathscr{D}(z, Z/S)$ (resp. $\mathscr{D}(z', Z'/S')$) is satisfied.*

Then, the following equality holds in $Z_{s'}$:
$$\langle Z' \rangle \otimes_{S'}^{\mathscr{S}} s' = (\langle Z \rangle \otimes_{S}^{\mathscr{S}} s) \otimes_k^{\flat} k'.$$

Proof According to Lemma 8.3.15, we have to prove the equality:

$$(8.3.22.1) \quad \sum_{z' \in Z_{s'}^{(0)}} m^{\mathscr{S}}(z', Z'/S').z' = \sum_{z \in Z_s^{(0)}} m^{\mathscr{S}}(z, Z/S).\langle \mathrm{Spec}\,(\kappa(z) \otimes_k k') \rangle_{Z_{s'}}.$$

As f is equidimensional of dimension n, we can assume according to Definition 8.3.18 that there exists a quasi-finite pseudo-dominant S-morphism $p : Z \longrightarrow \mathbf{A}_S^n$. For any generic point z of Z_s, $t = p(z)$ is the generic point of $\mathbf{A}_s^n$. Thus applying ($\mathscr{S}3$), we get:
$$m^{\mathscr{S}}(z, Z/S) = m^{\mathscr{S}}(z, Z/\mathbf{A}_S^n).$$

Consider the S' morphism $p' : Z' \longrightarrow \mathbf{A}_{Z'}^n$ obtained by base change. It is quasi-finite. As Z'/S' is equidimensional of dimension n, p' must be pseudo-dominant. For any generic point z' of $Z_{s'}$, $t' = p'(z')$ is the generic point of $\mathbf{A}_{s'}^n$ and as in the preceding paragraph, we get
$$m^{\mathscr{S}}(z', Z'/S') = m^{\mathscr{S}}(z', Z'/\mathbf{A}_{S'}^n).$$

Moreover, the residue field κ_t of t (resp. $\kappa_{t'}$ of t') is $k(t_1, ..., t_n)$ (resp. $k'(t_1, ..., t_n)$) and this implies $\mathrm{Spec}\left(\kappa(z) \otimes_{\kappa_t} \kappa_{t'}\right)$ is homeomorphic to $\mathrm{Spec}\left(\kappa(z) \otimes_k k'\right)$ and has the same geometric multiplicities. From this and the two preceding relations in (8.3.22.1), we reduce to the case $n = 0$ – indeed, according to [GD67, 14.4.1.1], $\mathbf{A}_S^n$ (resp. $\mathbf{A}_{S'}^n$) is geometrically unibranch at t (resp. t').

Assume now $n = 0$, so that f and f' are quasi-finite pseudo-dominant.

Let $\bar{k}$ be a separable closure of k and $\bar{k}'$ a separable closure of a composite of $\bar{k}$ and k'. It is sufficient to prove relation (8.3.22.1) after extension to $\bar{k}'$ (Lemma 8.1.19). Thus according to Lemma 8.3.17 and hypothesis (3), we can assume S and S' are integral strictly local schemes.

For any $z \in Z_s^{(0)}$, the extension $\kappa(z)/k$ is totally inseparable. Moreover, z corresponds to a unique point $z' \in Z_{s'}^{(0)}$ and we have to prove for any $z \in Z_s^{(0)}$:
$$m^{\mathscr{S}}(z', Z'/S') = m^{\mathscr{S}}(z, Z/S) . \mathrm{lg}(\kappa(z) \otimes_k k').$$

Let $S = \mathrm{Spec}\,(A)$, $K = \mathrm{Frac}(A)$ and $B = \mathscr{O}_{Z,z}$ (resp. $S' = \mathrm{Spec}\,(A')$, $K' = \mathrm{Frac}(A')$ and $B' = \mathscr{O}_{Z',z'}$). As B is quasi-finite dominant over A and A is henselian, B/A is necessarily finite dominant. The same is true for B'/A' and ($\mathscr{S}4$) gives the formulas:
$$m^{\mathscr{S}}(z, Z/S) = \frac{\dim_K(B \otimes_A K)}{[\kappa(z) : k]}, \quad m^{\mathscr{S}}(z', Z'/S') = \frac{\dim_{K'}(B' \otimes_{A'} K')}{[\kappa(z') : k']}.$$

As $B' \otimes_{A'} K' = (B \otimes_A K) \otimes_K K'$, the numerator of these two rationals are the same. To conclude, we are reduced to the easy relation
$$[\kappa(z') : k'] . \mathrm{lg}(\kappa(z) \otimes_k k') = [\kappa(z) : k].$$

Definition 8.3.23 Let S be a scheme and $\alpha = \sum_{i \in I} n_i . \langle Z_i \rangle_X$ be an S-cycle in standard form.

We say α/S is *pseudo-equidimensional over s* if it is pre-special and for any $i \in I$, the structural map $Z_i \longrightarrow S$ is equidimensional at the generic points of the fiber $Z_{i,s}$.

8 Relative cycles

Proposition 8.3.24 *Let S be a strictly local integral scheme with closed point s and residue field k and α be an S-cycle pseudo-equidimensional at s.*

Then for any extension $\operatorname{Spec}(k') \xrightarrow{s'} S$ *of s and any fat point* (R, k') *of S over s', the following relation holds:*

$$\alpha_{R,k'} = \left(\alpha \otimes_S^{\mathscr{L}} s\right) \otimes_k^{\flat} k'.$$

Proof We put $S' = \operatorname{Spec}(R)$ and denote by s' its closed point.

Reductions.– By additivity, we reduce to the case $\alpha = \langle Z \rangle$, Z is integral and the structural morphism $f : Z \longrightarrow S$ is equidimensional at the generic points of Z_s. Any generic point of $S'_{s'}$ dominates a generic point of Z_s so that we can argue locally at each generic point x of Z_s. Thus we can assume Z_s is irreducible with generic point x. Moreover, as Z is equidimensional at x, we can assume according to Definition 8.3.18 that there exists a quasi-finite pseudo-dominant S-morphism

(8.3.24.1) $$Z \xrightarrow{p} \mathbf{A}_S^n.$$

Note that S is geometrically unibranch at s. Thus, applying [GD67, 14.4.1] ("*critère de Chevalley*"), f is universally open at x. As S' is a trait whose closed point goes to s in S, it follows from [GD67, 14.3.7] that the base change $f' : Z' \longrightarrow S'$ of f along S'/S is pseudo-dominant.

Let T be an irreducible component of Z', with special fiber $T_{s'}$ and generic fiber $T_{K'}$ over S'. Then $T \longrightarrow S'$ is a dominant morphism of finite type. Thus, according to [GD67, 14.3.10], either $T_{s'} = \emptyset$ or $\dim(T_{s'}) = \dim(T_{K'})$. Moreover, the dimension of T_η is equal to the transcendental degree of the function field of T over K', which is equal to the transcendental degree of Z over K. This is n according to (8.3.24.1). Thus, in any case, T is equidimensional of dimension n over S' and this implies Z' is equidimensional of dimension n over S'. Moreover, either $T_{s'} = \emptyset$ or $\dim(T) = n + 1 = \dim(Z')$. Note this implies that for any generic point z' of $Z_{s'}$, the condition $\mathscr{D}(z', Z'/S')$ is satisfied.

Middle step.– We prove: $\alpha_{R,k} = \langle Z' \rangle \otimes_{S'}^{\mathscr{L}} s'$.
According to Lemma 8.3.16,

$$\alpha_{R,k} = \langle \overline{Z'_K} \rangle \otimes_R^{\flat} k' = \langle \overline{Z'_K} \rangle \otimes_{S'}^{\mathscr{L}} s'.$$

But the canonical map $\overline{Z'_K} \longrightarrow Z'$ is a birational universal homeomorphism so that we conclude this step by Lemma 8.3.12.

Final step.– We have only to point out that the conditions of Proposition 8.3.22 are fulfilled for the obvious square; this is precisely what we need. $\square$

Corollary 8.3.25 *Let S be a reduced scheme, s a point of S and α an S-cycle which is pseudo-equidimensional over s.*

Let $\bar{s}$ be a geometric point of S with image s in S and S′ be the strict localization of S at $\bar{s}$. We let $S' = \cup_{i \in I} S'_i$ be the irreducible components of S′ and α_i be the cycle made by the part of the cycle $\alpha \otimes_S^b S'$ whose points dominate S'_i.

Then the following conditions are equivalent:

(i) α/S is special at s;
(ii) the cycle $\alpha_\lambda \otimes_{S'_i}^{\mathscr{L}} \bar{s}$ does not depend on $i \in I$.

Moreover, when these conditions are fulfilled, $\alpha \otimes_S \bar{s} = \alpha_\lambda \otimes_{S'_i}^{\mathscr{L}} \bar{s}$.

Proof According to Corollary 8.3.8, we reduce to the case $S = S'$. Then this follows directly from the preceding proposition. □

Corollary 8.3.26 *Let S be a reduced scheme, geometrically unibranch at a point $s \in S$, and α an S-cycle. The following conditions are equivalent:*

(i) α/S is pseudo-equidimensional over s.
(ii) α/S is special at s.

Under these conditions, $\alpha \otimes_S s = \alpha \otimes_S^{\mathscr{L}} s$.

Remark 8.3.27 In particular, over a reduced geometrically unibranch scheme S, every cycle whose support is equidimensional over S is special.

Corollary 8.3.28 *Let S be a reduced scheme and $s \in S$ a point such that S is geometrically unibranch at s and $e(\mathscr{O}_{S,s}) = 1$. Then for any S-cycle α, the following conditions are equivalent:*

(i) α/S is pseudo-equidimensional over s.
(ii) α/S is Λ-universal at s.

Remark 8.3.29 In particular, over a regular scheme S, every cycle whose support is equidimensional over S is Λ-universal. Note also the following theorem:

Theorem 8.3.30 *Let S be an excellent scheme and $s \in S$ a point. The following conditions are equivalent:*

(i) S is regular at s.
(ii) S is geometrically unibranch at s and $e(\mathscr{O}_{S,s}) = 1$.
(iii) S is unibranch at s and $e(\mathscr{O}_{S,s}) = 1$.

Bibliographical references for the proof. We can assume S is the spectrum of an excellent local ring A with closed point s. The implication $(i) \Rightarrow (ii)$ follows from the fact that a normal local ring is geometrically unibranch (at its closed point) and from [Bou93, AC.VIII.§7, prop. 2]. $(ii) \Rightarrow (iii)$ is trivial. Concerning the implication $(iii) \Rightarrow (i)$, let $\hat{A}$ be the completion of the local ring A. We know from [Bou93, AC.VIII.108, ex. 24] that when $e(A) = 1$ and $\hat{A}$ is integral, A is regular. Note $e(A) = 1$ implies A is reduced. To conclude, we refer to [GD67, 7.8.3, (vii)], which established that if A is local excellent reduced, $\hat{A}$ is integral if and only if A is unibranch.

Finally, we get the following theorem already proved by Suslin and Voevodsky ([SV00b, 3.5.9]):

Theorem 8.3.31 *Let S be a scheme and s a point with residue field κ_s such that the local ring A of S at s is regular. Then for any equidimensional S-scheme Z and any generic point z of Z_s,*

$$m^{SV}(z, \langle Z \rangle \otimes_S s) = \sum_i (-1)^i \lg_A \mathrm{Tor}_i^A(\mathcal{O}_{Z,z}, \kappa_s).$$

Proof We reduce to the case $S = \mathrm{Spec}\,(A)$. Then Z is absolutely equidimensional, and we can apply Lemma 8.3.15 together with Corollary 8.3.26 to get that $m^{SV}(z, \langle Z \rangle \otimes_S s) = m^{\mathscr{S}}(z, Z/S)$. Then the result follows from a theorem of Serre [Ser75, IV.12, th. 1]. □

Remark 8.3.32 Let S be a regular scheme, X a smooth S-scheme and $\alpha \subset X$ an S-cycle whose support is equidimensional over S. Let s be a point of S and $i : X_s \longrightarrow X$ the closed immersion of the fiber of X at s. Then the cycle $i^*(\alpha)$ of [Ser75, V-28, par. 7] is well defined and we get:

$$\alpha \otimes_S s = i^*(\alpha).$$

9 Finite correspondences

9.0.1 In this section, $\mathscr{S}$ is the category of all noetherian schemes. We fix an admissible class $\mathscr{P}$ of morphisms in $\mathscr{S}$ and assume in addition that $\mathscr{P}$ is contained in the class of separated morphisms of finite type.

Consider two S-schemes X and Y. To clarify certain formulas, we will denote $X \times_S Y$ simply by XY and let $p_{XY}^X : XY \longrightarrow X$ be the canonical projection morphism. We fix a ring of coefficients $\Lambda \subset \mathbf{Q}$.

9.1 Definition and composition

9.1.1 Let S be a base scheme. For any $\mathscr{P}$-scheme X/S, we let $c_0(X/S, \Lambda)$ be the Λ-module comprising the finite and Λ-universal S-cycles with domain X.[88] Consider a morphism $f : Y \longrightarrow X$ of $\mathscr{P}$-schemes over S. Then the pushforward of cycles induces a well-defined morphism:

$$f_* : c_0(Y/S, \Lambda) \longrightarrow c_0(X/S, \Lambda).$$

Indeed, consider a cycle $\alpha \in c_0(Y/S)$. Let us denote by Z its support in Y and by $f(Z) \subset X$ the image of the latter under f. We consider these subsets as reduced

---

[88] With the notations of [SV00b], $c_0(X/S, \mathbf{Z}) = c_{equi}(X/S, 0)$ when S is reduced.

subschemes. Note that $f(Z)$ is separated and of finite type over S because X/S is noetherian, separated, and of finite type, by assumption 9.0.1. Because Z/S is proper, [GD61, 5.4.3(ii)] shows that $f(Z)$ is indeed proper over S. Thus, the cycle $f_*(\alpha)$ is Λ-universal according to Corollary 8.2.10. Finally, Z/S is finite, and we deduce that $f(Z)$ is quasi-finite, thus finite, over S. This implies the result.

Definition 9.1.2 Let X and Y be two $\mathscr{P}$-schemes over S.
A finite S-correspondence from X to Y with coefficients in Λ is an element of

$$c_S(X,Y)_\Lambda := c_0(X \times_S Y / X).$$

We denote such a correspondence by the symbol $X \bullet\!\xrightarrow{\alpha} Y$.

In the case $\Lambda = \mathbf{Z}$, we simply put $c_S(X,Y) := c_S(X,Y)_\mathbf{Z}$. Throughout the rest of this section, unless explicitly stated, any cycle and any finite S-correspondence are assumed to have coefficients in Λ.

Remark 9.1.3

1. According to properties (P7) and (P7') (*cf.* Section 8.1.41) of the pullback, $c_S(X,Y)_\Lambda$ commutes with finite sums in X and Y.
2. Consider $\alpha \in c_S(X,Y)_\Lambda$. Let Z be the support of α. Then, Z is finite pseudo-dominant over X (by Definition 8.1.20). This means that Z is finite equidimensional over X.
When X is regular (resp. X is reduced geometrically unibranch and $\operatorname{char}(X) \subset \Lambda^\times$), a cycle $\alpha \subset X \times_S Y$ written in standard form

$$\alpha = \sum_i n_i \langle Z_i \rangle_{X \times_S Y}$$

defines a finite S-correspondence from X to Y if and only if for any index $i \in I$, the scheme Z_i is finite equidimensional over X (*i.e.* finite and dominant over an irreducible component of X) – *cf.* Remark 8.3.29 (resp. Remark 8.3.27).
Moreover, in each respective case, $c_S(X,Y)_\Lambda$ is the free Λ-module generated by the closed integral subschemes Z of $X \times_S Y$ which are finite equidimensional over X.
3. By definition, we get an inclusion:

$$c_S(X,Y) \subset c_S(X,Y)_\Lambda$$

which induces an injective map:

$$c_S(X,Y) \otimes_\mathbf{Z} \Lambda \longrightarrow c_S(X,Y)_\Lambda.$$

According to Corollary 8.1.54, this map is a bijection. Indeed, given any finite Λ-linear S-correspondence $\alpha : X \bullet\!\longrightarrow Y$, applying the mentioned corollary, there exists an integer $N > 0$ such that $N.\alpha$ is $\mathbf{Z}$-universal, so in particular an element of $c_S(X,Y)$. If we assume that N is minimal, as α is Λ-universal by assumption,

9 Finite correspondences

N must be invertible in Λ. Therefore, $(N.\alpha) \otimes \frac{1}{N}$ belongs to $c_S(X,Y) \otimes_{\mathbb{Z}} \Lambda$ and is sent to α by the preceding map.

Given more generally inclusions of rings $\Lambda \subset \Lambda' \subset \mathbf{Q}$, we get an inclusion of groups
$$c_S(X,Y)_\Lambda \subset c_S(X,Y)_{\Lambda'}$$
which induces an injection:

(9.1.3.1) $\qquad c_S(X,Y)_\Lambda \otimes_\Lambda \Lambda' \longrightarrow c_S(X,Y)_{\Lambda'}$.

Applying Proposition 8.1.53 and the same argument as above, we get that this map is in fact surjective, and therefore a bijection.

Example 9.1.4

1. Let $f : X \longrightarrow Y$ be a morphism in $\mathscr{P}/S$. Because X/S is separated (assumption 9.0.1), the graph Γ_f of f is a closed subscheme of $X \times_S Y$. The canonical projection $\Gamma_f \longrightarrow X$ is an isomorphism. Thus $\langle\Gamma_f\rangle_{XY}$ is a Hilbert cycle over X. In particular, it is Λ-universal and also finite over X, thus it defines a finite S-correspondence from X to Y.
2. Let $f : Y \longrightarrow X$ be a finite S-morphism which is Λ-universal (as a morphism of the associated cycles). Then the graph Γ_f of f is closed in $X \times_S Y$ and the projection $\Gamma_f \longrightarrow X$ is isomorphic to f. Thus the cycle $\langle\Gamma_f\rangle_{XY}$ is a finite Λ-universal cycle over X which therefore defines a finite S-correspondence ${}^t f : X \bullet\!\!\longrightarrow Y$ called the *transpose* of the finite Λ-universal morphism f.

Suppose we are given finite S-correspondences $X \bullet\!\!\xrightarrow{\alpha} Y \bullet\!\!\xrightarrow{\beta} Z$. Consider the following diagram of cycles:

(9.1.4.1)
$$\begin{array}{ccc} \beta \otimes_Y \alpha & \longrightarrow & \beta \longrightarrow Z. \\ \downarrow & & \downarrow \\ \alpha & \longrightarrow & Y \\ \downarrow & & \\ X & & \end{array}$$

The pullback cycle is well-defined and has coefficients in Λ as β is Λ-universal over Y. Moreover, according to the definition of pullback (*cf.* Theorem 8.1.39) and Corollary 8.2.6, $\beta \otimes_Y \alpha$ is a finite Λ-universal cycle over X with domain XYZ. Note finally that according to Section 9.1.1, the pushforward of this latter cycle by p_{XYZ}^{XZ} is an element of $c_S(X,Z)_\Lambda$.

Definition 9.1.5 Using the preceding notations, we define the *composition product* of β and α as the finite S-correspondence
$$\beta \circ \alpha = p_{XYZ*}^{XZ}(\beta \otimes_Y \alpha) : X \bullet\!\!\longrightarrow Z.$$

Remark 9.1.6 In the case where S is regular and X, Y, Z are smooth over S, the composition product defined above agrees with the one defined in [Dég07, 4.1.16] in terms of the Tor-formula of Serre. In fact, this is a direct consequence of Theorem 8.3.31 after reduction to the case where α and β are represented by closed integral subschemes (see also point (2) of Remark 9.1.3).

We sum up the main properties of the composition for finite correspondences in the following proposition :

Proposition 9.1.7 *Let X, Y, Z be $\mathscr{P}$-schemes over S.*

1. *For any finite S-correspondences $X \bullet\xrightarrow{\alpha} Y \bullet\xrightarrow{\beta} Z \bullet\xrightarrow{\gamma} T$, we have $(\gamma \circ \beta) \circ \alpha = \gamma \circ (\beta \circ \alpha)$.*
2. *For any $X \bullet\xrightarrow{\alpha} Y \xrightarrow{g} Z$, $\langle \Gamma_g \rangle_{YZ} \circ \alpha = (1_X \times_S g)_*(\alpha)$.*
3. *For any $X \xrightarrow{f} Y \bullet\xrightarrow{\beta} Z$, $\beta \circ \langle \Gamma_f \rangle_{XY} = \beta \otimes_Y \langle X \rangle$.*
   Moreover, if f is flat, $\beta \circ \langle \Gamma_f \rangle_{XY} = (f \times_S 1_Z)^*(\beta)$ considering the flat pullback of cycles in the classical sense.
4. *For any $X \xleftarrow{f} Y \bullet\xrightarrow{\beta} Z$ such that f is finite Λ-universal, $\beta \circ {}^t f = (f \times_S 1_Z)_*(\beta)$.*
5. *For any $X \bullet\xrightarrow{\alpha} Y \xleftarrow{g} Z$ such that g is finite Λ-universal, ${}^t g \circ \alpha = \langle Z \rangle \otimes_Y \alpha$. If we suppose that g is finite flat, then ${}^t g \circ \alpha = (1_X \times_S g)^*(\alpha)$.*

Proof (1) Using respectively the projection formulas of Corollary 8.2.10 and Proposition 8.2.8, we obtain

$$(\gamma \circ \beta) \circ \alpha = p_{XYZT*}^{XT}\big((\gamma \otimes_Z \beta) \otimes_Y \alpha\big),$$
$$\gamma \circ (\beta \circ \alpha) = p_{XYZT*}^{XT}\big(\gamma \otimes_Z (\beta \otimes_Y \alpha)\big).$$

Thus this formula is a direct consequence of the associativity (Corollary 8.2.7).

(2) Let $\epsilon : \Gamma_g \longrightarrow Y$ and $p_{X\Gamma_g}^{XZ} : X\Gamma_g \longrightarrow XZ$ be the canonical projections. As ϵ is an isomorphism, we have tautologically $\langle Y \rangle = \epsilon_*(\langle \Gamma_g \rangle)$. The result follows from the computation:

$$(1_X \times_S g)_*(\alpha) = (1_X \times_S g)_*(\langle Y \rangle \otimes_Y \alpha) = (1_X \times_S g)_*(\epsilon_* \langle \Gamma_g \rangle \otimes_Y \alpha)$$
$$\stackrel{(*)}{=} (1_X \times_S g)_*(1_X \times_S \epsilon)_*(\langle \Gamma_g \rangle \otimes_Y \alpha) = p_{X\Gamma_g*}^{XZ}(\langle \Gamma_g \rangle \otimes_Y \alpha)$$
$$\stackrel{(*)}{=} p_{XYZ*}^{XZ}(\langle \Gamma_g \rangle_{YZ} \otimes_Y \alpha)$$

where the equalities labeled $(*)$ follow from the projection formula of Corollary 8.2.10.

(3) The first assertion follows from the projection formula of Proposition 8.2.8 and the fact that Γ_f is isomorphic to X:

$$\beta \circ \langle \Gamma_f \rangle_{XY} = p_{XYZ*}^{XZ}(\beta \otimes_Y \langle \Gamma_f \rangle_{XY}) = \beta \otimes_Y p_{XY*}^{X}(\langle \Gamma_f \rangle_{XY}) = \beta \otimes_Y \langle X \rangle.$$

The second assertion follows from Corollary 8.2.2.

(4) and (5): The proof of these assertions is nearly identical to that of (2) and (3) except that we use the projection formula of Proposition 8.2.8 (and do not need the commutativity of Corollary 8.2.3). □

As a corollary, we obtain that the composition of S-morphisms coincides with the composition of the associated graph considered as finite S-correspondences. For any S-morphism $f : X \longrightarrow Y$, we will still denote by $f : X \bullet\!\!\longrightarrow Y$ the finite S-correspondence equal to $\langle \Gamma_f \rangle_{XY}$. Note moreover that for any $\mathscr{P}$-scheme X/S, the identity morphism of X is the neutral element for the composition of finite S-correspondences.

Definition 9.1.8 We let $\mathscr{P}^{cor}_{\Lambda,S}$ be the category of $\mathscr{P}$-schemes over S with morphisms the finite S-correspondences and the composition product of Definition 9.1.5.

An object of $\mathscr{P}^{cor}_{\Lambda,S}$ will be denoted by $[X]$. The category $\mathscr{P}^{cor}_{\Lambda,S}$ is additive, and the direct sum is given by the disjoint union of $\mathscr{P}$-schemes over S. We have a canonical faithful functor

$$(9.1.8.1) \qquad \gamma : \mathscr{P}/S \longrightarrow \mathscr{P}^{cor}_{\Lambda,S}$$

which is the identity on objects and the graph on morphisms. We call it the *graph functor*.

9.1.9 Given an extension of rings $\Lambda \subset \Lambda' \subset \mathbf{Q}$, we get according to Remark 9.1.3(3) and the definition of composition of finite correspondences a functor of Λ'-linear categories:

$$(9.1.9.1) \qquad \mathscr{P}^{cor}_{\Lambda,S} \otimes_\Lambda \Lambda' \longrightarrow \mathscr{P}^{cor}_{\Lambda',S},$$

which is the identity on objects and the maps of the form (9.1.3.1) on morphisms. According to Remark 9.1.3(3), the latter maps are bijections and we get the following result about changing coefficients.

Proposition 9.1.10 *With the notations above, the functor* (9.1.9.1) *is an equivalence of categories.*

9.1.11 Given two S-morphisms $f : Y \longrightarrow X$ and $g : X' \longrightarrow X$ such that g is finite Λ-universal, we get from the previous proposition the equality of cycles in YX':

$$^tg \circ f = \langle X' \rangle \otimes_X \langle Y \rangle_{YX},$$

where Y is regarded as a closed subscheme of YX through the graph of f.

In particular, when either f or g is flat, we get (using property (P3) of Section 8.1.35 or Corollary 8.2.2):

$$^tg \circ f = \langle X' \times_X Y \rangle_{YX'}.$$

To state the next formulas (the generalized degree formulas), we introduce the following notion:

Definition 9.1.12 Let $f : X' \longrightarrow X$ be a finite pseudo-dominant morphism (recall Definition 8.1.2). For any generic point x of X, we define the degree of f at x as the integer:
$$\deg_x(f) = \sum_{x'/x} [\kappa_{x'} : \kappa_x],$$
where the sum runs over the generic points of X' lying above x.

Proposition 9.1.13 *Let X be a connected S-scheme and $f : X' \longrightarrow X$ be a finite S-morphism.*

If f is special then there exists an integer $d \in \mathbf{N}^*$ such that for any generic point x of X, $\deg_x(f) = d$.

Moreover, $f \circ {}^t f = d.1_X$.

We simply call d the *degree* of the finite special morphism f.

Proof Let Δ' be the diagonal of X'/S. For any generic point x of X, we let Δ_x be the diagonal of the corresponding irreducible component of X, considered as a closed subscheme of X. According to Proposition 9.1.7, and the definition of pushforwards, we get

$$\alpha := f \circ {}^t f = (f \times_S f)_*(\langle \Delta' \rangle_{X'X'}) = \sum_{x \in X^{(0)}} \deg_x(f).\langle \Delta_x \rangle_{XX}.$$

Considering generic points x, y of X, we prove $\deg_x(f) = \deg_y(f)$. By induction, we can reduce to the case where x and y have a common specialization s in X because X is connected and noetherian. Then, as α/X is special, we get by definition of the pullback (see more precisely Remark 8.1.44)

$$\alpha \otimes_S s = \deg_x(f).s = \deg_y(f).s,$$

as required. The remaining assertion then follows. □

Proposition 9.1.14 *Let $f : X' \longrightarrow X$ be an S-morphism which is finite, radicial and Λ-universal.*

Assume X is connected, and let d be the degree of f.

Then ${}^t f \circ f = d.1_{X'}$. In particular, if d is invertible in Λ, f is an isomorphism in $\mathscr{P}^{cor}_{\Lambda,S}$.

Proof According to Section 9.1.11, ${}^t f \circ f = \langle X' \rangle \otimes_X \langle X' \rangle$ as cycles in $X'X'$. Let x be the generic point of X and k be its residue field. Let $\{x'_i, i \in I\}$ be the set of generic points of X, and for any $i \in I$, k'_i be the residue field of x'_i. According to Lemma 8.2.1, we thus obtain:

$${}^t f \circ f = \sum_{(i,j) \in I^2} \langle \mathrm{Spec}\left(k'_i \otimes_k k'_j\right) \rangle_{X'X'}.$$

The result now follows by the definition of the degree and the fact that for any $i \in I$, k'_i/k is radicial. □

9.2 Monoidal structure

Fix a base scheme S. Let X, X', Y, Y' be $\mathscr{P}$-schemes over S.
Consider finite S-correspondences $\alpha : X \bullet\!\!\longrightarrow Y$ and $\alpha' : X' \bullet\!\!\longrightarrow Y'$. Then $\alpha X' := \alpha \otimes_X \langle XX' \rangle$ and $\alpha' X := \alpha' \otimes_{X'} \langle XX' \rangle$ are both finite Λ-universal cycles over XX'. Using stability by composition of finite Λ-universal morphisms (cf. Corollary 8.2.6), the cycle $(\alpha X') \otimes_{XX'} (\alpha' X)$ is finite Λ-universal over XX'.

Definition 9.2.1 Using the above notation, we define the *tensor product* of α and α' over S as the finite S-correspondence

$$\alpha \otimes_S^{tr} \alpha' = (\alpha X') \otimes_{XX'} (\alpha' X) : XX' \bullet\!\!\longrightarrow YY'.$$

Let us first remark that this tensor product is commutative (use commutativity of the pullback (Corollary 8.2.3)) and associative (use associativity of the pullback (Corollary 8.2.7)). Moreover, it is compatible with composition:

Lemma 9.2.2 *Suppose given finite S-correspondences: $\alpha : X \longrightarrow Y$, $\beta : Y \longrightarrow Z$, $\alpha' : X' \longrightarrow Y'$, $\beta' : Y' \longrightarrow Z'$. Then*

$$(\beta \circ \alpha) \otimes_S^{tr} (\beta' \circ \alpha') = (\beta \otimes_S^{tr} \beta') \circ (\alpha \otimes_S^{tr} \alpha').$$

Proof We put $\alpha X' = \alpha \otimes_X \langle XX' \rangle$, $\alpha' X = \alpha' \otimes_X \langle XX' \rangle$ and $\beta Y' = \beta \otimes_Y \langle YY' \rangle$, $\beta' Y = \beta' \otimes_Y \langle YY' \rangle$. We can compute the right-hand side of the above equation as follows:

$$p_{XX'YY'ZZ'*}^{XX'ZZ'}\Big((\beta Y' \otimes_{YY'} \beta' Y) \otimes_{YY'} (\alpha X' \otimes_{XX'} \alpha' X)\Big)$$

$$\stackrel{(1)}{=} p_{XX'YY'ZZ'*}^{XX'ZZ'}\Big((\beta Y' \otimes_{YY'} \beta' Y) \otimes_{YY'} (\alpha' X \otimes_{XX'} \alpha X')\Big)$$

$$\stackrel{(2)}{=} p_{XX'YY'ZZ'*}^{XX'ZZ'}\Big(\beta Y' \otimes_{YY'} ((\beta' Y \otimes_{YY'} \alpha' X) \otimes_{XX'} \alpha X')\Big)$$

$$\stackrel{(3)}{=} p_{XX'YY'ZZ'*}^{XX'ZZ'}\Big((\beta Y' \otimes_{YY'} \alpha X') \otimes_{XX'} (\beta' Y \otimes_{YY'} \alpha' X))\Big).$$

Equality (1) follows from commutativity (Corollary 8.2.3), equality (2) from associativity (Corollary 8.2.7) and equality (3) by both commutativity and associativity.

We note that using the projection formula of Corollary 8.2.10, the left-hand side is equal to

$$p_{XX'YY'ZZ'*}^{XX'ZZ'}\Big(\big((\beta \otimes_Y \alpha) \otimes_X \langle XX' \rangle\big) \otimes_{XX'} \big((\beta' \otimes_{Y'} \alpha') \otimes_{X'} \langle XX' \rangle\big)\Big).$$

We are left to observe that

$$(\beta \otimes_Y \alpha) \otimes_X \langle XX' \rangle = \big((\beta Y') \otimes_{YY'} \alpha\big) \otimes_X \langle XX' \rangle = \beta Y' \otimes_{YY'} \alpha X',$$

using transitivity (Proposition 8.2.4) and associativity (Corollary 8.2.7). The result thus follows by symmetry of the other part in the left-hand side. □

Definition 9.2.3 We define a symmetric monoidal structure on the category $\mathscr{P}^{cor}_{\Lambda,S}$ by putting $[X] \otimes^{tr}_S [Y] = [X \times_S Y]$ on objects and using the tensor product of the previous definition for morphisms.

9.2.4 Note that the functor $\gamma : \mathscr{P}/S \longrightarrow \mathscr{P}^{cor}_{\Lambda,S}$ is monoidal for the cartesian structure on the source category. Indeed, this is a consequence of property (P3) of the relative product (see Section 8.1.35) and the observation that for any morphisms $f : X \longrightarrow Y$ and $f' : X' \longrightarrow Y'$, $(\Gamma_f \times_S X') \times_{XX'} (\Gamma'_{f'} \times_S X) = \Gamma_{f \times_S f'}$.

9.3 Functoriality

Fix a morphism of schemes $f : T \longrightarrow S$. For any $\mathscr{P}$-scheme X/S, we put $X_T = X \times_S T$. For a pair of $\mathscr{P}$-schemes over S (resp. T-schemes) (X,Y), we put $XY = X \times_S Y$ (resp. $XY_T = X \times_T Y$).

9.3.a Base change

Consider a finite S-correspondence $\alpha : X \bullet\!\!\longrightarrow Y$. The cycle $\alpha \otimes_X \langle X_T \rangle$ defines a finite T-correspondence from X_T to Y_T denoted by α_T.

Lemma 9.3.1 *Consider finite S-correspondences $X \bullet\!\!\xrightarrow{\alpha} Y \bullet\!\!\xrightarrow{\beta} Y$.*
Then $(\beta \circ \alpha)_T = \beta_T \circ \alpha_T$.

Proof This follows easily using the projection formula of Corollary 8.2.10, the associativity formula of Corollary 8.2.7 and the transitivity formula of Proposition 8.2.4:

$$p^{XZ}_{XYZ*}(\beta \otimes_Y \alpha) \otimes_X \langle X_T \rangle = p^{XZ_T}_{XYZ_T*}\big((\beta \otimes_Y \alpha) \otimes_X \langle X_T \rangle\big)$$
$$= p^{XZ_T}_{XYZ_T*}\big(\beta \otimes_Y (\alpha \otimes_X \langle X_T \rangle)\big) = p^{XZ_T}_{XYZ_T*}\big((\beta \otimes_Y \langle Y_T \rangle) \otimes_{Y_T} (\alpha \otimes_X \langle X_T \rangle)\big).$$

Definition 9.3.2 Let $f : T \longrightarrow S$ be a morphism of schemes. Using the preceding lemma, we define the base change functor

$$f^* : \mathscr{P}^{cor}_{\Lambda,S} \longrightarrow \mathscr{P}^{cor}_{\Lambda,T}$$
$$[X/S] \longmapsto [X_T/T]$$
$$c_S(X,Y)_\Lambda \ni \alpha \longmapsto \alpha_T.$$

We sum up the basic properties of the base change for correspondences in the following lemma.

9 Finite correspondences

Lemma 9.3.3 *Take the notation and hypothesis of the previous definition.*

1. *The functor f^* is symmetric monoidal.*
2. *Let $f_0^* : \mathscr{P}/S \longrightarrow \mathscr{P}/T$ be the classical base change functor on $\mathscr{P}$-schemes over S. Then the following diagram is commutative:*

$$\begin{array}{ccc} \mathscr{P}/S & \xrightarrow{\gamma_S} & \mathscr{P}^{cor}_{\Lambda,S} \\ f_0^* \downarrow & & \downarrow f^* \\ \mathscr{P}/T & \xrightarrow{\gamma_T} & \mathscr{P}^{cor}_{\Lambda,T}. \end{array}$$

3. *Let $\sigma : T' \longrightarrow T$ be a morphism of schemes. Through the canonical isomorphisms $(X_T)_{T'} \simeq X_{T'}$, equality $(f \circ \sigma)^* = \sigma^* \circ f^*$ holds.*

Proof (1) This point follows easily using the associativity formula of Corollary 8.2.7 and the transitivity formulas of Proposition 8.2.4 and Corollary 8.2.6.
(2) This point follows from the fact that for any S-morphism $f : X \longrightarrow Y$, there is a canonical isomorphism $\Gamma_{f_T} \longrightarrow \Gamma_f \times_S T$.
(3) This point is a direct application of the transitivity (Proposition 8.2.4). □

Lemma 9.3.4 *Let $f : T \longrightarrow S$ be a universal homeomorphism. Then $f^* : \mathscr{P}^{cor}_{\Lambda,S} \longrightarrow \mathscr{P}^{cor}_{\Lambda,T}$ is fully faithful.*

Proof Let X and Y be $\mathscr{P}$-schemes over S. Then $X_T \longrightarrow X$ is a universal homeomorphism. Any generic point x of X corresponds uniquely to a generic point of X_T. Let m_x (resp. m'_x) be the geometric multiplicity of x in X (resp. X_T). Consider a finite S-correspondence $\alpha = \sum_{i \in I} n_i.z_i$. For each $i \in I$, let x_i be the generic point of X dominated by z_i. Then we get by definition:

$$f^*(\alpha) = \sum_{i \in I} m'_{x_i} \frac{n_i}{m_{x_i}} . z_i$$

and the lemma is clear. □

9.3.b Restriction

Consider a $\mathscr{P}$-morphism $p : T \longrightarrow S$. For any pair of T-schemes (X, Y), we denote by $\delta_{XY} : X \times_T Y \longrightarrow X \times_S Y$ the canonical closed immersion deduced by base change from the diagonal immersion of T/S.

Consider a finite T-correspondence $\alpha : X \bullet\!\!\!-\!\!\!\!\rightarrow Y$. The cycle $\delta_{XY*}(\alpha)$ is the cycle α considered as a cycle in $X \times_S Y$. It defines a finite S-correspondence from X to Y.

Lemma 9.3.5 *Let X, Y and Z be T-schemes. The following relations hold:*

1. *For any T-morphism $f : X \longrightarrow Y$, $\delta_{XY*}(\langle \Gamma_f \rangle_{XY_T}) = \langle \Gamma_f \rangle_{XY}$.*
2. *For all $\alpha \in c_T(X,Y)_\Lambda$ and $\beta \in c_T(Y,Z)_\Lambda$,*

$$\delta_{XZ*}(\beta \circ \alpha) = (\delta_{YZ*}(\beta)) \circ (\delta_{XY*}(\alpha)).$$

Proof The first assertion is obvious.

The second assertion is a consequence of the projection formulas of Proposition 8.2.8 and Corollary 8.2.10, and the functoriality of pushforwards:

$$(\delta_{YZ*}(\beta)) \circ (\delta_{XY*}(\alpha)) = p^{XZ}_{XYZ*}(\delta_{YZ*}(\beta) \otimes_Y \delta_{XY*}(\alpha))$$
$$= p^{XZ}_{XYZ*}\delta_{XYZ*}(\beta \otimes_Y \alpha) = \delta_{XZ*}p^{XZ_T}_{XYZ_T*}(\beta \otimes_Y \alpha).$$

Definition 9.3.6 Let $p : T \longrightarrow S$ be a $\mathscr{P}$-morphism.
Using the preceding lemma, we define a functor

$$p_\sharp : \mathscr{P}^{cor}_{\Lambda,T} \longrightarrow \mathscr{P}^{cor}_{\Lambda,S}$$
$$[X \longrightarrow T] \longmapsto [X \longrightarrow T \xrightarrow{p} S]$$
$$c_T(X,Y)_\Lambda \ni \alpha \longmapsto \delta_{XY*}(\alpha).$$

This functor enjoys the following properties:

Lemma 9.3.7 *Let $p : T \longrightarrow S$ be a $\mathscr{P}$-morphism.*

1. *The functor $p_\sharp$ is left adjoint to the functor p^*.*
2. *For any composable $\mathscr{P}$-morphisms $Z \xrightarrow{q} T \xrightarrow{p} S$, $(pq)_\sharp = p_\sharp q_\sharp$.*
3. *Let $p^0_\sharp : \mathscr{P}/T \longrightarrow \mathscr{P}/S$ be the functor induced by composition with p. Then the following diagram is commutative:*

$$\begin{array}{ccc} \mathscr{P}/T & \xrightarrow{\gamma_T} & \mathscr{P}^{cor}_{\Lambda,T} \\ p^0_\sharp \downarrow & & \downarrow p_\sharp \\ \mathscr{P}/S & \xrightarrow{\gamma_S} & \mathscr{P}^{cor}_{\Lambda,S}. \end{array}$$

Proof For point (1), we have to construct for schemes X/T and Y/S a natural isomorphism $c_S(p_\sharp X, Y)_\Lambda \simeq c_T(X, p^*Y)_\Lambda$. It is induced by the canonical isomorphism of schemes $(p_\sharp X) \times_S Y \simeq X \times_T (p^*Y)$.

Point (2) follows from the associativity of the pushforward functor on cycles. Note also that this identification is compatible with the transposition of the identification of Lemma 9.3.3(3) according to the adjunction property just obtained.

Point (3) is a reformulation of Lemma 9.3.5(2). □

9.3.c A finiteness property

9.3.8 We assume here that $\mathscr{P}$ is the class of all separated morphisms of finite type in $\mathscr{S}$.

Let I be a left filtering category and $(X_i)_{i \in I}$ be a projective system of separated S-schemes of finite type with affine dominant transition morphisms. We let $\mathscr{X}$ be the projective limit of $(X_i)_i$ and assume that $\mathscr{X}$ is Noetherian over S.

Proposition 9.3.9 *Let Y be a $\mathscr{P}$-scheme of finite type over S. Then the canonical morphism*
$$\varphi : \varinjlim_{i \in I^{op}} c_S(X_i, Y)_\Lambda \longrightarrow c_0(\mathscr{X} \times_S Y/\mathscr{X}, \Lambda)$$
is an isomorphism.

Proof Note that according to [AGV73, IV, 8.3.8(i)], we can assume that condition (2) of Section 8.3.5 is satisfied for $(X_i)_{i\in I}$. Thus conditions (1) to (4) of *loc. cit.* are satisfied. Then the surjectivity of φ follows from Proposition 8.3.9 and the injectivity from Proposition 8.3.6. □

9.4 The fibred category of correspondences

We can summarize the preceding constructions:

Proposition 9.4.1 *The 2-functor*
$$\mathscr{P}^{cor}_\Lambda : S \longmapsto \mathscr{P}^{cor}_{\Lambda,S}$$
equipped with the pullback defined in Definition 9.3.2 and with the tensor product of Definition 9.2.3 is a monoidal $\mathscr{P}$-fibred category such that the functor
$$\gamma : \mathscr{P} \longrightarrow \mathscr{P}^{cor}_\Lambda$$
(see (9.1.8.1)) is a morphism of monoidal $\mathscr{P}$-fibred category.

Proof According to Lemma 9.3.7, for any $\mathscr{P}$ morphisms p, p^* admits a left adjoint $p_\sharp$. We have checked that γ is symmetric monoidal and commutes with f^* and $p_\sharp$ (see respectively Section 9.2.4, Lemma 9.3.3 and Lemma 9.3.7). But γ is essentially surjective. Thus, to prove the properties ($\mathscr{P}$-BC) and ($\mathscr{P}$-PF) for the fibred category $\mathscr{P}^{cor}_\Lambda$, we are reduced to the case of $\mathscr{P}$, which is easy (see Example 1.1.28). This concludes the proof. □

Remark 9.4.2 Consider the definition above.

1. The category $\mathscr{P}^{cor}_\Lambda$ is Λ-linear. For any scheme S, $\mathscr{P}^{cor}_{\Lambda,S}$ is additive. For any finite family of schemes $(S_i)_{i\in I}$ which admits a sum S in $\mathscr{S}$, the canonical map
$$\mathscr{P}^{cor}_{\Lambda,S} \longrightarrow \bigoplus_{i \in I} \mathscr{P}^{cor}_{\Lambda,S_i}$$
is an isomorphism.
2. The functor $\gamma : \mathscr{P} \longrightarrow \mathscr{P}^{cor}_\Lambda$ is nothing else than the canonical geometric sections of $\mathscr{P}^{cor}_\Lambda$ (see Definition 1.1.35).

We will apply these definitions in the particular cases $\mathscr{P} = Sm$ (resp. $\mathscr{P} = \mathscr{S}^{ft}$), the class of smooth separated (resp. separated) morphisms of finite type. Note that we get a commutative square

$$Sm \xrightarrow{\gamma} \mathscr{S}m_\Lambda^{cor}$$
$$\downarrow \qquad \downarrow$$
$$\mathscr{S}^{ft} \xrightarrow{\gamma} \mathscr{S}_\Lambda^{ft,cor},$$

where the vertical maps are the obvious embeddings of monoidal Sm-fibred categories.

9.4.3 Consider the extensions of rings $\Lambda \subset \Lambda' \subset \mathbf{Q}$. The functors (9.1.9.1) for various schemes S in $\mathscr{S}$ are compatible with the operations of a $\mathscr{P}$-fibred category because they are only concerned with adding denominators in the coefficients of the finite correspondences considered. Thus they induce a morphism of monoidal $\mathscr{P}$-fibred categories over $\mathscr{S}$:

(9.4.3.1) $$\mathscr{P}_\Lambda^{cor} \otimes_\Lambda \Lambda' \longrightarrow \mathscr{P}_{\Lambda'}^{cor}.$$

According to Proposition 9.1.10, we immediately get the following result:

Proposition 9.4.4 *Then the morphism (9.4.3.1) is an equivalence of monoidal $\mathscr{P}$-fibred categories.*

Remark 9.4.5 The restriction of the category $\mathscr{P}_\mathbf{Z}^{cor}$ to the category of regular schemes was already defined in [Dég07]. Indeed, one can check using the comparison of Suslin–Voevodsky's multiplicities with Serre's intersection multiplicities (using Tor-formulas; *cf.* Theorem 8.3.31), that the operations τ^*, $\tau_\sharp$, and $\otimes^{tr}$ defined here coincide with that of [Dég07]. This remark extends Remark 9.1.6.

10 Sheaves with transfers

10.0.1 The category $\mathscr{S}$ is the category of noetherian schemes of finite dimension. We fix an admissible class $\mathscr{P}$ of morphisms in $\mathscr{S}$ satisfying the following assumptions:

(a) Any morphism in $\mathscr{P}$ is separated of finite type.
(b) Any étale separated morphism of finite type is in $\mathscr{P}$.

We fix a topology t on $\mathscr{S}$ which is $\mathscr{P}$-admissible and such that:

(c) For any scheme S, there is a class of covers $\mathscr{E}$ of the form $(p : W \longrightarrow S)$ with p a $\mathscr{P}$-morphism such that t is the topology generated by $\mathscr{E}$ and the covers of the form $(U \longrightarrow U \sqcup V, V \longrightarrow U \sqcup V)$ for any schemes U and V in $\mathscr{S}$.

We fix a ring of coefficients Λ. Whenever we speak of Λ-cycles (or the premotivic category $\mathscr{P}_\Lambda^{cor}$, etc...), we mean cycles with coefficients in the localization of $\mathbf{Z}$ with respect to invertible primes in Λ.

Note that in Sections 10.4 and 10.5, we will apply the conventions of Section 1.4 by replacing the class of smooth morphisms of finite type (resp. morphisms of finite type) there by the class of smooth separated morphisms of finite type (resp. separated morphisms of finite type).

10.1 Presheaves with transfers

We consider the additive category $\mathscr{P}^{cor}_{\Lambda,S}$ of Definition 9.1.8 and the graph functor $\gamma : \mathscr{P}/S \longrightarrow \mathscr{P}^{cor}_{\Lambda,S}$ of (9.1.8.1).

Definition 10.1.1 A *presheaf with transfers* F over S is an additive presheaf of Λ-modules over $\mathscr{P}^{cor}_{\Lambda,S}$. We denote by $\operatorname{PSh}\left(\mathscr{P}^{cor}_{\Lambda,S}\right)$ the corresponding category.

If X is a $\mathscr{P}$-scheme over S, we denote by $\Lambda^{tr}_S(X)$ the presheaf with transfers represented by X.

We denote by $\hat{\gamma}_*$ the functor

(10.1.1.1) $$\operatorname{PSh}\left(\mathscr{P}^{cor}_{\Lambda,S}\right) \longrightarrow \operatorname{PSh}(\mathscr{P}/S, \Lambda), F \longmapsto F \circ \gamma.$$

Note that $\operatorname{PSh}\left(\mathscr{P}^{cor}_{\Lambda,S}\right)$ is obviously a Grothendieck abelian category generated by the objects $\Lambda^{tr}_S(X)$ for a $\mathscr{P}$-scheme X/S. Moreover, the following proposition is straightforward:

Proposition 10.1.2 *There is an essentially unique Grothendieck abelian $\mathscr{P}$-premotivic category* $\operatorname{PSh}\left(\mathscr{P}^{cor}_{\Lambda}\right)$ *which is geometrically generated (cf. Definition 1.1.41), whose fiber over a scheme S is* $\operatorname{PSh}\left(\mathscr{P}^{cor}_{\Lambda,S}\right)$ *and such that the functor Λ^{tr}_S induces a morphism of additive monoidal $\mathscr{P}$-fibred categories.*

(10.1.2.1) $$\mathscr{P}^{cor}_{\Lambda} \longrightarrow \operatorname{PSh}\left(\mathscr{P}^{cor}_{\Lambda}\right).$$

Moreover, the functor (10.1.1.1) *induces a morphism of abelian $\mathscr{P}$-premotivic categories*

$$\hat{\gamma}^* : \operatorname{PSh}(\mathscr{P}, \Lambda) \rightleftarrows \operatorname{PSh}\left(\mathscr{P}^{cor}_{\Lambda}\right) : \hat{\gamma}_*.$$

Proof To help the reader, we recall the following consequence of Yoneda's lemma:

Lemma 10.1.3 *Let $F : (\mathscr{P}^{cor}_{\Lambda,S})^{op} \longrightarrow \Lambda\text{-mod}$ be a presheaf with transfers. Let $\mathscr{I}$ be the category of representable presheaves with transfers over F. Then the canonical map*

$$\varinjlim_{\Lambda^{tr}_S(X) \longrightarrow F} \Lambda^{tr}_S(X) \longrightarrow F$$

is an isomorphism. The limit is taken in $\operatorname{PSh}\left(\mathscr{P}^{cor}_{\Lambda,S}\right)$ *and runs over $\mathscr{I}$.* □

This lemma allows us to define the structural left adjoints of $\operatorname{PSh}\left(\mathscr{P}^{cor}_{\Lambda}\right)$ (recall f^*, $p_\sharp$ for p a $\mathscr{P}$-morphism and the tensor product) because they are indeed determined by (10.1.2.1). The existence of the structural right adjoints is formal.

The same lemma allows to get the adjunction $(\hat{\gamma}^*, \hat{\gamma}_*)$. □

Remark 10.1.4 Note that for any presheaf with transfers F over S, and any morphism $f : T \longrightarrow S$ (resp. $\mathscr{P}$-morphism $p : S \longrightarrow S'$), we get as usual $f_*F = F \circ f^*$ (resp. $p^*F = F \circ p_\sharp$) where the functor f^* (resp. $p_\sharp$) on the right-hand side is taken with respect to the $\mathscr{P}$-fibred category $\mathscr{P}^{cor}_{\Lambda}$.

Let us state the following lemma for future use.

Lemma 10.1.5 *Let $(S_\alpha)_{\alpha \in A}$ be a projective system of schemes in $\mathscr{S}$, with dominant affine transition maps, and such that $S = \varprojlim_{\alpha \in A} S_\alpha$ is representable in $\mathscr{S}$.*

Consider an index $\alpha_0 \in A$ and a presheaf with transfers F over S_{α_0}. For any index α/α_0, we denote by F_α (resp. F) the pullback of F_{α_0} over S_α (resp. S) in the sense of the premotivic structure on $\mathrm{PSh}\left(\mathscr{P}_\Lambda^{cor}\right)$.

Then the canonical map:

$$\varinjlim_{\alpha \in A/\alpha_0} F_\alpha(S_\alpha) \longrightarrow F(S)$$

is an isomorphism.

Proof The presheaf F_{α_0} can be written as an inductive limit of representable sheaves of the form $\Lambda_{S_{\alpha_0}}^{tr}(X_{\alpha_0})$ of a $\mathscr{P}$-scheme $X_{\alpha_0}/S_{\alpha_0}$. As the global section functor on presheaves with transfers commutes with inductive limits, we are reduced to the case where $F = \Lambda_{S_{\alpha_0}}^{tr}(X_{\alpha_0})$. In this case, the lemma follows directly from Proposition 9.3.9. □

10.2 Sheaves with transfers

Definition 10.2.1 A t-sheaf with transfers over S is a presheaf with transfers F such that the functor $F \circ \gamma_S$ is a t-sheaf. We denote by $\mathrm{Sh}_t\left(\mathscr{P}_{\Lambda,S}^{cor}\right)$ the full subcategory of $\mathrm{PSh}(\mathscr{P}_{\Lambda,S}^{cor}, \Lambda)$ of sheaves with transfers.

According to this definition, we get a canonical faithful functor

$$\gamma_* : \mathrm{Sh}_t\left(\mathscr{P}_{\Lambda,S}^{cor}\right) \longrightarrow \mathrm{Sh}_t(\mathscr{P}/S, \Lambda), F \mapsto F \circ \gamma.$$

Example 10.2.2 A particularly important case for us is the case when $t = Nis$ is the Nisnevich topology. According to Voevodsky's original definition, a Nisnevich sheaf with transfers will be called simply a *sheaf with transfers*.

Remark 10.2.3 Later on, in the case $\mathscr{P} = \mathscr{S}^{ft}$, we will use the notation $\underline{\Lambda}_S^{tr}(X)$ to denote the presheaf on the big site $\mathscr{S}_{\Lambda,S}^{ft,cor}$ represented by a separated S-scheme of finite type.

Proposition 10.2.4 *Let X be a $\mathscr{P}$-scheme over S.*
1. *The presheaf $\Lambda_S^{tr}(X)$ is an étale sheaf with transfers.*
2. *If $\mathrm{char}(X) \subset \Lambda^\times$, $\Lambda_S^{tr}(X)$ is a qfh-sheaf with transfers.*

Proof For point (1), we follow the proof of [Dég07, 4.2.4]: the computation of the pullback by an étale map is given in our context by point (3) of Proposition 9.1.7.

Moreover, the property for a cycle α/Y to be Λ-universal is étale-local on Y according to Corollary 8.3.8. For point (2), we refer to [SVoob, 4.2.7]. □

We can actually describe explicitly representable presheaves with transfers in the following case:

Proposition 10.2.5 *Let S be a scheme and X be a finite étale S-scheme. Then for any $\mathscr{P}$-scheme Y over S,*

$$\Gamma(Y, \Lambda_S^{tr}(X)) = \pi_0(Y \times_S X).\Lambda.$$

This readily follows from the next lemma:

Lemma 10.2.6 *Let $f : X \longrightarrow S$ be an étale separated morphism of finite type. Let $\pi_0^{finite}(X/S)$ be the set of connected components V of X such that $f(V)$ is equal to a connected component of S (i.e. f is finite over V).*
Then $c_0(X/S, \Lambda) = \pi_0^{finite}(X/S).\Lambda$.

Proof We can assume that S is reduced and connected.

We first treat the case where $X = S$. Consider a finite Λ-universal S-cycle α with domain S. Write $\alpha = \sum_{i \in I} n_i.\langle Z_i \rangle_S$ in standard form. By definition, Z_i dominates an irreducible component of S, thus Z_i is equal to that irreducible component.

Consider S_0 an irreducible component of S and an index $i \in I$ such that $S_0 \cap Z_i$ is not empty. Consider a point $s \in S_0 \cap Z_i$. We have obviously $\alpha_s = n_i.\langle \operatorname{Spec}(\kappa(s)) \rangle \neq 0$. Thus there exists a component of α which dominates S_0, i.e. $\exists j \in I$ such that $Z_j = S_0$. Moreover, computing α_s using alternatively Z_i and Z_j gives $n_i = n_j$. As S is noetherian, we see inductively that $\{Z_i | i \in I\}$ is the set of irreducible components of S and for any $i, j \in I$, $n_i = n_j$. Thus $c_0(S/S, \Lambda) = \mathbf{Z}$.

Consider now the case of an étale S-scheme X. By additivity of c_0, we can assume that X is connected. Consider the following canonical map:

$$c_0(X/S, \Lambda) \longrightarrow c_0(X \times_S X/X, \Lambda), \alpha \longmapsto \alpha \otimes_S^\flat X.$$

Note that considering the projection $p : X \times_S X \longrightarrow X$, by definition, $\alpha \otimes_S^\flat X = p^*(\alpha)$. Consider the diagonal $\delta : X \longrightarrow X \times_S X$ of X/S. Because X/S is étale and separated, δ is a direct factor of $X \times_S X$ and we can write $X \times_S X = X \sqcup U$. Because c_0 is additive,

$$c_0(X \times_S X/X, \Lambda) = c_0(X/X, \Lambda) \oplus c_0(U/X, \Lambda).$$

Moreover, the projection on the first factor is induced by the map δ^* on cycles. Because $\delta^* p^* = 1$, we deduce that a cycle in $c_0(X/S, \Lambda)$ corresponds uniquely to a cycle in $c_0(X/X, \Lambda)$. According to the preceding case, this latter group is the free group generated by the cycle $\langle X \rangle$. This latter cycle is Λ-universal over S, because X/S is flat. Thus, if X/S is finite, it is an element of $c_0(X/S, \Lambda)$ so that $c_0(X/S, \Lambda) = \Lambda$. Otherwise, no Λ-linear combination of $\langle X \rangle$ belongs to $c_0(X/S, \Lambda)$ so that $c_0(X/S, \Lambda) = 0$. □

10.3 Associated sheaf with transfers

10.3.1 Recall from Section 3.2.1 that we denote by $(\mathscr{P}/S)^{\amalg}$ the category of I-diagrams of objects in $\mathscr{P}/S$ indexed by a discrete category I. Given any simplicial object $\mathscr{X}$ of $(\mathscr{P}/S)^{\amalg}$, we will consider the complex $\Lambda_S^{tr}(\mathscr{X})$ of $\operatorname{PSh}\left(\mathscr{P}_{\Lambda,S}^{cor}\right)$ applying the definition of Section 5.1.8 to the Grothendieck $\mathscr{P}$-fibred category $\operatorname{PSh}(\mathscr{P})$.

Consider a t-cover $p : W \longrightarrow X$ in $\mathscr{P}/X$. We denote by W_X^n the n-fold product of W over X (in the category $\mathscr{P}/X$). We denote by $\check{S}(W/X)$ the Čech simplicial object of $\mathscr{P}_{\Lambda,S}^{cor}$ such that $\check{S}_n(W/X) = W_X^{n+1}$. The canonical morphism $\check{S}(W/X) \longrightarrow X$ is a t-hypercover according to Definition 3.2.1. We will call these particular types of t-hypercovers the *Čech t-hypercovers* of X.

Definition 10.3.2 We will say that the admissible topology t on $\mathscr{P}$ is *compatible with transfers* (resp. *weakly compatible with transfers*) if for any scheme S and any t-hypercover (resp. any Čech t-hypercover) $\mathscr{X} \longrightarrow X$ in the site $\mathscr{P}/S$, the canonical morphism of complexes

(10.3.2.1) $$\Lambda_S^{tr}(\mathscr{X}) \longrightarrow \Lambda_S^{tr}(X)$$

induces a quasi-isomorphism of the associated t-sheaves on $\mathscr{P}/S$.

Obviously, if t is compatible with transfers then it is weakly compatible with transfers.

Recall from Proposition 10.2.4 that, in the cases $t = Nis, \acute{e}t$, (10.3.2.1) is actually a morphism of complexes of t-sheaves with transfers. The following proposition is a generalization of [Voe96, 3.1.3] but its proof is in fact the same.

Proposition 10.3.3 *The Nisnevich (resp étale) topology t on $\mathscr{P}$ is weakly compatible with transfers.*

Proof We consider a t-cover $p : W \longrightarrow X$, the associated Čech hypercover $\mathscr{X} = \check{S}(W/X)$ of X and we prove that the map (10.3.2.1) is a quasi-isomorphism of t-sheaves. Recall that a point of $\mathscr{P}/S$ for the topology t is given by an essentially affine pro-object $(V_i)_{i \in I}$ of $\mathscr{P}/S$. Moreover, its projective limit $\mathscr{V}$ in the category of schemes is in particular a local henselian noetherian scheme.

It will be sufficient to check that the fiber of (10.3.2.1) at the point $(V_i)_{i \in I}$ is a quasi-isomorphism. Thus, according to Proposition 9.3.9, we can assume that $S = \mathscr{V}$ is a local henselian scheme and we are to reduce to proving that the complex of abelian groups

$$\ldots \longrightarrow c_0(W \times_X W/S, \Lambda) \longrightarrow c_0(W/S, \Lambda) \xrightarrow{p_*} c_0(X/S, \Lambda) \longrightarrow 0$$

is acyclic. We denote this complex by C.

Recall that the abelian group $c_0(X/S)$ is covariantly functorial in X with respect to separated morphisms of finite type $f : X' \longrightarrow X$ (*cf.* Section 9.1.1). Moreover, if f is an immersion, f_* is obviously injective.

Let $\mathscr{F}_0$ be the set of closed subschemes Z of X such that Z/S is finite. Given a closed subscheme Z in $\mathscr{F}_0$, we let C_Z be the complex of abelian groups

(10.3.3.1) $\quad \cdots \longrightarrow c_0(W_Z \times_Z W_Z/S, \Lambda) \longrightarrow c_0(W_Z/S, \Lambda) \xrightarrow{p_{Z*}} c_0(Z/S, \Lambda) \longrightarrow 0$

where p_Z is the t-cover obtained by pullback along $Z \longrightarrow X$. From what we have just recalled, we can identify C_Z with a subcomplex of C. The set $\mathscr{F}_0$ can be ordered by inclusion, and C is the union of its subcomplexes C_Z. If $\mathscr{F}_0$ is empty, then $C = 0$ and the proposition is clear. Otherwise, $\mathscr{F}_0$ is filtered and we can write:

$$C = \varinjlim_{Z \in \mathscr{F}_0} C_Z.$$

Thus, it will be sufficient to prove that C_Z is acyclic for any $Z \in \mathscr{F}_0$. Because S is henselian and Z is finite over S, Z is indeed a finite sum of local henselian schemes. This implies that the t-cover p_Z, which is in particular étale surjective, admits a splitting $s : Z \longrightarrow W_Z$. Then the complex (10.3.3.1) is contractible with contracting homotopy defined by the family of morphisms

$$(s \times_Z 1_{W_Z^n})_* : c_0(W_Z^n/S, \Lambda) \longrightarrow c_0(W_Z^{n+1}/S, \Lambda).$$

10.3.4 Considering an additive abelian presheaf G on $\mathscr{P}/S$, the natural transformation

$$X \longmapsto \mathrm{Hom}_{\mathrm{PSh}(\mathscr{P}/S)}(\hat{\gamma}_* \Lambda_S^{tr}(X), G)$$

defines a presheaf with transfers over S.[89] We will denote by G_τ its restriction to the site $\mathscr{P}/S$. Note that this definition can be applied in the case where G is a t-sheaf on $\mathscr{P}/S$, because under the assumptions of Section 10.0.1 on t, it is in particular an additive presheaf.

Definition 10.3.5 We will say that t is *mildly compatible with transfers* if for any scheme S and any t-sheaf F on $\mathscr{P}/S$, F_τ is a t-sheaf on $\mathscr{P}/S$.

If t is weakly compatible with transfers then it is mildly compatible with transfers.

Remark 10.3.6 Assume t is mildly compatible with transfers. Then for any scheme S, any t-cover $p : W \longrightarrow X$ in $\mathscr{P}/S$ induces a morphism

$$p_* : \Lambda_S^{tr}(W) \longrightarrow \Lambda_S^{tr}(X)$$

which is an epimorphism of the associated t-sheaves on $\mathscr{P}/S$. This means that for any correspondence $\alpha \in c_S(Y, X)$, there exists a t-cover $q : W' \longrightarrow Y$ and a correspondence $\alpha' \in c_S(W', W)$ making the following diagram commutative:

(10.3.6.1)
$$\begin{array}{ccc} W' & \xrightarrow{\hat{\alpha}} & W \\ {\scriptstyle q}\downarrow & & \downarrow{\scriptstyle p} \\ Y & \xrightarrow{\alpha} & X \end{array}$$

---

[89] Actually, this defines a right adjoint to the functor $\hat{\gamma}_*$.

Lemma 10.3.7 *Assume t is mildly compatible with transfers.*

Let S be a scheme and P^{tr} be a presheaf with transfers over S. We put $P = P^{tr} \circ \gamma$ as a presheaf on $\mathscr{P}/S$. We denote by F the t-sheaf associated with P and by $\eta : P \longrightarrow F$ the canonical natural transformation.

Then there exists a unique pair (F^{tr}, η^{tr}) such that:

1. *F^{tr} is a sheaf with transfers over S such that $F^{tr} \circ \gamma = F$.*
2. *$\eta^{tr} : P^{tr} \longrightarrow F^{tr}$ is a natural transformation of presheaves with transfers such that the induced transformation*

$$P = (P^{tr} \circ \gamma) \longrightarrow (F^{tr} \circ \gamma) = F$$

coincides with η.

Proof As a preliminary observation, we note that given a presheaf G on $\mathscr{P}/S$, the data of a presheaf with transfers G^{tr} such that $G^{tr} \circ \gamma = G$ is equivalent to the data for any $\mathscr{P}$-schemes X and Y of a bilinear product

(10.3.7.1) $$G(X) \otimes_{\mathbb{Z}} c_S(Y, X) \longrightarrow G(Y), \rho \otimes \alpha \mapsto \langle \rho, \alpha \rangle$$

such that:

(a) For any morphism $f : Y' \longrightarrow Y$ in $\mathscr{P}/S$, $f^* \langle \rho, \alpha \rangle = \langle \rho, \alpha \circ f \rangle$.
(b) For any morphism $f : X \longrightarrow X'$ in $\mathscr{P}/S$, if $\rho = f^*(\rho')$, then $\langle \rho, \alpha \rangle = \langle \rho', f \circ \alpha \rangle$.
(c) When $X = Y$, for any $\rho \in F(X)$, $\langle \rho, 1_X \rangle = \rho$.
(d) For any finite S-correspondence $\beta \in c_S(Z, Y)$, $\langle \langle \rho, \alpha \rangle, \beta \rangle = \langle \rho, \alpha \circ \beta \rangle$.

On the other hand, the data of products of the form (10.3.7.1) for any $\mathscr{P}$-schemes X and Y over S which satisfy the conditions (a) and (b) above is equivalent to the data of a natural transformation

$$\phi : G \longrightarrow G_\tau$$

by putting $\langle \rho, \alpha \rangle_\phi = [\phi_X(\rho)]_Y . \alpha$.

Applying this to the presheaf with transfers P^{tr}, we obtain a canonical natural transformation

$$\psi : P \longrightarrow P_\tau.$$

By assumption on t, F_τ is a t-sheaf. Thus, there exists a unique natural transformation ψ such that the following diagram commutes:

$$\begin{array}{ccc} P & \xrightarrow{\psi} & P_\tau \\ \eta \downarrow & & \downarrow \eta_\tau \\ F & \xrightarrow{\phi} & F_\tau. \end{array}$$

Thus we get products of the form (10.3.7.1) associated with ϕ which satisfies (a) and (b). The commutativity of the above diagram asserts they are compatible with the ones corresponding to P^{tr} and the unicity of the natural transformation ϕ implies the uniqueness statement of the lemma.

10 Sheaves with transfers

To finish the proof of the existence, we must show (c) and (d) for the product $\langle \cdot, \cdot \rangle_\phi$. Consider a pair $(\rho, \alpha) \in F(X) \times c_S(Y, X)$. Because F is the t-sheaf associated with P, there exists a t-cover $p : W \longrightarrow X$ and a section $\hat{\rho} \in P(W)$ such that $p^*(\rho) = \eta_W(\hat{\rho})$. Moreover, according to Remark 10.3.6, we get a t-cover $q : W' \longrightarrow Y$ and a correspondence $\hat{\alpha} \in c_S(W', W)$ making the diagram (10.3.6.1) commutative. Then we get using (a) and (b):

$$q^* \langle \rho, \alpha \rangle_\phi = \langle \rho, \alpha \circ q \rangle_\phi = \langle \rho, p \circ \hat{\alpha} \rangle_\phi = \langle p^* \rho, \hat{\alpha} \rangle_\phi = \langle \eta_W(\hat{\rho}), \hat{\alpha} \rangle_\phi = \langle \hat{\rho}, \hat{\alpha} \rangle_\psi.$$

Because $q^* : F(X) \longrightarrow F(W)$ is injective, we deduce easily from this principle the properties (c) and (d) and this concludes the proof. □

10.3.8 Let us consider the canonical adjunction

$$a_t^* : \mathrm{PSh}(\mathscr{P}/S, \Lambda) \rightleftarrows \mathrm{Sh}_t(\mathscr{P}/S, \Lambda) : \mathscr{O}_t$$

where $\mathscr{O}_t$ is the canonical forgetful functor.

We also denote by $\mathscr{O}_t^{tr} : \mathrm{Sh}_t\left(\mathscr{P}_{\Lambda,S}^{cor}\right) \longrightarrow \mathrm{PSh}\left(\mathscr{P}_{\Lambda,S}^{cor}\right)$ the obvious forgetful functor. Trivially, the following relation holds:

(10.3.8.1) $$\hat{\gamma}_* \, a_{t,*} = a_{t,*} \, \gamma_*.$$

Proposition 10.3.9 *Using the notations above, the following condition on the admissible topology t are equivalent:*

(i) t is mildly compatible with transfers.
(ii) For any scheme S, the functor $\mathscr{O}_t^{tr}$ admits a left adjoint

$$a_t^* : \mathrm{PSh}\left(\mathscr{P}_{\Lambda,S}^{cor}\right) \longrightarrow \mathrm{Sh}_t\left(\mathscr{P}_{\Lambda,S}^{cor}\right)$$

which is exact and such that the exchange transformation

(10.3.9.1) $$a_t^* \hat{\gamma}_* \longrightarrow \gamma_* a_t^*$$

induced by the identification (10.3.8.1) is an isomorphism.

Under these conditions, the following properties hold for any scheme S:

(iii) The category $\mathrm{Sh}_t\left(\mathscr{P}_{\Lambda,S}^{cor}\right)$ is a Grothendieck abelian category.
(iv) The functor γ_* commutes with all limits and colimits.

Proof The fact that (i) implies (ii) follows from the preceding lemma as we can put $a_t^{tr}(F) = F^{tr}$ with the notation of the lemma. The fact that this defines a functor, as well as the properties stated in (ii), follows from the uniqueness statement of *loc. cit.*

Let us assume (ii). Then (iii) follows formally from (ii), from the existence, adjunction property and exactness of a_t^*, as $\mathrm{PSh}\left(\mathscr{P}_{\Lambda,S}^{cor}\right)$ is a Grothendieck abelian category. Moreover, we deduce from the isomorphism (10.3.9.1) that γ_* is exact:

indeed, a_t^* and $\hat{\gamma}_*$ are exact. As γ_* commutes with arbitrary direct sums, we get (iv). From this point, we deduce the existence of a right adjoint $\gamma^!$ to the functor γ_*. Using again the isomorphism (10.3.9.1), we obtain for any t-sheaves F on $\mathscr{P}/S$ and any $\mathscr{P}$-scheme X/S a canonical isomorphism $F_\tau(X) = \gamma^! F(X)$. This proves (i). □

10.3.10 Under the assumption of the previous proposition, given any $\mathscr{P}$-scheme X/S, we will put $\Lambda_S^{tr}(X)_t = a_t^* \Lambda_S^{tr}(X)$. The above proposition shows that the family $\Lambda_S^{tr}(X)_t$ for $\mathscr{P}$-schemes X/S is a generating family in $\mathrm{Sh}_t\left(\mathscr{P}_{\Lambda,S}^{cor}\right)$. Moreover, we easily get the following corollary of the preceding proposition and Proposition 10.1.2:

Corollary 10.3.11 *Assume that t is mildly compatible with transfers.*

Then there exists an essentially unique Grothendieck abelian $\mathscr{P}$-premotivic category $\mathrm{Sh}_t(\mathscr{P}_\Lambda^{cor})$ which is geometrically generated, whose fiber over a scheme S is $\mathrm{Sh}_t\left(\mathscr{P}_{\Lambda,S}^{cor}\right)$ and such that the t-sheafification functor induces an adjunction of abelian $\mathscr{P}$-premotivic categories:

$$a_t^* : \mathrm{PSh}\left(\mathscr{P}_\Lambda^{cor}\right) \rightleftarrows \mathrm{Sh}_t\left(\mathscr{P}_\Lambda^{cor}\right) : \mathscr{O}_t^{tr}.$$

Moreover, the functor γ_* induces an adjunction of abelian $\mathscr{P}$-premotivic categories:

(10.3.11.1) $\qquad \gamma^* : \mathrm{Sh}_t(\mathscr{P},\Lambda) \rightleftarrows \mathrm{Sh}_t\left(\mathscr{P}_\Lambda^{cor}\right) : \gamma_*$.

Remark 10.3.12 Notice moreover that $\gamma^* a_t^* = a_t^* \hat{\gamma}^*$.

Proof In fact, using the exactness of a_t^*, given any sheaf F with transfers F over S, we get a canonical isomorphism

$$F = \varinjlim_{\Lambda_S^{tr}(X)_t \longrightarrow F} \Lambda_S^{tr}(X)_t$$

where the limit is taken in $\mathrm{Sh}_t\left(\mathscr{P}_{\Lambda,S}^{cor}\right)$ and runs over the representable t-sheaves with transfers over F. As in the proof of Proposition 10.1.2, this allows us to uniquely define the structural left adjoints of $\mathrm{Sh}_t(\mathscr{P}_\Lambda^{cor})$. The existence (and uniqueness) of the structural right adjoints then follows formally. The same remark allows us to construct the functor γ^*. □

Remark 10.3.13 Let us illustrate the meaning of the preceding corollary for a topology t which is compatible with transfers. Given a complex C with coefficients in the category $\mathrm{Sh}_t\left(\mathscr{P}_{\Lambda,S}^{cor}\right)$, the following conditions are equivalent:

(i) C is local (Definition 5.1.9),
(i') $\gamma_*(C)$ is local,
(i'') given any $\mathscr{P}$-scheme X/S and any integer $n \in \mathbf{Z}$, the canonical map

$$H^n(C(X)) \longrightarrow H_t^n(X, \gamma_*(C))$$

is an isomorphism,

(ii) C is t-flasque (Definition 5.1.9),
(ii') $\gamma_*(C)$ is t-flasque,
(ii") given any t-hypercover $p : \mathscr{X} \longrightarrow X$ in $\mathscr{P}/S$ and any integer $n \in \mathbf{Z}$, the canonical map
$$p^* : H^n(C(X)) \longrightarrow H^n(C(\mathscr{X}))$$
is an isomorphism.

More precisely, the equivalence of (i) and (ii) is the preceding corollary, while the equivalence of (i) and (i') (resp. (ii) and (ii')) follows from the existence of the adjunction (10.3.11.1) and the fact that γ_* is exact. The equivalence between (i') and (i") (resp. (ii') and (ii")) is a simple translation of Definition 5.1.9.

10.3.14 Recall from Definition 5.1.9 we say that the abelian $\mathscr{P}$-premotivic category $\mathrm{Sh}_t(\mathscr{P}_\Lambda^{cor})$ satisfies cohomological t-descent if for any scheme S, and any t-hypercover $\mathscr{X} \longrightarrow X$ in $\mathscr{P}/S$, the induced morphism of complexes in $\mathrm{Sh}_t\left(\mathscr{P}_{\Lambda,S}^{cor}\right)$
$$\Lambda_S^{tr}(\mathscr{X})_t \longrightarrow \Lambda_S^{tr}(X)_t$$
is a quasi-isomorphism. The preceding corollary thus gives the following one:

Corollary 10.3.15 *Assume t is mildly compatible with transfers. Then the following conditions are equivalent:*

(i) The topology t is compatible with transfers.
(ii) The abelian $\mathscr{P}$-premotivic category $\mathrm{Sh}_t(\mathscr{P}_\Lambda^{cor})$ satisfies cohomological t-descent.
(iii) The abelian $\mathscr{P}$-premotivic category $\mathrm{Sh}_t(\mathscr{P}_\Lambda^{cor})$ is compatible with t (see Definition 5.1.9).

Proof The equivalence of (i) and (ii) follows easily from the isomorphism (10.3.9.1). The equivalence of (ii) and (iii) is Proposition 5.1.26 applied to the adjunction (10.3.11.1), in view of Proposition 10.3.9(iv). □

10.3.16 Recall from Section 2.1.10 that a cd-structure P on $\mathscr{S}$ is the data of a family of commutative squares, called P-distinguished, of the form

(10.3.16.1)
$$\begin{array}{ccc} B & \xrightarrow{k} & Y \\ g \downarrow & Q & \downarrow f \\ A & \xrightarrow{i} & X \end{array}$$

which is stable under isomorphisms. Further, we will consider the following assumptions on P:

(a) P is complete, regular and bounded in the sense of [Voe10c].
(b) Any P-distinguished square as above is comprised of $\mathscr{P}$-morphisms and k is an immersion.

(c) Any square as above which is cartesian and such that $X = A \sqcup Y$, i and f being the obvious immersions, is P-distinguished.

Then the topology t_P associated with P (see Definition 2.1.10) is $\mathscr{P}$-admissible and satisfies assumption (c) of Section 10.0.1. Moreover, according to [Voe10c, 2.9], we obtain the following properties:

(d) A presheaf F on $\mathscr{P}/S$ is a t_P-sheaf if and only if $F(\varnothing) = 0$ and for any P-distinguished square (10.3.16.1) in $\mathscr{P}/S$, the sequence

$$0 \longrightarrow F(X) \xrightarrow{f^*+e^*} F(Y) \oplus F(A) \xrightarrow{k^*-g^*} F(B)$$

is exact.

(e) For any P-distinguished square (10.3.16.1) the sequence of representable presheaves on $\mathscr{P}/S$

$$0 \longrightarrow \Lambda_S(B) \xrightarrow{k_*-g_*} \Lambda_S(Y) \oplus \Lambda_S(A) \xrightarrow{f_*+e_*} \Lambda_S(X) \longrightarrow 0$$

becomes exact on the associated t_P-sheaves.

Proposition 10.3.17 *Consider a cd-structure P satisfying properties (a) and (b) above and assume that $t = t_P$ is the topology associated with P. Then the following conditions are equivalent:*

(i) The topology t is compatible with transfers.
(ii) The topology t is mildly compatible with transfers.
(iii) For any scheme S and any P-distinguished square (10.3.16.1) in $\mathscr{P}/S$, the short sequence of presheaves with transfers over S

$$0 \longrightarrow \Lambda_S^{tr}(B) \xrightarrow{k_*-g_*} \Lambda_S^{tr}(Y) \oplus \Lambda_S^{tr}(A) \xrightarrow{f_*+e_*} \Lambda_S^{tr}(X) \longrightarrow 0$$

becomes exact on the associated t-sheaves on $\mathscr{P}/S$.

Proof The implication (i) $\Rightarrow$ (ii) is obvious.

The implication (ii) $\Rightarrow$ (iii) follows from point (e) above and the following facts: γ^* is right exact (Corollary 10.3.11), $\gamma^* a_t = a_t^{tr} \hat{\gamma}^*$ (Remark 10.3.12), $k_* : \Lambda_S^{tr}(B) \longrightarrow \Lambda_S^{tr}(Y)$ is a monomorphism of presheaves with transfers (use Proposition 9.1.7(2) and the fact k is an immersion from assumption (b)).

Assume (iii). Then we obtain (ii) as a direct consequence of the point (d) above. Thus, to prove (i), we have only to prove that the abelian $\mathscr{P}$-premotivic category $\mathrm{Sh}_t(\mathscr{P}_\Lambda^{cor})$ satisfies cohomological t-descent according to Corollary 10.3.15.

Let S be a scheme. Recall that the category $\mathrm{D}\big(\mathrm{Sh}_t(\mathscr{P}/S,\Lambda)\big)$ has a canonical DG-structure (see for example Section 5.0.2). The cohomological t-descent for $\mathrm{Sh}_t\big(\mathscr{P}_{\Lambda,S}^{cor}\big)$ can be reformulated by saying that for any complex K of t-sheaves on $\mathscr{P}/S$, and any t-hypercover $\mathscr{X} \longrightarrow X$, the canonical map of $\mathrm{D}(\Lambda\text{-mod})$

$$\mathbf{R}\,\mathrm{Hom}^\bullet_{\mathrm{D}(\mathrm{Sh}_t(\mathscr{P}/S,\Lambda))}(\gamma_*\Lambda_S^{tr}(X)_t, K) \longrightarrow \mathbf{R}\,\mathrm{Hom}^\bullet_{\mathrm{D}(\mathrm{Sh}_t(\mathscr{P}/S,\Lambda))}(\gamma_*\Lambda_S^{tr}(\mathscr{X})_t, K)$$

10 Sheaves with transfers

is an isomorphism. Recall also that there is an injective model structure on the category $C(\mathrm{Sh}_t(\mathscr{P}/S, \Lambda))$ for which every object is cofibrant, with quasi-isomorphisms as weak equivalences (see [CD09, 2.1] for more details). Replacing K by a fibrant resolution for the injective model structure, we get for any simplicial objects $\mathscr{X}$ of $\mathscr{P}/S^{\mathrm{II}}$ that:

$$\mathrm{R}\mathrm{Hom}^{\bullet}_{\mathrm{D}(\mathrm{Sh}_t(\mathscr{P}/S, \Lambda))}(\gamma_* \Lambda_S^{tr}(\mathscr{X})_t, K) = \mathrm{Hom}^{\bullet}_{\mathrm{D}(\mathrm{Sh}_t(\mathscr{P}/S, \Lambda))}(\gamma_* \Lambda_S^{tr}(\mathscr{X})_t, K).$$

Thus it is sufficient to prove that the presheaf

$$E : \mathscr{P}/S^{op} \longrightarrow C(\Lambda\text{-}\mathrm{mod}), X \longmapsto \mathrm{Hom}^{\bullet}_{\mathrm{D}(\mathrm{Sh}_t(\mathscr{P}/S, \Lambda))}(\gamma_* \Lambda_S^{tr}(X)_t, K)$$

satisfies t-descent in the sense of Definition 3.2.5.

We derive from (iii) that E sends a P-distinguished square to a homotopy cartesian square in $D(\Lambda\text{-}\mathrm{mod})$. Thus the assertion follows from the arguments on t-descent from [Voe10b, Voe10c]. □

Remark 10.3.18 It follows from Remark 10.3.13 that under the equivalent conditions (i), (ii), (iii) of the above corollary, the admissible topology $t = t_P$ is bounded in $\mathrm{Sh}_t(\mathscr{P}_\Lambda^{cor})$ in the sense of Definition 5.1.28. Over a scheme S, a bounded generating family is given by the following complexes of $\mathrm{Sh}_t(\mathscr{P}_{\Lambda,S}^{cor})$:

$$\cdots \longrightarrow 0 \longrightarrow \Lambda_S^{tr}(B) \xrightarrow{k_*-g_*} \Lambda_S^{tr}(Y) \oplus \Lambda_S^{tr}(A) \xrightarrow{f_*+e_*} \Lambda_S^{tr}(X) \longrightarrow 0 \longrightarrow \cdots$$

induced by a P-distinguished square of the form (10.3.16.1) – see also Example 5.1.29.

We conclude this section with the compatibility of certain sheaves with transfers with projective limits of schemes. This will be the key point to establish continuity for motivic complexes.

Proposition 10.3.19 *Let t be one of the topologies Nis, ét, cdh.*

Let $(S_\alpha)_{\alpha \in A}$ be a projective system of schemes in $\mathscr{S}$, with dominant affine transition maps, and such that $S = \varprojlim_{\alpha \in A} S_\alpha$ is representable in $\mathscr{S}$.

Consider an index $\alpha_0 \in A$ and a t-sheaf with transfers F over $\mathscr{S}_{\Lambda, S_0}^{ft, cor}$. For any index α/α_0, we denote by F_α (resp. F) the pullback of F_{α_0} over S_α (resp. S) in the sense of the premotivic structure on $\mathrm{Sh}_t(\mathscr{P}_\Lambda^{cor})$ (obtained in Corollary 10.3.11).

Then the canonical map:

$$\varinjlim_{\alpha \in A/\alpha_0} F_\alpha(S_\alpha) \longrightarrow F(S)$$

is an isomorphism.

Proof We consider the forgetful functor: $\mathscr{O}_t^{tr} : \mathrm{Sh}_t\left(\mathscr{S}_\Lambda^{ft, cor}\right) \longrightarrow \mathrm{PSh}\left(\mathscr{S}_\Lambda^{ft, cor}\right)$. It is fully faithful and it commutes with the global section functor. We want to prove the proposition by using Lemma 10.1.5. Thus it is sufficient to prove that, for any morphism $f : X \longrightarrow S$ in $\mathscr{S}$, the functor $\mathscr{O}_t^{tr}$ commutes with f^*. In other words, the pullback functor $\hat{f}^*$ for presheaves with transfers on $\mathscr{S}_\Lambda^{ft,cor}$ preserves t-sheaves with transfers: for any t-sheaf with transfers F over S, $\hat{f}^*(F)$ is a t-sheaf with transfers.

Let us first treat the case where f is separated of finite type. Then $\hat{f}^*$ admits a left adjoint $\hat{f}_\sharp$ which preserves t-covers. Thus the property is clear.

In the general case, we write f as a projective limit of morphisms of schemes $(f_\alpha : X_\alpha \longrightarrow S)_{\alpha \in A}$ such that the transition morphisms of the projective scheme $(X_\alpha)_{\alpha \in A}$ are affine and dominant and each f_α is separated of finite type.[90] To check that $\hat{f}^*(F)$ is a t-sheaf, we consider a t-cover $p : W \longrightarrow X$ of an S-scheme separated of finite type. Because of our choice of topology t, there exists an index α_1/α_0 such that $p : W \longrightarrow X$ can be lifted as a t-cover $p_1 : W_{\alpha_1} \longrightarrow X_{\alpha_1}$ over S_{α_1}. Using Lemma 10.1.5 again, we are now reduced to proving that for any α/α_1, $\hat{f}_{\alpha_1}^*(F)$ satisfies the t-sheaf property with respect to the pullback of p_1 over S_α/S_{α_1}. This follows from the first case treated. □

Remark 10.3.20 The previous proposition generalizes [Dég07, Prop. 2.19].

10.4 Examples

10.4.1 Assume that t is the Nisnevich topology. According to Lemma 10.3.3 and Proposition 10.3.17, t is then compatible with transfers. With the notation of Corollary 10.3.11, we get the following definition:

Definition 10.4.2 We denote by

$$\mathrm{Sh}^{tr}(-,\Lambda), \quad \underline{\mathrm{Sh}}^{tr}(-,\Lambda)$$

the respective abelian premotivic and generalized abelian premotivic categories defined in Corollary 10.3.11 in the respective cases $\mathscr{P} = \mathscr{S}m$, $\mathscr{P} = \mathscr{S}^{ft}$.

From now on, the objects of $\mathrm{Sh}^{tr}(S,\Lambda)$ (resp. $\underline{\mathrm{Sh}}^{tr}(S,\Lambda)$) are called *sheaves with transfers over S* (resp. *generalized sheaves with transfers over S*).

Let X be a separated S-scheme of finite type. We let $\underline{\Lambda}_S^{tr}(X)$ be the generalized sheaf with transfers represented by X (*cf.* Proposition 10.2.4). If X is S-smooth, we denote by $\Lambda_S^{tr}(X)$ its restriction to $\mathscr{S}m_{\Lambda,S}^{cor}$ – *i.e.* the sheaf with transfers over S represented by X.

An important property of sheaves with transfers is that the abelian premotivic category $\mathrm{Sh}^{tr}(-,\Lambda)$ (resp. $\underline{\mathrm{Sh}}^{tr}(-,\Lambda)$) is compatible with the Nisnevich topology on $\mathscr{S}m$ (resp. $\mathscr{S}^{ft}$) according to Proposition 10.3.17. Note moreover that it is compactly geometrically generated.

---

[90] Write the $\mathscr{O}_S$-algebra $f_*(\mathscr{O}_X)$ as the filtered union of its finite type sub-$\mathscr{O}_S$-algebras, ordered by inclusion.

10.4.3 We also obtained an adjunction (resp. generalized adjunction) of premotivic abelian categories

$$\gamma^* : \mathrm{Sh}(\mathscr{S}m, \Lambda) \rightleftarrows \underline{\mathrm{Sh}}^{tr}(-, \Lambda) : \gamma_*,$$

$$\gamma^* : \mathrm{Sh}\left(\mathscr{S}^{ft}, \Lambda\right) \rightleftarrows \underline{\mathrm{Sh}}^{tr}(-, \Lambda) : \gamma_*.$$

Note that in each case γ_* is conservative and exact according to Proposition 10.3.9(iv).

Remark 10.4.4 An important application of the existence of the pair of adjoint functors (γ^*, γ_*) is the following computation: given any complex K of sheaves with transfers over S and any smooth S-scheme X,

$$\begin{aligned}\mathrm{Hom}_{D(\underline{\mathrm{Sh}}^{tr}(S,\Lambda))}(\Lambda_S^{tr}(X), K[n]) &= \mathrm{Hom}_{D(\underline{\mathrm{Sh}}^{tr}(S,\Lambda))}(\mathbf{L}\gamma^*\Lambda_S(X), K[n]) \\ &= \mathrm{Hom}_{D(\mathrm{Sh}(\mathscr{S}m,\Lambda))}(\Lambda_S(X), \gamma_*(K)[n]) \\ &= H^n_{Nis}(X, \gamma_*(K)).\end{aligned}$$

This is a generalization of [VSF00, chap. 5, 3.1.9] to unbounded complexes and arbitrary bases.

10.4.5 Let S be a scheme. Consider the inclusion functor $\varphi : \mathscr{S}m_{\Lambda,S}^{cor} \longrightarrow \mathscr{S}_{\Lambda,S}^{ft,cor}$. It induces a functor

$$\varphi^* : \underline{\mathrm{Sh}}^{tr}(S, \Lambda) \longrightarrow \mathrm{Sh}^{tr}(S, \Lambda), F \longmapsto F \circ \varphi$$

which is obviously exact and commutes with arbitrary direct sums. In particular, it commutes with arbitrary colimits.

Lemma 10.4.6 *With the notations above, the functor φ^* admits a left adjoint $\varphi_!$ such that for any smooth S-scheme X, $\varphi_!(\Lambda_S^{tr}(X)) = \underline{\Lambda}_S^{tr}(X)$. The functor $\varphi_!$ is fully faithful.*

In other words, we have defined an enlargement of premotivic abelian categories (*cf.* Definition 1.4.13)

(10.4.6.1) $\qquad \varphi_! : \mathrm{Sh}^{tr}(-, \Lambda) \longrightarrow \underline{\mathrm{Sh}}^{tr}(-, \Lambda) : \varphi^*.$

Proof Let F be a sheaf with transfers. Let $\{X/F\}$ be the category of representable sheaves $\Lambda_S^{tr}(X)$ over F for a smooth S-scheme X. We put

$$\varphi_!(F) = \varinjlim_{\{X/F\}} \underline{\Lambda}_S^{tr}(X).$$

The adjunction property of $\varphi_!$ is immediate from the Yoneda lemma. We prove that for any sheaf with transfers F, the unit adjunction morphism $F \longrightarrow \varphi^*\varphi_!(F)$ is an isomorphism. As already remarked, φ^* commutes with colimits so that we are restricted to the case where $F = \Lambda_S^{tr}(X)$, which follows by definition. $\square$

10.4.7 Assume now that $t = cdh$ is the cdh-topology, and $\mathscr{P} = \mathscr{S}^{ft}$ is the class of separated morphisms of finite type. Recall the topology t is associated with the *lower cd-structure* – see Example 2.1.11. Then the assumptions of Proposition 10.3.17 with respect to the lower cd-structure are fulfilled according to [SVoob, 4.3.3] combined with [SVoob, 4.2.9]. Thus we get the following result:

Proposition 10.4.8 *The admissible topology cdh on $\mathscr{S}^{ft}$ is compatible with transfers.*

As a corollary, we get a generalized premotivic abelian category whose fiber over a scheme S is the category $\underline{Sh}^{tr}_{cdh}(S, \Lambda)$ of cdh-sheaves with transfers on $\mathscr{S}^{ft}$. It is compatible with the cdh-topology. Moreover, the restriction of a_{cdh} to $\underline{Sh}^{tr}(S, \Lambda)$ induces a morphism of generalized premotivic categories; we get the following commutative diagram of such morphisms:

$$\begin{array}{ccc} \underline{Sh}(-, \Lambda) & \xrightarrow{a^*_{cdh}} & \underline{Sh}_{cdh}(-, \Lambda) \\ \gamma^* \downarrow & & \downarrow \gamma^*_{cdh} \\ \underline{Sh}^{tr}(-, \Lambda) & \xrightarrow{a^*_{cdh}} & \underline{Sh}^{tr}_{cdh}(-, \Lambda). \end{array}$$

10.5 Comparison results

10.5.a Change of coefficients

10.5.1 Assume the topology t is mildly compatible with transfers and consider a localization Λ' of Λ.

Then the morphism (9.4.3.1) of $\mathscr{P}$-premotivic categories extends to an adjunction of abelian $\mathscr{P}$-premotivic categories:

$$(10.5.1.1) \qquad \mathrm{Sh}_t\left(\mathscr{P}^{cor}_{\Lambda}\right) \otimes_{\Lambda} \Lambda' \rightleftarrows \mathrm{Sh}_t\left(\mathscr{P}^{cor}_{\Lambda'}\right).$$

Proposition 9.4.4 immediately yields the following result:

Proposition 10.5.2 *Consider the above notations. Then the adjunction (10.5.1.1) is an equivalence of $\mathscr{P}$-premotivic categories.*

Remark 10.5.3 Remark 9.4.5 can be extended to sheaves with transfers: for any regular scheme S, the category $\mathrm{Sh}^{tr}(S, \mathbf{Z}) = \mathrm{Sh}_{Nis}\left(\mathscr{S}m^{cor}_{\mathbf{Z},S}\right)$ defined here coincides with that defined in [Dég07], as well as its operations of a $\mathscr{P}$-premotivic category when restricted to regular schemes.

Remark 10.5.4 In a previous version of this text, the preceding proposition was obtained under restrictive hypotheses. We have been able to remove these unnecessary assumptions thanks to point (3) of Remark 9.1.3, which is a consequence of Proposition 8.1.53 (bound on the denominators of intersection multiplicities).

10.5.b Representable qfh-sheaves

10.5.5 Let us denote by $\mathrm{Sh}_{qfh}(S, \Lambda)$ the category of *qfh*-sheaves of Λ-modules over $\mathscr{S}^{ft}/S$. Note that for an S-scheme X, the Λ-presheaf represented by X is not a sheaf for the *qfh*-topology. We denote the associated sheaf by $\underline{\Lambda}_S^{qfh}(X)$. We let a_{qfh} be the associated *qfh*-sheaf functor. Recall that for any S-scheme X, the graph functor (10.4.3) induces a morphism of sheaves

$$\underline{\Lambda}_S(X) \xrightarrow{\gamma_{X/S}} \underline{\Lambda}_S^{tr}(X).$$

We recall the following theorem of Suslin and Voevodsky (see [SV00b, 4.2.7+4.2.12]):

Theorem 10.5.6 *Let S be a scheme such that $\mathrm{char}(S) \subset \Lambda^\times$. Then, for any S-scheme X, the application of a_{qfh} to the map $\gamma_{X/S}$ gives an isomorphism in $\mathrm{Sh}_{qfh}(S, \Lambda)$:*

$$\underline{\Lambda}_S^{qfh}(X) \xrightarrow{\gamma_{X/S}^{qfh}} \underline{\Lambda}_S^{tr}(X).$$

10.5.7 Assume $\mathrm{char}(S) \subset \Lambda^\times$. Using the previous theorem, we associate to any *qfh*-sheaf $F \in \mathrm{Sh}_{qfh}(S, \Lambda)$ a presheaf with transfers

$$\rho(F) : X \longmapsto \mathrm{Hom}_{\mathrm{Sh}_{qfh}(S,\Lambda)}(\underline{\Lambda}_S^{tr}(X), F).$$

We obviously get $\gamma^* \rho(F) = F$ as a presheaf over $\mathscr{S}^{ft}/S$ so that $\rho(F)$ is a sheaf with transfers and we have defined a functor

$$\rho : \mathrm{Sh}_{qfh}(S, \Lambda) \longrightarrow \underline{\mathrm{Sh}}^{tr}(S, \Lambda).$$

For any S-scheme X, $\rho(\underline{\Lambda}_S^{qfh}(X)) = \underline{\Lambda}_S^{tr}(X)$ according to the previous proposition.

Corollary 10.5.8 *Assume $\mathrm{char}(S) \subset \Lambda^\times$. Let $f : X' \longrightarrow X$ be a morphism of S-schemes.*

If f is a universal homeomorphism, then the map $f_* : \underline{\Lambda}_S^{tr}(X') \longrightarrow \underline{\Lambda}_S^{tr}(X)$ is an isomorphism in $\underline{\mathrm{Sh}}^{tr}(S, \Lambda)$.

Proof Indeed, according to [Voe96, 3.2.5], $\Lambda_S^{qfh}(X') \longrightarrow \Lambda_S^{qfh}(X)$ is an isomorphism in $\mathrm{Sh}_{qfh}(S, \Lambda)$ and the result follows by applying the functor ρ. □

10.5.c qfh-sheaves and transfers

Proposition 10.5.9 *Assume $\mathrm{char}(S) \subset \Lambda^\times$. Any qfh-sheaf of Λ-modules over the category of S-schemes of finite type is naturally endowed with a unique structure of a sheaf with transfers, and any morphism of such qfh-sheaves is a morphism of sheaves with transfers.*

In particular, the qfh-sheafification functor defines a left exact functor left adjoint to the forgetful functor $\rho : \mathrm{Sh}_{qfh}(S, \Lambda) \longrightarrow \underline{\mathrm{Sh}}^{tr}(S, \Lambda)$ introduced in Section 10.5.7.

Proof It follows from Theorem 10.5.6 that the category of Λ-linear finite correspondences is canonically equivalent to the full subcategory of the category of *qfh*-sheaves of Λ-modules spanned by the objects of shape $\Lambda_S^{qfh}(X)$ for X separated of finite type over S. The first assertion is thus an immediate consequence of Theorem 10.5.6 and of the (additive) Yoneda lemma. The fact that the *qfh*-sheafification functor defines a left adjoint to the restriction functor ρ is then obvious, while its left exactness is a consequence of the facts that it is left exact (at the level of sheaves without transfers) and that forgetting transfers defines a conservative and exact functor from the category of Nisnevich sheaves with transfers to the category of Nisnevich sheaves. □

Recall the following theorem:

Theorem 10.5.10 *Assume Λ is a $\mathbf{Q}$-algebra. Let F be an étale Λ-sheaf on $\mathscr{S}^{ft}/S$. Then for any S-scheme X, and any integer i, the canonical map*

$$H^i_{Nis}(X, F) \longrightarrow H^i_{\acute{e}t}(X, F)$$

is an isomorphism.

Proof Using the compatibility of étale cohomology with projective limits of schemes, we are reduced to proving that $H^i_{\acute{e}t}(X, F) = 0$ whenever X is henselian local and $i > 0$. Let k be the residue field of X, G its absolute Galois group and F_0 the restriction of F to $\mathrm{Spec}(k)$. Then F_0 is a G-module and according to [AGV73, 8.6], $H^i_{\acute{e}t}(X, F) = H^i(G, F_0)$. As G is profinite, this group must be torsion so that it vanishes by assumption. □

Remark 10.5.11 The preceding theorem also follows formally from Theorem 3.3.23.

Proposition 10.5.12 *Assume Λ is a $\mathbf{Q}$-algebra. Let S be an excellent scheme and F be a qfh-sheaf of Λ-modules on $\mathscr{S}^{ft}/S$. Then for any geometrically unibranch S-scheme X of finite type, and any integer i, the canonical map*

$$H^i_{Nis}(X, F) \longrightarrow H^i_{qfh}(X, F)$$

is an isomorphism.

Proof According to Theorem 10.5.10, $H^i_{Nis}(X, F) = H^i_{\acute{e}t}(X, F)$. Let $p : X' \longrightarrow X$ be the normalization of X. As X is an excellent geometrically unibranch scheme, p is a finite universal homeomorphism. It follows from [AGV73, VII, 1.1] that $H^i_{\acute{e}t}(X, F) = H^i_{\acute{e}t}(X', F)$ and from [Voe96, 3.2.5] that $H^i_{qfh}(X, F) = H^i_{qfh}(X', F)$. Thus we can assume that X is normal, and the result is now exactly [Voe96, 3.4.1]. □

Corollary 10.5.13 *Assume Λ is a $\mathbf{Q}$-algebra. Let S be an excellent scheme.*

1. *Let X be a geometrically unibranch S-scheme of finite type. For any point x of X, the local henselian scheme X_x^h is a point for the category of sheaves $\mathrm{Sh}_{qfh}(S, \Lambda)$ (i.e. evaluating at X_x^h defines an exact functor).*
2. *The family of points X_x^h of the previous type is a conservative family for $\mathrm{Sh}_{qfh}(S, \Lambda)$.*

Proof The first point follows from the previous proposition. For any excellent scheme X, the normalization morphism $X' \to X$ is a qfh-cover. Thus the category $\mathrm{Sh}_{qfh}(S,\Lambda)$ is equivalent to the category of qfh-sheaves on the site comprising geometrically unibranch S-schemes of finite type. This implies the second assertion. $\square$

10.5.14 Given any scheme S, we introduce the following composite functor using the notations of Sections 10.5.7 and 10.4.5:

$$\psi^* : \mathrm{Sh}_{qfh}(S,\Lambda) \xrightarrow{\rho} \underline{\mathrm{Sh}}^{tr}(S,\Lambda) \xrightarrow{\varphi^*} \mathrm{Sh}^{tr}(S,\Lambda).$$

Theorem 10.5.15 *Assume Λ is a **Q**-algebra and let S be a geometrically unibranch excellent scheme. Considering the above notation, the following conditions are true:*

(i) *For any S-scheme X of finite type, $\psi^*(\Lambda_S^{qfh}(X)) = \Lambda_S^{tr}(X)$.*
(ii) *The functor ψ^* admits a left adjoint $\psi_!$.*
(iii) *For any smooth S-scheme X, $\psi_!(\Lambda_S^{tr}(X)) = \Lambda_S^{qfh}(X)$.*
(iv) *The functor ψ^* is exact and preserves colimits.*
(v) *The functor $\psi_!$ is fully faithful.*

According to property (iii), the functor $\psi_!$ commutes with pullbacks. In other words, we have defined an enlargement of abelian premotivic categories (*cf.* Definition 1.4.13) over the category of (noetherian) geometrically unibranch schemes:

(10.5.15.1) $$\psi_! : \mathrm{Sh}^{tr}(-,\Lambda) \rightleftarrows \mathrm{Sh}_{qfh}(-,\Lambda) : \psi^*$$

Proof Point (i) follows from Theorem 10.5.6. Recall the enlargement of (10.4.6.1):

$$\varphi_! : \mathrm{Sh}^{tr}(-,\Lambda) \longrightarrow \underline{\mathrm{Sh}}^{tr}(-,\Lambda) : \varphi^*.$$

We define the functor $\psi_!$ as the composite:

$$\mathrm{Sh}^{tr}(S,\Lambda) \xrightarrow{\varphi_!} \underline{\mathrm{Sh}}^{tr}(S,\Lambda) \xrightarrow{\gamma^*} \underline{\mathrm{Sh}}(S,\Lambda) \xrightarrow{a_{qfh}} \mathrm{Sh}_{qfh}(S,\Lambda).$$

According to the properties of the functors in this composite, $\psi_!$ is exact and preserves colimits. Moreover, for any smooth S-scheme X, as $\Lambda_S^{tr}(X)$ is a qfh-sheaf over $\mathscr{S}^{ft}/S$ according to Proposition 10.2.4, $\psi_!(\Lambda_S^{tr}(X)) = \Lambda_S^{qfh}(X)$ which proves (iii). Property (ii) follows from (iii) and the fact that $\psi_!$ commutes with colimits, while the sheaves $\Lambda_S^{tr}(X)$ for X/S smooth generate $\mathrm{Sh}^{tr}(S,\Lambda)$.

For any smooth S-scheme X, $\Gamma(X;\psi^*(F)) = F(X)$. Thus the exactness of ψ^* follows from Corollary 10.5.13 — here we use the assumption that S is geometrically unibranch and excellent, as it implies that X satisfies the same properties, so that we can apply the mentioned corollary. As ψ^* obviously preserves direct sums, we get (iv).

To check that for any sheaf with transfers F the unit map $F \longrightarrow \psi^*\psi_!(F)$ is an isomorphism, we thus are reduced to the case where $F = \Lambda_S^{tr}(X)$ for a smooth S-scheme X, which follows from (i) and (iii). □

11 Motivic complexes

11.0.1 In this section, $\mathscr{S}$ is the category of noetherian finite-dimensional schemes. It is adequate in the sense of Section 2.0.1. Given a scheme S, we denote by $\mathscr{S}m_S$ the category of smooth separated S-schemes of finite type. It is admissible in the sense of Section 1.0.1.

We fix a ring of coefficients Λ.

11.1 Definition and basic properties

11.1.a Premotivic categories

According to Proposition 10.3.17 and Corollary 10.3.15, the abelian premotivic category $\mathrm{Sh}^{tr}(-,\Lambda)$ constructed in Definition 10.4.2 is compatible with the Nisnevich topology. Thus we can apply to it the general definitions of Section 5. This gives the following definition:

Definition 11.1.1 We define the (Λ-linear) category of *motivic complexes* (resp. *stable motivic complexes* or simply *motives*) following Definition 5.2.16 (resp. Definition 5.3.22) as

$$\mathrm{DM}_\Lambda^{e\!f\!f} = \mathrm{D}_{\mathbf{A}^1}^{e\!f\!f}\left(\mathrm{Sh}^{tr}(-,\Lambda)\right)$$
$$\text{resp. } \mathrm{DM}_\Lambda = \mathrm{D}_{\mathbf{A}^1}\left(\mathrm{Sh}^{tr}(-,\Lambda)\right).$$

Given a scheme S, we will put: $\mathrm{DM}^{e\!f\!f}(S,\Lambda) = \mathrm{DM}_\Lambda^{e\!f\!f}(S)$, $\mathrm{DM}(S,\Lambda) = \mathrm{DM}_\Lambda(S)$.

11.1.2 Let us unfold the preceding definition. Given a scheme S in $\mathscr{S}$, the triangulated category $\mathrm{DM}^{e\!f\!f}(S,\Lambda)$ is equal to the $\mathbf{A}^1$-localization of the derived category $\mathrm{D}(\mathrm{Sh}^{tr}(S,\Lambda))$ of the category of sheaves with transfers over S.

Given a smooth S-scheme X of finite type, we have denoted by $\Lambda_S^{tr}(X)$ the sheaf with transfers represented by X over S. We will view this sheaf as an object of $\mathrm{DM}^{e\!f\!f}(S,\Lambda)$, as a complex concentrated in degree 0, and call it the effective motivic complex associated with X/S.

Recall the following operations as part of the premotivic structure:

- Given any morphism $f : T \longrightarrow S$ in $\mathscr{S}$, there exists an adjunction of the form:

$$\mathbf{L}f^* : \mathrm{DM}^{e\!f\!f}(S,\Lambda) \rightleftarrows \mathrm{DM}^{e\!f\!f}(T,\Lambda) : \mathbf{R}f_*\,.$$

11 Motivic complexes

- Given a separated smooth morphism of finite type $f : T \longrightarrow S$ in $\mathscr{S}$, there exists an adjunction of the form:

$$\mathbf{L}f_\sharp : \mathrm{DM}^{e\!f\!f}(S,\Lambda) \rightleftarrows \mathrm{DM}^{e\!f\!f}(T,\Lambda) : f^* = \mathbf{L}f^*.$$

- Given any noetherian finite-dimensional scheme S, the category $\mathrm{DM}^{e\!f\!f}(S,\Lambda)$ is symmetric closed monoidal.

These operations are subject to the properties of a premotivic category: functoriality, smooth base change formula, smooth projection formula – see Section 1 for more details. By construction, the triangulated premotivic category $\mathrm{DM}^{e\!f\!f}_\Lambda$ satisfies the homotopy property and the Nisnevich descent properties.

By construction (cf. (5.3.23.2)), we get an adjunction of triangulated premotivic categories

$$(11.1.2.1) \qquad \Sigma^\infty : \mathrm{DM}^{e\!f\!f}_\Lambda \rightleftarrows \mathrm{DM}_\Lambda : \Omega^\infty.$$

Considering the *Tate motivic complex*

$$(11.1.2.2) \qquad \Lambda^{tr}_S(1) := \Lambda^{tr}_S(\mathbf{P}^1_S/\{1\}),$$

the object $\Sigma^\infty(\Lambda^{tr}_S(1))$ is $\otimes$-invertible in $\mathrm{DM}(S,\Lambda)$ and this property uniquely characterizes the homotopy category $\mathrm{DM}(S,\Lambda)$ – see Remark 5.3.29. Given a smooth separated S-scheme X of finite type, we put:

$$M_S(X) := \Sigma^\infty \Lambda^{tr}_S(X)$$

and simply call it the motive associated with X/S. Usually we denote by $\mathbb{1}_S$ the unit of the monoidal category $\mathrm{DM}(S,\Lambda)$.

By construction, the premotivic category DM_Λ satisfies the homotopy, stability and Nisnevich descent properties (see Section 5.3.23).

Example 11.1.3

- Let k be a perfect field. Then $\mathrm{DM}^{e\!f\!f}(k,\mathbf{Z})$ contains as a full subcategory the category $\mathrm{DM}^{e\!f\!f}_-(k)$ defined by Voevodsky (cf. [VSF00, Chap. 5]). This is the content of the proof of [VSF00, Chap. 5, Prop. 3.2.3]. Indeed, recall from Section 5.2.18 that $\mathrm{DM}^{e\!f\!f}(k,\mathbf{Z})$ is equivalent to the full subcategory of $D(\mathrm{Sh}^{tr}(k,\mathbf{Z}))$ comprising the complexes which are $\mathbf{A}^1$-local. Over a perfect field, Theorem 3.1.12 of [VSF00, Chap. 5] implies that a complex of sheaves with transfers is $\mathbf{A}^1$-local if and only if its homotopy sheaves are $\mathbf{A}^1$-invariant.
- Let S be a regular scheme. The triangulated categories $\mathrm{DM}^{e\!f\!f}(S,\mathbf{Z})$ and $\mathrm{DM}(S,\mathbf{Z})$ introduced here coincide with those constructed in [CD09]. The same is true concerning the operations of premotivic triangulated categories (see Remark 10.5.3).

11.1.4 Let Λ' be a localization of Λ. The premotivic adjunction

(11.1.4.1) $$\mathrm{Sh}^{tr}(-,\Lambda) \otimes_\Lambda \Lambda' \rightleftarrows \mathrm{Sh}^{tr}(-,\Lambda')$$

obtained as a particular case of (10.5.1.1) gives the following adjunctions of triangulated premotivic categories:

(11.1.4.2) $$\begin{aligned} \mathrm{DM}_\Lambda \otimes_\Lambda \Lambda' &\rightleftarrows \mathrm{DM}_{\Lambda'}, \\ \mathrm{DM}_\Lambda^{\mathit{eff}} \otimes_\Lambda \Lambda' &\rightleftarrows \mathrm{DM}_{\Lambda'}^{\mathit{eff}}. \end{aligned}$$

Proposition 10.5.2 immediately yields the following result:

Proposition 11.1.5 *The premotivic adjunctions* (11.1.4.2) *are equivalences of triangulated premotivic categories.*

In other words, for any scheme S, the triangulated monoidal category $\mathrm{DM}(S,\Lambda')$ (resp. $\mathrm{DM}^{\mathit{eff}}(S,\Lambda')$) is the naive localization of the category $\mathrm{DM}(S,\Lambda)$ (resp. $\mathrm{DM}^{\mathit{eff}}(S,\Lambda)$) with respect to integers invertible in Λ'.

11.1.b Constructible and geometric motives

11.1.6 The premotivic triangulated category $\mathrm{DM}_\Lambda^{\mathit{eff}}$ is geometrically generated: given any scheme S, the essentially small set $\mathscr{G}_S^{\mathit{eff}}$ of motivic complexes of the form $\Lambda_S^{tr}(X)$ for a smooth separated S-scheme X of finite type form a set of generators in the triangulated category $\mathrm{DM}^{\mathit{eff}}(S,\Lambda)$.

Similarly, the premotivic triangulated category DM_Λ is $\mathbf{Z}$-generated, where $\mathbf{Z}$ is the set of twists corresponding to the Tate twist: given any scheme S, the essentially small set $\mathscr{G}_S$ of motives of the form $M_S(X)(n)$ for a smooth separated S-scheme X of finite type and an integer $n \in \mathbf{Z}$ form a set of generators in the triangulated category $\mathrm{DM}(S,\Lambda)$.

Following the general conventions on premotivic triangulated categories (Definition 1.4.9), we define the notion of constructibility for motives as follows:

Definition 11.1.7 Given any scheme S, we define the category of *constructible motives* (resp. *constructible motivic complexes*) over S as the thick triangulated subcategory of $\mathrm{DM}(S,\Lambda)$ (resp. $\mathrm{DM}^{\mathit{eff}}(S,\Lambda)$) generated by $\mathscr{G}_S$ (resp. $\mathscr{G}_S^{\mathit{eff}}$). We denote it by $\mathrm{DM}_c(S,\Lambda)$ (resp. $\mathrm{DM}_c^{\mathit{eff}}(S,\Lambda)$).

Remark 11.1.8 Recall that $\mathrm{DM}_{c,\Lambda}$ (resp. $\mathrm{DM}_{c,\Lambda}^{\mathit{eff}}$) is an Sm-fibred monoidal subcategory of DM_Λ (resp. $\mathrm{DM}_\Lambda^{\mathit{eff}}$) over $\mathscr{S}$. In other words, constructible motives (resp. motivic complexes) are stable under the operations f^*, $p_\sharp$ (for p smooth) and tensor products. This is obvious from the definitions.

11.1.9 Let S be a scheme. Consider the triangulated subcategory $\mathscr{V}_S$ of $\mathrm{K}^b(\mathscr{S}m_{\Lambda,S}^{cor})$ generated by complexes of one the following forms:

1. for any distinguished square
$$\begin{array}{ccc} W & \xrightarrow{k} & V \\ g \downarrow & \downarrow f \\ U & \xrightarrow{j} & X \end{array}$$
of smooth S-schemes,

$$[W] \xrightarrow{g_*-k_*} [U] \oplus [V] \xrightarrow{j^*+f^*} [X],$$

2. for any smooth S-scheme X, $p : \mathbf{A}_X^1 \longrightarrow X$ the canonical projection.

$$[\mathbf{A}_X^1] \xrightarrow{p_*} [X].$$

Definition 11.1.10 We define the category $\mathrm{DM}_{gm}^{eff}(S,\Lambda)$ of *geometric effective motives* over S as the pseudo-abelian envelope of the triangulated category

$$\mathrm{K}^b(\mathscr{S}m_{\Lambda,S}^{cor})/\mathscr{V}_S.$$

We define the category $\mathrm{DM}_{gm,\Lambda}(S)$ of *geometric motives* over S as the triangulated category obtained from $\mathrm{DM}_{gm}^{eff}(S,\Lambda)$ by formally inverting the Tate complex

$$[\mathbf{P}_S^1] \longrightarrow [S].$$

Remark 11.1.11 The categories of geometric motives (resp. effective geometric motives) over an arbitrary base, as defined here, already appears in the work of Ivorra [Ivo07, sec. 1.3].

11.1.12 According to this definition, we can construct for any scheme S a commutative diagram of functors:

(11.1.12.1)
$$\begin{array}{ccc} \mathrm{DM}_{gm}^{eff}(S,\Lambda) & \longrightarrow & \mathrm{DM}^{eff}(S,\Lambda) \\ \downarrow & & \downarrow \Sigma^\infty \\ \mathrm{DM}_{gm}(S,\Lambda) & \longrightarrow & \mathrm{DM}(S,\Lambda), \end{array}$$

where the right vertical map is the left adjoint of (11.1.2.1).

Recall from Remark 10.3.18 that the Nisnevich topology is bounded in $\mathrm{Sh}^{tr}(-,\Lambda)$. Thus, as a corollary of Proposition 5.2.38, Corollary 5.2.39 and Corollary 5.3.42 we get the following result:

Theorem 11.1.13 *The horizontal functors of the square* (11.1.12.1) *are fully faithful and their essential images consist of constructible objects in the sense of Definition 11.1.7.*

Given any motive (resp. motivic complex) $\mathscr{M}$ over S, the following conditions are equivalent:

(i) $\mathscr{M}$ is geometric (i.e. in the image of the horizontal map of diagram (11.1.12.1)*),*

(ii) $\mathcal{M}$ *is constructible,*
(iii) $\mathcal{M}$ *is compact.*

The triangulated category $\mathrm{DM}(S,\Lambda)$ *(resp.* $\mathrm{DM}^{\mathit{eff}}(S,\Lambda)$*) is compactly generated. More precisely, the objects of the set of generators* $\mathcal{G}_S$ *(resp.* $\mathcal{G}_S^{\mathit{eff}}$*) defined in Section 11.1.6 are compact.*

Remark 11.1.14 If $S = \mathrm{Spec}\,(k)$ is the spectrum of a perfect field, then the categories $\mathrm{DM}_{gm}(S,\Lambda)$ and $\mathrm{DM}_{gm}^{\mathit{eff}}(S,\Lambda)$ coincide with the categories introduced by Voevodsky in [VSF00, chap. 5, Sec. 2.1]. The above theorem is a generalization of [VSF00, chap. 5, Th. 3.2.6] to an arbitrary base (and the non-effective case).

11.1.c Enlargement, descent and continuity

11.1.15 We can apply the definitions of Section 5 to the generalized abelian premotivic category $\underline{\mathrm{Sh}}^{tr}(-,\Lambda)$ constructed in Definition 10.4.2

Definition 11.1.16 We define the (Λ-linear) category of *generalized motivic complexes* (resp. *generalized motives*) following Definition 5.3.22 (resp. Definition 5.2.16) as

$$\underline{\mathrm{DM}}_\Lambda^{\mathit{eff}} = \mathrm{D}_{\mathbf{A}^1}^{\mathit{eff}}\left(\underline{\mathrm{Sh}}^{tr}(-,\Lambda)\right)$$

$$\text{resp. } \underline{\mathrm{DM}}_\Lambda = \mathrm{D}_{\mathbf{A}^1}\left(\underline{\mathrm{Sh}}^{tr}(-,\Lambda)\right).$$

11.1.17 The advantage of this definition is that any separated S-scheme X of finite type defines a generalized motivic complex, given by the sheaf with transfers $\underline{\Lambda}_S^{tr}(X)$ viewed as a complex concentrated in degree 0 (see Definition 10.4.2).
  The category $\underline{\mathrm{DM}}_\Lambda^{\mathit{eff}}$, as a generalized premotivic category, admits the following operations:

- Given any morphism $f : T \longrightarrow S$ in $\mathscr{S}$, there exists an adjunction of the form:

$$\mathbf{L}f^* : \underline{\mathrm{DM}}^{\mathit{eff}}(S,\Lambda) \rightleftarrows \underline{\mathrm{DM}}^{\mathit{eff}}(T,\Lambda) : \mathbf{R}f_*\,.$$

- Given a separated morphism $f : T \longrightarrow S$ of finite type in $\mathscr{S}$ (not necessarily smooth), there exists an adjunction of the form:

$$\mathbf{L}f_\sharp : \underline{\mathrm{DM}}^{\mathit{eff}}(S,\Lambda) \rightleftarrows \underline{\mathrm{DM}}^{\mathit{eff}}(T,\Lambda) : f^* = \mathbf{L}f^*\,.$$

- Given any noetherian finite-dimensional scheme S, the category $\underline{\mathrm{DM}}^{\mathit{eff}}(S,\Lambda)$ is symmetric closed monoidal.

These operations satisfy the properties of a generalized premotivic category, for which we refer the reader to Section 1.4.

11 Motivic complexes

As in the non-generalized case, we get from the general construction (see (5.3.23.2)) an adjunction of triangulated generalized premotivic categories

(11.1.17.1) $\quad\Sigma^\infty : \underline{DM}^{e\!f\!f}_\Lambda \rightleftarrows \underline{DM}_\Lambda : \Omega^\infty$.

To any separated S-scheme X of finite type, we associate a generalized motive as:

$$\underline{M}_S(X) := \Sigma^\infty \underline{\Lambda}^{tr}_S(X).$$

By construction, the generalized premotivic category $\underline{DM}^{e\!f\!f}_\Lambda$ (resp. $\underline{DM}_\Lambda$) satisfies the homotopy property and the Nisnevich descent property (resp. and stability property).

11.1.18 By virtue of Remark 10.3.18, the Nisnevich topology is bounded in $\underline{Sh}^{tr}(-,\Lambda)$. Therefore, as a corollary of Proposition 5.2.38 (resp. Corollary 5.2.39), we obtain in particular that $\underline{DM}^{e\!f\!f}(S,\Lambda)$ (resp. $\underline{DM}(S,\Lambda)$) is compactly generated, with the essentially small family of objects $\underline{\Lambda}^{tr}_S(X)$ (resp. $\underline{M}_S(X)(n)$) for a separated S-scheme of finite type X (resp. and an integer $n \in \mathbf{Z}$) as compact generators.

Recall that for any scheme S, the obvious restriction functor

$$\varphi^* : \mathrm{Sh}^{tr}(S,\Lambda) \longrightarrow \underline{Sh}^{tr}(S,\Lambda)$$

admits a left adjoint $\varphi_!$ which is fully faithful (Lemma 10.4.6). Moreover, the adjoint pair $(\varphi_!, \varphi^*)$ satisfies the assumptions of Proposition 6.1.4, so that applying Corollary 6.1.9 gives the following proposition:

Proposition 11.1.19 *Given any scheme S, the adjoint pair $(\varphi_!, \varphi^*)$ can be derived and induces the following pair of adjoint functors*

(11.1.19.1) $\quad\begin{aligned}\varphi_! &: \mathrm{DM}(S,\Lambda) \rightleftarrows \underline{DM}(S,\Lambda) : \varphi^* \\ \text{resp. } \varphi_! &: \mathrm{DM}^{e\!f\!f}(S,\Lambda) \rightleftarrows \underline{DM}^{e\!f\!f}(S,\Lambda) : \varphi^*,\end{aligned}$

such that $\varphi_!$ is fully faithful.

More generally, the family of these adjunctions for a noetherian finite-dimensional scheme S defines an enlargement of premotivic categories (Definition 1.4.13).

The abuse of notation is justified because of the following essentially commutative diagram of functors:

(11.1.19.2) $\quad\begin{array}{ccc}\mathrm{DM}^{e\!f\!f}_\Lambda & \xrightarrow{\Sigma^\infty} & \mathrm{DM}_\Lambda \\ \varphi_! \downarrow & & \downarrow \varphi_! \\ \underline{DM}^{e\!f\!f}_\Lambda & \xrightarrow{\Sigma^\infty} & \underline{DM}_\Lambda.\end{array}$

Recall that, given a smooth separated S-scheme X, we have the relation:

$$\varphi_!(M_S(X)) = \underline{M}_S(X).$$

Remark 11.1.20 Beware that the functor φ^* is far from being conservative. The following example was suggested by V. Vologodsky: let Z be a nowhere dense closed subscheme of S. Then $\varphi^*(\underline{M}_S(Z)) = 0$. In fact, one can see that $DM(S, \Lambda)$ is a localization of the category $\underline{DM}(S, \Lambda)$ with respect to the objects $\mathscr{M}$ such that $\varphi^*(\mathscr{M}) = 0$.

11.1.21 With rational coefficients, the preceding proposition can be refined. Recall that the *qfh*-sheafification functor (10.5.9) induces by Section 5.3.28 a premotivic adjunction
$$\underline{\alpha}^* : \underline{DM}_\mathbf{Q} \rightleftarrows \underline{DM}_{qfh,\mathbf{Q}} : \underline{\alpha}_* .$$

Theorem 11.1.22 *If S is a geometrically unibranch excellent noetherian scheme of finite dimension then the composite functor*
$$\underline{\alpha}^* \varphi_! : DM(S, \mathbf{Q}) \longrightarrow \underline{DM}_{qfh,\mathbf{Q}}(S)$$
is fully faithful.

Proof Note that $DM^{eff}(S, \mathbf{Q})$ and $D^{eff}_{\mathbf{A}^1}(\mathrm{Sh}_{qfh}(S, \mathbf{Q}))$ are compactly generated; see Example 5.1.29 and Proposition 5.2.38. Hence this corollary follows from Theorem 10.5.15 and Proposition 6.1.8. □

Remark 11.1.23 Recall this theorem can be rephrased by saying that motives over S satisfy *qfh*-descent – see Remark 5.2.11 and more generally Section 3. In the next section, we will give applications of this fact to motivic cohomology.

Theorem 11.1.24 *The following assertions hold:*

1. *The triangulated premotivic categories DM^{eff}_Λ and DM_Λ are weakly continuous (Definition 4.3.2).*
2. *The generalized triangulated premotivic categories $\underline{DM}^{eff}_\Lambda$ and $\underline{DM}_\Lambda$ are weakly continuous.*

Proof Note that Proposition 10.3.19 shows precisely that the generalized premotivic abelian category $\underline{\mathrm{Sh}}^{tr}(-, \Lambda)$ satisfies Property (wC) of Section 5.1.35. Therefore, the assertion (2) follows from Propositions 5.2.41 and 5.3.44.

Moreover, the assertion (1) follows from Corollary 6.1.12 given the enlargement obtained in Proposition 11.1.19. □

Example 11.1.25 From the previous theorem and Proposition 4.3.4, we obtain in particular that for any pro-scheme $(S_\alpha)_{\alpha \in A}$ with affine and dominant transition map such that $S = \varprojlim_{\alpha \in A} S_\alpha$ is noetherian finite-dimensional, there exist canonical equivalences of categories:
$$2\text{-}\varinjlim_\alpha \left(DM^{eff}_{gm,\Lambda}(S_\alpha) \right) \longrightarrow DM^{eff}_{gm,\Lambda}(S),$$
$$2\text{-}\varinjlim_\alpha \left(DM_{gm,\Lambda}(S_\alpha) \right) \longrightarrow DM_{gm,\Lambda}(S).$$

This result generalizes [Ivo07, 4.16].

11.2 Motivic cohomology

11.2.a Definition and functoriality

Definition 11.2.1 Let S be a scheme and $(n,m) \in \mathbf{Z}^2$ be a pair of integers. We define the *motivic cohomology* of S in degree n and twist m with coefficients in Λ as the Λ-module
$$H^{n,m}_{\mathscr{M}}(S,\Lambda) = \mathrm{Hom}_{\mathrm{DM}(S,\Lambda)}\left(\mathbb{1}_S, \mathbb{1}_S(m)[n]\right).$$
Assuming $m \geq 0$, we define the *effective motivic cohomology* of S in degree n and twist m with coefficients in Λ as the Λ-module
$$H^{n,m}_{\mathscr{M},\mathit{eff}}(S,\Lambda) = \mathrm{Hom}_{\mathrm{DM}^\mathit{eff}(S,\Lambda)}\left(\Lambda^{tr}_S, \Lambda^{tr}_S(m)[n]\right).$$

Motivic cohomology (resp. effective motivic cohomology) is contravariant with respect to morphisms of schemes and the monoidal structure on DM_Λ (resp. $\mathrm{DM}^{\mathit{eff}}_\Lambda$) defines a ring structure compatible with pullbacks: given two cohomology classes
$$\alpha: \mathbb{1}_S \longrightarrow \mathbb{1}_S(m)[n], \alpha': \mathbb{1}_S \longrightarrow \mathbb{1}_S(m')[n'],$$
one simply puts
$$\alpha.\alpha' = \alpha \otimes_S \alpha'.$$
The link between motivic cohomology and effective motivic cohomology is provided by Proposition 5.3.39. Given any scheme S and any pair of integers $(n,m) \in \mathbf{Z}^2$, one has a canonical identification:
$$H^{n,m}_{\mathscr{M}}(S,\Lambda) = \varinjlim_{r >> 0} \mathrm{Hom}_{\mathrm{DM}^\mathit{eff}(S,\Lambda)}\left(\Lambda^{tr}_S(r), \Lambda^{tr}_S(m+r)[n]\right).$$

11.2.2 Let Λ' be a localization of Λ. Then using the left adjoint of the premotivic adjunction (11.1.4.2), we get a canonical morphism
$$H^{n,m}_{\mathscr{M}}(S,\Lambda) \otimes_\Lambda \Lambda' \longrightarrow H^{n,m}_{\mathscr{M}}(S,\Lambda').$$
It is obviously compatible with pullbacks and the product structure. According to Proposition 11.1.5, this map is an isomorphism.

Example 11.2.3 Let k be a perfect field. Given any smooth separated k-scheme S of finite type, with structural morphism f, and any pair of integers $(n,m) \in \mathbf{Z}^2$, motivic cohomology as defined in the previous definition coincides with motivic cohomology as defined by Voevodsky in [VSF00, chap. 5] according to the following computation and Remark 11.1.14:
$$H^{n,m}_{\mathscr{M}}(X,\mathbf{Z}) = \mathrm{Hom}_{\mathrm{DM}(X,\mathbf{Z})}(\mathbb{1}_X, \mathbb{1}_X(m)[n]) = \mathrm{Hom}_{\mathrm{DM}(X,\mathbf{Z})}(\mathbb{1}_X, f^*(\mathbb{1}_k)(m)[n])$$
$$= \mathrm{Hom}_{\mathrm{DM}(k,\mathbf{Z})}(\mathbf{L}f_\sharp(\mathbb{1}_X), \mathbb{1}_k(m)[n]) = \mathrm{Hom}_{\mathrm{DM}(k,\mathbf{Z})}(M_k(X), \mathbb{1}_k(m)[n])$$
$$= \mathrm{Hom}_{\mathrm{DM}_{gm}(k,\mathbf{Z})}(M_k(X), \mathbb{1}_k(m)[n]).$$

In particular, it coincides with higher Chow groups (cf. [Voe02a]) according to the following formula:
$$H^{n,m}_{\mathcal{M}}(X, \mathbf{Z}) = CH^m(X, 2m - n).$$
Recall in particular the following computations:
$$H^{n,m}_{\mathcal{M}}(X, \mathbf{Z}) = \begin{cases} \mathbf{Z}^{\pi_0(X)} & \text{if } n = m = 0, \\ \mathbf{G}_m(X) & \text{if } n = m = 1, \\ CH^m(X) & \text{if } n = 2m, \\ 0 & \text{if } m < 0, n > \min(m + \dim(X), 2m), \end{cases}$$
where $CH^m(X)$ is the usual Chow group of m-codimensional cycles in X.

Note we will extend the identification of motivic cohomology as defined in the previous definition with the general version defined by Voevodsky – [Voe98] – in Section 11.2.c.

11.2.4 Consider a separated morphism $p : X \longrightarrow S$ of finite type. Recall from the $\mathscr{S}^{ft}$-fibred structure of $\underline{\mathrm{DM}}_\Lambda$ that $\underline{M}_S(X) = \mathbf{L}p_\sharp p^*(\mathbb{1}_S)$. Using the adjunction property of the pair $(\mathbf{L}p_\sharp, p^*)$, we easily get:

(11.2.4.1)
$$H^{n,m}_{\mathcal{M}}(X, \Lambda) = \mathrm{Hom}_{\underline{\mathrm{DM}}(X, \Lambda)}\left(\mathbb{1}_X, \mathbb{1}_X(m)[n]\right) = \mathrm{Hom}_{\underline{\mathrm{DM}}(X, \Lambda)}\left(\mathbb{1}_X, \mathbb{1}_X(m)[n]\right)$$
$$= \mathrm{Hom}_{\underline{\mathrm{DM}}(S, \Lambda)}\left(\underline{M}_S(X), \mathbb{1}_S(m)[n]\right).$$

In particular, given any finite S-correspondence $\alpha : X \bullet\!\!\!-\!\!\!\rightarrow Y$ between separated S-schemes of finite type, we obtain a pullback
$$\alpha^* : H^{n,m}_{\mathcal{M}}(Y, \Lambda) \longrightarrow H^{n,m}_{\mathcal{M}}(X, \Lambda)$$
which is, among other properties, functorial with respect to composition of finite S-correspondences and extends the natural contravariant functoriality of motivic cohomology.

In particular, given any finite Λ-universal morphism $f : Y \longrightarrow X$, we obtain a pushout
$$f_* : H^{n,m}_{\mathcal{M}}(Y, \Lambda) \longrightarrow H^{n,m}_{\mathcal{M}}(X, \Lambda)$$
by considering the transpose of the graph of f.

Proposition 11.2.5 *Let $f : Y \longrightarrow X$ be a finite Λ-universal morphism of schemes. Assume X is connected and let $d > 0$ be the degree of f (cf. Definition 9.1.12). Then for any element $x \in H^{n,m}_{\mathcal{M}}(X, \Lambda)$, $f_* f^*(x) = d.x$.*

This is a simple application of Proposition 9.1.13. We leave to the reader the exercise of determining projection and base change formulas for this pushout.

Example 11.2.6 Let $f : Y \longrightarrow X$ be a finite morphism. Recall that f is Λ-universal in the following particular cases:

- f is flat (see Example 8.1.50);
- X is regular and f sends the generic points of Y to generic points of X (see Corollary 8.3.28).

In particular, motivic cohomology is covariant with respect to this kind of finite morphism.

Another important application of generalized motives is obtained using Corollary 10.5.8:

Proposition 11.2.7 *Let $f : X' \longrightarrow X$ be a separated universal homeomorphism of finite type. Assume that $\mathrm{char}(X) \subset \Lambda^\times$. Then the pullback functor*

$$H^{n,m}_{\mathscr{M}}(X, \Lambda) \longrightarrow H^{n,m}_{\mathscr{M}}(X', \Lambda)$$

is an isomorphism.

Remark 11.2.8 The preceding considerations hold similarly for the effective motivic cohomology.

Example 11.2.9 In characteristic 0, motivic cohomology (effective and non-effective) is invariant under semi-normalization ([Swa80]).

When restricted to an excellent geometrically unibranch scheme X, motivic cohomology (effective and non-effective) is invariant under normalization. Indeed, the normalization $X' \longrightarrow X$ of such a scheme is a universal homeomorphism ([GD67, IV_0, 23.2.2]) of finite type.

11.2.b Effective motivic cohomology in weight 0 and 1

11.2.10 Let S be a scheme and X a smooth S-scheme. For any subscheme Y of X, we denote by $\Lambda_S^{tr}(X/Y)$ the cokernel of the canonical morphism of sheaves with transfers $\Lambda_S^{tr}(Y) \longrightarrow \Lambda_S^{tr}(X)$. As this morphism is a monomorphism, we obtain a canonical distinguished triangle in $\mathrm{DM}^{\mathit{eff}}(S, \Lambda)$

$$\Lambda_S^{tr}(Y) \longrightarrow \Lambda_S^{tr}(X) \longrightarrow \Lambda_S^{tr}(X/Y) \longrightarrow \Lambda_S^{tr}(X)[1].$$

Using this notation and according to Definition 2.4.17, the Tate motivic complex is defined as: $\Lambda_S^{tr}(1) = \Lambda_S^{tr}(\mathbf{P}_S^1/\{\infty\})[-2]$.

The following computation is classical:

$$\Lambda_S^{tr}(1) = \Lambda_S^{tr}(\mathbf{P}_S^1/\mathbf{A}_S^1)[-2] = \Lambda_S^{tr}(\mathbf{A}_S^1/\mathbf{G}_m)[-2];$$

the first identification follows from homotopy invariance and the second one by Nisnevich descent (cf. Proposition 5.2.13).

Proposition 11.2.11 *Suppose S is a normal scheme.
Then the sheaf on Sm_S represented by $\mathbf{G}_m$ admits a canonical structure of a sheaf with transfers and there is a canonical isomorphism in $DM^{eff}(S, \Lambda)$:*

$$\mathbf{G}_m \otimes_{\mathbf{Z}} \Lambda \xrightarrow{\simeq} \Lambda_S^{tr}(1)[1].$$

Proof Let U be an open subscheme of $\mathbf{A}_S^1$ and X be a smooth S-scheme. Note that X is normal according to [GD67, 18.10.7]. Consider a cycle

$$\alpha = \sum_i n_i . \langle Z_i \rangle$$

of $X \times_S U$ with $n_i \in \Lambda$ and Z_i irreducible finite and dominant over an irreducible component of X. Then Z_i is a divisor in $X \times_S U$ and according to [GD67, 21.14.3], it is flat over X. In other words, α is a Hilbert cycle, which implies it is Λ-universal (Example 8.1.50). As a consequence, we obtain the equality

$$H^i \Gamma(X; \underline{C}^* \Lambda_S^{tr}(U)) = H_{-i}^{sing}(X \times_S U/X) \otimes_{\mathbf{Z}} \Lambda,$$

where the functor $\underline{C}^*$ is the associated Suslin singular complex (see (5.2.32.1)) and the right-hand side is the Suslin homology of $X \times_S U/X$ (*cf.* [SVoob]).

Suppose in addition that X and U are affine and let $Z = \mathbf{P}_S^1 - U$. According to a theorem of Suslin and Voevodsky (*cf.* [SVoob, th. 3.1]),

$$H_{-i}^{sing}(X \times_S U/X) = \begin{cases} \operatorname{Pic}(X \times_S \mathbf{P}_S^1, X \times_S Z) & \text{if } i = 0 \\ 0 & \text{otherwise;} \end{cases}$$

the group on the left-hand side is the *relative Picard group*. In particular, the complex $\underline{C}^* \Lambda_S^{tr}(U)$, regarded as a complex of presheaves with transfers, is concentrated in cohomological degree 0 and its 0-th cohomology is the presheaf $X \mapsto \operatorname{Pic}(X \times_S \mathbf{P}_S^1, X \times_S Z) \otimes_{\mathbf{Z}} \Lambda$.

Consider the following exact sequence of presheaves with transfers:

$$0 \longrightarrow \Lambda_S^{tr}(\mathbf{G}_m) \longrightarrow \Lambda_S^{tr}(\mathbf{A}_S^1) \longrightarrow \tilde{\Lambda}_S^{tr}(\mathbf{A}_S^1/\mathbf{G}_m) \longrightarrow 0.$$

Applying the functor $\underline{C}^*$ to it, relatively to the category of complexes of presheaves with transfers, we obtain a distinguished triangle in $D(\operatorname{PSh}^{tr}(S, \Lambda))$:

$$\underline{C}^* \Lambda_S^{tr}(\mathbf{G}_m) \longrightarrow \underline{C}^* \Lambda_S^{tr}(\mathbf{A}_S^1) \longrightarrow \underline{C}^* \tilde{\Lambda}_S^{tr}(\mathbf{A}_S^1/\mathbf{G}_m) \longrightarrow \underline{C}^* \Lambda_S^{tr}(\mathbf{G}_m)[1].$$

Taking the associated long exact sequence of cohomology presheaves, we obtain that the complex of presheaves with transfers $\underline{C}^* \tilde{\Lambda}_S^{tr}(\mathbf{A}_S^1/\mathbf{G}_m)$ is concentrated in cohomological degree 0 and -1, and we get an exact sequence of presheaves:

$$0 \longrightarrow \hat{H}^{-1}[\underline{C}^* \tilde{\Lambda}_S^{tr}(\mathbf{A}_S^1/\mathbf{G}_m)] \longrightarrow \hat{H}^0[\underline{C}^* \Lambda_S^{tr}(\mathbf{G}_m)] \longrightarrow \hat{H}^0[\underline{C}^* \Lambda_S^{tr}(\mathbf{A}_S^1)]$$
$$\longrightarrow \hat{H}^0[\underline{C}^* \tilde{\Lambda}_S^{tr}(\mathbf{A}_S^1/\mathbf{G}_m)] \longrightarrow 0.$$

11 Motivic complexes

By definition of the relative Picard group, given any smooth (affine) scheme X, we get an exact sequence of abelian groups:

(11.2.11.1) $\quad 0 \longrightarrow \mathbf{G}_m(X) \longrightarrow \operatorname{Pic}(X \times_S \mathbf{P}_S^1, X_0 \sqcup X_\infty) \longrightarrow \operatorname{Pic}(X \times_S \mathbf{P}_S^1, X_0) \longrightarrow 0.$

Thus we deduce that:

$$\hat{H}^0[\underline{C}^* \tilde{\Lambda}_S^{tr}(\mathbf{A}_S^1/\mathbf{G}_m)] = 0,$$
$$\hat{H}^{-1}[\underline{C}^* \tilde{\Lambda}_S^{tr}(\mathbf{A}_S^1/\mathbf{G}_m)] = \mathbf{G}_m \otimes_{\mathbf{Z}} \Lambda.$$

This gives in particular a canonical isomorphism:

$$\underline{C}^* \tilde{\Lambda}_S^{tr}(\mathbf{A}_S^1/\mathbf{G}_m)[-1] \simeq \mathbf{G}_m \otimes_{\mathbf{Z}} \Lambda$$

in $D(\operatorname{PSh}^{tr}(S, \Lambda))$. Taking its image in $\mathrm{DM}^{\mathit{eff}}(S, \Lambda)$ we obtain a canonical isomorphism which can be written as:

$$\underline{C}^* \Lambda_S^{tr}(\mathbf{A}_S^1/\mathbf{G}_m)[-1] \simeq \mathbf{G}_m \otimes_{\mathbf{Z}} \Lambda.$$

Thus the result follows because, according to Lemma 5.2.35, the canonical map

$$\Lambda_S^{tr}(\mathbf{A}_S^1/\mathbf{G}_m) \longrightarrow \underline{C}^* \Lambda_S^{tr}(\mathbf{A}_S^1/\mathbf{G}_m)$$

is an isomorphism in $\mathrm{DM}^{\mathit{eff}}(S, \Lambda)$. □

Remark 11.2.12 In the course of the proof, a canonical structure of a sheaf with transfers over S on $\mathbf{G}_m$ has naturally appeared – described by the exact sequence (11.2.11.1). This structure is classical (see [MVW06, Ex. 2.4]). One can describe it as follows.

Let X and Y be smooth S-schemes. Assume X is connected (thus irreducible as it is normal). Let Z be a closed integral subscheme Z of $X \times_S Y$ which is finite surjective over X. Then Z/X corresponds to an extension of function fields L/K. The norm map of L/K induces a morphism of abelian groups: $N_{Z/X} : \mathbf{G}_m(Z) \longrightarrow \mathbf{G}_m(X)$. Then we associate with Z, viewed as a finite correspondence from X to Y, the following morphism:

$$\mathbf{G}_m(Y) \xrightarrow{p^*} \mathbf{G}_m(Z) \xrightarrow{N_{Z/X}} \mathbf{G}_m(X),$$

where $p : Y \longrightarrow Z$ is the natural projection.

The following proposition is well known to the expert. We include a proof for completeness.

Proposition 11.2.13 *For any regular scheme X and any integer $i \geq 0$,*

$$H_{Nis}^i(X, \mathbf{G}_m) = \begin{cases} \mathscr{O}_X(X)^\times & \text{if } i = 0, \\ \operatorname{Pic}(X) & \text{if } i = 1, \\ 0 & \text{otherwise,} \end{cases}$$

where $\mathrm{Pic}(X)$ is the Picard group of X.

Proof Let Y be any étale scheme over X. We let $C^0(V)$ be the abelian group comprising the invertible rational functions on V and $C^1(V)$ be the group of 1-codimensional cycles in V. Classically, one associates with any rational function f on V its Weil divisor $\mathrm{div}(f) \in C^1(V)$. Recall, when V is integral with function field K, $f \in K$, one puts:

$$\mathrm{div}_V(f) = \sum_{x \in V^{(1)}} v_x(f).x;$$

the sum runs over the points of codimension 1 in V and $v_x(f)$ is the valuation of f corresponding to the valuation ring $\mathcal{O}_{X,x}$.

According to this definition, we get a complex:

$$0 \longrightarrow \mathbf{G}_m(V) \longrightarrow C^0(V) \xrightarrow{\mathrm{div}_V} C^1(V).$$

This sequence is functorial with respect to pullback of étale X-schemes. Thus we have defined a morphism of presheaves on $X_{\acute{e}t}$:

$$\pi : \mathbf{G}_m \longrightarrow C^*.$$

Given any Nisnevich distinguished square Q (Example 2.1.11), one can check easily that the image of Q by C^0 (resp. C^1) is cocartesian. As a consequence C^* is a complex of Nisnevich sheaves which satisfies the Brown–Gersten property – *i.e.* it is Nisnevich flasque in the sense of Definition 5.1.9 according to Proposition 5.2.13 applied to the derived category of Nisnevich sheaves over X.

On the other hand, π is a quasi-isomorphism of Nisnevich sheaves over S: indeed it is well known that for any regular local ring A, the sequence

$$0 \longrightarrow A^\times \longrightarrow \mathrm{Frac}(A)^\times \xrightarrow{\mathrm{div}_A} Z^1(A) \longrightarrow 0$$

is exact. This is an easy consequence of the fact that A is a unique factorization domain – the classical Auslander–Buchsbaum theorem, (e.g. [Mat70, 20.3]).

In particular, we get $H^i(X, \mathbf{G}_m) = H^i(C^*(X))$ and this concludes the proof. □

The following theorem is a generalization of a well-known computation of Voevodsky for smooth schemes over a perfect field. The second case is a corollary of the two preceding propositions.

Theorem 11.2.14 *Let S be a scheme and $n \in \mathbf{Z}$ an integer. The following computation holds:*

1.
$$H^{n,0}_{\mathcal{M},\mathit{eff}}(S, \Lambda) = \mathrm{Hom}_{\mathrm{DM}^{\mathit{eff}}(S)}(\Lambda^{tr}_S, \Lambda^{tr}_S[n]) = \begin{cases} \Lambda^{\pi_0(S)} & \textit{if } n = 0 \\ 0 & \textit{otherwise;} \end{cases}$$

2. *if S is regular,*

11 Motivic complexes

$$H^{n,1}_{\mathscr{M},eff}(S,\Lambda) = \mathrm{Hom}_{\mathrm{DM}^{eff}(S)}(\Lambda^{tr}_S, \Lambda^{tr}_S(1)[n]) = \begin{cases} \mathscr{O}_S(S)^\times \otimes_{\mathbf{Z}} \Lambda & \text{if } n = 1 \\ \mathrm{Pic}(S) \otimes_{\mathbf{Z}} \Lambda & \text{if } n = 2 \\ 0 & \text{otherwise}. \end{cases}$$

Proof For the first case, according to Proposition 10.2.5, the sheaf Λ^{tr}_S is Nisnevich local and $\mathbf{A}^1$-local as a complex of sheaves. It is obviously acyclic for the Nisnevich topology. Thus, the result follows using again Proposition 10.2.5 in the case $n = 0$.

Consider the second case. According to Proposition 11.2.13, the sheaf $\mathbf{G}_m$ on Sm_S is $\mathbf{A}^1$-local. Thus according to Proposition 11.2.11 $\mathbf{G}_m \otimes \Lambda[-1]$ is an $\mathbf{A}^1$-resolution of $\Lambda^{tr}_S(1)$. In particular,

$$\mathrm{Hom}_{\mathrm{DM}^{eff}(S)}(\Lambda^{tr}_S, \Lambda^{tr}_S(1)[n]) = \mathrm{Hom}_{D(\mathrm{Sh}^{tr}(S,\Lambda))}(\Lambda^{tr}_S, \mathbf{G}_m \otimes \Lambda[n-1])$$
$$= H^{n-1}_{Nis}(S, \mathbf{G}_m) \otimes \Lambda$$

where the second identification follows from Remark 10.4.4. The conclusion follows from another application of Proposition 11.2.13. □

The following corollary is a (very) weak cancellation result in $\mathrm{DM}^{eff}(S)$:

Corollary 11.2.15 *Let S be a regular scheme. Then*

$$\mathbf{R}Hom(\Lambda^{tr}_S(1), \Lambda^{tr}_S(1)) = \Lambda^{tr}_S.$$

Moreover, if $m = 0$ or $m = 1$, for any integer $n > m$,

$$\mathbf{R}Hom(\Lambda^{tr}_S(n), \Lambda^{tr}_S(m)) = 0.$$

Proof We consider the first assertion. Any smooth S-scheme is regular. Hence, it is sufficient to prove that for any connected regular scheme S, for any integer $n \in \mathbf{Z}$,

$$\mathrm{Hom}_{\mathrm{DM}^{eff}(S)}(\Lambda^{tr}_S(1), \Lambda^{tr}_S(1)[n]) = \begin{cases} \Lambda & \text{if } n = 0 \\ 0 & \text{otherwise}. \end{cases}$$

Using the exact triangle

(11.2.15.1) $\qquad \Lambda^{tr}_S(\mathbf{G}_m) \longrightarrow \Lambda^{tr}_S(\mathbf{A}^1) \longrightarrow \Lambda^{tr}_S(1)[2] \xrightarrow{+1}$

and the second case of the previous theorem, we obtain the following long exact sequence

$$\cdots \longrightarrow \mathrm{Hom}(\Lambda^{tr}_S(\mathbf{A}^1), \Lambda^{tr}_S(1)[n]) \longrightarrow \mathrm{Hom}(\Lambda^{tr}_S(\mathbf{G}_m), \Lambda^{tr}_S(1)[n])$$
$$\longrightarrow \mathrm{Hom}(\Lambda^{tr}_S(1), \Lambda^{tr}_S(1)[n-1]) \longrightarrow \mathrm{Hom}(\Lambda^{tr}_S(\mathbf{A}^1), \Lambda^{tr}_S(1)[n+1]) \longrightarrow \cdots$$

Then the result follows using the previous theorem and the fact that

$$\mathrm{Pic}(\mathbf{A}^1 \times S) = \mathrm{Pic}(\mathbf{G}_m \times S)$$

whenever S is regular.

For the last assertion, we are reduced to proving that if S is a regular scheme, for any integers $n > 0$ and i,
$$\mathrm{Hom}_{\mathrm{DM}^{e\!f\!f}(S)}(\Lambda_S^{tr}(n), \Lambda_S^{tr}[i]) = 0.$$
This is obviously implied by the case $n = 1$.

Consider the distinguished triangle (11.2.15.1) again. Then the long exact sequence attached to the cohomological functor $\mathrm{Hom}_{\mathrm{DM}^{e\!f\!f}(S,\Lambda)}(-, \Lambda_S^{tr})$ and applied to this triangle together with the first case of the previous theorem yields the result. □

11.2.c The motivic cohomology ring spectrum

11.2.16 According to Definition 10.4.2 and Section 10.4.3, we have an adjunction of abelian premotivic categories
$$\gamma^* : \mathrm{Sh}(-, \Lambda) \rightleftarrows \mathrm{Sh}^{tr}(-, \Lambda) : \gamma_*$$
such that γ_* is conservative and exact. According to Section 5.3.28, it induces an adjunction of triangulated premotivic categories

(11.2.16.1) $$\mathbf{L}\gamma^* : \mathrm{D}_{\mathbf{A}^1, \Lambda} \rightleftarrows \mathrm{DM}_\Lambda : \mathbf{R}\gamma_*.$$

Composing with the premotivic adjunction between the stable homotopy category and the $\mathbf{A}^1$-derived homotopy category (5.3.35.1), we finally get a canonical premotivic adjunction:

(11.2.16.2) $$\varphi^* : \mathrm{SH} \rightleftarrows \mathrm{DM}_\Lambda : \varphi_*.$$

Recall that, because φ^* is monoidal, φ_* is weakly monoidal. In particular, for any scheme S, one gets canonical morphisms
$$\mathbb{1}_S \longrightarrow \varphi_*(\mathbb{1}_S), \quad \varphi_*(\mathbb{1}_S) \wedge \varphi_*(\mathbb{1}_S) \longrightarrow \varphi_*(\mathbb{1}_S)$$
which gives a structure of a commutative monoid to the spectrum $\varphi_*(\mathbb{1}_S)$, *i.e.* a ring spectrum.

Definition 11.2.17 Given any scheme S, one defines the *motivic cohomology ring spectrum* over S with coefficients in Λ as the commutative ring spectrum:
$$H_{\mathcal{M}, S}^\Lambda := \varphi_*(\mathbb{1}_S).$$

The properties of the functor φ_* immediately imply that the ring spectrum $H_{\mathcal{M}, S}^\Lambda$ represents motivic cohomology. One now easily checks that this ring spectrum coincides with the original one of Voevodsky (see [Voe98, sec. 6.1]) in the case $\Lambda = \mathbf{Z}$. Therefore, our definition of motivic cohomology (with $\mathbf{Z}$-coefficients) agrees with that given by Voevodsky in *loc. cit.*

11 Motivic complexes

11.2.18 Consider a localization Λ' of Λ. Then one gets an essentially commutative diagram of right adjoints of premotivic adjunctions:

$$\begin{array}{ccc}
D_{\mathbf{A}^1}(S,\Lambda) \otimes_\Lambda \Lambda' & \longleftarrow & DM(S,\Lambda) \otimes_\Lambda \Lambda' \\
\uparrow & & \uparrow \\
SH(S) & (1) & (2) \\
\downarrow & & \downarrow \\
D_{\mathbf{A}^1}(S,\Lambda') & \longleftarrow & DM(S,\Lambda')
\end{array}$$

where the map (1) is the canonical equivalence (see Proposition 5.3.37) and the map (2) is the equivalence from (11.1.4.2). Note in particular that (2) is monoidal (as its reciprocal equivalence is monoidal as the left adjoint of a premotivic adjunction). Thus this essentially commutative diagram defines a canonical morphism of ring spectra:

(11.2.18.1) $$H^\Lambda_{\mathcal{M},S} \otimes_\Lambda \Lambda' \longrightarrow H^{\Lambda'}_{\mathcal{M},S}.$$

As a corollary of Proposition 11.1.5, we get the following result:

Proposition 11.2.19 *The map* (11.2.18.1) *is an isomorphism of ring spectra.*

Remark 11.2.20 In a previous version of this text, we only get the above result in particular cases. The main argument for the general case obtained above can be traced back to Proposition 8.1.53.

11.2.21 Let $f : T \longrightarrow S$ be a morphism of schemes. Recall from the structure of the premotivic adjunction (φ^*, φ_*) defined above that we get an exchange morphism:

$$f^* \varphi_* \longrightarrow \varphi_* f^*.$$

Applying this natural transformation to the unit object $\mathbb{1}_S$ of $DM(S,\Lambda)$, one gets a canonical morphism of ring spectra:

$$\tau_f : f^*(H^\Lambda_{\mathcal{M},S}) \longrightarrow H^\Lambda_{\mathcal{M},T}.$$

Note that this shows the collection $(H^\Lambda_{\mathcal{M},S})$ is a section of the fibred category SH. Recall also the following conjecture of Voevodsky ([Voe02b, conj. 17]):

Conjecture 11.2.22 (Voevodsky) *For any morphism f as above, the map τ_f is an isomorphism.*

Remark 11.2.23 At least, Voevodsky formulated this conjecture in the case where $\Lambda = \mathbf{Z}$. According to the preceding proposition, this implies the case of any coefficient ring $\Lambda \subset \mathbf{Q}$. We will solve affirmatively a particular case of this conjecture in Corollary 16.1.7 when $\Lambda = \mathbf{Q}$. We will see below that this conjecture of Voevodsky is strongly related to the behavior of the six operations in DM_Λ; see Proposition 11.4.7. References for other known cases of variants of the conjecture may be found in Remark 11.4.8.

11.3 Orientation and purity

11.3.1 For any scheme S, we let $\mathbf{P}_S^\infty$ be the ind-scheme

$$S \longrightarrow \mathbf{P}_S^1 \longrightarrow \cdots \longrightarrow \mathbf{P}_S^n \longrightarrow \mathbf{P}_S^{n+1} \longrightarrow$$

comprising the obvious closed immersions. This ind-scheme has a comultiplication given by the Segre embedding

$$\mathbf{P}_S^\infty \times_S \mathbf{P}_S^\infty \longrightarrow \mathbf{P}_S^\infty.$$

Define $\Lambda_S^{tr}(\mathbf{P}^\infty) = \varinjlim \Lambda_S^{tr}(\mathbf{P}^n)$. Applying Theorem 11.2.14 in the case $S = \mathrm{Spec}\,(\mathbf{Z})$, we obtain a canonical isomorphism:

$$\mathrm{Hom}_{\mathrm{DM}^{e\!f\!f}(\mathrm{Spec}\,(\mathbf{Z}),\Lambda)}(\Lambda^{tr}(\mathbf{P}^\infty), \Lambda^{tr}(1)[2]) = \mathrm{Pic}(\mathbf{P}^\infty) \otimes_{\mathbf{Z}} \Lambda,$$

whose aim is a free Λ-algebra of power series in one variable. Considering the canonical dual invertible sheaf on $\mathbf{P}^\infty$, we obtain a canonical formal generator of this Λ-algebra and thus a morphism $\mathrm{DM}^{e\!f\!f}(\mathrm{Spec}\,(\mathbf{Z}),\Lambda)$:

$$c_1 : \Lambda^{tr}(\mathbf{P}^\infty) \longrightarrow \Lambda^{tr}(1)[2].$$

For any scheme S, considering the canonical projection $f : S \longrightarrow \mathrm{Spec}\,(\mathbf{Z})$, we obtain by pullback along f a morphism of $\mathrm{DM}^{e\!f\!f}(S,\Lambda)$

$$c_{1,S} : \Lambda_S^{tr}(\mathbf{P}_S^\infty) \longrightarrow \Lambda_S^{tr}(1)[2].$$

Consider $\mathbf{G}_m$ as a sheaf of groups over Sm_S. Following [MV99, part 4], we introduce its classifying space $B\mathbf{G}_m$ as a simplicial sheaf over Sm_S. From proposition 1.16 of *loc. cit.*, we get $\mathrm{Hom}_{\mathscr{H}_\bullet^s(S)}(S_+, B\mathbf{G}_m) = \mathrm{Pic}(S)$. Moreover, in the homotopy category of pointed simplicial sheaves $\mathscr{H}_\bullet(S)$, we have a canonical isomorphism $B\mathbf{G}_m \simeq \mathbf{P}_S^\infty$ (*cf. loc. cit.*, prop. 3.7). Thus, finally, we obtain a canonical map of pointed sets

$$\mathrm{Pic}(S) = \mathrm{Hom}_{\mathscr{H}_\bullet^s(S)}(S_+, B\mathbf{G}_m) \longrightarrow \mathrm{Hom}_{\mathscr{H}_\bullet(S)}(S_+, \mathbf{P}^\infty)$$
$$\longrightarrow \mathrm{Hom}_{\mathrm{DM}^{e\!f\!f}(S,\Lambda)}(\Lambda_S^{tr}, \Lambda_S^{tr}(\mathbf{P}^\infty/*)) \longrightarrow \mathrm{Hom}_{\mathrm{DM}^{e\!f\!f}(S,\Lambda)}(\Lambda_S^{tr}, \Lambda_S^{tr}(\mathbf{P}^\infty)).$$

Definition 11.3.2 Consider the above notations. We define the first motivic Chern class as the following composite morphism

$$c_1 : \mathrm{Pic}(S) \longrightarrow \mathrm{Hom}_{\mathrm{DM}^{e\!f\!f}(S,\Lambda)}(\Lambda_S^{tr}, \Lambda_S^{tr}(\mathbf{P}_S^\infty)) \xrightarrow{(c_{1,S})_*} \mathrm{Hom}_{\mathrm{DM}^{e\!f\!f}(S,\Lambda)}(\Lambda_S^{tr}, \Lambda_S^{tr}(1)[2])$$
$$\longrightarrow \mathrm{Hom}_{\mathrm{DM}(S,\Lambda)}(\mathbb{1}_S, \mathbb{1}_S(1)[2]) = H_{\mathscr{M}}^{2,1}(S,\Lambda).$$

The first motivic Chern class is evidently compatible with pullback.

Remark 11.3.3 Beware that the map

$$\mathrm{Pic}(S) \longrightarrow \mathrm{Hom}_{\mathrm{DM}^{\mathit{eff}}(S,\Lambda)}(\Lambda_S^{tr}, \Lambda_S^{tr}(\mathbf{P}_S^\infty))$$

defined above is not necessarily a morphism of abelian groups. However, the composite:

$$\mathrm{Pic}(S) \longrightarrow \mathrm{Hom}_{\mathrm{DM}^{\mathit{eff}}(S,\Lambda)}(\Lambda_S^{tr}, \Lambda_S^{tr}(\mathbf{P}_S^\infty)) \xrightarrow{(c_{1,S})_*} \mathrm{Hom}_{\mathrm{DM}^{\mathit{eff}}(S,\Lambda)}(\Lambda_S^{tr}, \Lambda_S^{tr}(1)[2])$$

is the isomorphism of Theorem 11.2.14 when S is normal. In particular, it is a morphism of abelian groups in this case. We will give an argument below for the general case.

11.3.4 The triangulated category $\mathrm{DM}(S, \Lambda)$ thus satisfies all the axioms of [Dég08, §2.1] (see also Section 2.3.1 of *loc. cit.* in the regular case). In particular, we derive from the main results of *loc. cit.* the following facts:

1. Let $p : P \longrightarrow S$ be a projective bundle of rank n. Let $c : \mathbb{1}_S \longrightarrow \mathbb{1}_S(1)[2]$ be the first Chern class of the canonical line bundle on P. Then the map

$$M_S(P) \xrightarrow{\Sigma_i \, p \otimes c^i} \bigoplus_{i=0}^{n} \mathbb{1}_S(i)[2i]$$

is an isomorphism. This gives the projective bundle theorem in motivic cohomology for any base scheme.

One deduces using the method of Grothendieck that motivic cohomology possesses Chern classes of vector bundles which satisfy all the usual properties (see the remark below for additivity).

2. Let $i : Z \longrightarrow X$ be a closed immersion between smooth separated S-schemes of finite type. Assume i has pure codimension c and let j be the complementary open immersion. Then there is a canonical *purity isomorphism*:

$$\mathfrak{p}_{X,Z} : M_S(X/X - Z) \longrightarrow M_S(Z)(c)[2c].$$

This defines in particular the *Gysin triangle*

$$M_S(X - Z) \xrightarrow{j_*} M_S(X) \xrightarrow{i^*} M_S(Z)(c)[2c] \xrightarrow{\partial_{X,Z}} M_S(X - Z)[1].$$

3. Let $f : Y \longrightarrow X$ be a projective morphism between smooth separated S-schemes of finite type. Assume f has pure relative dimension d. Then there is an associated *Gysin morphism*

$$f^* : M_S(X) \longrightarrow M_S(Y)(d)[2d]$$

functorial in f. We refer the reader to *loc. cit.* for various formulas involving the Gysin morphism (projection formula, excess intersection,...)

Note in particular that we deduce from this Gysin morphism the following map in motivic cohomology:

$$f_* : H_{\mathcal{M}}^{n,i}(Y, \Lambda) \longrightarrow H_{\mathcal{M}}^{n+2d, i+d}(X, \Lambda).$$

4. For any smooth projective S-scheme X, the premotive $M_S(X)$ admits a *strong dual*. If X has pure relative dimension d over S, the strong dual of $M_S(X)$ is $M_S(X)(-d)[-2d]$.

Remark 11.3.5 According to *loc. cit.*, there exists for any scheme S a formal group law $F_S(x, y)$ with coefficients in the graded ring $H_{\mathcal{M}}^{2*,*}(S, \Lambda)$. If one consider the Segre embedding

$$\Sigma : \mathbf{P}_S^\infty \longrightarrow \mathbf{P}_S^\infty \times_S \mathbf{P}_S^\infty$$

one has: $F_S(x, y) = \sigma^*(1)$ through the isomorphism:

$$H_{\mathcal{M}}^{2*,*}(\mathbf{P}_S^\infty \times_S \mathbf{P}_S^\infty, \Lambda) \simeq H_{\mathcal{M}}^{2*,*}(S, \Lambda)[[x, y]],$$

which results from the projective bundle formula in motivic cohomology.

According to Remark 11.3.3, whenever S is normal, one gets $F_S(x, y) = x + y$. In particular, $F_{\mathrm{Spec}(\mathbf{Z})}(x, y) = x + y$. On the other hand, according to the above definition of $F_S(x, y)$, $F_S(x, y)$ is compatible with pullback. Thus one deduces that $F_S(x, y) = x + y$ for any scheme S.

11.3.6 According to the properties that we have previously proved, motivic cohomology, and in particular the bigraded part $H_{\mathcal{M}}^{2n,n}(X, \mathbf{Z})$, possesses many of the desired properties of a generalized Chow theory for regular schemes (see [BGI71, XIV, §8]).

Note in particular that the existence of Chern classes allows us to define a Chern character:

$$K_0(X) \otimes_{\mathbf{Z}} \mathbf{Q} \xrightarrow{ch} H_{\mathcal{M}}^{2*,*}(X, \mathbf{Z}) \otimes \mathbf{Q} \simeq H_{\mathcal{M}}^{2*,*}(X, \mathbf{Q}),$$

where the final isomorphism follows from Section 11.2.2. In particular, we will prove in the next section (Corollary 16.1.7) that, when X is regular, this map is an isomorphism, as expected.

Remark 11.3.7 Among the good properties of motivic cohomology is the fact that it is defined, with its ring structure and natural functoriality, for arbitrary schemes. On the other hand, even when X is regular, one cannot presently describe the cohomology group $H_{\mathcal{M}}^{2n,n}(X, \mathbf{Z})$ in terms of classes of n-codimensional cycles in X modulo an appropriate equivalence relation.

Let us however mention the two following interesting facts:

1. Let X be a scheme of finite type over $\mathrm{Spec}\,(\mathbf{Z})$ and X_p be its fiber over a primer p. Then one has a pullback map:

$$H_{\mathcal{M}}^{2n,n}(X, \mathbf{Z}) \longrightarrow H_{\mathcal{M}}^{2n,n}(X_p, \mathbf{Z}), \sigma \longmapsto \sigma_p.$$

When X is an arithmetic scheme (regular and flat over $\mathbf{Z}$) with good reduction at p, the target is the Chow group of n-codimensional cycles (see Example 11.2.3).

11 Motivic complexes

Then σ_p should be thought of as the specialization of its generic fiber (which lies in $H_{\mathcal{M}}^{2n,n}(X_{\mathbf{Q}}, \mathbf{Z}) = CH^n(X_{\mathbf{Q}})$ according to Example 11.2.3). This construction should coincide with other specialization maps in the arithmetic case (see for example [Ful98, §20.3]).

2. Let X be a smooth S-scheme. Then any n-codimensional closed subscheme Z of X which is smooth over S defines, using the Gysin morphism, an element

$$[Z] = i_*(1) \in H_{\mathcal{M}}^{2n,n}(X, \mathbf{Z})$$

which should be called the fundamental class of X. One can extract from [Dég08] some expected properties of these fundamental classes (e.g. the relation to Chern classes, pullback properties such as compatibility with base change).

In particular, any S-point of X defines an element of $H_{\mathcal{M}}^{2d,d}(X, \mathbf{Z})$ where d is the dimension of X (assumed of pure dimension). In particular, the group $H_{\mathcal{M}}^{2d,d}(X, \mathbf{Z})$ is close to a group of cycles in X of relative dimension 0 over S.

11.3.8 We conclude this series of remarks on motivic cohomology with the following construction that the reader might enjoy.

Let S be any scheme and $\mathscr{P}_S$ be the category of smooth projective S-schemes. Given any scheme X and Y in $\mathscr{P}_S$, one can use the group

$$H_{\mathcal{M}}^{2d,d}(X \times_S Y, \Lambda),$$

where d is the relative dimension of Y as a group of correspondences using the properties obtained so far from motivic cohomology. In particular, one can mimic the construction of the category of Chow motives over a field k using the category $\mathscr{P}_S$ and these correspondences. One obtains an additive monoidal category $\mathrm{Chow}'(S, \Lambda)$ of *strong Chow motives*.

According to the duality property of motives (Section 11.3.4, point 4) one also obtains a canonical isomorphism

$$\mathrm{Hom}_{\mathrm{DM}(S,\Lambda)}(M_S(X), M_S(Y)) = H_{\mathcal{M}}^{2d,d}(X \times_S Y, \Lambda).$$

Thus one deduces a canonical full embedding of monoidal categories:

$$\mathrm{Chow}'(S, \Lambda) \longrightarrow \mathrm{DM}_{gm}(S, \Lambda),$$

which extends the well-known case when S is a perfect field.

Remark 11.3.9 Beware that, with rational coefficients, a sharper notion of Chow motives – in more precise terms, these are motives *of weight zero* – has been introduced recently (see [Héb11], [Bon14]).

11.4 The six functors

11.4.1 Recall that according to Definition 10.4.2 and Section 10.4.3, we have an adjunction of abelian premotivic categories

$$\gamma^* : \mathrm{Sh}(-,\Lambda) \rightleftarrows \mathrm{Sh}^{tr}(-,\Lambda) : \gamma_*$$

such that γ_* is exact and conservative. Moreover, for any scheme S, any smooth S-schemes X, Y and any open immersion $j : U \longrightarrow X$, the canonical map:

$$j_* : c_S(Y,U) \longrightarrow c_S(Y,X)$$

is obviously a monomorphism. Thus the abelian premotivic category $\mathrm{Sh}^{tr}(-,\Lambda)$ satisfies the assumptions (i)–(iv) of Section 6.3.1. In particular, we deduce from Corollaries 6.3.12 and 6.3.15 the following theorem:

Proposition 11.4.2 *The premotivic triangulated category* DM_Λ *satisfies the support property. Moreover, for any scheme S and any closed immersion* $i : Z \longrightarrow X$ *between smooth S-schemes,* DM_Λ *satisfies the localization property with respect to i, (Loc_i).*

An important corollary of this proposition is that given any separated morphism $f : Y \longrightarrow X$ of finite type, one can construct an adjunction of triangulated categories:

$$f_! : \mathrm{DM}(Y,\Lambda) \rightleftarrows \mathrm{DM}(X,\Lambda) : f^!$$

such that $f_! = f_*$ when f is proper (see Section 2.2). We will elaborate on this fact at the end of this section.

11.4.3 Note that in particular, the premotivic category DM_Λ satisfies the weak localization property (wLoc). According to the premotivic adjunction (11.2.16.2) and the existence of the first Chern class in motivic cohomology (Definition 11.3.2), one can apply Example 2.4.40 to the premotivic triangulated category DM_Λ (which satisfies the Nisnevich separation property by construction). This implies in particular that DM_Λ is oriented as a premotivic triangulated category (Definition 2.4.38).

In particular, one can apply Corollary 2.4.43 to DM_Λ and get the following result:

Proposition 11.4.4 *Any smooth projective morphism f is* DM_Λ-*pure: the canonical purity map (2.4.39.3)*

$$f_\# \longrightarrow f_!(d)[2d]$$

is an isomorphism, where d is the relative dimension of f.

In particular, DM_Λ is weakly pure. The only property of the premotivic triangulated category DM_Λ that we cannot prove is the localization property for general closed immersions. However, the properties we have seen so far allow us to construct the six operations and establish some further properties that are already of interest. Let us summarize this formalism, from Theorem 2.2.14 together with Lemma 2.4.23:

11 Motivic complexes

Theorem 11.4.5 *For any separated morphism of finite type $f : Y \longrightarrow X$, there exists an essentially unique pair of adjoint functors*

$$f_! : \mathrm{DM}(Y, \Lambda) \rightleftarrows \mathrm{DM}(X, \Lambda) : f^!$$

such that:

1. *There exists a structure of a covariant (resp. contravariant) 2-functor on $f \mapsto f_!$ (resp. $f \mapsto f^!$).*
2. *There exists a natural transformation $\alpha_f : f_! \longrightarrow f_*$ which is an isomorphism when f is proper. Moreover, α is a morphism of 2-functors.*
3. *For any smooth projective morphism $f : X \longrightarrow S$ of relative dimension d, there are canonical natural isomorphisms*

$$\mathfrak{p}_f^! : f_\sharp \longrightarrow f_!(d)[2d],$$
$$\mathfrak{p}_f^{\prime !} : f^* \longrightarrow f^!(-d)[-2d],$$

which are dual to each other.
4. *For any cartesian square:*

$$\begin{array}{ccc} Y' & \xrightarrow{f'} & X' \\ g' \downarrow & \Delta & \downarrow g \\ Y & \xrightarrow{f} & X, \end{array}$$

such that f is separated of finite type, there exist natural transformations

$$g^* f_! \xrightarrow{\sim} f'_! g'^*,$$
$$g'_* f'^! \xrightarrow{\sim} f^! g_*,$$

which are isomorphisms in the following cases:
   - *g is smooth;*
   - *f is projective and smooth.*
5. *For any smooth projective morphism $f : Y \longrightarrow X$, there exist natural isomorphisms*

$$Ex(f_1^*, \otimes) : (f_! K) \otimes_X L \xrightarrow{\sim} f_!(K \otimes_Y f^* L),$$
$$Hom_X(f_!(L), K) \xrightarrow{\sim} f_* Hom_Y(L, f^!(K)),$$
$$f^! Hom_X(L, M) \xrightarrow{\sim} Hom_Y(f^*(L), f^!(M)).$$

Remark 11.4.6 As an example of application, let us recall the construction of the general trace map (from [AGV73]) in the case of a smooth projective morphism $f : Y \longrightarrow X$ of relative dimension d. It is the following composite map:

$$f_*f^* \xrightarrow{\alpha_f^{-1}} f_! f^* \xrightarrow{p_f'^t} f_! f^!(d)[2d] \xrightarrow{ad'(f_!,f^!)} 1(d)[2d].$$

This allows one to recover the Gysin map associated with f, already constructed in Section 11.3.4, as well as the duality property for the motive $M_X(Y)$.

We will reformulate Voevodsky's conjecture 11.2.22 in terms of the six operations as follows.

Proposition 11.4.7 *We fix a base scheme S as well as a ring of coefficients Λ. The following assertions are equivalent:*

(i) for any S-schemes X and Y and any morphism of finite type $f : X \longrightarrow Y$, the canonical map $\tau_f : f^*(H^\Lambda_{\mathscr{M},X}) \longrightarrow H^\Lambda_{\mathscr{M},Y}$ is invertible;
(ii) for any S-scheme X, the canonical functor

$$\mathrm{Ho}(H^\Lambda_{\mathscr{M},X}\text{-mod}) \longrightarrow \mathrm{DM}(X,\Lambda)$$

is an equivalence of categories, and $\mathrm{DM}(-,\Lambda)$ is a motivic category over S-schemes;
(iii) the premotivic category $\mathrm{DM}(-,\Lambda)$ has the localization property for S-schemes;
(iv) $\mathrm{DM}(-,\Lambda)$ is a motivic category over S-schemes.

Proof The fact that properties (iii) and (iv) are equivalent is obvious, since the only missing property that is not known for DM_Λ to be a motivic category is the localization property. Condition (iv) is obviously a consequence of condition (ii).

Keeping track of the notations introduced in Section 11.2.16, we shall observe that the forgetful functor

$$\varphi_* : \mathrm{DM}_\Lambda \longrightarrow \mathrm{SH}$$

commutes with the operator $j_\sharp$, for any open immersion j, as follows. Since the forgetful functor from $\mathrm{D}_{\mathbf{A}^1,\Lambda}$ to SH is conservative and commutes with $j_\sharp$ for any open immersion j, it is sufficient to prove that the functor $\mathbf{R}\gamma_* : \mathrm{DM}_\Lambda \longrightarrow \mathrm{D}_{\mathbf{A}^1,\Lambda}$ has the same property, which is precisely Proposition 6.3.11.

Let us check that condition (i) (i.e. Voevodsky's conjecture 11.2.22) is a consequence of condition (iv). Let us assume that (iv) holds true, and that we have a morphism of finite type $f : X \longrightarrow Y$. The property that the canonical map

$$\tau_f : f^*(H^\Lambda_{\mathscr{M},X}) \longrightarrow H^\Lambda_{\mathscr{M},Y}$$

is invertible is local for the Zariski topology on X and on Y, so that we may assume that f is affine. Since the map τ_f is invertible for f smooth, we observe from there that it is sufficient to prove that τ_f is invertible when f is a closed immersion. Let $j : U \longrightarrow Y$ be the open immersion complement to f. Assuming (iv), there is a homotopy cofiber sequence of the form

$$j_\sharp \mathbb{1}_U \longrightarrow \mathbb{1}_Y \longrightarrow f_* \mathbb{1}_X$$

in $\mathrm{DM}(Y,\Lambda)$, the image of which is isomorphic to the homotopy cofiber sequence

11 Motivic complexes

$$j_\sharp H^\Lambda_{\mathcal{M},U} \longrightarrow H^\Lambda_{\mathcal{M},Y} \longrightarrow f_* H^\Lambda_{\mathcal{M},X}$$

in SH(Y), since the functor φ_* commutes with $j_\sharp$ (as recalled above) and with f_* (for obvious reasons). But the localization property in SH implies that the homotopy cofiber of the map $j_\sharp H^\Lambda_{\mathcal{M},U} \longrightarrow H^\Lambda_{\mathcal{M},Y}$ is $f_* f^* H^\Lambda_{\mathcal{M},Y}$. Since the functor f_* is conservative in SH (being fully faithful), this shows that the map τ_f is invertible.

Let us assume that condition (i) is true. Since the forgetful functor φ_* is conservative and commutes with i_* for any closed immersion i, in order to prove that condition (iv) holds, i.e. that DM_Λ has the localization property, it is sufficient to prove that condition (ii) of Corollary 2.3.18 is satisfied in DM_Λ. We observe furthermore that, for any smooth and projective morphism of S-schemes $p : X \longrightarrow Y$ everywhere of relative dimension d, the functor φ_* commutes with $p_\sharp$. Indeed, for any object M in $\mathrm{DM}(X,\Lambda)$, we have:

$$\begin{aligned}\varphi_* p_\sharp(M) &\simeq \varphi_* p_*(M)(d)[2d] \\ &\simeq p_* \varphi_*(M)(d)[2d] \\ &\simeq p_!(Th_X(T_f) \otimes \varphi_*(M)) \\ &\simeq p_\sharp \varphi_*(M)\end{aligned}$$

(where the identification $Th_X(T_f) \otimes \varphi_*(M) \simeq \varphi_*(M)(d)[2d]$ comes from the orientation on $\varphi_*(M)$ induced by its $H^\Lambda_{\mathcal{M},X}$-module structure). This implies that the functor φ_* commutes with $f_\sharp$ for any smooth morphism of S-schemes f. Indeed, this is a local condition with respect to the Zariski topology both on the source and on the target of f, so that it is sufficient to check the case where f is quasi-projective. Since the case where f is an open immersion is already known, and since we just discussed the case where f is a smooth and projective, this proves our claim. Finally, we observe that, given a closed immersion $i : Z \longrightarrow X$ as well as a smooth morphism $f : Y \longrightarrow X$, the diagram

$$g_\sharp H^\Lambda_{\mathcal{M}, f^{-1}(X-Z)} \longrightarrow f_\sharp H^\Lambda_{\mathcal{M},Y} \longrightarrow i_* i^* H^\Lambda_{\mathcal{M},Y}$$

is a homotopy cofiber sequence, where $g : f^{-1}(X - Z) \longrightarrow X$ is the restriction of f. Since the functor φ_* is conservative and commutes with $f_\sharp$, $g_\sharp$ and i_*, this proves that DM_Λ has the localization property, by Corollary 2.3.18.

If condition (i) is true, then, by virtue of Proposition 7.2.13, there is a morphism of premotivic categories

$$\alpha^* : \mathrm{Ho}(H^\Lambda_{\mathcal{M}}\text{-}\mathrm{mod}) \rightleftarrows \mathrm{DM}_\Lambda : \alpha_* \, .$$

Furthermore, Proposition 7.2.18 implies that, under condition (i), $\mathrm{Ho}(H^\Lambda_{\mathcal{M}}\text{-}\mathrm{mod})$ is a motivic category (in particular, has the localization property). We just saw that DM_Λ is a motivic category as well. To prove that the functor α^* is an equivalence of categories, by virtue of Corollary 1.3.20, it is sufficient to prove that, for any smooth morphism $f : X \longrightarrow Y$, the unit map

$$f_\sharp H^\Lambda_{\mathscr{M},X} \longrightarrow \alpha_* \alpha^* f_\sharp H^\Lambda_{\mathscr{M},X} \simeq \alpha_* f_\sharp \alpha^* H^\Lambda_{\mathscr{M},X}$$

is invertible. Since the operators α_* and $f_\sharp$ commute (when we forget the $H^\Lambda_{\mathscr{M},X}$-module structure, α_* is just φ_*), it is sufficient to check this property when f is the identity. But the map

$$H^\Lambda_{\mathscr{M},X} \longrightarrow \alpha_* \alpha^* H^\Lambda_{\mathscr{M},X}$$

is invertible (in fact the identity), by definition. $\square$

Remark 11.4.8 A variant of Voevodsky's conjecture would be that the map $\tau_f : f^*(H^\Lambda_{\mathscr{M},X}) \longrightarrow H^\Lambda_{\mathscr{M},Y}$ is invertible for *regular* S-schemes. We invite the reader to check that this version of the conjecture may be reformulated as in Proposition 11.4.7 (restricting ourselves to regular S-schemes, obviously), essentially with the same proof. Evidence for this weaker form of the conjecture is given by the fact that over any field of exponent characteristic p, it is true with $\Lambda = \mathbf{Z}[1/p]$; see [CD15]. A variant consists in replacing DM_Λ by its *cdh*-local version. In equal characteristic zero, this is proved for possibly singular schemes in [CD15] (in characteristic $p > 0$, this also holds up to p-torsion). The *cdh*-local version of $H^\Lambda_{\mathscr{M}}$ should be isomorphic to Spitzweck's motivic cohomology spectrum [Spi18].

Part IV
Beilinson motives and algebraic K-theory

12 Stable homotopy theory of schemes

12.1 Ring spectra

Consider a base scheme S.
Recall that a *ring spectrum* E over S is a monoid object in the monoidal category $\operatorname{SH}(S)$. We say that E is *commutative* if it is commutative as a monoid in the symmetric monoidal category $\operatorname{SH}(S)$. In what follows, we will assume that all our ring spectra are commutative without mentioning it.

The premotivic category is $\mathbf{Z}^2$-graded where the first index refers to the simplicial sphere and the second one to the Tate twist. According to our general convention, a cohomology theory representable in SH is $\mathbf{Z}^2$-graded accordingly: given such a ring spectrum E, for any smooth S-scheme X, and any integers $(i, n) \in \mathbf{Z}^2$, we get a bigraded ring:
$$E^{n,i}(X) = \operatorname{Hom}_{\operatorname{SH}(S)}\left(\Sigma^{\infty} X_+, E(i)[n]\right).$$
When X is a pointed smooth S-scheme, it defines a pointed sheaf of sets still denoted by X and we denote by $\tilde{E}^{n,i}(X)$ the corresponding cohomology ring.

The *coefficient ring* associated with E is the cohomology of the base $E^{**} := E^{**}(S)$. The ring $E^{**}(X)$ (resp. $\tilde{E}^{**}(X)$) is in fact an E^{**}-algebra.

12.1.1 We say E is a *strict ring spectrum* if there exists a monoid in the category of symmetric Tate spectra E' and an isomorphism of ring spectra $E \simeq E'$ in $\operatorname{SH}(S)$. In this case, a module M over the monoid E in the monoidal category $\operatorname{SH}(S)$ will be said to be *strict* if there exists an E'-module M' in the category of symmetric Tate spectra, as well as an isomorphism of E-modules $M \simeq M'$ in $\operatorname{SH}(S)$.

12.2 Orientation

12.2.1 Consider the infinite tower
$$\mathbf{P}_S^1 \longrightarrow \mathbf{P}_S^2 \longrightarrow \cdots \longrightarrow \mathbf{P}_S^n \longrightarrow \cdots$$
of schemes pointed at infinity. We denote by $\mathbf{P}_S^{\infty}$ the limit of this tower as a pointed Nisnevich sheaf of sets and by $\iota : \mathbf{P}_S^1 \longrightarrow \mathbf{P}_S^{\infty}$ the induced inclusion. Recall the following definition, classically translated from topology:

Definition 12.2.2 Let E be a ring spectrum over S. An *orientation* of E is a cohomology class c in $\tilde{E}^{2,1}(\mathbf{P}_S^{\infty})$ such that $\iota^*(c)$ is sent to the unit of the coefficient ring of E by the canonical isomorphism $\tilde{E}^{2,1}(\mathbf{P}_S^1) = E^{0,0}$.

We then say that (E, c) is an *oriented ring spectrum*. We shall also say that E is *orientable* if there exists an orientation c.

According to [MV99, 1.16 and 3.7], we get a canonical map for any smooth S-scheme X

$$\mathrm{Pic}(X) = H^1(X, \mathbf{G}_m) \longrightarrow \mathrm{Hom}_{\mathcal{H}_\bullet(S)}(X_+, \mathbf{P}^\infty) \longrightarrow \mathrm{Hom}_{\mathrm{SH}(S)}(\Sigma^\infty X_+, \Sigma^\infty \mathbf{P}^\infty)$$

(the first map is an isomorphism whenever S is regular (or even geometrically unibranch)). Given this map, an orientation c of a ring spectrum E defines a map of sets

$$c_{1,X} : \mathrm{Pic}(X) \longrightarrow E^{2,1}(X)$$

which is natural in X (and from its construction in [MV99], one can check that $c = c_{1,\mathbf{P}_S^\infty}(\mathcal{O}(1))$). Usually, we put $c_1 = c_{1,X}$.

Example 12.2.3

1. The original example of an oriented ring spectrum is the algebraic cobordism spectrum MGL introduced by Voevodsky (cf. [Voe98]).
2. According to Definition 11.3.2, the motivic cohomology ring spectrum $H^\Lambda_{\mathcal{M},S}$ defined in Definition 11.2.17 is an oriented ring spectrum.
3. Consider a triangulated premotivic category $\mathcal{T}$ which satisfies the weak localization property (wLoc) and such that there exists an adjunction of triangulated premotivic categories:

$$\varphi^* : \mathrm{SH} \rightleftarrows \mathcal{T} : \varphi_*.$$

Recall that φ^* is symmetric monoidal. Thus, its right adjoint is weakly symmetric monoidal and for the spectrum

$$H_{\mathcal{T},S} := \varphi_*(\mathbb{1}_S)$$

admits a (commutative) ring structure.

Then $\mathcal{T}$ is oriented in the sense of Definition 2.4.38 if and only if the ring spectrum $H_{\mathcal{T},S}$ is oriented in the sense of Definition 12.2.2 – see Example 2.4.40. Moreover, an orientation of $\mathcal{T}$ is equivalent to the data of orientations $H_{\mathcal{T},S}$ for any scheme S which is stable under pullbacks (on cohomology).

Remark 12.2.4 When E is a strict ring spectrum, the category E-mod satisfies the axioms of [Dég08, 2.1] (see example 2.12 of *loc. cit.*).

Recall the following result, which first appeared in [Vez01]:

Proposition 12.2.5 (Morel) *Let (E,c) be an oriented ring spectrum. Then:*

$$E^{**}(\mathbf{P}_S^\infty) = E^{**}[[c]] \quad \text{and} \quad E^{**}(\mathbf{P}_S^\infty \times \mathbf{P}_S^\infty) = E^{**}[[x,y]],$$

where x (resp. y) is the pullback of c along the first (resp. second) projection.

Remark 12.2.6 When E is a strict ring spectrum, this is [Dég08, 3.2] according to Remark 12.2.4. The proof follows an argument of Morel ([Dég08, lemma 3.3]) and the arguments of *op. cit.*, p. 634, can be easily used to obtain the proposition arguing directly for the cohomology functor $X \mapsto E^{*,*}(X)$.

12 Stable homotopy theory of schemes

12.2.7 Recall that the Segre embedding

$$\mathbf{P}_S^n \times \mathbf{P}_S^m \longrightarrow \mathbf{P}_S^{n+m+nm}$$

defines a map

$$\sigma : \mathbf{P}_S^\infty \times \mathbf{P}_S^\infty \longrightarrow \mathbf{P}_S^\infty.$$

It gives the structure of an H-group to $\mathbf{P}_S^\infty$ in the homotopy category $\mathscr{H}(S)$. Consider the hypotheses of the previous proposition. Then the pullback along σ in E-cohomology induces a map

$$E^{**}[[c]] \xrightarrow{\sigma^*} E^{**}[[x,y]]$$

and following Quillen, we check that the formal power series $\sigma^*(c)$ defines a formal group law over the ring E^{**}.

Definition 12.2.8 Let (E, c) be an oriented ring spectrum and consider the previous notation.

The formal group law $F_E(x, y) := \sigma^*(c)$ will be called the formal group law associated to (E, c).

Recall the formal group law has the form:

$$F_E(x, y) = x + y + \sum_{i+j>0} a_{ij} . x^i y^j$$

with $a_{ij} = a_{ji}$ in $E^{-2i-2j,-i-j}$.

The coefficients a_{ij} describe the failure of additivity of the first Chern class c_1. Indeed, assuming the previous definition, we get the following result:

Proposition 12.2.9 *Let X be a smooth S-scheme.*

1. *For any line bundle L/X, the class $c_1(L)$ is nilpotent in $E^{**}(X)$.*
2. *Suppose X admits an ample line bundle. For any line bundles L, L' over X,*

$$c_1(L_1 \otimes L_2) = F_E(c_1(L_1), c_1(L_2)) \in E^{2,1}(X).$$

We refer to [Dég08, 3.8] in the case where E is strict; the proof is the same in the general case.

Recall the following theorem of Vezzosi (cf. [Vez01, 4.3]):

Theorem 12.2.10 (Vezzosi) *Let (E, c) be an oriented spectra over S, with formal group law F_E. Then there exists a bijection between the following sets:*

(i) Orientation classes c' of E.
(ii) Morphisms of ring spectra $MGL \longrightarrow E$ in $SH(S)$.
(iii) Pairs (F, φ) where F is a formal group law over E^{**} and φ is a power series over E^{**} which defines an isomorphism of formal group laws: φ is invertible as a power series and $F_E(\varphi(x), \varphi(y)) = F(x, y)$.

12.3 Rational category

In what follows, we shall frequently use the equivalence of premotivic categories (see Section 5.3.35)
$$\mathrm{SH}_\mathbf{Q} \rightleftarrows D_{\mathbf{A}^1,\mathbf{Q}},$$
and shall identify freely any rational spectrum over a scheme S with an object of $D_{\mathbf{A}^1}(S,\mathbf{Q})$.

13 Algebraic K-theory

13.1 The K-theory spectrum

We consider the spectrum KGL_S which represents homotopy invariant K-theory in SH(S) according to Voevodsky (see [Cis13], [Voe98, 6.2], [Rio10, 5.2] and [PPR09]). It is characterized by the following properties:

(K1) For any morphism of schemes $f : T \longrightarrow S$, there is an isomorphism $f^*KGL_S \simeq KGL_T$ in SH(T).

(K2) For any regular scheme S and any integer n, there exists an isomorphism
$$\mathrm{Hom}_{\mathrm{SH}(S)}(\mathbb{1}_S[n], KGL_S) \longrightarrow K_n(S)$$
(where the right-hand side is Quillen's algebraic K-theory as defined by Thomason and Trobaugh, [TT90], in the case where S does not admit an ample family) such that, for any morphism $f : T \longrightarrow S$ of regular schemes, the following diagram is commutative:

$$\begin{array}{ccccc}
\mathrm{Hom}(\mathbb{1}_S[n], KGL_S) & \longrightarrow & \mathrm{Hom}(f^*\mathbb{1}_S[n], f^*KGL_S) & = & \mathrm{Hom}(\mathbb{1}_T[n], KGL_T) \\
\downarrow & & & & \downarrow \\
K_n(S) & & \xrightarrow{\quad f^* \quad} & & K_n(T)
\end{array}$$

(where the lower horizontal map is the pullback in Quillen's algebraic K-theory along the morphism f and the upper horizontal map is obtained by using the functor $f^* : \mathrm{SH}(S) \longrightarrow \mathrm{SH}(T)$ and the identification (K1)).

(K3) For any scheme S, there exists a unique structure of a commutative monoid on KGL_S which is compatible with base change – using the identification (K1) – and induces the canonical ring structure on $K_0(S)$.

Thus, according to (K1) and (K3), the collection of ring spectra KGL_S for schemes S form an absolute ring spectrum. As usual, when no confusion can arise, we will not indicate the base in the notation KGL.

13 Algebraic K-theory

Note that (K1) implies formally that the isomorphism of (K2) can be extended for any smooth S-scheme X (with S regular), giving a natural isomorphism:

$$\mathrm{Hom}_{\mathrm{SH}(S)}(\Sigma^\infty X_+[n], KGL) \longrightarrow K_n(X).$$

13.2 Periodicity

13.2.1 Recall from the construction the following property of the spectrum KGL:

(K4) the spectrum KGL is a $\mathbf{P}^1$-periodic spectrum in the sense that there exists a canonical isomorphism

$$KGL \xrightarrow{\sim} \mathbf{R}Hom\left(\Sigma^\infty \mathbf{P}^1_S, KGL\right) = KGL(-1)[-2].$$

As usual, $\mathbf{P}^1_S$ is pointed by the infinite point.

This isomorphism, classically called the Bott isomorphism, is characterized uniquely by the fact that its adjoint isomorphism (obtained by tensoring with $\mathbb{1}_S(1)[2]$) is equal to the composite

(13.2.1.1) $$\gamma_u : KGL(1)[2] \xrightarrow{1 \otimes u} KGL \wedge KGL \xrightarrow{\mu} KGL,$$

where $u : \Sigma^\infty \mathbf{P}^1 \longrightarrow KGL$ corresponds to the class $[\mathcal{O}(1)] - 1$ in $\tilde{K}_0(\mathbf{P}^1)$ through the isomorphism (K2) and μ is the structural map of monoid from (K3).

Using the isomorphism of (K4), the property (K1) can be extended as follows: for any smooth S-scheme X and any integers $(i,n) \in \mathbf{Z}^2$, there is a canonical isomorphism:

(13.2.1.2) $$KGL^{n,i}(X) \xrightarrow{\sim} K_{2i-n}(X).$$

Remark 13.2.2 The element u is invertible in the ring $KGL^{*,*}(S)$. Its inverse is the Bott element $\beta \in KGL^{2,1}(S)$. If we choose as an orientation of the ring spectrum KGL (cf. Definition 12.2.2) the class

$$\beta.([\mathcal{O}(1)] - 1) \in KGL^{2,1}(\mathbf{P}^\infty),$$

the corresponding formal group law is the multiplicative formal group law:

$$F(x,y) = x + y + \beta^{-1}.xy.$$

13.3 Modules over algebraic K-theory

Theorem 13.3.1 (Østvær, Röndigs, Spitzweck) *The spectrum KGL can be represented canonically by a cartesian monoid KGL', as well as by a homotopy cartesian commutative monoid KGL^β in the fibred model category of symmetric $\mathbf{P}^1$-spectra, in such a way that there exists a morphism of monoids $KGL' \longrightarrow KGL^\beta$ which is a termwise stable $\mathbf{A}^1$-equivalence.*

Proof For any noetherian scheme of finite dimension S, one has a strict commutative ring spectrum KGL_S^β which is canonically isomorphic to KGL_S in $\mathrm{SH}(S)$ as ring spectra; see [RSØ10]. One can check that the objects KGL_S^β do form a commutative monoid over the diagram of all noetherian schemes of finite dimension (i.e. a commutative monoid in the category of sections of the fibred category of $\mathbf{P}^1$-spectra over the category of noetherian schemes of finite dimension), either by hand, by following the explicit construction of *loc. cit.*, or by modifying its construction very slightly as follows: one can perform *mutatis mutandis* the construction of *loc. cit.* in the $\mathbf{P}^1$-stabilization of the $\mathbf{A}^1$-localization of the model category of Nisnevich simplicial sheaves over (any essentially small adequate subcategory of) the category of all noetherian schemes of finite dimension, and get an object KGL^β, whose restriction to each of the categories Sm/S is the object KGL_S^β. From this point of view, we clearly have canonical maps $f^*(KGL_S^\beta) \longrightarrow KGL_T^\beta$ for any morphism of schemes $f : T \longrightarrow S$. The object KGL^β is homotopy cartesian, as the composed map

$$\mathbf{L}f^*(KGL_S) \simeq \mathbf{L}f^*(KGL_S^\beta) \longrightarrow f^*(KGL_S^\beta) \longrightarrow KGL_T^\beta \simeq KGL_T$$

is an isomorphism in $\mathrm{SH}(T)$. Consider now a cofibrant resolution

$$KGL'_{\mathrm{Spec}(\mathbf{Z})} \longrightarrow KGL^\beta_{\mathrm{Spec}(\mathbf{Z})}$$

in the model category of monoids of the category of symmetric $\mathbf{P}^1$-spectra over $\mathrm{Spec}(\mathbf{Z})$; see Theorem 7.1.3. Then, we define, for each noetherian scheme of finite dimension S, the $\mathbf{P}^1$-spectrum KGL'_S as the pullback of $KGL'_{\mathrm{Spec}(\mathbf{Z})}$ along the map $f : S \longrightarrow \mathrm{Spec}(\mathbf{Z})$. As the functor f^* is a left Quillen functor, the object KGL'_S is cofibrant (both as a monoid and as a $\mathbf{P}^1$-spectrum), so that we get, by construction, a termwise cofibrant cartesian strict $\mathbf{P}^1$-ring spectrum KGL', as well as a morphism $KGL' \longrightarrow KGL^\beta$ which is a termwise stable $\mathbf{A}^1$-equivalence. □

13.3.2 For each noetherian scheme of finite dimension S, one can consider the model categories of modules over KGL'_S and KGL_S^β, respectively; see Theorem 7.2.2. The change of scalars functor along the stable $\mathbf{A}^1$-equivalence $KGL'_S \longrightarrow KGL_S^\beta$ defines a left Quillen equivalence, whence an equivalence of homotopy categories:

$$\mathrm{Ho}(KGL'_S\text{-}\mathrm{mod}) \simeq \mathrm{Ho}(KGL_S^\beta\text{-}\mathrm{mod}).$$

13 Algebraic K-theory

Definition 13.3.3 We define the premotivic triangulated category of KGL-modules over $\mathscr{S}$ as the fibred triangulated category whose fiber over a scheme S in $\mathscr{S}$ is defined as:
$$KGL\text{-}\mathrm{mod}(S) := \mathrm{Ho}(KGL_S^\beta\text{-}\mathrm{mod}).$$

13.3.4 By definition, for any smooth S-scheme X, we have a canonical isomorphism
$$\mathrm{Hom}_{\mathrm{SH}(S)}(\Sigma^\infty(X_+), KGL[n]) \simeq \mathrm{Hom}_{KGL}(KGL_S(X), KGL[n])$$
(where $KGL_S(X) = KGL_S \wedge_S^{\mathbf{L}} \Sigma^\infty(X_+)$, while Hom_{KGL} stands for $\mathrm{Hom}_{KGL\text{-}\mathrm{mod}(S)}$).

According to (K1) and (K3), for any regular scheme X, we thus get a canonical isomorphism:

(13.3.4.1) $\qquad \epsilon_S : \mathrm{Hom}_{KGL}(KGL_S[n], KGL_S) \xrightarrow{\sim} K_n(S).$

Using Bott periodicity (K4), and the compatibility with base change, this isomorphism can be extended for any smooth S-scheme X and any pair $(n,m) \in \mathbf{Z}^2$:

(13.3.4.2) $\qquad \epsilon_{X/S} : \mathrm{Hom}_{KGL}(KGL_S(X), KGL_S(m)[n]) \xrightarrow{\sim} K_{2m-n}(X).$

Corollary 13.3.5 *The premotivic triangulated category $KGL\text{-}\mathrm{mod}$ forms a motivic category, and the functors*
$$\mathrm{SH}(S) \longrightarrow KGL\text{-}\mathrm{mod}(S), \quad M \mapsto KGL_S \wedge_S^{\mathbf{L}} M$$
for a scheme S in $\mathscr{S}$ define a morphism of motivic categories
$$\mathrm{SH} \longrightarrow KGL\text{-}\mathrm{mod}$$
over the category of noetherian schemes of finite dimension.

Proof This follows from the preceding theorem and from Propositions 7.2.13 and 7.2.18. □

13.4 K-theory with support

13.4.1 Consider a closed immersion $i : Z \longrightarrow S$ with complementary open immersion $j : U \longrightarrow S$. Assume S is regular.

We use the definition of [Gil81, 2.13] for the K-theory of S with support in Z denoted by $K_*^Z(S)$. In other words, we define $K^Z(S)$ as the homotopy fiber of the restriction map
$$\mathbf{R}\Gamma(S, KGL_S) = K(S) \longrightarrow K(U) = \mathbf{R}\Gamma(U, KGL_U),$$
and put: $K_n^Z(S) = \pi_n(K^Z(S))$.

Applying the derived global section functor $\mathbf{R}\Gamma(S,-)$ to the homotopy fiber sequence

(13.4.1.1) $$i_! i^! KGL_S \longrightarrow KGL_S \longrightarrow j_* j^* KGL_S,$$

we get a homotopy fiber sequence

(13.4.1.2) $$\mathbf{R}\Gamma(S, i_! i^! KGL_S) \longrightarrow \mathbf{R}\Gamma(S, KGL_S) \longrightarrow \mathbf{R}\Gamma(U, KGL_S)$$

from which we deduce an isomorphism in the stable homotopy category of S^1-spectra:

(13.4.1.3) $$\mathbf{R}\Gamma(Z, i^! KGL_S) = \mathbf{R}\Gamma(S, i_! i^! KGL_S) \simeq K^Z(S).$$

We thus get the following property:

(K6) There is a canonical isomorphism

$$\mathrm{Hom}_{\mathrm{SH}(S)}\left(\mathbf{1}_S[n], i_! i^! KGL_S\right) \longrightarrow K_n^Z(S)$$

which satisfies the following compatiblities:
(K6a) the following diagram is commutative:

$$\begin{array}{ccccc}
\mathrm{Hom}(\mathbf{1}[n+1], j_*j^* KGL_S) & \longrightarrow & \mathrm{Hom}(\mathbf{1}[n], i_! i^! KGL_S) & \longrightarrow & \mathrm{Hom}(\mathbf{1}[n], KGL_S) \\
\downarrow & & \downarrow & & \downarrow \\
K_{n+1}(U) & \longrightarrow & K_n^Z(S) & \longrightarrow & K_n(S),
\end{array}$$

where the upper horizontal arrows are induced by the localization sequence (13.4.1.1), and the lower row is the canonical sequence of K-theory with support. The extreme left and right vertical maps are the isomorphisms of (K2);
(K6b) for any morphism $f : Y \longrightarrow S$ of regular schemes, $k : T \longrightarrow Y$ the pullback of i along f, the following diagram is commutative:

$$\begin{array}{ccccc}
\mathrm{Hom}(\mathbf{1}[n], i_! i^! KGL_S) & \to & \mathrm{Hom}(f^*\mathbf{1}[n], f^* i_! i^! KGL_S) & \to & \mathrm{Hom}(\mathbf{1}[n], k_! k^! KGL_Y) \\
\downarrow & & & & \downarrow \\
K_n^Z(S) & & \xrightarrow{f^*} & & K_n^T(Y),
\end{array}$$

where the lower horizontal map is given by the functoriality of relative K-theory (induced by the funtoriality of K-theory) and the upper row is obtained using the functor f^* of SH, the canonical exchange morphism $f^* i_! i^! \longrightarrow k_! k^! f^*$ and the identification (K1).

This property can be extended to the motivic category $\mathrm{Ho}(KGL\text{-mod})$ and we get a canonical isomorphism

13 Algebraic K-theory

(13.4.1.4) $\quad \epsilon_i : \mathrm{Hom}_{KGL}(KGL_S[n], i_!i^! KGL_S) \xrightarrow{\sim} K_n^Z(S)$

satisfying the analog of (K6a) and (K6b).

13.5 The fundamental class

13.5.1 Consider a cartesian square of regular schemes

$$\begin{array}{ccc} Z' & \xrightarrow{k} & S' \\ {\scriptstyle g}\downarrow & & \downarrow{\scriptstyle f} \\ Z & \xrightarrow{i} & S \end{array}$$

with i a closed immersion. We will say that this square is *Tor-independent* if Z and S' are Tor-independent over S in the sense of [BGI71, III, 1.5]: for any $i > 0$, $\mathrm{Tor}_i^S(\mathcal{O}_Z, \mathcal{O}_{S'}) = 0.$[91]

In this case, when we assume in addition that all the schemes in the previous square are regular and that i is a closed immersion, we get from [TT90, 3.18][92] the formula

$$f^* i_* = k_* g^* : K_*(Z) \longrightarrow K_*(S')$$

in Quillen K-theory. An important point for us is that there exists a *canonical homotopy* between these morphisms at the level of the Waldhausen spectra.[93] According to the localization theorem of Quillen [Qui73, 3.1], we get:

Theorem 13.5.2 (Quillen) *For any closed immersion* $i : Z \longrightarrow S$ *between regular schemes, there exists a canonical isomorphism*

$$\mathfrak{q}_i : K_n^Z(S) \longrightarrow K_n(Z).$$

Moreover, this isomorphism is functorial with respect to the Tor-independent squares as above, with i a closed immersion and all the schemes regular.

Remark 13.5.3 Under the condition of this theorem, the following diagram is commutative by construction:

$$\begin{array}{c} K_n^Z(S) \\ {\scriptstyle \mathfrak{q}_i}\downarrow \searrow \\ K_n(Z) \xrightarrow[i_*]{} K_n(S) \end{array}$$

where the non-labeled map is the canonical one.

---

[91] For example, when i is a regular closed immersion of codimension 1, this happens if and only if the above square is transversal.

[92] When all the schemes in the square admit ample line bundles, we can refer to [Qui73, 2.11].

[93] In Quillen's proof, one can also trace back a canonical homotopy with the restriction mentioned in the preceding footnote.

Definition 13.5.4 Let $i : Z \longrightarrow S$ be a closed immersion between regular schemes. We define the *fundamental class* associated with i as the morphism of KGL-modules:
$$\eta_i : i_* KGL_Z \longrightarrow KGL_S$$
defined by the image of the unit element 1 through the following morphism:
$$K_0(Z) \xrightarrow{q_i^{-1}} K_0^Z(S) \xrightarrow{\epsilon_i^{-1}} \mathrm{Hom}(KGL_S, i_! i^! KGL_S) = \mathrm{Hom}(i_* KGL_Z, KGL_S).$$

We also denote by $\eta'_i : KGL_Z \longrightarrow i^! KGL_S$ the morphism obtained by adjunction.

Remark 13.5.5 The fundamental class has the following functoriality properties.

(1) By definition, and applying Remark 13.5.3, the composite map
$$KGL_S \longrightarrow i_* i^*(KGL_S) = i_* KGL_Z \xrightarrow{\eta_i} KGL_S$$
corresponds via the isomorphism ϵ_S to $i_*(1) \in K_0(S)$. According to [BGI71, Exp. VII, 2.7], this class is equal to $\lambda_{-1}(N_i)$, where N_i is the conormal sheaf of the regular immersion i.

(2) In the situation of a Tor-independent square as in Section 13.5.1, observe that $f^* \eta_i = \eta_k$ through the canonical exchange isomorphism $f^* i_* = k_* g^*$ — apply the functoriality of ϵ_i from (K6b) and that of q_i.

(3) Using the identification $i^! i_* = 1$, we get $\eta'_i = i^! \eta_i$. Consider a cartesian square as in Section 13.5.1 and assume f is smooth. Then the square is Tor-independent and we get $g^* \eta'_i = \eta'_k$ using the exchange isomorphism $g^* i^! = k^! f^*$.

13.6 Absolute purity for K-theory

Proposition 13.6.1 *For any closed immersion $i : Z \longrightarrow S$ between regular schemes, the following diagram is commutative:*

$$\begin{array}{ccc} \mathrm{Hom}_{KGL}(KGL_Z[n], KGL_Z) & \xrightarrow{\eta'_i} & \mathrm{Hom}_{KGL}(KGL_Z[n], i^! KGL_S) \\ {\scriptstyle \epsilon_Z} \downarrow & (*) & \downarrow {\scriptstyle \epsilon_i} \\ K_n(Z) & \xrightarrow{q_i^{-1}} & K_n^Z(S). \end{array}$$

Proof In this proof, we denote by $[-,-]$ the bifunctor $\mathrm{Hom}_{KGL}(-,-)$.
Step 1: We assume that $i : Z \longrightarrow S$ admits a retraction $p : S \longrightarrow Z$.
Consider a KGL-linear map $\alpha : KGL_Z[n] \longrightarrow KGL_Z$. Then, $\eta'_i(\alpha)$ corresponds by adjunction to the composition
$$i_* KGL_Z[n] \xrightarrow{i_*(\alpha)} i_* KGL_Z \xrightarrow{\eta_i} KGL_S.$$

13 Algebraic K-theory

Applying the projection formula for the motivic category Ho(KGL-mod), we get:

$$i_*(\alpha) = i_*(1 \otimes i^*p^*(\alpha)) = i_*(1) \otimes p^*(\alpha).$$

Here 1 stands for the identity morphism of the KGL-module KGL_Z. This shows that $\eta'_i(\alpha)$ corresponds by adjunction to the composite map:

$$\eta_i \otimes p^*(\alpha) : i_*KGL_Z[n] = i_*KGL_Z[n] \otimes KGL_S \longrightarrow KGL_S \otimes KGL_S = KGL_S$$

(the tensor product is the KGL-linear one). By assumption, $i_* : K_*(Z) \longrightarrow K_*(S)$ admits a retraction, which implies the canonical map $\mathscr{O}_i : K_*^Z(S) \longrightarrow K_*(S)$ admits a retraction (*cf.* Remark 13.5.3). To check that the diagram (∗) is commutative, we can thus compose with $\mathscr{O}_i$.

Recall the first point of Remark 13.5.5: applying property (K6a) and the fact the isomorphism $\epsilon_S : [KGL_S[n], KGL_S] \longrightarrow K_n(S)$ is compatible with the algebra structures, we are finally reduced to proving that

$$i_*(\alpha) = i_*(1).p^*(\alpha) \in K_n(S).$$

This follows from the projection formula in K-theory (see [Qui73, 2.10] and [TT90, 3.17]).

Step 2: We shall reduce the general case to Step 1. We consider the following deformation to the normal cone diagram: let D be the blow-up of $\mathbf{A}_S^1$ in the closed subscheme $\{0\} \times Z$, P be the projective completion of the normal bundle of Z in S and s be the canonical section of P/Z; we get the following diagram of regular schemes:

(13.6.1.1)
$$\begin{array}{ccccc} Z & \xrightarrow{s_1} & \mathbf{A}_Z^1 & \xleftarrow{s_0} & Z \\ {\scriptstyle i}\downarrow & & \downarrow & & \downarrow {\scriptstyle s} \\ S & \longrightarrow & D & \longleftarrow & P \end{array}$$

where s_0 (resp. s_1) is the zero (resp. unit) section of $\mathbf{A}_Z^1$ over Z. These squares are cartesian and Tor-independent in the sense of Section 13.5.1. The maps s_0 and s_1 induce isomorphisms in K-theory because Z is regular. Thus, the second point of Remark 13.5.5 allows us to reduce to the case of the immersion s which was done in Step 1. □

13.6.2 Consider a cartesian square

$$\begin{array}{ccc} T & \xrightarrow{k} & X \\ {\scriptstyle g}\downarrow & & \downarrow {\scriptstyle f} \\ Z & \xrightarrow{i} & S \end{array}$$

such that S and Z are regular, i is a closed immersion and f is smooth. In this case, the following diagram is commutative

$$\begin{array}{ccc}
\operatorname{Hom}_{KGL}(KGL_Z(T)[n], KGL_Z) & \xrightarrow{\eta'_i} & \operatorname{Hom}_{KGL}(KGL_Z(T)[n], i^! KGL_S) \\
\| & & \| \\
\operatorname{Hom}_{KGL}(KGL_T[n], KGL_T) & \xrightarrow{\eta'_k} & \operatorname{Hom}_{KGL}(KGL_T[n], k^! KGL_X)
\end{array}$$

using the adjunction $(g_\sharp, g_*)$, the exchange isomorphism $g^* i^! \simeq k^! f^*$ (which uses relative purity for smooth morphisms) and the third point of Remark 13.5.5. In particular, the preceding proposition has the following consequences:

Theorem 13.6.3 (Absolute purity) *For any closed immersion $i : Z \longrightarrow S$ between regular schemes, the map*

$$\eta'_i : KGL_Z \longrightarrow i^! KGL_S$$

is an isomorphism in the category $\operatorname{Ho}(KGL\text{-}\operatorname{mod})(Z)$ *(or in* $\operatorname{SH}(Z)$*)*.

Corollary 13.6.4 *Given a cartesian square as above, for any pair $(n, m) \in \mathbf{Z}^2$, the following diagram is commutative:*

$$\begin{array}{ccc}
\operatorname{Hom}(KGL_S(X), i_* KGL_Z(m)[n]) & \xrightarrow{\eta_i} & \operatorname{Hom}(KGL_S(X), KGL_S(m)[n]) \\
\| & & \\
\operatorname{Hom}(KGL_Z(T), KGL_Z(m)[n]) & & \sim \;\Big\downarrow \epsilon_{X/S} \\
{\scriptstyle \epsilon_{T/Z}} \Big\downarrow \sim & & \\
K_{2m-n}(T) & \xrightarrow{k_*} & K_{2m-n}(X)
\end{array}$$

where the vertical maps are the isomorphisms (13.3.4.2).

13.7 Trace maps

13.7.1 Let S be a regular scheme. Let Y be a smooth S-scheme. The obvious map $\operatorname{Pic}(Y) \longrightarrow K_0(Y)$ together with the canonical maps

$$K_0(Y) \xrightarrow{\sim} \operatorname{Hom}_{KGL}(KGL_S(Y), KGL_S) \xrightarrow{\beta_*} \operatorname{Hom}_{KGL}(KGL_S(Y), KGL_S(1)[2])$$

defines Chern classes in the category $\operatorname{Ho}(KGL\text{-}\operatorname{mod})(S)$; they corresponds to the orientation defined in Remark 13.2.2.

Let $p : P \longrightarrow S$ be a projective bundle of rank n. Let $v = [\mathcal{O}(1)] - 1$ in $K_0(P)$. It corresponds to a map $\mathfrak{v} : KGL_S(P) \longrightarrow KGL_S$. According to [Dégo8, 3.2] and our choice of Chern classes, the map

13 Algebraic K-theory

$$KGL_S(P) \xrightarrow{\Sigma_i \beta^i . v^i \boxtimes p_*} \bigoplus_{0 \le i \le n} KGL_S(i)[2i]$$

is an isomorphism. As β is invertible, it follows that the map

(13.7.1.1) $$\varphi_{P/S} : KGL_S(P) \xrightarrow{\Sigma_i v^i \boxtimes p_*} \bigoplus_{0 \le i \le n} KGL_S$$

is an isomorphism as well. Using this formula, the map $\text{Hom}(\varphi_{P/S}, KGL_S)$ is equal to the isomorphism of Quillen's projective bundle theorem in K-theory (cf. [Qui73, 4.3]):

$$f_{P/S} : \bigoplus_{i=0}^n K_*(S) \longrightarrow K_*(P), (S_0, ..., S_n) \longmapsto \sum_i p^*(S_i).v^i.$$

Let $p_* : K_*(P) \longrightarrow K_*(S)$ be the pushout by the projective morphism p. According to the projection formula, it is $K_*(S)$-linear. In particular, it is determined by the $n+1$-uple $(a_0, ..., a_n)$ where $a_i = p_*(v^i) \in K_0(S)$ through the isomorphism $f_{P/S}$. Let $\mathfrak{a}_i : KGL_S \longrightarrow KGL_S$ be the map corresponding to a_i.

Definition 13.7.2 Consider the previous notations. We define the *trace map* associated with the projection $p : P \longrightarrow S$ as the morphism of KGL-modules $\text{Tr}_p^{KGL} : p_*(KGL_P) \longrightarrow KGL_S$ determined as the composite

$$p_*(KGL_P) = \mathbf{R}Hom(KGL_S(P), KGL_S) \xrightarrow{(\varphi_{P/S}^*)^{-1}} \bigoplus_{i=0}^n KGL_S \xrightarrow{(\mathfrak{a}_0, ..., \mathfrak{a}_n)} KGL_S.$$

From this definition, it follows that Tr_p represents the push-forward by p in K-theory:

$$\begin{array}{c} \text{Hom}_{KGL}(KGL_S[n], p_*KGL_P) \xrightarrow{\text{Tr}_{p_*}^{KGL}} \text{Hom}_{KGL}(KGL_S[n], KGL_S) \\ \parallel \\ \text{Hom}_{KGL}(KGL_P[n], KGL_P) \\ \epsilon_P \downarrow \qquad \qquad \qquad \qquad \qquad \qquad \qquad \downarrow \epsilon_S \\ K_n(P) \xrightarrow{\qquad p_* \qquad} K_n(S). \end{array}$$

Consider moreover a cartesian square:

$$\begin{array}{ccc} Q & \xrightarrow{q} & P \\ g \downarrow & & \downarrow p \\ Y & \xrightarrow{f} & S \end{array}$$

such that f is smooth. From the projective base change theorem, we get $f^* p_* p^* = q_* q^* g^*$. Using this identification, we easily obtain that $f^* \text{Tr}_p^{KGL} = \text{Tr}_q^{KGL}$. Thus, we conclude that the map

$$\operatorname{Hom}_{KGL}(KGL_S(Y)[n], p_*KGL_P) \xrightarrow{\operatorname{Tr}_p^{KGL}} \operatorname{Hom}_{KGL}(KGL_S(Y)[n], KGL_S)$$

represents the usual pushout map

$$q_* : K_n(Q) \longrightarrow K_n(Y)$$

through the canonical isomorphisms (13.3.4.2).

13.7.3 Consider a projective morphism $f : T \longrightarrow S$ between regular schemes and choose a factorization

$$T \xrightarrow{i} P \xrightarrow{p} S$$

where i is a closed immersion and p is the projection of a projective bundle. Let us define a morphism

$$\operatorname{Tr}_{(p,i)}^{KGL} : f_*KGL_T = p_*i_*KGL_T \xrightarrow{p_*\eta_i} p_*KGL_P \xrightarrow{\operatorname{Tr}_p^{KGL}} KGL_S.$$

According to Corollary 13.6.4 and the previous paragraph, for any cartesian square

$$\begin{array}{ccc} Y & \xrightarrow{g} & X \\ b \downarrow & & \downarrow a \\ T & \xrightarrow{f} & S \end{array}$$

such that a is smooth, the following diagram is commutative.

(13.7.3.1)
$$\begin{array}{ccc}
\operatorname{Hom}_{KGL}(KGL_S(X), f_*KGL_T(m)[n]) & \xrightarrow{\operatorname{Tr}_{(p,i)_*}^{KGL}} & \operatorname{Hom}_{KGL}(KGL_S(X), KGL_S(m)[n]) \\
\| & & \\
\operatorname{Hom}_{KGL}(KGL_T(Y), KGL_Z(m)[n]) & & \simeq \downarrow \epsilon_{X/S} \\
\epsilon_{Y/T} \downarrow \simeq & & \\
K_{2m-n}(Y) & \xrightarrow{g_*} & K_{2m-n}(X).
\end{array}$$

Definition 13.7.4 Considering the above notations, we define the *trace map* associated to f as the morphism

$$\operatorname{Tr}_f^{KGL} = \operatorname{Tr}_{(p,i)}^{KGL} : f_*f^*KGL_S \longrightarrow KGL_S.$$

Remark 13.7.5 By definition, the trace map $\operatorname{Tr}_f^{KGL}$ is a morphism of KGL-modules. As a consequence, the map obtained by adjunction

$$\eta'_f : KGL_T \simeq f^*KGL_S \longrightarrow f^!KGL_S$$

13 Algebraic K-theory

is also a morphism of KGL-modules. This implies that the morphism η'_f (and thus also Tr_f^{KGL}) is completely determined by the element

$$\eta'_f \in \mathrm{Hom}_{KGL}(KGL_T, f^! KGL_S) \simeq \mathrm{Hom}_{\mathrm{SH}(T)}(\mathbb{1}_T, f^! KGL_S).$$

Moreover, as p is smooth, there is a canonical isomorphism $p^! KGL_S \simeq KGL_P$ (by relative purity for p and by periodicity; see [Rio10, lemma 6.1.3.3]). From there, we deduce from Theorem 13.6.3 that we have a canonical isomorphism

$$f^! KGL_S \simeq i^! KGL_P \simeq KGL_T.$$

This implies that we have an isomorphism:

$$\mathrm{Hom}_{\mathrm{SH}(T)}(\mathbb{1}_T, f^! KGL_S) \simeq K_0(T).$$

Hence, the map η'_f is completely determined by a class in $K_0(T)$. The problem of the functoriality of trace maps in the motivic category $\mathrm{Ho}(KGL\text{-}\mathrm{mod})$ is thus a matter of functoriality of this element η'_f in K_0, which can be translated faithfully to the problem of the functoriality of pushforwards for K_0.

However, the only property of trace maps we shall use here is the following.

Proposition 13.7.6 *Let* $f : T \longrightarrow S$ *be a finite flat morphism of regular schemes such that the $\mathcal{O}_S$-module $f_* \mathcal{O}_T$ is (globally) free of rank d. Then the composite map*

$$KGL_S \longrightarrow f_* f^* KGL_S \xrightarrow{\mathrm{Tr}_f^{KGL}} KGL_S$$

is equal to $d.1_{KGL_S}$ in $\mathrm{Ho}(KGL\text{-}\mathrm{mod})(S)$ (whence in $\mathrm{SH}(S)$ as well).

Proof Let φ be the composite map of $\mathrm{Ho}(KGL\text{-}\mathrm{mod})(S)$

$$KGL_S \longrightarrow f_* f^* KGL_S \xrightarrow{\mathrm{Tr}_f} KGL_S.$$

As φ is KGL_S-linear by construction, it corresponds to an element

$$\varphi \in \mathrm{Hom}_{KGL}(KGL_S, KGL_S) \simeq \mathrm{Hom}_{\mathrm{SH}(S)}(\mathbb{1}_S, KGL_S) \simeq K_0(S).$$

According to the commutative diagram (13.7.3.1), if we apply the global sections functor $\mathrm{Hom}_{\mathrm{SH}(S)}(\mathbb{1}_S, -)$ to φ, we obtain through the evident canonical isomorphisms the composition of the usual pullback and pushforward by f in K-theory:

$$K_0(S) \xrightarrow{f^*} K_0(T) \xrightarrow{f_*} K_0(S).$$

With these notations, the element of $K_0(S)$ corresponding to φ is the pushforward of $1_T = f^*(1_S)$ by f, while the element corresponding to the identity of KGL_S is of course 1_S. Under our assumptions on f, it is obvious that we have the identity $f_*(1_T) = d.1_S \in K_0(S)$. This means that φ is d times the identity of KGL_S. □

# 14 Beilinson motives

14.1 The γ-filtration

14.1.1 We denote by $KGL_{\mathbf{Q}}$ the $\mathbf{Q}$-localization of the absolute ring spectrum KGL, considered as a cartesian section of $D_{\mathbf{A}^1,\mathbf{Q}}$. From [Rio10, 5.3.10], this spectrum has the following property:

(K5) For any scheme S, there exists a canonical decomposition, called the *Adams decomposition*
$$KGL_{\mathbf{Q},S} \simeq \bigoplus_{i \in \mathbf{Z}} KGL_S^{(i)}$$
compatible with base change and such that for any regular scheme S, the isomorphism of (K2) induces an isomorphism:
$$\mathrm{Hom}_{D_{\mathbf{A}^1}(S,\mathbf{Q})}\left(\mathbf{Q}_S(X)[n], KGL_S^{(i)}\right) \simeq K_n^{(i)}(X) := Gr_\gamma^i K_n(X)_{\mathbf{Q}},$$
where the right-hand side is the i-th graded piece of the γ-filtration on K-theory groups.

We will denote by
$$\pi_i : KGL_{\mathbf{Q},S} \longrightarrow KGL_S^{(i)}$$
$$\text{resp. } \iota_i : KGL_S^{(i)} \longrightarrow KGL_{\mathbf{Q},S}$$

the projection (resp. inclusion) defined by the decomposition (K3) and we put $p_i = \iota_i \pi_i$ for the corresponding projector on $KGL_{\mathbf{Q},S}$.

Definition 14.1.2 (Riou) We define the *Beilinson motivic cohomology spectrum* as the rational Tate spectrum $H_{\mathrm{B},S} = KGL_S^{(0)}$.

Remark 14.1.3 Note that, by definition, for any morphism of schemes $f : T \longrightarrow S$, we have $f^*H_{\mathrm{B},S} \simeq H_{\mathrm{B},T}$.

Lemma 14.1.4 *The isomorphism γ_u of (13.2.1.1) is homogeneous of degree $+1$ with respect to the graduation (K5). In other words, for any integer $i \in \mathbf{Z}$, the following composite map is an isomorphism*
$$KGL^{(i)}(1)[2] \xrightarrow{\iota_i} KGL_{\mathbf{Q}}(1)[2] \xrightarrow{\gamma_u} KGL_{\mathbf{Q}} \xrightarrow{\pi_i} KGL^{(i+1)}.$$

For any integer $i \in \mathbf{Z}$, we thus get a canonical isomorphism

(14.1.4.1) $$H_{\mathrm{B}}(i)[2i] \xrightarrow{\sim} KGL^{(i)}.$$

14 Beilinson motives

Proof It is sufficient to check that, for $j \neq i + 1$,

$$\begin{cases} p_j \circ \gamma_u \circ p_i = 0, \\ p_j \circ \gamma_u^{-1} \circ p_i = 0 \end{cases}$$

in $\mathrm{Hom}_{D_{A^1}(S,Q)}(KGL_Q, KGL_Q)$. But according to [Rio10, 5.3.1 and 5.3.6], we have only to check these equalities for the induced endomorphism of K_0 (regarded as a presheaf on the category of smooth schemes over $\mathrm{Spec}\,(Z)$). This follows then from the compatibility of the projective bundle isomorphism with the γ-filtration; see [BGI71, Exp. VI, 5.6]. □

14.1.5 Recall from [NSØ09] that KGL_Q is canonically isomorphic (with respect to the orientation in Remark 13.2.2) to the universal oriented rational ring spectrum with multiplicative formal group law introduced in [NSØ09]. The isomorphism of the preceding corollary shows in particular that H_B is obtained from KGL_Q by killing the elements β^n for $n \neq 0$. In particular, this shows that H_B is canonically isomorphic to the spectrum denoted by LQ in [NSØ09], which corresponds to the universal additive formal group law over Q. This implies that H_B has a natural structure of a (commutative) ring spectrum.

Proposition 14.1.6 *The multiplication map*

$$\mu : H_B \otimes H_B \longrightarrow H_B$$

is an isomorphism.

This trivially implies that the following map is an isomorphism:

(14.1.6.1) $\qquad 1 \otimes \eta : H_B \longrightarrow H_B \otimes H_B$.

Proof It is enough to treat the case $S = \mathrm{Spec}\,(Z)$. We will prove that the projector

$$\psi : H_B \otimes H_B \xrightarrow{\mu} H_B \xrightarrow{1 \otimes \eta} H_B \otimes H_B$$

is an isomorphism (in which case it is in fact the identity). We do that for the isomorphic ring spectrum LQ.

Let $H^{top}Q$ be the topological spectrum representing rational singular cohomology. In the terminology of [NSØ09], LQ is a Tate spectrum representing the Landweber exact cohomology which corresponds to the Adams graded MU_*-algebra Q obtained by killing every generator of the Lazard ring MU_*. The corresponding topological spectrum is of course $H^{top}Q$.

According to [NSØ09, 9.2], the spectrum $E = LQ \otimes LQ$ is a Landweber exact spectrum corresponding to the MU_*-algebra $Q \otimes_{MU_*} Q = Q$. In particular, the corresponding topological spectrum is simply $H^{top}Q$. Thus, according to [NSØ09, 9.7], applied with $F = E = LQ \otimes LQ$, we get an isomorphism of Q-vector spaces

$$\mathrm{End}(L\mathbf{Q} \otimes L\mathbf{Q}) = \mathrm{Hom}_{\mathbf{Q}}(\mathbf{Q}, E_{**}) = \mathbf{Q}.$$

Thus $\psi = \lambda.\mathrm{Id}$ for $\lambda \in \mathbf{Q}$. But $\lambda = 0$ is excluded because ψ is a projector on a non-trivial factor, and this concludes the proof. □

14.2 Definition

Definition 14.2.1 Let S be any scheme.

We say that an object E of $D_{\mathbf{A}^1}(S, \mathbf{Q})$ is H_B-*acyclic* if $H_B \otimes E = 0$ in $D_{\mathbf{A}^1}(S, \mathbf{Q})$. A morphism of $D_{\mathbf{A}^1}(S, \mathbf{Q})$ is an H_B-*equivalence* if its cone is H_B-acyclic (or, equivalently, if its tensor product with H_B is an isomorphism).

An object M of $D_{\mathbf{A}^1}(S, \mathbf{Q})$ is H_B-*local* if, for any H_B-acyclic object E, the group $\mathrm{Hom}(E, M)$ vanishes.

We denote by $\mathrm{DM}_B(S)$ the Verdier quotient of $D_{\mathbf{A}^1}(S, \mathbf{Q})$ by the localizing subcategory comprising H_B-acyclic objects (i.e. the localization of $D_{\mathbf{A}^1}(S, \mathbf{Q})$ by the class of H_B-equivalences).

The objects of $\mathrm{DM}_B(S)$ are called the *Beilinson motives*.

Proposition 14.2.2 *An object E of $D_{\mathbf{A}^1}(S, \mathbf{Q})$ is H_B-acyclic if and only if we have $KGL_{\mathbf{Q}} \otimes E = 0$.*

Proof This follows immediately from property (K5) (see Section 14.1.1) and Lemma 14.1.4. □

Proposition 14.2.3 *The localization functor $D_{\mathbf{A}^1}(S, \mathbf{Q}) \longrightarrow \mathrm{DM}_B(S)$ admits a fully faithful right adjoint whose essential image in $D_{\mathbf{A}^1}(S, \mathbf{Q})$ is the full subcategory spanned by H_B-local objects. More precisely, there is a left Bousfield localization of the stable model category of symmetric Tate spectra $\mathrm{Sp}(S, \mathbf{Q})$ by a small set of maps whose homotopy category is precisely $\mathrm{DM}_B(S)$.*

Proof For each smooth S-scheme X and any integers $n, i \in \mathbf{Z}$, we have a functor with values in the category of $\mathbf{Q}$-vector spaces

$$F_{X,n,i} = \mathrm{Hom}_{D_{\mathbf{A}^1}(S, \mathbf{Q})}(\Sigma^\infty Q_S(X), H_B \otimes (-)(i)[n]) : \mathrm{Sp}(S, \mathbf{Q}) \longrightarrow \mathbf{Q}\text{-}\mathrm{mod}$$

which preserves filtered colimits. We define the class of H_B-weak equivalences as the class of maps of $\mathrm{Sp}(S, \mathbf{Q})$ whose image under $F_{X,n,i}$ is an isomorphism for all X and n, i. By virtue of [Bekoo, Prop. 1.15 and 1.18], we can apply Smith's theorem [Bekoo, Theorem 1.7] (with the class of cofibrations of $\mathrm{Sp}(S, \mathbf{Q})$), which implies the proposition. □

Remark 14.2.4 We shall often make the abuse of considering $\mathrm{DM}_B(S)$ as a full subcategory in $D_{\mathbf{A}^1, \mathbf{Q}}(S)$, with an implicit reference to the preceding proposition.

Note that H_B-acyclic objects are stable under the operations f^*, $f_\sharp$ and $\otimes$, so that applying Corollary 5.2.5, we obtain a premotivic category DM_B together with a premotivic adjunction:

14 Beilinson motives

(14.2.4.1) $$\beta^* : D_{\mathbf{A}^1,\mathbf{Q}} \rightleftarrows DM_B : \beta_*.$$

Proposition 14.2.5 *The spectrum $H_{B,S}$ is H_B-local and the unit map $\eta_{H_B} : \mathbb{1} \longrightarrow H_{B,S}$ is an H_B-equivalence in $D_{\mathbf{A}^1}(S, \mathbf{Q})$.*

Proof The unit map $\eta : \mathbb{1}_S \longrightarrow H_{B,S}$ is an H_B-equivalence by Corollary 14.1.6.

Consider a rational spectrum E over S such that $E \otimes H_B = 0$ and a map $f : E \longrightarrow H_B$. It follows trivially from the commutative diagram

$$\begin{array}{ccc} E & \xrightarrow{f} & H_{B,S} \\ {\scriptstyle 1\otimes\eta}\downarrow & {\scriptstyle 1\otimes\eta}\downarrow & \searrow \\ E \otimes H_{B,S} & \xrightarrow{f\otimes 1} & H_{B,S} \otimes H_{B,S} \xrightarrow{\mu} H_{B,S} \end{array}$$

that $f = 0$, which shows that $H_{B,S}$ is H_B-local. □

Corollary 14.2.6 *The family of ring spectra $H_{B,S}$ comes from a cofibrant cartesian commutative monoid (Section 7.2.10) of the symmetric monoidal fibred model category of Tate spectra over the category of schemes.*

Proof By virtue of Proposition 14.2.5 and of Corollary 7.1.9, there exists a cofibrant commutative monoid in the model category of symmetric Tate spectra over $\mathrm{Spec}\,(\mathbf{Z})$ which is canonically isomorphic to $H_{B,\mathbf{Z}}$ in $D_{\mathbf{A}^1}(\mathrm{Spec}\,(\mathbf{Z}), \mathbf{Q})$ (as commutative ring spectra). For a morphism of schemes $f : S \longrightarrow \mathrm{Spec}\,(\mathbf{Z})$, we can then define $H_{B,S}$ as the pullback of $H_{B,\mathbf{Z}}$ (at the level of the model categories); using Proposition 7.1.11, we see that this defines a cofibrant cartesian commutative monoid on the fibred category of spectra which is isomorphic to $H_{B,S}$ as commutative ring spectra in $D_{\mathbf{A}^1}(S, \mathbf{Q})$. □

14.2.7 From now on, we shall assume that H_B is given by a cofibrant cartesian commutative monoid of the symmetric monoidal fibred model category of Tate spectra over the category of schemes. By virtue of Propositions 7.2.11 and 7.2.18, we get the motivic category $\mathrm{Ho}(H_B\text{-}\mathrm{mod})$ of H_B-modules, together with a commutative diagram of morphisms of premotivic categories

$$\begin{array}{ccc} D_{\mathbf{A}^1,\mathbf{Q}} & \xrightarrow{H_B \otimes (-)} & \mathrm{Ho}(H_B\text{-}\mathrm{mod}) \\ {\scriptstyle \beta}\searrow & & \nearrow{\scriptstyle \varphi} \\ & DM_B & \end{array}$$

(any H_B-acyclic object becomes null in the homotopy category of H_B-modules by definition, so that $H_B \otimes (-)$ factors uniquely through DM_B by the universal property of localization).

Proposition 14.2.8 *The forgetful functor $U : \mathrm{Ho}(H_B\text{-}\mathrm{mod})(S) \longrightarrow D_{\mathbf{A}^1}(S, \mathbf{Q})$ is fully faithful.*

Proof We have to prove that, for any $H_{\mathcal{B},S}$-module M, the map

$$H_{\mathcal{B},S} \otimes M \longrightarrow M$$

is an isomorphism in $D_{\mathbf{A}^1,\mathbf{Q}}(S)$. As this is a natural transformation between exact functors which commute with small sums, and as $D_{\mathbf{A}^1,\mathbf{Q}}$ is a compactly generated triangulated category, it is sufficient to check this for $M = H_{\mathcal{B},S} \otimes E$, with E a (compact) object of $D_{\mathbf{A}^1,\mathbf{Q}}(S)$ (see Proposition 7.2.7). In this case, this follows immediately from the isomorphism (14.1.6.1). □

Theorem 14.2.9 *The functor* $\mathrm{DM}_{\mathcal{B}}(S) \longrightarrow \mathrm{Ho}(H_{\mathcal{B},S}\text{-}\mathrm{mod})$ *is an equivalence of triangulated monoidal categories.*

Proof This follows formally from the preceding proposition by definition of $\mathrm{DM}_{\mathcal{B}}$ (see for instance [GZ67, Chap. I, Prop. 1.3]). □

Remark 14.2.10 The preceding theorem shows that the premotivic category of $H_{\mathcal{B}}$-modules $\mathrm{Ho}(H_{\mathcal{B}}\text{-}\mathrm{mod})$ as well as the morphism $D_{\mathbf{A}^1,\mathbf{Q}} \longrightarrow \mathrm{Ho}(H_{\mathcal{B}}\text{-}\mathrm{mod})$ are completely independent of the choice of the strictification of the (commutative) monoid structure on $H_{\mathcal{B}}$ given by Corollary 14.2.6.

Corollary 14.2.11 *The premotivic category* $\mathrm{DM}_{\mathcal{B}} \simeq \mathrm{Ho}(H_{\mathcal{B}}\text{-}\mathrm{mod})$ *is a* $\mathbf{Q}$-*linear motivic category.*

Proof It follows from Proposition 7.2.18 and Theorem 14.2.9 that $\mathrm{DM}_{\mathcal{B}}$ satisfies the homotopy, stability and localization properties (because this is true for $D_{\mathbf{A}^1,\mathbf{Q}}$ by Corollary 6.2.2). It is also well generated because it is a localization of $D_{\mathbf{A}^1,\mathbf{Q}}$. Thus we can apply Remark 2.4.47 to conclude the proof. □

Remark 14.2.12 One can also prove that $\mathrm{DM}_{\mathcal{B}}$ is motivic much more directly: this follows from the fact that $D_{\mathbf{A}^1,\mathbf{Q}}$ is motivic and that the six Grothendieck operations preserve $H_{\mathcal{B}}$-acyclic objects, so that all the properties of $D_{\mathbf{A}^1,\mathbf{Q}}$ induce their analogs on $\mathrm{DM}_{\mathcal{B}}$ by the 2-universal property of localization (we leave this as an easy exercise for the reader).

Definition 14.2.13 For a scheme X, we define its Beilinson motivic cohomology by the formula:

$$H_{\mathcal{B}}^q(X, \mathbf{Q}(p)) = \mathrm{Hom}_{\mathrm{DM}_{\mathcal{B}}(X)}(\mathbb{1}_X, \mathbb{1}_X(p)[q]).$$

In fact, according to the preceding corollary, the cohomology theory defined above is represented by the ring spectrum $H_{\mathcal{B}}$. In particular, we can now justify the terminology of Beilinson motives:

Corollary 14.2.14 *For any regular scheme* X, *we have a canonical isomorphism*

$$H_{\mathcal{B}}^q(X, \mathbf{Q}(p)) \simeq Gr_\gamma^p K_{2p-q}(X)_{\mathbf{Q}}.$$

14 Beilinson motives

14.2.15 Recall from Section 14.1.5 that $H_{\mathcal{B},S}$ is canonically oriented for any scheme S. Moreover, these orientations are compatible with pullbacks with respect to S. This means in particular that the motivic triangulated category $\mathrm{DM}_{\mathcal{B}}$ is oriented (see Example 12.2.3).

In particular, the fibred category $\mathrm{DM}_{\mathcal{B}}$ satisfies the usual Grothendieck six functors formalism. We refer the reader to Theorem 2.4.50 for the precise statement.

It was remarked in Section 14.1.5 that $H_{\mathcal{B},S}$ is the universal oriented ring spectrum with additive formal group law over S. This property can be expressed by the following nice description of Beilinson motives:

Corollary 14.2.16 *Let E be a rational spectrum over S. The following conditions are equivalent:*

(i) E is a Beilinson motive (i.e. is in the essential image of the right adjoint of the localization functor $D_{\mathbf{A}^1,\mathbf{Q}} \longrightarrow \mathrm{DM}_{\mathcal{B}}$);
(ii) E is $H_{\mathcal{B}}$-local;
(iii) the map $\eta \otimes 1_E : E \longrightarrow H_{\mathcal{B}} \otimes E$ is an isomorphism;
(iv) E is an $H_{\mathcal{B}}$-module in $D_{\mathbf{A}^1,\mathbf{Q}}$;
(v) E admits a strict $H_{\mathcal{B}}$-module structure.

If, in addition, E is a commutative ring spectrum, these conditions are equivalent to the following:

(Ri) E is orientable;
(Rii) E is an $H_{\mathcal{B}}$-algebra;
(Riii) E admits a unique structure of an $H_{\mathcal{B}}$-algebra;

Moreover, if E is a strict commutative ring spectrum, these conditions are equivalent to the following:

(Riv) there exists a morphism of commutative monoids $H_{\mathcal{B}} \longrightarrow E$ in the stable model category of Tate spectra;
(Rv) there exists a unique morphism $H_{\mathcal{B}} \longrightarrow E$ in the homotopy category of commutative monoids of the category of Tate spectra.

Proof The equivalence between statements (i)–(v) follows immediately from Theorem 14.2.9. If E is a ring spectrum, the equivalence with (Ri), (Rii) and R(iii) is a consequence of Theorem 12.2.10 and of the fact that $MGL_{\mathbf{Q}}$ is $H_{\mathcal{B}}$-local; see [NSØ09, Cor. 10.6]. It remains to prove the equivalence with (Riv) and (Rv). Then, E is $H_{\mathcal{B}}$-local if and only if the map $E \longrightarrow H_{\mathcal{B}} \otimes E$ is an isomorphism. But this map can be seen as a morphism of strict commutative ring spectra (using the model structure of Theorem 7.1.8 applied to the model category of Tate spectra) whose target is clearly an $H_{\mathcal{B}}$-algebra, so that (Riv) is equivalent to (ii). It remains to check that there is at most one strict $H_{\mathcal{B}}$-algebra structure on E (up to homotopy), which follows from the fact that $H_{\mathcal{B}}$ is the initial object in the homotopy category of commutative monoids of the model category given by Theorem 7.1.8 applied to the model structure of Proposition 14.2.3. □

Corollary 14.2.17 *One has the following properties:*

1. *The ring structure on the spectrum H_B is given by the following structural maps (with the notations of Section 14.1.1).*

$$H_\mathrm{B} \otimes H_\mathrm{B} \xrightarrow{\iota_0 \otimes \iota_0} KGL_\mathbf{Q} \otimes KGL_\mathbf{Q} \xrightarrow{\mu_{KGL}} KGL_\mathbf{Q} \xrightarrow{\pi_0} H_\mathrm{B},$$

$$\mathbf{Q} \xrightarrow{\eta_{KGL}} KGL_\mathbf{Q} \xrightarrow{\pi_0} H_\mathrm{B}.$$

2. *The map $\iota_0 : H_\mathrm{B} \longrightarrow KGL_\mathbf{Q}$ is compatible with the monoid structures.*
3. *Let $H_\mathrm{B}[t,t^{-1}] = \bigoplus_{i \in \mathbf{Z}} H_\mathrm{B}(i)[2i]$ be the free H_B-algebra generated by one invertible generator t of bidegree $(2,1)$. Then the section $u : \mathbf{Q}(1)[2] \longrightarrow KGL_\mathbf{Q}$ induces an isomorphism of H_B-algebras:*

$$\gamma'_u : H_\mathrm{B}[t,t^{-1}] \longrightarrow KGL_\mathbf{Q}.$$

Proof Property (1) follows from properties (2) and (3). Property (2) is a trivial consequence of the previous corollary. Using the isomorphisms (14.1.4.1) of Lemma 14.1.4, we get a canonical isomorphism

$$H_{\mathrm{B},S}[t,t^{-1}] \xrightarrow{\sim} \bigoplus_{i \in \mathbf{Z}} KGL^{(i)}.$$

Through this isomorphism, the map γ'_u corresponds to the Adams decomposition (i.e. to the isomorphism (K5) of Section 14.1.1) from which we deduce property (3). □

Remark 14.2.18 One deduces easily, from the preceding proposition and from Corollary 14.1.6, another proof of the fact that $KGL_\mathbf{Q}$ is a strict commutative ring spectrum.

The isomorphism (3) is in fact compatible with the grading of each term: the factor $H_\mathrm{B}.t^i$ is sent to the factor $KGL^{(i)}$. Recall also the parameter t corresponds to the unit β^{-1} in $KGL^{*,*}$.

Corollary 14.2.19 *The Adams decomposition is compatible with the monoid structure on $KGL_\mathbf{Q}$: for any integers i,j,l such that $l \neq i+j$, the following map is zero:*

$$KGL^{(i)} \otimes KGL^{(j)} \xrightarrow{t_i \otimes t_j} KGL_\mathbf{Q} \otimes KGL_\mathbf{Q} \xrightarrow{\mu} KGL_\mathbf{Q} \xrightarrow{\pi_l} KGL^{(l)}.$$

14.2.20 Let R be a $\mathbf{Q}$-algebra with structural morphism φ. Recall from Section 5.3.36 that we get an adjunction of premotivic triangulated categories:

$$\varphi^* : \mathrm{D}_{\mathbf{A}^1,\mathbf{Q}} \longrightarrow \mathrm{D}_{\mathbf{A}^1,R} : \varphi_*.$$

Moreover, for any object M and N of $\mathrm{D}_{\mathbf{A}^1,\mathbf{Q}}(S)$, the canonical map

(14.2.20.1) $$\mathrm{Hom}(M,N) \otimes_\mathbf{Q} R \longrightarrow \mathrm{Hom}(\varphi^*(M),\varphi^*(N))$$

is an isomorphism provided M is compact or R is a finite $\mathbf{Q}$-vector space.

14 Beilinson motives

In particular, the ring spectrum $KGL_R := \varphi^*(KGL_\mathbf{Q})$ represents Quillen's algebraic K-theory with coefficients in R over regular schemes. We can repeat Definition 14.2.1 with R-coefficients and this gives the category $\mathrm{DM}_B(S, R)$ of Beilinson motives with R-coefficients together with an adjunction:

$$\varphi^* : \mathrm{DM}_B \longrightarrow \mathrm{DM}_B(-, R) : \varphi_*.$$

Moreover, using the canonical map (14.2.20.1) and the fact it is an isomorphism when M is a constructible Beilinson motive, we immediately extend all the properties proved so far from $\mathbf{Q}$-coefficients to R-coefficients.

14.3 Motivic proper descent

Recall from Definition 4.3.2 we have defined the notion of continuity for a triangulated premotivic category which is the homotopy category of a premotivic model category, such as the triangulated motivic category DM_B — in this case, the notion of continuity is relative to the Tate twist.

Proposition 14.3.1 *The motivic triangulated category* DM_B *is continuous.*

Proof We consider the adjunction (14.2.4.1). According to Theorem 14.2.9, the functor β_* commutes with pullbacks by arbitrary morphisms. Thus the continuity property for DM_B follows from the continuity property for $\mathrm{D}_{\mathbf{A}^1, \mathbf{Q}}$ which was established in Example 6.1.13. □

We will give the main applications of continuity in the section on constructible Beilinson motives. Recall from Proposition 4.3.9 the following corollary of the continuity property of the motivic category DM_B:

Corollary 14.3.2 *Let X be a scheme, and consider an X-scheme Y of finite type. Given a point $x \in X$, we denote by X_x^h the spectrum of the local henselian ring of X at the point x. Let $a_x : Y \times_X X_x^h \longrightarrow Y$ be the canonical map. Then the family of functors*

$$\mathrm{DM}_B(Y) \longrightarrow \mathrm{DM}_B(Y \times_X X_x^h), \quad E \longmapsto a_x^*(E)$$

is conservative.

As the reader might expect, this proposition is very useful when reducing global properties of the motivic category DM_B to local properties. This is in particular illustrated by the following proposition.

Theorem 14.3.3 *The motivic category* DM_B *is separated (on the category of noetherian schemes of finite dimension).*

Proof According to Proposition 2.3.9, it is sufficient to check that, for any finite surjective morphism $f : T \longrightarrow S$, the pullback functor

$$f^* : \mathrm{DM}_B(S) \longrightarrow \mathrm{DM}_B(T)$$

is conservative.

We argue by induction on the dimension of S.

Let us first treat the case where $\dim(S) = 0$. Using the localization property, we can assume that S and T are reduced (cf. Proposition 2.3.6). Then S is a disjoint sum of spectra of fields. In particular, f is not only finite surjective but also flat. Moreover, it is also globally free. It will be sufficient to prove that, for any Beilinson motive E over S, the adjunction map

$$E \longrightarrow f_* f^*(E)$$

is a monomorphism in DM_B. Using the projection formula in DM_B applied to the finite morphism f (point (5) of Theorem 2.4.50), this latter map is isomorphic to

$$\left(H_B \longrightarrow f_* f^*(H_B)\right) \otimes 1_E.$$

We are finally reduced to proving that the map $H_{B,S} \longrightarrow f_* f^* H_{B,S}$ is a monomorphism in DM_B (any monomorphism of a triangulated category splits). As $H_{B,S}$ is a direct factor of $KGL_{\mathbf{Q},S}$, it is sufficient to find a retraction of the adjunction map

$$KGL_{\mathbf{Q},S} \longrightarrow f_* f^* KGL_{\mathbf{Q},S},$$

and this follows from Proposition 13.7.6.

Let us finally solve the induction process. Applying the preceding proposition, we can assume that S is local henselian. Let s be the closed point of S and U be the open complement. Let f_s (resp. f_U) be the pullback of f above s (resp. U). Using the localization property of DM_B and the base change isomorphisms (point (4) of Theorem 2.4.50), it is sufficient to treat the case of the finite morphisms f_U and f_s. The case of f_U follows by the induction hypothesis while the case of f_s follows from the case treated previously. This completes the induction process. □

According respectively to Proposition 3.3.33 and Theorem 3.3.37, we deduce from the preceding proposition the following result:

Theorem 14.3.4

1. *The motivic category* DM_B *satisfies étale descent.*
2. *The motivic category* DM_B *satisfies h-descent when restricted to quasi-excellent schemes.*

Recall this means that for any étale hypercover (resp. h-hypercover of a quasi-excellent scheme) $p : \mathscr{X} \longrightarrow X$ and for any Beilinson motive E over X, the map

$$p^* : \mathbf{R}\Gamma(X,E) \longrightarrow \mathbf{R}\Gamma(\mathscr{X},E) = \mathbf{R}\varprojlim_n \mathbf{R}\Gamma(\mathscr{X}_n,E)$$

is an isomorphism in the derived category of the category of **Q**-vector spaces (see Corollary 3.2.17, taking into account Definition 3.2.20).

14.4 Motivic absolute purity

Theorem 14.4.1 (Absolute purity) *Let $i : Z \longrightarrow S$ be a closed immersion between regular schemes. Assume i has pure codimension n. Then, considering the notations of Section 14.1.1, Definition 13.5.4, and the identification (14.1.4.1), the composed map*

$$H_{\mathrm{B},Z} \xrightarrow{\iota_0} KGL_{\mathbf{Q},Z} \xrightarrow{\eta'_i} i^! KGL_{\mathbf{Q},S} \xrightarrow{\pi_n} i^! H_{\mathrm{B},S}(n)[2n]$$

is an isomorphism.

This isomorphism, or equivalently the map obtained by adjunction:

$$i_*(H_{\mathrm{B},Z}) \longrightarrow H_{\mathrm{B},S}(n)[2n]$$

is called the *fundamental class* associated with i. In fact, this is a canonical class in the Beilinson motivic cohomology of X with support in Z of bidegree $(2n, n)$.

Remark 14.4.2 It follows from Remark 13.5.5 that the fundamental class in Beilinson motivic cohomology is compatible with pullback with respect to Tor-independent square.

Proof We have only to check that the above composition induces an isomorphism after applying the functor $\mathrm{Hom}(\mathbf{Q}_S(X), -(a)[b])$ for a smooth S-scheme X and a pair of integers $(a, b) \in \mathbf{Z}^2$. Using Remark 13.5.5(3), this composition is compatible with smooth base change, and we can assume $X = S$. Let us consider the projector

$$p_a : K_r^Z(S)_{\mathbf{Q}} = K_r(S/S - Z)_{\mathbf{Q}} \longrightarrow K_r(S/S - Z)_{\mathbf{Q}}$$

induced by $\pi_a \circ \iota_a : KGL_{\mathbf{Q}} \longrightarrow KGL_{\mathbf{Q}}$, and denote by $K_r^{(a)}(S/S-Z)$ (with $r = 2a-b$) its image. By virtue of Proposition 13.6.1, we only have to check that the following composite is an isomorphism:

$$\rho_i : K_r^{(a)}(Z) \xrightarrow{\iota_a} K_r(Z)_{\mathbf{Q}} \xrightarrow{q_i^{-1}} K_r(S/S - Z)_{\mathbf{Q}} \xrightarrow{\pi_a} K_r^{(a+n)}(S/S - Z).$$

From Theorem 13.5.2, the morphism ρ_i is functorial with respect to Tor-independent cartesian squares of regular schemes (*cf.* Section 13.5.1). Thus, using again the deformation diagram (13.6.1.1), we get a commutative diagram

$$\begin{array}{ccccc}
K_r^{(a)}(Z) & \longrightarrow & K_r^{(a)}(\mathbf{A}_Z^1) & \longleftarrow & K_r^{(a)}(Z) \\
\downarrow{\rho_i} & & \downarrow & & \downarrow{\rho_s} \\
K_r^{(a+n)}(S/S - Z) & \longrightarrow & K_r^{(a+n)}(D/D - \mathbf{A}_Z^1) & \longleftarrow & K_r^{(a+n)}(P/P - Z)
\end{array}$$

in which any of the horizontal maps is an isomorphism (as a direct factor of an isomorphism). Thus, we are reduced to the case of the closed immersion $s : Z \longrightarrow P$, the canonical section of the projectivization of a vector bundle E (where E is the normal bundle of the closed immersion i). Moreover, as the assertion is local on Z, we may assume E is a trivial vector bundle.

Let $p : P \longrightarrow Z$ be the canonical projection and $j : P - Z \longrightarrow P$ the obvious open immersion. Considering the element $v' := ([\mathcal{O}(1)] - 1)$ of $K_0(P)$, we let v be its projection on the first graded part of the γ-filtration, $v \in K_0^{(1)}(P)$.

Recall that, according to the projective bundle formula, the horizontal lines in the following commutative diagram are split short exact sequences:

$$\begin{array}{ccccccccc} 0 & \longrightarrow & K_r(P/P-Z)_{\mathbb{Q}} & \xrightarrow{\nu} & K_r(P)_{\mathbb{Q}} & \xrightarrow{j^*} & K_r(P-Z)_{\mathbb{Q}} & \longrightarrow & 0 \\ & & \downarrow & & \downarrow & & \downarrow & & \\ 0 & \longrightarrow & K_r^{(a+n)}(P/P-Z) & \xrightarrow{\nu'} & K_r^{(a+n)}(P) & \longrightarrow & K_r^{(a+n)}(P-Z) & \longrightarrow & 0. \end{array}$$

By assumption on E, v^n lies in the kernel of j^* and the diagram allows us to identify the graded piece $K_r^{(a+n)}(P/P-Z)$ with the submodule of $K_r^{(a+n)}(P)$ of the form $K_r^{(a)}(Z).v^n$.

On the other hand, $j^* s_* = 0$: there exists a unique element $\epsilon \in K_0(Z)$ such that $s_*(1) = p^*(\epsilon).v^n$ in $K_0(P)$. From the relation $p_* s_*(1) = 1$, we obtain that ϵ is a unit in $K_0(Z)$, with inverse the element $p_*(v^n)$. By virtue of [BGI71, Exp. VI, Cor. 5.8], $p_*(v^n)$ belongs to the 0-th γ-graded part of $K_0(P)_{\mathbb{Q}}$ so that the same holds for its inverse ϵ. In the end, for any element $z \in K_r(Z)$, we get the following expression:

$$s_*(z) = s_*(1.s^* p^*(z)) = s_*(1).p^*(z) = p^*(\epsilon.z).v^n.$$

Thus, the commutative diagram

$$K_r^{(a)}(Z) \longrightarrow K_r(Z)_{\mathbb{Q}} \xrightarrow{q_s^{-1}} K_r(P/P-Z)_{\mathbb{Q}} \longrightarrow K_r^{(a+n)}(P/P-Z)$$

with s_* to $K_r(P)_{\mathbb{Q}}$ via ν, and $K_r(P)_{\mathbb{Q}} \longrightarrow K_r^{(n)}(P)$ via ν'

implies that the isomorphism q_s^{-1} preserves the γ-filtration (up to a shift by n). Hence, it induces an isomorphism on the graded pieces by functoriality. $\square$

15 Constructible Beilinson motives

15.1 Definition and basic properties

In this section, we apply the general results of Section 4 to the triangulated motivic category DM_B. Let us first recall the definition of constructibility (Definition 4.2.1) which corresponds to the Tate twist.

15 Constructible Beilinson motives

Definition 15.1.1 Given any scheme S, we define the category $DM_{B,c}(S)$ of *constructible Beilinson motives* over S as the thick triangulated subcategory of $DM_B(S)$ generated by the motives of the form $M_S(X)(i)$ for a smooth S-scheme X and an integer $i \in \mathbf{Z}$.

Remark 15.1.2 Constructible Beilinson motives play, within the class of all Beilinson motives, the same role as complexes of étale sheaves with bounded cohomology and constructible cohomology sheaves play in the class of all complexes of étale sheaves (in the case of torsion coefficients prime to the residue characteristics). This fact will be even more striking after Theorems 15.2.1 and 15.2.4.

15.1.3 Recall from Corollary 6.2.2 that $D_{\mathbf{A}^1,\mathbf{Q}}$ is compactly generated by Tate twists. According to Theorem 14.2.9, the same is true for the motivic category DM_B. Thus Proposition 1.4.11 gives the following criterion of constructibility for Beilinson motives:

Proposition 15.1.4 *Given any base scheme S, a Beilinson motive $\mathscr{M}$ over S is constructible if and only if it is compact.*

Remark 15.1.5 In the sequel, we will give several concrete descriptions of the category of constructible Beilinson motives (see Corollaries 16.1.6 and 16.2.16).

Recall from Proposition 14.3.1 that DM_B is continuous (with respect to the Tate twist). Proposition 4.3.4 thus implies the following properties of constructible Beilinson motives:

Proposition 15.1.6 *Let $(S_\alpha)_{\alpha \in A}$ be a pro-object of noetherian finite-dimensional schemes with affine transition maps and such that the scheme $S = \varprojlim_{\alpha \in A} S_\alpha$ is noetherian of finite dimension. Then the canonical functor:*

(15.1.6.1) $$2\text{-}\varinjlim_\alpha DM_{B,c}(S_\alpha) \longrightarrow DM_{B,c}(S)$$

is an equivalence of monoidal triangulated categories.

Example 15.1.7 Under the assumptions of the above proposition, for any pair of integers (p, q), the canonical map

$$\varinjlim_\alpha H_B^q(S_\alpha, \mathbf{Q}(p)) \longrightarrow H_B^q(S, \mathbf{Q}(p))$$

is an isomorphism.[94]

---

[94] This result is to be compared with [Qui73, Sec. 7, 2.2] — it concerns homotopy invariant K-theory rather than K-theory.

15.2 The Grothendieck six functors formalism and duality

The motivic triangulated category DM_B is separated (Theorem 14.3.3) and weakly pure (see Definition 4.2.20; this follows directly from Theorem 14.4.1). Thus the abstract Theorem 4.2.29 gives the finiteness theorem, which we state here explicitly to help the reader:

Theorem 15.2.1 *The triangulated subcategory $DM_{B,c}$ of DM_B is stable under the following operations:*

1. *f^* for any morphism of schemes f.*
2. *f_* for any morphism $f : Y \longrightarrow X$ of finite type such that X is quasi-excellent (resp. any proper morphism f).*
3. *$f_!$ for any separated morphism of finite type f.*
4. *$f^!$ for any morphism $f : Y \longrightarrow X$ of finite type such that X is quasi-excellent.*
5. *$\otimes_X$ for any scheme X.*
6. *Hom_X for any quasi-excellent scheme X.*

To be more precise, points (1) and (5) are obvious, the condition of point (2) not in parentheses is the hardest fact and follows from Theorem 4.2.24, point (3) as well as the parenthesized condition of point (2) is Corollary 4.2.12, point (4) is Corollary 4.2.28 and point (6) is Corollary 4.2.25.

15.2.2 Let B be an excellent scheme such that $\dim(B) \leq 2$. Recall that B satisfies wide resolution of singularities up to quotient singularities (see Definition 4.1.9 and the result of De Jong recalled in Corollary 4.1.11). Thus according to Corollary 4.4.3, we get the following description of constructible Beilinson motives:

Proposition 15.2.3 *Let S be a separated B-scheme of finite type, and $T \subset S$ a closed subscheme. Then the triangulated category $DM_{B,c}(S)$ is the smallest triangulated category of $DM_B(S)$ which contains motives of the form*

$$f_*(\mathbb{1}_X)(n)$$

where n is an integer and $f : X \longrightarrow S$ is a projective morphism such that X is regular connected and $f^{-1}(T)_{red}$ is either empty, or X of the support of a strict normal crossing divisor.

The main motivation to introduce the notion of constructibility is Grothendieck duality. We obtain this duality from the theoretical result on motivic triangulated categories, more precisely Corollary 4.4.24:

Theorem 15.2.4 *Let B be an excellent scheme such that $\dim(B) \leq 2$ and S be a regular separated B-scheme of finite type.*
Then for any separated morphism $f : X \longrightarrow S$ of finite type, the premotive $f^!(\mathbb{1}_S)$ is a dualizing object of $DM_{B,c}(X)$. In fact, if we put $D_X(M) := Hom_X(M, f^!(\mathbb{1}_S))$ for any constructible Beilinson motive M, the following properties hold:

(a) *For any separated S-scheme of finite type X, the functor D_X preserves constructible objects.*

(b) *For any separated S-scheme of finite type X, the natural map*

$$M \longrightarrow D_X(D_X(M))$$

is an isomorphism for any constructible Beilinson motive M.

(c) *For any separated S-scheme of finite type X, and for any Beilinson motive M and N over X, if N is constructible then we have a canonical isomorphism*

$$D_X(M \otimes_X D_X(N)) \simeq Hom_X(M, N).$$

(d) *For any morphism between separated S-schemes of finite type $f : Y \longrightarrow X$, we have natural isomorphisms*

$$D_Y(f^*(M)) \simeq f^!(D_X(M)),$$
$$f^*(D_X(M)) \simeq D_Y(f^!(M)),$$
$$D_X(f_!(N)) \simeq f_*(D_Y(N)),$$
$$f_!(D_Y(N)) \simeq D_X(f_*(N)),$$

where M (resp. N) is a constructible Beilinson motive over X (resp. Y).

15.2.5 Let R be a $\mathbf{Q}$-algebra.[95] We define the premotivic triangulated category of constructible Beilinson motives with coefficients in R as the category of constructible objects of the category $DM_B(-, R)$ defined in Section 14.2.20.

According to *loc. cit.*, for any constructible Beilinson motives with coefficients in $\mathbf{Q}$, we get an isomorphism:

$$\mathrm{Hom}_{DM_{B,c}(S)}(M, N) \otimes_{\mathbf{Q}} R \longrightarrow \mathrm{Hom}_{DM_{B,c}(S,R)}\left(\mathbf{L}\varphi^*(M), \mathbf{L}\varphi^*(N)\right).$$

It is straightforward to see that this isomorphism allows us to extend all the results proved so far for Beilinson motives with coefficient in $\mathbf{Q}$ to the case of R-coefficients.

16 Comparison theorems

16.1 Comparison with Voevodsky motives

16.1.1 We consider the premotivic adjunction of Section 11.4.1

(16.1.1.1) $$\gamma^* : D_{\mathbf{A}^1, \mathbf{Q}} \rightleftarrows DM_{\mathbf{Q}} : \gamma_*.$$

---

[95] The examples we have in mind are: $R = E$ is a number field, $R = \mathbf{C}$, $R = \mathbf{Q}_l, \bar{\mathbf{Q}}_l$ for a prime l.

For a scheme S, $\gamma_*(\mathbb{1}_S)$ is a (strict) commutative ring spectrum, and, for any object M of $\mathrm{DM}_\mathbb{Q}(S)$, $\gamma_*(M)$ is naturally endowed with the structure of a $\gamma_*(\mathbb{1}_S)$-module. On the other hand, as we have the projective bundle formula in $\mathrm{DM}_\mathbb{Q}(S)$ (Section 11.3.4), $\gamma_*(\mathbb{1}_S)$ is orientable (Theorem 12.2.10), which implies that, for any object M of $\mathrm{DM}_\mathbb{Q}(S)$, $\gamma_*(M)$ is an $H_{\mathrm{B},S}$-module, whence is H_{B}-local (Corollary 14.2.16). As consequence, we get a canonical factorization of (16.1.1.1):

(16.1.1.2) $\qquad \mathrm{D}_{\mathbb{A}^1,\mathbb{Q}} \xrightarrow{\beta^*} \mathrm{DM}_{\mathrm{B}} \xrightarrow{\varphi^*} \mathrm{DM}_\mathbb{Q}$.

Consider the commutative diagram of premotivic categories

(16.1.1.3) $\qquad \begin{array}{ccc} \mathrm{D}_{\mathbb{A}^1,\mathbb{Q}} & \xrightarrow{\gamma^*} & \mathrm{DM}_\mathbb{Q} \\ \rho_\sharp \downarrow & & \downarrow \psi_\sharp \\ \underline{\mathrm{D}}_{\mathbb{A}^1,\mathbb{Q}} & \xrightarrow{\underline{\gamma}^*} & \underline{\mathrm{DM}}_\mathbb{Q} \end{array}$

in which the two vertical maps are the canonical enlargements, and, in particular, are fully faithful (see Proposition 6.1.8).

Let t denote either the qfh-topology or the h-topology. We also have the following commutative triangle

(16.1.1.4) $\qquad \underline{\mathrm{D}}_{\mathbb{A}^1,\mathbb{Q}} \xrightarrow{\underline{\gamma}^*} \underline{\mathrm{DM}}_\mathbb{Q} \xrightarrow{\underline{\alpha}^*} \underline{\mathrm{DM}}_{t,\mathbb{Q}}$
with $\underline{a}^*$ underneath

in which both $\underline{a}^*$ and $\underline{\alpha}^*$ are induced by the t-sheafification functor; see Example 5.3.31 and Section 11.1.21. We obtain from (16.1.1.2), (16.1.1.3), and (16.1.1.4) the commutative diagram of premotivic categories below, in which $\chi_\sharp = \varphi^* \underline{\alpha}^* \psi_\sharp$:

(16.1.1.5) $\qquad \begin{array}{ccc} \mathrm{D}_{\mathbb{A}^1,\mathbb{Q}} & \xrightarrow{\beta^*} & \mathrm{DM}_{\mathrm{B}} \\ \rho_\sharp \downarrow & & \downarrow \chi_\sharp \\ \underline{\mathrm{D}}_{\mathbb{A}^1,\mathbb{Q}} & \xrightarrow{\underline{a}^*} & \underline{\mathrm{DM}}_{t,\mathbb{Q}}. \end{array}$

From now on, we shall fix an excellent noetherian scheme of finite dimension S.

Theorem 16.1.2 *We have canonical equivalences of categories*

$$\mathrm{DM}_{\mathrm{B}}(S) \simeq \mathrm{DM}_{qfh,\mathbb{Q}}(S) \simeq \mathrm{DM}_{h,\mathbb{Q}}(S)$$

(recall that, for $t = qfh, h$, $\mathrm{DM}_{t,\mathbb{Q}}(S)$ stands for the localizing subcategory of $\underline{\mathrm{DM}}_{t,\mathbb{Q}}(S)$, spanned by the objects of shape $\Sigma^\infty \mathbf{Q}_S(X)(n)$, where X runs over the family of smooth S-schemes, and $n \leq 0$ is an integer; see Example 5.3.31).

Proof Let t denote the qfh-topology or the h-topology. We shall prove that the functor

$$\chi_\sharp : \mathrm{DM}_\mathrm{B}(S) \longrightarrow \underline{\mathrm{DM}}_{t,\mathbf{Q}}(S)$$

is fully faithful, and that its essential image is precisely $\mathrm{DM}_{t,\mathbf{Q}}$. The functor

$$\beta_* : \mathrm{DM}_\mathrm{B} \longrightarrow D_{\mathbf{A}^1,\mathbf{Q}}(S)$$

is fully faithful, so that its composition with its left adjoint β^* is canonically isomorphic to the identity. In particular, we get isomorphisms of functors:

$$\chi_\sharp \simeq \chi_\sharp \beta^* \beta_* \simeq \underline{a}^* \rho_\sharp \beta_* .$$

The right adjoint of $\underline{a}^*$ is fully faithful, and its essential image consists of the objects of $\underline{D}_{\mathbf{A}^1,\mathbf{Q}}(S)$ which satisfy t-descent (Proposition 5.3.30). On the other hand, the functor $\rho_\sharp$ is fully faithful, and an object of $D_{\mathbf{A}^1,\mathbf{Q}}(S)$ satisfies t-descent if and only if its image under $\rho_\sharp$ satisfies t-descent (Corollary 6.1.11). By virtue of Theorem 14.3.4, this immediately implies that $\chi_\sharp$ is fully faithful. Let $\mathrm{DM}_{t,\mathbf{Q}}'(S)$ be the localizing subcategory of $DM_{t,\mathbf{Q}}(S)$ spanned by the objects of shape $\Sigma^\infty \mathbf{Q}(X)(n)$, where X runs over the family of smooth S-schemes, and $n \leq 0$ is an integer (Example 5.3.31). We know that $\mathrm{DM}_{t,\mathbf{Q}}(S)$ is compactly generated (see Example 5.1.29, Proposition 5.2.38 and Corollary 5.3.40), and that $\chi_\sharp$ is a fully faithful exact functor which preserves small sums as well as compact objects from $\mathrm{DM}_\mathrm{B}(S)$ to $\mathrm{DM}_{t,\mathbf{Q}}(S)$. As, by construction, there exists a generating family of compact objects of $\mathrm{DM}_{t,\mathbf{Q}}(S)$ in the essential image of $\chi_\sharp$, this implies that $\chi_\sharp$ induces an equivalence of triangulated categories $\mathrm{DM}_\mathrm{B}(S) \simeq \mathrm{DM}_{t,\mathbf{Q}}(S)$ (see Corollary 1.3.20). □

Let us underline the following result, which completes Corollary 14.2.16:

Theorem 16.1.3 *Let E be an object of $D_{\mathbf{A}^1}(S, \mathbf{Q})$. The following conditions are equivalent:*

(i) E is a Beilinson motive;
(ii) E satisfies h-descent;
(iii) E satisfies qfh-descent;

Proof We already know that condition (i) implies condition (ii) (second point of Theorem 14.3.4), and condition (ii) obviously implies condition (iii). It is thus sufficient to prove that condition (iii) implies condition (i). If E satisfies qfh-descent, then $\rho_\sharp(E)$ satisfies qfh-descent in $\underline{\mathrm{DM}}(S, \mathbf{Q})$ as well. The commutativity of (16.1.1.4) implies then that $\rho_\sharp(E)$ belongs to the essential image of $\underline{\gamma}_*$ (the right adjoint of $\underline{\gamma}^*$). As $\rho_\sharp$ is fully faithful, the commutativity of (16.1.1.3) thus implies that E itself belongs to the essential image of γ_* (the right adjoint to γ^*). In particular, E is then a module over the ring spectrum $\gamma_*(\mathbb{1}_S)$, which is itself an H_B-algebra. The result follows by Corollary 14.2.16. □

Theorem 16.1.4 *If S is excellent and geometrically unibranch, then the comparison functor*
$$\varphi^* : \mathrm{DM}_{\mathrm{B}}(S) \longrightarrow \mathrm{DM}_{\mathbf{Q}}(S)$$
is an equivalence of triangulated monoidal categories.

Proof If S is geometrically unibranch, then we know that the composed functor
$$\mathrm{DM}_{\mathbf{Q}}(S) \xrightarrow{\psi_\sharp} \underline{\mathrm{DM}}_{\mathbf{Q}}(S) \xrightarrow{\alpha^*} \underline{\mathrm{DM}}_{qfh,\mathbf{Q}}(S)$$
is fully faithful (Theorem 11.1.22). The commutative diagram
$$\mathrm{DM}_{\mathrm{B}}(S) \xrightarrow{\varphi^*} \mathrm{DM}_{\mathbf{Q}}(S) \xrightarrow{\alpha^*\psi_\sharp} \underline{\mathrm{DM}}_{qfh,\mathbf{Q}}(S)$$
$$\underset{\chi_\sharp}{\longrightarrow}$$

and Theorem 16.1.2 imply that φ^* is fully faithful. As φ^* is exact, preserves small sums as well as compact objects, and as $\mathrm{DM}_{\mathbf{Q}}(S)$ has a generating family of compact objects in the essential image of φ^*, the functor φ^* has to be an equivalence of categories (Corollary 1.3.20). □

Remark 16.1.5 A version of the preceding theorem (the one obtained by replacing DM_{B} by $\mathrm{Ho}(H_{\mathrm{B}}\text{-}\mathrm{mod})$) was already known in the case where S is the spectrum of a perfect field; see [RØ08a, theorem 68]. The proof used de Jong's resolution of singularities by alterations and Poincaré duality in a crucial way. The proof of the preceding theorem we gave here relies on proper descent but does not use any kind of resolution of singularities.

The preceding theorem allows us to give the following description of constructible Beilinson motives over geometrically unibranch schemes:

Corollary 16.1.6 *For any geometrically unibranch scheme S, the functor φ^* induces an equivalence of triangulated monoidal categories:*
$$\mathrm{DM}_{\mathrm{B},c}(S) \xrightarrow{\sim} \mathrm{DM}_{gm}(S,\mathbf{Q}),$$
where the right-hand side is the $\mathbf{Q}$-linear version of the category of geometric (Voevodsky) motives (Definition 11.1.10).

Note that we also applied Proposition 11.1.5 to get this corollary.

We finally point out the following important fact about Voevodsky's motivic cohomology spectrum $H_{\mathcal{M},S} = \gamma_*(\mathbb{1}_S)$ with rational coefficients:

Corollary 16.1.7

1. For any geometrically unibranch excellent scheme S, the canonical map
$$H_{\mathrm{B},S} \longrightarrow H^{\mathbf{Q}}_{\mathcal{M},S}$$
is an isomorphism of ring spectra.

2. For any morphism $f : T \longrightarrow S$ of excellent geometrically unibranch schemes, the canonical map
$$f^* H^Q_{\mathcal{M},S} \longrightarrow H^Q_{\mathcal{M},T}$$
is an isomorphism of ring spectra.

The second part is the last conjecture of Voevodsky's paper [Voe02b] with rational coefficients (and geometrically unibranch schemes) – see also Section 11.2.21.

Proof The first part is a trivial consequence of the previous theorem, and the second follows from the first, as the Beilinson motivic cohomology spectrum is stable under pullbacks. □

16.2 Comparison with Morel motives

16.2.1 Let S be a scheme. The permutation isomorphism

(16.2.1.1) $\qquad \tau : Q(1)[1] \otimes^L_Q Q(1)[1] \longrightarrow Q(1)[1] \otimes^L_Q Q(1)[1]$

satisfies the equation $\tau^2 = 1$ in $D_{A^1}(S, Q)$. Hence it defines an element ϵ in $\mathrm{End}_{D_{A^1}(S,Q)}(Q)$ which also satisfies the relation $\epsilon^2 = 1$. We define two projectors

(16.2.1.2) $\qquad e_+ = \dfrac{1-\epsilon}{2} \quad \text{and} \quad e_- = \dfrac{1+\epsilon}{2}.$

As the triangulated category $D_{A^1}(S, Q)$ is pseudo-abelian, we can define two objects by the formulæ:

(16.2.1.3) $\qquad Q_+ = \mathrm{Im}\, e_+ \quad \text{and} \quad Q_- = \mathrm{Im}\, e_-.$

Then for an object M of $D_{A^1}(S, Q)$, we set

(16.2.1.4) $\qquad M_+ = Q_+ \otimes^L_Q M \quad \text{and} \quad M_- = Q_- \otimes^L_Q M.$

It is obvious that for any objects M and N of $D_{A^1}(S, Q)$, one has

(16.2.1.5) $\qquad \mathrm{Hom}_{D_{A^1}(S,Q)}(M_i, N_j) = 0 \quad \text{for } i, j \in \{+, -\} \text{ with } i \neq j.$

Denote by $D_{A^1}(S, Q)_+$ (resp. $D_{A^1}(S, Q)_-$) the full subcategory of $D_{A^1}(S, Q)$ comprising objects which are isomorphic to some M_+ (resp. some M_-) for an object M in $D_{A^1}(S, Q)$. Then (16.2.1.5) implies that the direct sum functor $(M_+, M_-) \mapsto M_+ \oplus M_-$ induces an equivalence of triangulated categories

(16.2.1.6) $\qquad (D_{A^1}(S, Q)_+) \times (D_{A^1}(S, Q)_-) \simeq D_{A^1}(S, Q).$

We shall call $D_{\mathbf{A}^1}(S,\mathbf{Q})_+$ the *category of Morel motives over S*. The aim of this section is to compare this category with $\mathrm{DM}_{\mathrm{B}}(S)$ (see Theorem 16.2.13). This will consist essentially of proving that $\mathbf{Q}_+$ is nothing else than Beilinson's motivic spectrum H_{B} (which was announced by Morel in [Moro6]). The main ingredients of the proof are the description of $\mathrm{DM}_{\mathrm{B}}(S)$ as a full subcategory of $D_{\mathbf{A}^1}(S,\mathbf{Q})$, the homotopy t-structure on $D_{\mathbf{A}^1}(S,\mathbf{Q})$, and Morel's computation of the endomorphism ring of the motivic sphere spectrum in terms of Milnor–Witt K-theory [Moro3, Moro4a, Moro4b, Mor12].

16.2.2 For a little while, we shall assume that S is the spectrum of a field k.
Recall that the *algebraic Hopf fibration* is the map
$$\mathbf{A}^2 - \{0\} \longrightarrow \mathbf{P}^1, \quad (x,y) \longmapsto [x,y].$$
This defines, by desuspension, a morphism
$$\eta : \mathbf{Q}(1)[1] \longrightarrow \mathbf{Q}$$
in $D_{\mathbf{A}^1}(S,\mathbf{Q})$; see [Moro3, 6.2] (recall that we identify $D_{\mathbf{A}^1}(S,\mathbf{Q})$ with $\mathrm{SH}_{\mathbf{Q}}(S)$ and that, under this identification, $\mathbf{Q}(1)[1]$ corresponds to $\Sigma^\infty(\mathbf{G}_m)$).

Lemma 16.2.3 *We have $\eta = \epsilon\eta$ in $\mathrm{Hom}_{D_{\mathbf{A}^1}(S,\mathbf{Q})}(\mathbf{Q}(1)[1],\mathbf{Q})$.*

Proof See [Moro3, 6.2.3]. □

16.2.4 Recall the *homotopy t-structure* on $D_{\mathbf{A}^1}(S,\mathbf{Q})$; see [Moro3, 5.2]. To remain close to the conventions of *loc. cit.*, we shall adopt homological notations, so that, for any object M of $D_{\mathbf{A}^1}(S,\mathbf{Q})$, we have the following truncation triangle
$$\tau_{>0}M \longrightarrow M \longrightarrow \tau_{\leq 0}M \longrightarrow \tau_{>0}M[1].$$
We shall write H_0 for the zeroth homology functor in the sense of this t-structure. This t-structure can be described in terms of generators, as in [Ayo07a, definition 2.2.41]: the category $D_{\mathbf{A}^1}(S,\mathbf{Q})_{\geq 0}$ is the smallest full subcategory of $D_{\mathbf{A}^1}(S,\mathbf{Q})$ which contains the objects of shape $\mathbf{Q}_S(X)(m)[m]$ for X smooth over S, $m \in \mathbf{Z}$, and which satisfies the following stability conditions:

(a) $D_{\mathbf{A}^1}(S,\mathbf{Q})_{\geq 0}$ is stable under suspension; i.e. for any object M in $D_{\mathbf{A}^1}(S,\mathbf{Q})_{\geq 0}$, $M[1]$ is in $D_{\mathbf{A}^1}(S,\mathbf{Q})_{\geq 0}$;
(b) $D_{\mathbf{A}^1}(S,\mathbf{Q})_{\geq 0}$ is closed under extensions: for any distinguished triangle
$$M' \longrightarrow M \longrightarrow M'' \longrightarrow M'[1],$$
if M' and M'' are in $D_{\mathbf{A}^1}(S,\mathbf{Q})_{\geq 0}$, so is M;
(c) $D_{\mathbf{A}^1}(S,\mathbf{Q})_{\geq 0}$ is closed under small sums.

With this description, it is easy to see that $D_{\mathbf{A}^1}(S,\mathbf{Q})_{\geq 0}$ is also closed under tensor products (because the class of generators has this property). The category

$D_{\mathbf{A}^1}(S,\mathbf{Q})_{\leq 0}$ is the full subcategory of $D_{\mathbf{A}^1}(S,\mathbf{Q})$ which consists of objects M such that
$$\mathrm{Hom}_{D_{\mathbf{A}^1}(S,\mathbf{Q})}(\mathbf{Q}_S(X)(m)[m+n], M) \simeq 0$$
for X/S smooth, $m \in \mathbf{Z}$, and $n > 0$; see [Ayo07a, 2.1.72].

Note that the heart of the homotopy t-structure is symmetric monoidal, with tensor product $\otimes^h$ defined by the formula:
$$F \otimes^h G = H_0(F \otimes_S^{\mathbf{L}} G)$$
(the unit object is $H_0(\mathbf{Q})$).

We shall still write $\eta : H_0(\mathbf{Q}(1)[1]) \longrightarrow H_0(\mathbf{Q})$ for the map induced by the algebraic Hopf fibration.

Proposition 16.2.5 *Tensoring by $\mathbf{Q}(n)[n]$ defines a t-exact endofunctor of $D_{\mathbf{A}^1}(S,\mathbf{Q})$ for any integer n.*

Proof As tensoring by $\mathbf{Q}(n)[n]$ is an equivalence of categories, it is sufficient to prove this for $n \geq 0$. This is then a particular case of [Ayo07a, 2.2.51]. □

Proposition 16.2.6 *For any smooth S-scheme X of dimension d, and for any object M of $D_{\mathbf{A}^1}(S,\mathbf{Q})$, the map*
$$\mathrm{Hom}(\mathbf{Q}_S(X), M) \longrightarrow \mathrm{Hom}(\mathbf{Q}_S(X), M_{\leq n})$$
is an isomorphism for $n > d$.

Proof Using [Mor03, lemma 5.2.5], it is sufficient to prove the analog for the homotopy t-structure on $D^{e\!f\!f}_{\mathbf{A}^1,\mathbf{Q}}(S)$, which follows from [Mor05, lemma 3.3.3]. □

Proposition 16.2.7 *The homotopy t-structure is non-degenerate. Even better, for any object M of $D_{\mathbf{A}^1}(S,\mathbf{Q})$, we have canonical isomorphisms*
$$\mathbf{L}\varinjlim_n \tau_{>n} M \simeq M \quad \text{and} \quad \mathbf{R}\varprojlim_n \tau_{>n} M \simeq 0,$$
as well as isomorphisms
$$\mathbf{L}\varinjlim_n \tau_{\leq n} M \simeq 0 \quad \text{and} \quad M \simeq \mathbf{R}\varprojlim_n \tau_{\leq n} M .$$

Proof The first assertion is a direct consequence of Propositions 16.2.5 and 16.2.6 (because the objects of shape $\mathbf{Q}_S(X)(m)[i]$, for X/S smooth, and $m, i \in \mathbf{Z}$, form a generating family). As the objects $\mathbf{Q}_S(X)(m)[m+n]$ are compact in $D_{\mathbf{A}^1}(S,\mathbf{Q})$, the category $D_{\mathbf{A}^1}(S,\mathbf{Q})_{\leq 0}$ is closed under small sums. As $D_{\mathbf{A}^1}(S,\mathbf{Q})_{\geq 0}$ is also closed under small sums, we deduce easily that the truncation functors $\tau_{>0}$ and $\tau_{\leq 0}$ preserve small sums, which implies that the homology functor H_0 has the same property. Moreover, if
$$C_0 \longrightarrow \cdots \longrightarrow C_n \longrightarrow C_{n+1} \longrightarrow \cdots$$

is a sequence of maps in $D_{\mathbf{A}^1}(S,\mathbf{Q})$, then $C = \mathbf{L}\varinjlim_n C_n$ fits in a distinguished triangle of shape

$$\bigoplus_n C_n \xrightarrow{1-s} \bigoplus_n C_n \longrightarrow C \longrightarrow \bigoplus_n C_n[1],$$

where s is the map induced by the maps $C_n \longrightarrow C_{n+1}$. This implies that, for any integer i, we have

$$\varinjlim_n H_i(C_n) \simeq H_i(C)$$

(where the colimit is taken in the heart of the homotopy t-structure). As the homotopy t-structure is non-degenerate, this proves the two formulas

$$\mathbf{L}\varinjlim_n \tau_{>n} M \simeq M \quad \text{and} \quad \mathbf{L}\varinjlim_n \tau_{\leq n} M \simeq 0.$$

Let X be a smooth S-scheme of finite type, and p, q be some integers. To prove that the map

$$\mathrm{Hom}(\mathbf{Q}_S(X)(m)[i], M) \longrightarrow \mathrm{Hom}(\mathbf{Q}_S(X)(m)[i], \mathbf{R}\varprojlim_n \tau_{\leq n} M)$$

is bijective, we may assume that $m = 0$ (replacing M by $M(-m)[-m]$ and i by $i - m$, and using Proposition 16.2.5). Consider the Milnor short exact sequence below, with $A = \mathbf{Q}_S(X)[i]$ (in which the first map is injective, but we will not use it):

$$\varprojlim_n{}^1 \mathrm{Hom}(A[1], \tau_{\leq n} M) \longrightarrow \mathrm{Hom}(A, \mathbf{R}\varprojlim_n \tau_{\leq n} M) \longrightarrow \varprojlim_n \mathrm{Hom}(A, \tau_{\leq n} M) \longrightarrow 0.$$

Using Proposition 16.2.6, as $\varprojlim{}^1$ of a constant functor vanishes, we get that the map

$$\mathrm{Hom}(A, M) \longrightarrow \mathrm{Hom}(A, \mathbf{R}\varprojlim_n \tau_{\leq n} M)$$

is an isomorphism. This gives the isomorphism

$$M \simeq \mathbf{R}\varprojlim_n \tau_{\leq n} M.$$

Using the previous isomorphism, and by contemplating the homotopy limit of the homotopy cofiber sequences

$$\tau_{>n} M \longrightarrow M \longrightarrow \tau_{\leq n} M,$$

we deduce the isomorphism $\mathbf{R}\varprojlim_n \tau_{>n} M \simeq 0$. $\square$

16 Comparison theorems

Lemma 16.2.8 *We have $H_B \in D_{\mathbf{A}^1}(S,\mathbf{Q})_{\geq 0}$, so that we have a canonical map*

$$H_B \longrightarrow H_0(H_B)$$

in $D_{\mathbf{A}^1}(S,\mathbf{Q})$. In particular, for any object M in the heart of the homotopy t-structure, if M is endowed with an action of the monoid $H_0(H_B)$, then M has a natural structure of an H_B-module in $D_{\mathbf{A}^1}(S,\mathbf{Q})$.

Proof As H_B is isomorphic to the motivic cohomology spectrum in the sense of Voevodsky (Corollary 16.1.7), the first assertion is the first assertion of [Mor03, theorem 5.3.2]. Therefore, the truncation triangle for H_B gives a triangle

$$\tau_{>0} H_B \longrightarrow H_B \longrightarrow H_0(H_B) \longrightarrow \tau_{>0} H_B[1],$$

which gives the second assertion. For the third assertion, consider an object M in the heart of the homotopy t-structure, endowed with an action of $H_0(H_B)$. Note that $D_{\mathbf{A}^1}(S,\mathbf{Q})_{\geq 0}$ is closed under tensor products, so that $H_B \otimes_S^{\mathbf{L}} M$ is in $D_{\mathbf{A}^1}(S,\mathbf{Q})_{\geq 0}$. Hence we have natural maps

$$H_B \otimes_S^{\mathbf{L}} M \longrightarrow H_0(H_B \otimes_S^{\mathbf{L}} M) \longrightarrow H_0(H_0(H_B) \otimes_S^{\mathbf{L}} M) = H_0(H_B) \otimes^h M.$$

Then the structural map $H_0(H_B) \otimes^h M \longrightarrow M$ defines a map $H_B \otimes_S^{\mathbf{L}} M \longrightarrow M$ which gives the expected action (observe that, as we already know that H_B-modules form a thick subcategory of $D_{\mathbf{A}^1}(S,\mathbf{Q})$ (Proposition 14.2.8), we don't even need to check all the axioms of an internal module: it is sufficient to check that the unit $\mathbf{Q} \longrightarrow H_B$ induces a section $M \longrightarrow H_B \otimes_S^{\mathbf{L}} M$ of the map constructed above). □

Lemma 16.2.9 *We have the following exact sequence in the heart of the homotopy t-structure.*

$$H_0(\mathbf{Q}(1)[1]) \xrightarrow{\eta} H_0(\mathbf{Q}) \longrightarrow H_0(H_B) \longrightarrow 0.$$

Proof Using the equivalence of categories from the heart of the homotopy t-structure to the category of homotopy modules in the sense of [Mor03, definition 5.2.4], by virtue of Corollary 16.1.7 and [Mor03, theorem 5.3.2], we know that $H_0(H_B)$ corresponds to the homotopy module $\underline{K}_*^M \otimes \mathbf{Q}$ associated with Milnor K-theory, while $H_0(\mathbf{Q})$ corresponds to the homotopy module $\underline{K}_*^{MW} \otimes \mathbf{Q}$ associated with Milnor–Witt K-theory (which follows easily from [Mor12, theorems 2.11, 6.13 and 6.40]). Considering $\underline{K}_*^M$ and $\underline{K}_*^{MW}$ as unramified sheaves in the sense of Morel [Mor12], this lemma is then a reformulation of the isomorphism

$$K_*^{MW}(F)/\eta \simeq K_*^M(F)$$

for any field F; see [Mor12, remark 2.2]. □

Proposition 16.2.10 *We have $H_{B+} \simeq H_B$, and the induced map $\mathbf{Q}_+ \longrightarrow H_B$ gives a canonical isomorphism $H_0(\mathbf{Q}_+) \simeq H_0(H_B)$.*

Proof The map $\epsilon(1)[1] : \mathbf{Q}(1)[1] \longrightarrow \mathbf{Q}(1)[1]$ can be described geometrically as the morphism associated with the pointed morphism

$$\iota : \mathbf{G}_m \longrightarrow \mathbf{G}_m, \quad t \mapsto t^{-1}$$

(see the second assertion of [Moro3, lemma 6.1.1]). In the decomposition

$$K_1(\mathbf{G}_m) \simeq k[t,t^{-1}]^\times \simeq k^\times \oplus \mathbf{Z},$$

the map ι induces multiplication by -1 on $\mathbf{Z}$. Using the periodicity isomorphism $KGL(1)[2] \simeq KGL$, we get the identifications:

$$K_1(\mathbf{G}_m) \supset \mathrm{Hom}_{\mathrm{SH}(k)}(\Sigma^\infty(\mathbf{G}_m)[1], KGL) \simeq \mathrm{Hom}_{KGL}(KGL, KGL) \simeq K_0(k) \simeq \mathbf{Z}.$$

Therefore, ϵ acts as multiplication by -1 on the spectrum $KGL_\mathbf{Q}$, whence on H_B as well. This means precisely that $H_{\mathrm{B}+} \simeq H_\mathrm{B}$. By Lemma 16.2.3, the class 2η vanishes in $\mathbf{Q}_+$, so that, applying the (t-exact) functor $M \mapsto M_+$ to the exact sequence of Lemma 16.2.9, we get an isomorphism $H_0(\mathbf{Q}_+) \simeq H_0(H_{\mathrm{B}+}) \simeq H_0(H_\mathrm{B})$. □

Corollary 16.2.11 *For any object M in the heart of the homotopy t-structure, M_+ is a Beilinson motive.*

Proof The object M is an $H_0(\mathbf{Q})$-module, so that M_+ is an $H_0(\mathbf{Q}_+)$-module. By virtue of Proposition 16.2.10, M_+ is then a module over $H_0(H_\mathrm{B})$, so that, by Lemma 16.2.8, M_+ is naturally endowed with an action of H_B. □

Remark 16.2.12 Until now, we have not really used the fact we are in a $\mathbf{Q}$-linear context (replacing H_B by Voevodsky's motivic spectrum, we just needed 2 to be invertible in the preceding corollary). However, the following result really uses $\mathbf{Q}$-linearity (because, in the proof, we regard $\mathrm{DM}_\mathrm{B}(S)$ as a full subcategory of $\mathrm{D}_{\mathbf{A}^1}(S,\mathbf{Q})$; see Proposition 14.2.3).

Theorem 16.2.13 *For any noetherian scheme of finite dimension S, the map $\mathbf{Q}_+ \longrightarrow H_\mathrm{B}$ is an isomorphism in $\mathrm{D}_{\mathbf{A}^1}(S,\mathbf{Q})$. As a consequence, we have a canonical equivalence of triangulated monoidal categories*

$$\mathrm{D}_{\mathbf{A}^1}(S,\mathbf{Q})_+ \simeq \mathrm{DM}_\mathrm{B}(S).$$

This theorem has already been proved by Morel when S is the spectrum of a perfect field – where the left-hand side is the rational category of Voevodsky motives. Morel announced that the category $\mathrm{D}_{\mathbf{A}^1}(S,\mathbf{Q})_+$ should be *the* category of rational motives and this theorem confirm his insight.

Proof Observe that, if ever $\mathbf{Q}_+ \simeq H_\mathrm{B}$, we have $\mathrm{D}_{\mathbf{A}^1}(S,\mathbf{Q})_+ \simeq \mathrm{DM}_\mathrm{B}(S)$: this follows from the fact that an object M of $\mathrm{D}_{\mathbf{A}^1}(S,\mathbf{Q})$ belongs to $\mathrm{D}_{\mathbf{A}^1}(S,\mathbf{Q})_+$ (resp. to $\mathrm{DM}_\mathrm{B}(S)$) if and only if there exists an isomorphism $M \simeq M_+$ (resp. $M \simeq H_\mathrm{B} \otimes_S^\mathrm{L} M$; see Corollary 14.2.16). It is thus sufficient to prove the first assertion.

As both $\mathbf{Q}_+$ and H_B are stable under pullback, it is sufficient to treat the case where $S = \mathrm{Spec}(\mathbf{Z})$. Using Corollary 14.3.2, we may replace S by any of its henselisations, so that, by the localization property, it is sufficient to treat the case where S is the spectrum of a (perfect) field k.

We shall prove directly that, for any object M of $D_{A^1}(S, \mathbf{Q})$, M_+ is an H_B-module (or, equivalently, is H_B-local). Note that $DM_B(S)$ is closed under homotopy limits and homotopy colimits in $D_{A^1}(S, \mathbf{Q})$: indeed the inclusion functor $DM_B \longrightarrow D_{A^1, \mathbf{Q}}$ has a left adjoint which preserves a family of compact generators, whence it also has a left adjoint (Proposition 1.3.19). By virtue of Proposition 16.2.7, we may thus assume that M is bounded with respect to the homotopy t-structure. As $DM_B(S)$ is certainly closed under extensions in $D_{A^1}(S, \mathbf{Q})$, we may even assume that M belongs to the heart of the homotopy t-structure. The result follows by Corollary 16.2.11. □

Corollary 16.2.14 *For any noetherian scheme of finite dimension S, if -1 is a sum of squares in all the residue fields of S, then $\mathbf{Q}_- \simeq 0$ in $D_{A^1}(S, \mathbf{Q})$, and we have a canonical equivalence of triangulated monoidal categories*

$$D_{A^1}(S, \mathbf{Q}) \simeq DM_B(S).$$

Proof It is sufficient to prove that, under this assumption, $\mathbf{Q}_- \simeq 0$. As in the preceding proof, we may replace S by any of its henselisations (Proposition 4.3.9), so that, by the localization property (and by induction on the dimension), it is sufficient to treat the case where S is the spectrum of a field k. We have to check that, if -1 is a sum of squares in k, then we have $\epsilon = -1$. Using [Moro3, remark 6.3.5 and lemma 6.3.7], we see that, if k is of characteristic 2, we always have $\epsilon = -1$, while, if the characteristic of k is distinct from 2, we have a morphism of rings

$$GW(k) \longrightarrow \mathrm{Hom}_{D_{A^1, \mathbf{Q}}(\mathrm{Spec}(k))}(\mathbf{Q}, \mathbf{Q}),$$

where $GW(k)$ denotes the Grothendieck–Witt ring[96] over k. This morphism sends the class of the quadratic form $-X^2$ to $-\epsilon$ and this proves the result. (For a more precise version of this, with integral coefficients, see [Mor12, proposition 2.13].) □

16.2.15 Recall from Example 5.3.43 that we can describe the category $D_{A^1, c}(S, \mathbf{Q})$ of compact objects of $D_{A^1}(S, \mathbf{Q})$ as the triangulated monoidal category obtained from

$$\left(K^b \left(\mathbf{Q}(Sm/S) \right) / (BG_S \cup \mathcal{T}_{A_S^1}) \right)^{\natural}$$

by formally inverting the Tate twist. The operation ϵ still acts on this category and the decomposition into $+$ and $-$ parts of a motive respects constructibility as this is a decomposition by direct factors. The preceding theorem gives the following description of constructible Beilinson motives:

Corollary 16.2.16 *For any noetherian scheme of finite dimension S, there is a canonical equivalence of triangulated monoidal categories*

$$DM_{B, c}(S) \simeq D_{A^1, c}(S, \mathbf{Q})_+.$$

---

[96] *i.e.* the Grothendieck group of quadratic forms.

When -1 is a sum of squares in all the residue fields of S, this equivalence can be written:

$$\mathrm{DM}_{\mathrm{B},c}(S) \simeq \mathrm{D}_{\mathbf{A}^1,c}(S,\mathbf{Q}).$$

16.2.17 Consider the **Q**-linear *étale motivic category* $\mathrm{D}_{\mathbf{A}^1,\acute{e}t}(-,\mathbf{Q})$, defined by

$$\mathrm{D}_{\mathbf{A}^1,\acute{e}t}(S,\mathbf{Q}) = \mathrm{D}_{\mathbf{A}^1}(\mathrm{Sh}_{\acute{e}t}(Sm/S,\mathbf{Q}))$$

(see Example 5.3.31). The étale sheafification functor induces a morphism of motivic categories

(16.2.17.1) $\qquad \mathrm{D}_{\mathbf{A}^1}(S,\mathbf{Q}) \longrightarrow \mathrm{D}_{\mathbf{A}^1,\acute{e}t}(S,\mathbf{Q}).$

We shall prove the following result, as an application of Theorem 16.2.13.

Theorem 16.2.18 *For any noetherian scheme of finite dimension S, there is a canonical equivalence of categories*

$$\mathrm{DM}_{\mathrm{B}}(S) \simeq \mathrm{D}_{\mathbf{A}^1,\acute{e}t}(S,\mathbf{Q}).$$

As for Theorem 16.2.13, the idea of this result is due to F. Morel, who already proved it at least in the case of a base field.

In order to prove the above theorem, we shall study the behavior of the decomposition (16.2.1.3) in $\mathrm{D}_{\mathbf{A}^1,\acute{e}t}(S,\mathbf{Q})$:

Lemma 16.2.19 *We have $\mathbf{Q}_- \simeq 0$ in $\mathrm{D}_{\mathbf{A}^1,\acute{e}t}(S,\mathbf{Q})$.*

Proof Proceeding as in the proof of Theorem 16.2.13, we may assume that S is the spectrum of a perfect field k. By étale descent, we see that we may replace k by any of its finite extensions. In particular, we may assume that -1 is a sum of squares in k. But then, by virtue of Corollary 16.2.14, $\mathbf{Q}_- \simeq 0$ in $\mathrm{D}_{\mathbf{A}^1}(S,\mathbf{Q})$, so that, by functoriality, $\mathbf{Q}_- \simeq 0$ in $\mathrm{D}_{\mathbf{A}^1,\acute{e}t}(S,\mathbf{Q})$. $\square$

Proof (of Theorem 16.2.18) Note that the functor (16.2.17.1) has a fully faithful right adjoint, whose essential image consists of objects of $\mathrm{D}_{\mathbf{A}^1}(S,\mathbf{Q})$ which satisfy étale descent. As any Beilinson motive satisfies étale descent (first point of Theorem 14.3.4), $\mathrm{DM}_{\mathrm{B}}(S)$ can be regarded naturally as a full subcategory of $\mathrm{D}_{\mathbf{A}^1,\acute{e}t}(S,\mathbf{Q})$. On the other hand, by virtue of the preceding lemma, any object of $\mathrm{D}_{\mathbf{A}^1}(S,\mathbf{Q})$ which satisfies étale descent belongs to $\mathrm{D}_{\mathbf{A}^1}(S,\mathbf{Q})_+$. Hence, by Theorem 16.2.13, any object of $\mathrm{D}_{\mathbf{A}^1}(S,\mathbf{Q})$ which satisfies étale descent is a Beilinson motive. This completes the proof. $\square$

Remark 16.2.20 If S is excellent, and if all the residue fields of S are of characteristic zero, one can prove Theorem 16.2.18 independently of Morel's theorem: this then follows directly from a descent argument, namely from Corollary 3.3.38 and from Theorem 16.1.3.

Corollary 16.2.21 *For any regular noetherian scheme of finite dimension S, we have a canonical isomorphism*

$$\mathrm{Hom}_{\mathrm{D}_{\mathbf{A}^1,\acute{e}t}(S,\mathbf{Q})}(\mathbf{Q}_S, \mathbf{Q}_S(p)[q]) \simeq Gr_\gamma^p K_{2p-q}(S)_\mathbf{Q} .$$

Proof This follows immediately from Theorem 16.2.18, by definition of DM_{B} (Corollary 14.2.14). □

Corollary 16.2.22 *For any geometrically unibranch excellent noetherian scheme of finite dimension S, there is a canonical equivalence of symmetric monoidal triangulated categories*

$$\mathrm{D}_{\mathbf{A}^1,\acute{e}t}(S,\mathbf{Q}) \simeq \mathrm{DM}(S,\mathbf{Q}) .$$

Proof This follows from Theorems 16.1.4 and 16.2.18. □

Remark 16.2.23 The preceding corollary extends immediately to the case of coefficients in a $\mathbf{Q}$-algebra R (cf. Example 5.3.36 for the left-hand side and Section 14.2.20 for the right-hand side).

Corollary 16.2.24 *Let S be an excellent noetherian scheme of finite dimension. An object of* $\mathrm{D}_{\mathbf{A}^1}(S,\mathbf{Q})$ *satisfies h-descent if and only if it satisfies étale descent.*

Proof This follows from Theorems 16.1.3 and 16.2.18. □

17 Realizations

17.1 Tilting

17.1.1 Let $\mathscr{M}$ be a stable perfect symmetric monoidal Sm-fibred combinatorial model category over an adequate category of $\mathscr{S}$-schemes $\mathscr{S}$, such that $\mathrm{Ho}(\mathscr{M})$ is motivic, with generating set of twists τ.

Consider a homotopy cartesian commutative monoid $\mathscr{E}$ in $\mathscr{M}$. Then $\mathscr{E}$-mod is an Sm-fibred model category, such that $\mathrm{Ho}(\mathscr{E}\text{-mod})$ is motivic, and we have a morphism of motivic categories (see Propositions 7.2.13 and 7.2.18)

$$\mathrm{Ho}(\mathscr{M}) \longrightarrow \mathrm{Ho}(\mathscr{E}\text{-mod}), \quad M \mapsto \mathscr{E} \otimes^\mathbf{L} M .$$

In practice, all the realization functors are obtained in this way (at least over fields), which can be formulated as follows (for simplicity, we shall work here in a $\mathbf{Q}$-linear context, but, if we are ready to consider higher categorical constructions, there is no reason to make such an assumption).

17.1.2 Consider a quasi-excellent noetherian scheme S of finite dimension, as well as two stable symmetric monoidal Sm-fibred combinatorial model categories $\mathscr{M}$ and $\mathscr{M}'$ over the category of S-schemes of finite type such that $\mathrm{Ho}(\mathscr{M})$ and $\mathrm{Ho}(\mathscr{M}')$ are motivic (as triangulated premotivic categories). We also assume that both $\mathrm{Ho}(\mathscr{M})$ and $\mathrm{Ho}(\mathscr{M}')$ are **Q**-linear and separated.

Consider a Quillen adjunction

(17.1.2.1) $$\varphi^* : \mathscr{M} \rightleftarrows \mathscr{M}' : \varphi_*,$$

inducing a morphism of Sm-fibred categories

(17.1.2.2) $$\mathbf{L}\varphi^* : \mathrm{Ho}(\mathscr{M}) \longrightarrow \mathrm{Ho}(\mathscr{M}').$$

We consider both $\mathrm{Ho}(\mathscr{M})$ and $\mathrm{Ho}(\mathscr{M}')$ as endowed with their Tate twists, which defines two motivic subcategories of constructible objects $\mathrm{Ho}(\mathscr{M})_c$ and $\mathrm{Ho}(\mathscr{M}')_c$, respectively. The functor $\mathbf{L}\varphi^*$ preserves constructible objects, and thus defines a morphism of premotivic categories

(17.1.2.3) $$\mathbf{L}\varphi^* : \mathrm{Ho}(\mathscr{M})_c \longrightarrow \mathrm{Ho}(\mathscr{M}')_c.$$

Proposition 17.1.3 *Under the assumptions of Section 17.1.2, if, for any regular S-scheme of finite type X, and for any integers p and q, the map*

$$\mathrm{Hom}_{\mathrm{Ho}(\mathscr{M})(X)}(\mathbb{1}_X, \mathbb{1}_X(p)[q]) \longrightarrow \mathrm{Hom}_{\mathrm{Ho}(\mathscr{M}')(X)}(\mathbb{1}_X, \mathbb{1}_X(p)[q])$$

is bijective, then the morphism (17.1.2.3) *is an equivalence of premotivic categories. Moreover, if both $\mathrm{Ho}(\mathscr{M})$ and $\mathrm{Ho}(\mathscr{M}')$ are compactly generated by their Tate twists, then the morphism* (17.1.2.2) *is an equivalence of motivic categories.*

Proof Note first that, for any separated S-scheme of finite type X, and for any integers p and q, the map

$$\mathrm{Hom}_{\mathrm{Ho}(\mathscr{M})(X)}(\mathbb{1}_X, \mathbb{1}_X(p)[q]) \longrightarrow \mathrm{Hom}_{\mathrm{Ho}(\mathscr{M}')(X)}(\mathbb{1}_X, \mathbb{1}_X(p)[q])$$

is bijective. Indeed, it is equivalent to prove that the maps

$$\mathbf{R}\Gamma(X, \mathbb{1}_X(p)) \longrightarrow \mathbf{R}\Gamma(X, \varphi^*(\mathbb{1}_X)(p))$$

are isomorphisms in the derived category of **Q**-vector spaces: by h-descent (Theorem 3.3.37), and by virtue of Gabber's weak uniformization Theorem 4.1.2, it is sufficient to treat the case where X is regular, which is done by assumption. Let T be an S-scheme of finite type. To prove that the functor

$$\mathbf{L}\varphi^* : \mathrm{Ho}(\mathscr{M})_c(T) \longrightarrow \mathrm{Ho}(\mathscr{M}')_c(T)$$

is fully faithful, it is sufficient to choose two small families $\mathfrak{A}$ and $\mathfrak{B}$ of objects of $\mathrm{Ho}(\mathscr{M})(T)$ such that the thick subcategory generated by $\mathfrak{A}$ (by $\mathfrak{B}$, respectively) contains $\mathrm{Ho}(\mathscr{M})(T)$, and to check that the maps

$$\mathrm{Hom}_{\mathrm{Ho}(\mathscr{M})(T)}(A, B) \longrightarrow \mathrm{Hom}_{\mathrm{Ho}(\mathscr{M}')(T)}(\varphi^*(A), \varphi^*(B))$$

17 Realizations

are bijective, where A and B run over $\mathfrak{A}$ and $\mathfrak{B}$, respectively. By virtue of Proposition 4.2.13, it is thus sufficient to prove that, for any separated smooth morphism $f : X \longrightarrow T$, for any projective morphism $g : Y \longrightarrow T$, and for any integers p and q, the map

$$\mathrm{Hom}_{\mathrm{Ho}(\mathscr{M})}(\mathbf{L}f_\sharp(\mathbb{1}_X), \mathbf{R}g_*(\mathbb{1}_Y)(p)[q]) \longrightarrow \mathrm{Hom}_{\mathrm{Ho}(\mathscr{M}')}(\mathbf{L}f_\sharp(\mathbb{1}_X), \mathbf{R}g_*(\mathbb{1}_Y)(p)[q])$$

is an isomorphism. Consider the pullback square

$$\begin{array}{ccc} X \times_T Y & \xrightarrow{pr_2} & Y \\ {\scriptstyle pr_1}\downarrow & & \downarrow{\scriptstyle g} \\ X & \xrightarrow{f} & T. \end{array}$$

From Proposition 2.4.53, the functor φ^* commutes with $f_!$ when f is a separated morphism of finite type. One then easily concludes using this fact and the isomorphisms (obtained by adjunction and smooth (or proper) base change)

$$\begin{aligned}
\mathrm{Hom}(\mathbf{L}f_\sharp(\mathbb{1}_X), \mathbf{R}g_*(\mathbb{1}_Y)(p)[q]) &\simeq \mathrm{Hom}(\mathbb{1}_X, \mathbf{L}f^* \mathbf{R}g_*(\mathbb{1}_X)(p)[q]) \\
&\simeq \mathrm{Hom}(\mathbb{1}_X, \mathbf{R}pr_{1,*} \mathbf{L}pr_2^*(\mathbb{1}_X)(p)[q]) \\
&\simeq \mathrm{Hom}(\mathbb{1}_X, \mathbf{R}pr_{1,*}(\mathbb{1}_{X \times_T Y})(p)[q]), \\
&\simeq \mathrm{Hom}(\mathbb{1}_{X \times_T Y}, \mathbb{1}_{X \times_S Y}(p)[q]),
\end{aligned}$$

that (17.1.2.3) is fully faithful and that $\mathrm{Ho}(\mathscr{M}')_c(T)$ is the thick subcategory generated by the image by $\mathbf{L}\varphi^*$ of constructible objects of $\mathrm{Ho}(\mathscr{M})(T)$. In other words, the functor (17.1.2.3) is an equivalence of categories.

If, moreover, both $\mathrm{Ho}(\mathscr{M})$ and $\mathrm{Ho}(\mathscr{M}')$ are compactly generated by their Tate twists, then the sum-preserving exact functor

$$\mathbf{L}\varphi^* : \mathrm{Ho}(\mathscr{M})(T) \longrightarrow \mathrm{Ho}(\mathscr{M}')(T)$$

is an equivalence at the level of compact objects, hence is an equivalence of categories (Corollary 1.3.20). $\square$

17.1.4 Under the assumptions of Section 17.1.2, assume that $\mathscr{M}$ and $\mathscr{M}'$ are strongly $\mathbb{Q}$-linear (Definition 7.1.4), left proper, tractable, satisfy the monoid axiom, and have cofibrant unit objects. Let $\mathscr{E}'$ be a fibrant resolution of $\mathbb{1}$ in $\mathscr{M}'(\mathrm{Spec}\,(k))$. By virtue of Theorem 7.1.8, we may assume that $\mathscr{E}'$ is a fibrant and cofibrant commutative monoid in $\mathscr{M}'$. Then $\mathbf{R}\varphi_*(\mathbb{1}) = \varphi_*(\mathscr{E}')$ is a commutative monoid in $\mathscr{M}$. Let $\mathscr{E}$ be a cofibrant resolution of $\varphi_*(\mathscr{E}')$ in $\mathscr{M}(\mathrm{Spec}\,(k))$. Using Theorem 7.1.8, we may assume that $\mathscr{E}$ is a fibrant and cofibrant commutative monoid, and that the map

$$\mathscr{E} \longrightarrow \mathbf{R}\varphi_*(\mathscr{E}')$$

is a morphism of commutative monoids (and a weak equivalence by construction). We can view $\mathscr{E}$ and $\mathscr{E}'$ as cartesian commutative monoids in $\mathscr{M}$ and $\mathscr{M}'$, respectively (by considering their pullbacks along morphisms of finite type $f : X \longrightarrow \operatorname{Spec}(k)$). We obtain the essentially commutative diagram of left Quillen functors below (in which the lower horizontal map is the functor induced by φ^* and by the change of scalars functor along the map $\varphi^*(\mathcal{E}) \longrightarrow \mathcal{E}'$):

(17.1.4.1)
$$\begin{array}{ccc} \mathcal{M} & \longrightarrow & \mathcal{M}' \\ \downarrow & & \downarrow \\ \mathcal{E}\text{-mod} & \longrightarrow & \mathcal{E}'\text{-mod} \end{array}$$

where $\mathcal{E}$-mod and $\mathcal{E}'$-mod are respectively the model premotivic categories of $\mathcal{E}$-modules and $\mathcal{E}'$-modules (see Proposition 7.2.11).

Note furthermore that the right-hand vertical left Quillen functor is a Quillen equivalence by construction (identifying $\mathcal{M}'(X)$ with $\mathbb{1}_X$-modules, and using the fact that the morphism of monoids $\mathbb{1}_X \longrightarrow \mathcal{E}'_X$ is a weak equivalence in $\mathcal{M}'(X)$).

Theorem 17.1.5 *Consider the assumptions of Section 17.1.4, with $S = \operatorname{Spec}(k)$ the spectrum of a field k. We suppose furthermore that one of the following conditions is satisfied.*

(i) The field k is perfect.
(ii) The motivic categories $\operatorname{Ho}(\mathcal{M})$ *and* $\operatorname{Ho}(\mathcal{M}')$ *are continuous and semi-separated.*

Then the morphism

$$\operatorname{Ho}(\mathcal{E}\text{-mod})_c \longrightarrow \operatorname{Ho}(\mathcal{E}'\text{-mod})_c \simeq \operatorname{Ho}(\mathcal{M}')_c$$

is an equivalence of motivic categories. Under these identifications, the morphism (17.1.2.3) *corresponds to the change of scalar functor*

$$\operatorname{Ho}(\mathcal{M})_c \longrightarrow \operatorname{Ho}(\mathcal{M}')_c \simeq \operatorname{Ho}(\mathcal{E}\text{-mod})_c, \quad M \longmapsto \mathcal{E} \otimes^{\mathbf{L}} M.$$

If moreover both $\operatorname{Ho}(\mathcal{M})$ *and* $\operatorname{Ho}(\mathcal{M}')$ *are compactly generated by their Tate twists, then these identifications extend to non-constructible objects, so that, in particular, the morphism* (17.1.2.2) *corresponds to the change of scalar functor*

$$\operatorname{Ho}(\mathcal{M}) \longrightarrow \operatorname{Ho}(\mathcal{M}') \simeq \operatorname{Ho}(\mathcal{E}\text{-mod}), \quad M \longmapsto \mathcal{E} \otimes^{\mathbf{L}} M.$$

Remark 17.1.6 This theorem can be thought as (a part of) a *tilting theory* for motivic (homotopy) categories. Note that the theorem above readily implies that the morphism of motivic categories

$$\varphi^* : \operatorname{Ho}(\mathcal{M})_c \longrightarrow \operatorname{Ho}(\mathcal{M}')$$

commutes with the six operations (because, by virtue of Theorem 4.4.25, the functor $M \longmapsto \mathcal{E} \otimes^{\mathbf{L}} M$ has this property, as well as the inclusion $\operatorname{Ho}(\mathcal{M}')_c \subset \operatorname{Ho}(\mathcal{M}')$).

Proof For any regular k-scheme of finite type X, and for any integers p and q, the map

17 Realizations

$$\mathrm{Hom}_{\mathrm{Ho}(\mathscr{M})(X)}(\mathbb{1}_X, \mathscr{E}_X(p)[q]) \longrightarrow \mathrm{Hom}_{\mathrm{Ho}(\mathscr{M}')(X)}(\mathbb{1}_X, \mathscr{E}'_X(p)[q])$$

is bijective: this is easy to check whenever X is smooth over k, which proves the assertion under condition (i), while, under condition (ii), we see immediately from Proposition 4.3.16 that we may assume condition (i). The first assertion is then a special case of the first assertion of Proposition 17.1.3. Similarly, by Proposition 7.2.7, the second assertion follows from the second assertion of Proposition 17.1.3.□

Example 17.1.7 Let $\mathscr{M}$ be the stable Sm-fibred model category of Tate spectra, so that $\mathrm{Ho}(\mathscr{M}) = \mathrm{D}_{\mathbf{A}^1, \mathbf{Q}}$, and write $\mathscr{M}_{\mathrm{B}}$ for the left Bousfield localization of $\mathscr{M}$ by the class of H_{B}-equivalences (see Proposition 14.2.3), so that $\mathrm{Ho}(\mathscr{M}_{\mathrm{B}}) = \mathrm{DM}_{\mathrm{B}}$.

Let k be a field of characteristic zero, endowed with an embedding $\sigma : k \longrightarrow \mathbf{C}$. Given a complex analytic manifold X, let $\mathscr{M}_{an}(X)$ be the category of complexes of sheaves of $\mathbf{Q}$-vector spaces on the smooth analytic site of X (i.e. on the category of smooth analytic X-manifolds, endowed with the Grothendieck topology corresponding to open coverings), endowed with its local model structure (see [Ayo07b, 4.4.16] and [Ayo10]). We shall write $\mathscr{M}^{eff}_{Betti}(X)$ for the stable left Bousfield localization of $\mathscr{M}_{an}(X)$ by the maps of shape $\mathbf{Q}(U \times \mathbf{D}^1) \longrightarrow \mathbf{Q}(U)$ for any analytic smooth $X(\mathbf{C})$-manifold U (where $\mathbf{D}^1$ denotes the closed unit disc). We define at last $\mathscr{M}_{Betti}(X)$ as the stable model category of analytic $\mathbf{Q}(1)[1]$-spectra in $\mathscr{M}^{eff}_{Betti}(X)$, where $\mathbf{Q}(1)[1]$ stands for the cokernel of the map $\mathbf{Q} \longrightarrow \mathbf{Q}(\mathbf{A}^{1,an} - \{0\})$ induced by $1 \in \mathbf{C}$; see [Ayo10, section 1].

Given a k-scheme of finite type X, we shall write

(17.1.7.1) $$\mathrm{D}_{Betti}(X) := \mathrm{Ho}(\mathscr{M}_{Betti}(X))$$

(where the topological space $X(\mathbf{C})$ is endowed with its canonical analytic structure). According to [Ayo10, 1.8 and 1.10], there exist canonical equivalences of categories

(17.1.7.2) $$\mathrm{D}_{Betti}(X) \simeq \mathrm{Ho}(\mathscr{M}^{eff}_{Betti}(X)) \simeq \mathrm{D}(X(\mathbf{C}), \mathbf{Q}),$$

where $\mathrm{D}(X(\mathbf{C}), \mathbf{Q})$ stands for the (unbounded) derived category of the abelian category of sheaves of $\mathbf{Q}$-vector spaces on the small site of $X(\mathbf{C})$. By virtue of [Ayo10, section 2], there exists a symmetric monoidal left Quillen morphism of monoidal Sm-fibred model categories over the category of k-schemes of finite type

(17.1.7.3) $$An^* : \mathscr{M} \longrightarrow \mathscr{M}_{Betti},$$

which induces a morphism of motivic categories over the category of k-schemes of finite type. Hence $\mathbf{R}An_*(\mathbb{1})$ is a ring spectrum in $\mathrm{D}_{\mathbf{A}^1, \mathbf{Q}}(\mathrm{Spec}(k))$ which represents Betti cohomology of smooth k-schemes. As D_{Betti} satisfies étale descent, it follows from Corollary 3.3.38 that it satisfies h-descent, from which, by virtue of Theorem 16.1.3, the morphism (17.1.7.3) defines a left Quillen functor

(17.1.7.4) $$An^* : \mathscr{M}_{\mathrm{B}} \longrightarrow \mathscr{M}_{Betti},$$

hence gives rise to a morphism of motivic categories

(17.1.7.5) $$DM_B \longrightarrow D_{Betti},$$

the *Betti realization functor* of Beilinson motives.

Applying Theorem 17.1.5 to (17.1.7.4), we obtain a commutative ring spectrum $\mathscr{E}_{Betti} = \mathbf{R}An_*(\mathbb{1})$ which represents Betti cohomology of smooth k-schemes, such that the restriction of the functor (17.1.7.5) to constructible objects corresponds to the change of scalars functor $M \mapsto \mathscr{E}_{Betti} \otimes^{\mathbf{L}} M$:

(17.1.7.6) $$DM_{B,c}(X) \longrightarrow Ho(\mathscr{E}_{Betti}\text{-}\mathrm{mod})_c(X) \simeq D_c^b(X(\mathbf{C}), \mathbf{Q}).$$

It should be pointed out that, here, $D_c^b(X(\mathbf{C}), \mathbf{Q})$ means the derived category of sheaves which are *constructible of geometric origin* (i.e. constructible in the algebraic sense, and not in the analytic sense).

In other words, once Betti cohomology of smooth k-schemes is known, one can reconstruct canonically the bounded derived categories of constructible sheaves of geometric origin on $X(\mathbf{C})$ for any k-scheme of finite type X, from the theory of mixed motives. We expect all the realization functors to be of this shape (which should follow from (some variant of) Theorem 17.1.5): the (absolute) cohomology of smooth k-schemes with constant coefficients determines the derived categories of constructible sheaves of geometric origin over any k-scheme of finite type. For instance, the geometric part of the theory of variations of mixed Hodge structures should be obtained from Deligne cohomology, regarded as a ring spectrum in $DM_B(k)$ (or, more precisely, in $\mathscr{M}_B(k)$). Work in progress of Brad Drew [Dre13, Dre18] goes in this direction.

17.2 Mixed Weil cohomologies

Let S be an excellent (regular) noetherian scheme of finite dimension, and $\mathbf{K}$ a field of characteristic zero, called the *field of coefficients*.

17.2.1 Let E be a Nisnevich sheaf of commutative differential graded $\mathbf{K}$-algebras (i.e. E is a commutative monoid in the category of sheaves of complexes of $\mathbf{K}$-vector spaces). We shall write

$$H^n(X, E) = \mathrm{Hom}_{D_{\mathbf{A}^1, \mathbf{Q}}^{\mathit{eff}}(X)}(\mathbf{Q}_X, E[n])$$

for any smooth S-scheme of finite type X, and for any integer n (note that, if E satisfies Nisnevich descent and is $\mathbf{A}^1$-homotopy invariant, which we can always assume, using Theorem 7.1.8, then $H^n(X, E) = H^n(E(X))$).

We introduce the following axioms:

W1 *Dimension.*— $H^i(S, E) \simeq \begin{cases} \mathbf{K} & \text{if } i = 0, \\ 0 & \text{otherwise.} \end{cases}$

17 Realizations

W2 *Stability.—* $\dim_K H^i(\mathbf{G}_m, E) = \begin{cases} 1 & \text{if } i = 0 \text{ or } i = 1, \\ 0 & \text{otherwise.} \end{cases}$

W3 *Künneth formula.—* For any smooth S-schemes X and Y, the exterior cup product induces an isomorphism

$$\bigoplus_{p+q=n} H^p(X, E) \otimes_K H^q(Y, E) \xrightarrow{\sim} H^n(X \times_S Y, E).$$

W3' *Weak Künneth formula.—* For any smooth S-scheme X, the exterior cup product induces an isomorphism

$$\bigoplus_{p+q=n} H^p(X, E) \otimes_K H^q(\mathbf{G}_m, E) \xrightarrow{\sim} H^n(X \times_S \mathbf{G}_m, E).$$

17.2.2 Under assumptions W1 and W2, we will call any non-zero element $c \in H^1(\mathbf{G}_m, E)$ a *stability class*. Note that such a class corresponds to a non-trivial map

$$c : \mathbf{Q}_S(1) \longrightarrow E$$

in $\mathrm{D}^{\mathit{eff}}_{\mathbf{A}^1, \mathbf{Q}}(S)$ (using the decomposition $\mathbf{Q}(\mathbf{G}_m) = \mathbf{Q} \oplus \mathbf{Q}(1)[1]$). In particular, possibly after replacing E by a fibrant resolution (so that E is homotopy invariant and satisfies Nisnevich descent), such a stability class can be lifted to an actual map of complexes of presheaves. Such a lift will be called a *stability structure* on E.

Definition 17.2.3 A sheaf of commutative differential graded $\mathbf{K}$-algebras E as above is a *mixed Weil cohomology* (resp. a *stable cohomology*) if it satisfies the properties W1, W2 and W3 (resp. W1, W2 and W3') stated above.

Proposition 17.2.4 *Let E be a stable cohomology. There exists a (commutative) ring spectrum $\mathscr{E}$ in $\mathrm{DM}_{\mathrm{B}}(S)$ with the following properties.*

(i) *For any smooth S-scheme X, and any integer i, there is a canonical isomorphism of $\mathbf{K}$-vector spaces*

$$H^i(X, E) \simeq \mathrm{Hom}_{\mathrm{DM}_{\mathrm{B}}(S)}(M_S(X), \mathscr{E}[i]).$$

(ii) *Any choice of a stability structure on E defines a map $\mathbf{Q}(1) \longrightarrow \mathscr{E}$ in $\mathrm{DM}_{\mathrm{B}}(S)$, which induces an $\mathscr{E}$-linear isomorphism $\mathscr{E}(1) \simeq \mathscr{E}$.*

Proof One explicitly defines the commutative ring spectrum $\mathscr{E}$ as follows. First, by virtue of Theorem 7.1.8, we may assume that E is a Nisnevich sheaf of commutative differential graded algebras and is fibrant for the $\mathbf{A}^1$-local projective model structure: for any smooth S-scheme X, the two maps

$$H^n(E(X)) \longrightarrow H^n_{Nis}(X, E) \longrightarrow H^n_{Nis}(X \times \mathbf{A}^1, E)$$

are isomorphisms for any $n \in \mathbf{Z}$. Let L be the constant Nisnevich sheaf of complexes of $\mathbf{K}$-vector spaces associated to the kernel of the map induced by $S = \{1\} \subset \mathbf{G}_m$:

$$L = \ker\bigl(E(\mathbf{G}_m) \xrightarrow{1^*} E(S)\bigr).$$

We remark that L is cofibrant, and one defines

$$\mathscr{E}_n = Hom(L^{\otimes n}, E),$$

this sheaf being endowed with an action of the symmetric group on n letters by permuting the factors on $L^{\otimes n}$. We then have canonical pairings

$$Hom(L^{\otimes m}, E) \otimes_{\mathbf{Q}} Hom(L^{\otimes n}, E) \longrightarrow Hom(L^{\otimes m+n}, E \otimes_{\mathbf{Q}} E) \longrightarrow Hom(L^{\otimes m+n}, E)$$

which turn the collection $\mathscr{E} = \{\mathscr{E}_n\}_{n \geq 0}$ into a commutative monoid in the category of symmetric sequences of sheaves of complexes of $\mathbf{Q}$-vector spaces; see Definition 5.3.7. On the other hand, we remark that L is the constant sheaf associated to $\Gamma(S, Hom(\mathbf{Q}(1)[1], E))$, from which we deduce that there is a natural map

$$L \longrightarrow Hom(\mathbf{Q}(1)[1], E)$$

which can be transposed into a canonical map

$$\mathbf{Q}(1)[1] \longrightarrow Hom(L, E) = \mathscr{E}_1.$$

This defines a canonical structure of a commutative monoid in the category of symmetric $\mathbf{Q}(1)[1]$-spectra on the symmetric sequence $\mathscr{E}$ (see Remark 5.3.10).[97]

By virtue of [CD12, Proposition 2.1.6], for any smooth S-scheme X, and any integer i, there is a canonical isomorphism of $\mathbf{K}$-vector spaces

$$H^i(X, E) \simeq \mathrm{Hom}_{D_{\mathbf{A}^1, \mathbf{Q}}(S)}(M_S(X), \mathscr{E}[i]),$$

and any choice of a stability structure on E defines an isomorphism $\mathscr{E}(1) \simeq \mathscr{E}$. Moreover, [CD12, corollary 2.2.8] and Theorem 12.2.10 assert that this ring spectrum $\mathscr{E}$ is oriented, so that, by Corollary 14.2.16, $\mathscr{E}$ is an H_B-module, i.e. belongs to $DM_B(S)$. □

17.2.5 Given a stable cohomology E and its associated ring spectrum $\mathscr{E}$, we can view $\mathscr{E}$ as a cartesian commutative monoid: we define, for an S-scheme X, with structural map $f : X \longrightarrow S$:

$$\mathscr{E}_X = \mathbf{L} f^*(\mathscr{E})$$

(which means that we take a cofibrant replacement $\mathscr{E}'$ of $\mathscr{E}$ in the model category of commutative monoids of the category of Tate spectra, and define $\mathscr{E}_X = f^*(\mathscr{E}')$), and put

---

[97] Here, we work with $\mathbf{Q}(1)[1]$-spectra. However, the paper [CD12] is written in the language of symmetric $\mathbf{Q}(1)$-spectra. We leave as an exercise to the reader the task of the translation, which consists in checking that the functor $\{\mathscr{E}_n\}_{n \geq 0} \longmapsto \{\mathscr{E}_n[n]\}_{n \geq 0}$ is a symmetric monoidal left Quillen equivalence from symmetric $\mathbf{Q}(1)[1]$-spectra to symmetric $\mathbf{Q}(1)$-spectra, which is also a right Quillen functor (and thus, in particular, preserves and detects stable $\mathbf{A}^1$-equivalences).

17 Realizations

(17.2.5.1) $\quad D(X, \mathscr{E}) := \mathrm{Ho}(\mathscr{E}\text{-mod})(X) = \mathrm{Ho}(\mathscr{E}_X\text{-mod})$.

We thus have realization functors

(17.2.5.2) $\quad \mathrm{DM}_\mathrm{B}(X) \longrightarrow D(X, \mathscr{E}), \quad M \longmapsto \mathscr{E}_X \otimes_X^\mathrm{L} M$

which commute with the six operations of Grothendieck if ever S is the spectrum of a field (Theorem 4.4.25). Furthermore, $D(-, \mathscr{E})$ is a motivic category which is **Q**-linear (in fact **K**-linear), separated, and continuous.

For an S-scheme X, define

$$H^q(X, E(p)) = \mathrm{Hom}_{\mathrm{DM}_\mathrm{B}(X)}(\mathbf{Q}_X, \mathscr{E}(p)[q]) \simeq \mathrm{Hom}_{D(X, \mathscr{E})}(\mathscr{E}_X, \mathscr{E}_X(p)[q])$$

(this notation is compatible with Section 17.2.1 by virtue of Proposition 17.2.4).

Corollary 17.2.6 *Any stable cohomology (in particular, any mixed Weil cohomology) extends naturally to S-schemes of finite type, and this extension satisfies cohomological h-descent (in particular, étale descent as well as proper descent).*

Proof This follows immediately from the construction above and from Theorem 14.3.4. □

17.2.7 We denote by $D^\vee(S, \mathscr{E})$ the localizing subcategory of $D(S, \mathscr{E})$ generated by its rigid objects (i.e. by the objects which have strong duals). For instance, for any smooth and proper S-scheme X, $\mathscr{E}(X) = \mathscr{E} \otimes_S^\mathrm{L} M_S(X)$ belongs to $D^\vee(S, \mathscr{E})$; see Proposition 2.4.31.

If we denote by $D(\mathbf{K})$ the (unbounded) derived category of the abelian category of **K**-vector spaces, we get the following interpretation of the Künneth formula.

Theorem 17.2.8 *If E is a mixed Weil cohomology, then the functor*

$$\mathbf{R}\,\mathrm{Hom}_\mathscr{E}(\mathscr{E}, -) : D^\vee(S, \mathscr{E}) \longrightarrow D(\mathbf{K})$$

is an equivalence of symmetric monoidal triangulated categories.

Proof This is [CD12, theorem 2.6.2]. □

Theorem 17.2.9 *If S is the spectrum of a field, then $D^\vee(S, \mathscr{E}) = D(S, \mathscr{E})$.*

Proof This follows from Corollary 4.4.17. □

Remark 17.2.10 It is not reasonable to expect the analog of Theorem 17.2.9 to hold whenever S is of dimension > 0; see (the proof of) [CD12, corollary 3.2.7]. Heuristically, for higher dimensional schemes X, the rigid objects of $D(X, \mathscr{E})$ are extensions of some kind of locally constant sheaves (in the ℓ-adic setting, these correspond to $\mathbf{Q}_\ell$-*faisceaux lisses*).

Corollary 17.2.11 *If E is a mixed Weil cohomology, and if S is the spectrum of a field, then the functor*

$$R\operatorname{Hom}_{\mathscr{E}}(\mathscr{E},-) : D(S,\mathscr{E}) \longrightarrow D(\mathbf{K})$$

is an equivalence of symmetric monoidal triangulated categories.

Remark 17.2.12 This result can be thought of as a *tilting theory* for the spectra associated with mixed Weil cohomologies.

17.2.13 Assume that E is a mixed Weil cohomology, and that S is the spectrum of a field k. For each k-scheme of finite type X, denote by $D_c(X,\mathscr{E})$ the category of constructible objects of $D(X,\mathscr{E})$: by definition, this is the thick triangulated subcategory of $D(X,\mathscr{E})$ generated by objects of shape $\mathscr{E}(Y) = \mathscr{E} \otimes_X^L M_X(Y)$ for Y smooth over X (we can drop Tate twists because of Proposition 17.2.4 (ii)). The category $D_c(X,\mathscr{E})$ also coincides with the category of compact objects in $D(X,\mathscr{E})$; see Proposition 1.4.11. Write $D^b(\mathbf{K})$ for the bounded derived category of the abelian category of finite-dimensional $\mathbf{K}$-vector spaces. Note that $D^b(\mathbf{K})$ is canonically equivalent to the homotopy category of perfect complexes of $\mathbf{K}$-modules, i.e. to the category of compact objects of $D(\mathbf{K})$.

Corollary 17.2.14 *Under the assumptions of Section 17.2.13, we have a canonical equivalence of symmetric monoidal triangulated categories*

$$D_c(\operatorname{Spec}(k),\mathscr{E}) \simeq D^b(\mathbf{K}).$$

Proof This follows from Corollary 17.2.11 and from the fact that equivalences of categories preserve compact objects. □

Corollary 17.2.15 *Under the assumptions of Section 17.2.13, if E' is another $\mathbf{K}$-linear stable cohomology with associated ring spectrum $\mathscr{E}'$, any morphism of presheaves of commutative differential $\mathbf{K}$-algebras $E \longrightarrow E'$ inducing an isomorphism $H^1(\mathbf{G}_m, E) \simeq H^1(\mathbf{G}_m, E')$ gives a canonical isomorphism $\mathscr{E} \simeq \mathscr{E}'$ in the homotopy category of commutative ring spectra. In particular, we get canonical equivalences of categories*

$$D(X,\mathscr{E}) \simeq D(X,\mathscr{E}')$$

for any k-scheme of finite type X (and these are compatible with the six operations of Grothendieck, as well as with the realization functors).

Proof This follows from Theorem 17.2.9 and from [CD12, Theorem 2.6.5]. □

The preceding corollary can be stated in the following way: if $\mathscr{E}$ and $\mathscr{E}'$ are two (strict) commutative ring spectra associated to $\mathbf{K}$-linear mixed Weil cohomologies defined on smooth k-schemes E and E', respectively, then any morphism $\mathscr{E} \longrightarrow \mathscr{E}'$ in the homotopy category of (commutative) monoids in the model category of $\mathbf{K}$-linear Tate spectra is invertible. Moreover, $\mathscr{E}$ is isomorphic to $\mathscr{E}'$ if and only if E is isomorphic to E' (in the appropriate homotopy categories of commutative monoids). To be more precise (and more general), this last assertion follows immediately from Corollary 17.2.15 and from the following result.

Proposition 17.2.16 *Let $\mathbf{E}$ be a commutative monoid in the $\mathbf{A}^1$-stable model category of sheaves of complexes of symmetric $\mathbf{Q}(1)[1]$-spectra over the Nisnevich smooth site of k. Suppose that there exists an isomorphism $\mathbf{E}(1) \simeq \mathbf{E}$ in the homotopy category of $\mathbf{E}$-modules and that*

$$H^n(\operatorname{Spec}(k), \mathbf{E}) = \begin{cases} \mathbf{K} & \text{if } n = 0, \\ 0 & \text{otherwise.} \end{cases}$$

Then $E = \mathbf{R}\Gamma(-, \mathbf{E})$ is a stable cohomology theory, and the commutative ring spectrum $\mathcal{E}$ associated to E by Proposition 17.2.4 is canonically isomorphic to $\mathbf{E}$ in the homotopy category of (strict) commutative ring spectra.

Proof By virtue of Theorem 7.1.8, we may assume that $\mathbf{E}$ is (cofibrant and) fibrant. The ring spectrum $\mathbf{E}$ is defined by a symmetric sequence of complexes of Nisnevich sheaves of $\mathbf{K}$-vector spaces $\mathbf{E}_n$, $n \geq 0$, (endowed with an action of the symmetric group on n-letters), together with maps $\sigma_n : \mathbf{E}_n(1)[1] \longrightarrow \mathbf{E}_{n+1}$ inducing quasi-isomorphisms

$$\mathbf{E}_n \xrightarrow{\sim} \operatorname{Hom}(\mathbf{K}(1)[1], \mathbf{E}_{n+1})$$

as well as pairings

$$\mathbf{E}_m \otimes_{\mathbf{K}} \mathbf{E}_n \longrightarrow \mathbf{E}_{m+n}$$

satisfying a few compatibilities. In particular,

$$E = \mathbf{R}\Gamma(-, \mathbf{E}) = \mathbf{E}_0$$

is naturally endowed with the structure of a Nisnevich sheaf of commutative differential graded algebras which satisfies Nisnevich descent and $\mathbf{A}^1$-homotopy invariance. Moreover, for any integer $n \geq 0$, the Nisnevich sheaf of complexes of $\mathbf{K}$-vector spaces $\mathbf{E}_n$ also has the properties of Nisnevich descent and of $\mathbf{A}^1$-homotopy invariance, and is naturally endowed with the structure of an E-module. It is clear that E is a stable cohomology theory, so that (the proof of) Proposition 17.2.4 provides a commutative ring spectrum $\mathcal{E}$ associated to it. With the notations introduced in the proof of Proposition 17.2.4, we know that $\mathcal{E}$ is comprised of the symmetric sequence $\{\mathcal{E}_n = \operatorname{Hom}(L^{\otimes n}, E)\}_{n \geq 0}$, where L is the constant sheaf associated to $\Gamma(S, \operatorname{Hom}(\mathbf{K}(1)[1], E))$. Let us define $\mathscr{L} = L(1)[1]$. We define a new symmetric sequence $\underline{\mathbf{E}}$ by the formula

$$\underline{\mathbf{E}}_n = \operatorname{Hom}(\mathscr{L}^{\otimes n}, \mathbf{E}_n), \quad n \geq 0,$$

where the symmetric group acts through the diagonal $\mathfrak{S}_n \longrightarrow \mathfrak{S}_n \times \mathfrak{S}_n$ by permutation of the factors on $\mathscr{L}^{\otimes n}$ and by the structural action on $\mathbf{E}_n$. We see that $\underline{\mathbf{E}}$ is a commutative monoid in the category of symmetric sequences with the pairings defined by the tensor product map

$$\operatorname{Hom}(\mathscr{L}^{\otimes m}, \mathbf{E}_m) \otimes_{\mathbf{K}} \operatorname{Hom}(\mathscr{L}^{\otimes n}, \mathbf{E}_n) \longrightarrow \operatorname{Hom}(\mathscr{L}^{\otimes m+n}, \mathbf{E}_m \otimes_{\mathbf{K}} \mathbf{E}_n)$$

composed with the multiplication of **E**:

$$Hom(\mathcal{L}^{\otimes m+n}, \mathbf{E}_m \otimes_\mathbf{K} \mathbf{E}_n) \longrightarrow Hom(\mathcal{L}^{\otimes m+n}, \mathbf{E}_{m+n}).$$

Finally, we can compose the transposition of the map $\sigma_1 : E(1)[1] \longrightarrow \mathscr{E}_1$ with the structural map $\mathbf{K}(1)[1] \longrightarrow Hom(L,E) = \mathscr{E}_1$, to obtain:

$$\mathbf{K}(1)[1] \longrightarrow Hom(L,E) \longrightarrow Hom(L, Hom(\mathbf{K}(1)[1], \mathbf{E}_1) \simeq Hom(\mathcal{L}, \mathbf{E}_1) = \underline{\mathbf{E}}_1.$$

This defines a structure of a commutative ring spectrum on $\underline{\mathbf{E}}$. Note that L is chain homotopy equivalent to $\mathbf{K}[-1]$, so that the functors $Hom(L^{\otimes n}, -)$ preserve quasi-isomorphisms (more precisely, L is concentrated in cohomological degree 1, and its first cohomology sheaf is the constant sheaf associated to the **K**-vector space of dimension one $H^1(\mathbf{G}_m, \mathbf{E}))$. Therefore, one has a quasi-isomorphism of commutative monoids of **K**-linear Tate spectra $\mathscr{E} \longrightarrow \underline{\mathbf{E}}$, defined by the canonical maps

$$Hom(L^{\otimes n}, E) \longrightarrow Hom(L^{\otimes n}, Hom(\mathbf{K}(n)[n], \mathbf{E}_n) \simeq Hom(\mathcal{L}^{\otimes n}, \mathbf{E}_n).$$

It remains to produce a quasi-isomorphism of commutative monoids of Tate spectra $\mathbf{E} \longrightarrow \underline{\mathbf{E}}$. We have a structural map $\mathbf{K}(1)[1] \longrightarrow Hom(L,E)$ which can be transposed into a map

$$\mathcal{L} = L(1)[1] \longrightarrow E = \mathbf{E}_0.$$

As E is a commutative monoid and each $\mathbf{E}_n$ is an E-module, we have natural maps

$$\mathcal{L}^{\otimes n} \otimes_\mathbf{K} \mathbf{E}_n \longrightarrow E^{\otimes n} \otimes_\mathbf{K} \mathbf{E}_n \longrightarrow E \otimes_\mathbf{K} \mathbf{E}_n \longrightarrow \mathbf{E}_n$$

which can be transposed into $\mathfrak{S}_n$-equivariant maps

$$\mathbf{E}_n \longrightarrow Hom(\mathcal{L}^{\otimes n}, \mathbf{E}_n) = \underline{\mathbf{E}}_n.$$

These define a morphism of commutative monoids of **K**-linear Tate spectra $\mathbf{E} \longrightarrow \underline{\mathbf{E}}$. It remains to check that the maps $\mathbf{E}_n \longrightarrow \underline{\mathbf{E}}_n$ are quasi-isomorphisms for each $n \geq 0$. As $Hom(\mathbf{K}(n)[n], E) \simeq \mathbf{E}_n$, we can replace $\mathbf{E}_n$ by $Hom(L^{\otimes n}, E)$. The case $n = 1$ is then a reformulation of Proposition 17.2.4 (ii), and the general case follows by an obvious induction. □

Theorem 17.2.17 *Under the assumptions of Section 17.2.13, the six operations of Grothendieck preserve constructibility in the motivic category* $D(-, \mathscr{E})$, *as defined in Section 17.2.5.*

Proof Observe that the motivic category $D(-, \mathscr{E})$ is **Q**-linear and separated (because $DM_\mathbb{B}$ is so, see Proposition 7.2.18), as well as τ-compatible (because by Proposition 4.4.16, it is even τ-dualizable, which is stronger than τ-compatible; see Definition 4.4.13). The result follows by Theorem 4.2.29. □

17 Realizations

17.2.18 As a consequence, we have, for any k-scheme of finite type X, a realization functor
$$\mathrm{DM}_{\mathrm{B},c}(X) \longrightarrow \mathrm{D}_c(X, \mathscr{E})$$
and we deduce from Theorem 4.4.25 that it preserves all of the Grothendieck six operations. For $X = \mathrm{Spec}(k)$, by virtue of Corollary 17.2.14, this corresponds to a symmetric monoidal exact realization functor
$$R : \mathrm{DM}_{\mathrm{B},c}(\mathrm{Spec}(k)) \longrightarrow \mathrm{D}^b(\mathbf{K}).$$

This leads to a finiteness result:

Corollary 17.2.19 *Under the assumptions of Section 17.2.13, for any k-scheme of finite type X, and for any objects M and N in $\mathrm{D}_c(X, \mathscr{E})$, $\mathrm{Hom}_{\mathscr{E}}(M, N[n])$ is a finite-dimensional $\mathbf{K}$-vector space, and it is trivial for all but a finite number of values of n.*

Proof Let $f : X \longrightarrow \mathrm{Spec}(k)$ be the structural map. By virtue of Theorem 17.2.17, as M and N are constructible, the object $\mathbf{R}f_* \mathbf{R}Hom_X(M, N)$ is constructible as well, i.e. is a compact object of $\mathrm{D}(\mathrm{Spec}(k), \mathscr{E})$. But $\mathbf{R}\mathrm{Hom}_{\mathscr{E}}(M, N)$ is nothing else than the image of $\mathbf{R}f_* \mathbf{R}Hom_X(M, N)$ by the equivalence of categories given by Corollary 17.2.11. Hence $\mathbf{R}\mathrm{Hom}_{\mathscr{E}}(M, N)$ is a compact object of $\mathrm{D}(\mathbf{K})$, which means that it belongs to $\mathrm{D}^b(\mathbf{K})$. □

17.2.20 For a $\mathbf{K}$-vector space V and an integer n, define
$$V(n) = \begin{cases} V \otimes_{\mathbf{K}} \mathrm{Hom}_{\mathbf{K}}(H^1(\mathbf{G}_m, E)^{\otimes n}, \mathbf{K}) & \text{if } n > 0, \\ V \otimes_{\mathbf{K}} H^1(\mathbf{G}_m, E)^{\otimes(-n)} & \text{if } n \leq 0. \end{cases}$$

Any choice of a generator in $\mathbf{K}(-1) = H^1(\mathbf{G}_m, E) \simeq H^2(\mathbf{P}^1_k, E)$ defines a natural isomorphism $V(n) \simeq V$ for any integer n. We have canonical isomorphisms
$$H^q(X, E(p)) \simeq H^q(X, E)(p)$$
(using the fact that the equivalence of Corollary 17.2.14 is monoidal). The realization functors (17.2.5.2) induce in particular cycle class maps
$$cl_X : H^q_{\mathrm{B}}(X, \mathbf{Q}(p)) \longrightarrow H^q(X, E)(p)$$
(and similarly for cohomology with compact support, for homology, and for Borel–Moore homology).

Example 17.2.21 Let k be a field of characteristic zero. We then have a mixed Weil cohomology E_{dR} defined by the algebraic de Rham complex
$$E_{dR}(X) = \Omega^*_{A/k}$$
for any smooth affine k-scheme of finite type $X = \mathrm{Spec}(A)$ (algebraic de Rham cohomology of smooth k-schemes of finite type is obtained by Zariski descent); see

[CD12, 3.1.5]. We obtain from Proposition 17.2.4 a commutative ring spectrum $\mathscr{E}_{dR}$, and, for a k-scheme of finite type X, we define

$$D_{dR}(X) = D_c(X, \mathscr{E}_{dR}).$$

We thus get a motivic category D_{dR}, and we have a natural definition of algebraic de Rham cohomology of k-schemes of finite type, given by

$$H_{dR}^n(X) = \mathrm{Hom}_{D_{dR}(X)}(\mathscr{E}_{dR,X}, \mathscr{E}_{dR,X}[n]).$$

This definition coincides with the usual one: this is true by definition for separated smooth k-schemes of finite type, while the general case follows from h-descent (Corollary 17.2.6) and from de Jong's Theorem (Corollary 4.1.11) (or resolution of singularities à la Hironaka). We have, by construction, a *de Rham realization functor*

$$R_{dR} : \mathrm{DM}_{\mathrm{B},c}(X) \longrightarrow D_{dR}(X)$$

which preserves the six operations of Grothendieck (Theorem 4.4.25). In particular, we have cycle class maps

$$H_{\mathrm{B}}^q(X, \mathbf{Q}(p)) \longrightarrow H_{dR}^q(X)(p).$$

Note that, for any field extension k'/k, we have natural isomorphisms

$$H_{dR}^n(X) \otimes_k k' \simeq H_{dR}^n(X \times_{\mathrm{Spec}(k)} \mathrm{Spec}(k')).$$

Example 17.2.22 Let k be a field of characteristic zero, which is algebraically closed and complete with respect to some valuation (archimedean or not). We can then define a stable cohomology $E_{dR,an}$ as analytic de Rham cohomology of X^{an}, for any smooth k-scheme of finite type X; see [CD12, 3.1.7]. As above, we get a ring spectrum $\mathscr{E}_{dR,an}$, and for any k-scheme of finite type, a category of coefficients

$$D_{dR,an}(X) = D_c(X, \mathscr{E}_{dR,an}),$$

which allows us to define the analytic de Rham cohomology of any k-scheme of finite type X by

$$H_{dR,an}^n(X) = \mathrm{Hom}_{D_{dR,an}(X)}(\mathscr{E}_{dR,an,X}, \mathscr{E}_{dR,an,X}[n]).$$

We also have a realization functor

$$R_{dR,an} : \mathrm{DM}_{\mathrm{B},c}(X) \longrightarrow D_{dR,an}(X)$$

which preserves the six operations of Grothendieck.
We then have a morphism of stable cohomologies

$$E_{dR} \longrightarrow E_{dR,an}$$

17 Realizations

which happens to be a quasi-isomorphism locally for the Nisnevich topology (this is Grothendieck's theorem in the case where K is archimedean, and Kiers's theorem in the case where K is non-archimedean; anyway, one obtains this directly from Corollary 17.2.15). This induces a canonical isomorphism

$$\mathscr{E}_{dR} \simeq \mathscr{E}_{dR,an}$$

in the homotopy category of commutative ring spectra. In particular, $E_{dR,an}$ is a mixed Weil cohomology, and, for any k-scheme of finite type, we have natural equivalences of categories

$$\mathrm{D}_{dR}(X) \longrightarrow \mathrm{D}_{dR,an}(X), \quad M \mapsto \mathscr{E}_{dR,an} \otimes^{\mathbf{L}}_{\mathscr{E}_{dR}} M$$

which commute with the six operations of Grothendieck and are compatible with the realization functors.

Note that, in the case $k = \mathbf{C}$, $\mathscr{E}_{dR,an}$ coincides with Betti cohomology (after tensorization by $\mathbf{C}$), so that we have canonical fully faithful functors

$$\mathrm{D}_{Betti,c}(X) \otimes_{\mathbf{Q}} \mathbf{C} \longrightarrow \mathrm{D}_{dR,an}(X)$$

which are compatible with the realization functors. More precisely, it follows from Proposition 17.2.16 that the Betti spectrum $\mathscr{E}_{Betti}$, obtained by applying Theorem 17.1.5 to Ayoub's realization functor (17.1.7.4), is the spectrum associated to $\mathbf{Q}$-linear Betti cohomology, regarded as a mixed Weil cohomology, from Proposition 17.2.4. Therefore, the holomorphic Poincaré Lemma, together with Corollary 17.2.15, provide an isomorphism

$$\mathscr{E}_{Betti} \otimes_{\mathbf{Q}} \mathbf{C} \simeq \mathscr{E}_{dR,an}$$

in the homotopy category of commutative monoids of the model category of $\mathbf{C}$-linear Tate spectra. We thus have triangulated equivalences of categories

$$\mathrm{D}^b_c(X(\mathbf{C}), \mathbf{C}) \simeq \mathrm{Ho}(\mathscr{E}_{Betti} \otimes_{\mathbf{Q}} \mathbf{C}\text{-mod})_c(X) \simeq \mathrm{D}_{dR,an}(X)$$

which commute with the six operations as well as with the realization functors. In particular, for X smooth, by the Riemann–Hilbert correspondence, $\mathrm{D}_{dR,an}(X)$ is equivalent to the bounded derived category of analytic regular holonomic $\mathscr{D}$-modules on X which are constructible of geometric origin. But we can go backwards: as proved by Brad Drew in his thesis [Dre13], using Corollary 17.2.15, one can actually show directly (i.e. motivically) that, for X smooth, $\mathrm{D}_{dR,an}(X)$ is equivalent to the bounded derived category of analytic regular holonomic $\mathscr{D}$-modules on X which are constructible of geometric origin, and thus give a new algebraic proof of the Riemann–Hilbert correspondence.

Example 17.2.23 Let V be a complete discrete valuation ring of mixed characteristic with perfect residue field k and field of functions K. The Monsky–Washnitzer complex defines a stable cohomology E_{MW} over smooth V-schemes of finite type, defined by

$$E_{MW}(X) = \Omega^*_{A^\dagger/V} \otimes_V K$$

for any affine smooth V-scheme $X = \operatorname{Spec}(A)$ (the case of a smooth V-scheme of finite type is obtained by Zariski descent); see [CD12, 3.2.3]. Let $\mathscr{E}_{MW}$ be the corresponding ring spectrum in $\operatorname{DM}_B(\operatorname{Spec}(V))$, and write $j : \operatorname{Spec}(K) \longrightarrow \operatorname{Spec}(V)$ and $i : \operatorname{Spec}(k) \longrightarrow \operatorname{Spec}(V)$ for the canonical immersions. As we obviously have $j^*\mathscr{E}_{MW} = 0$ (the Monsky–Washnitzer cohomology of a smooth V-scheme with empty special fiber vanishes), we have a canonical isomorphism

$$\mathscr{E}_{MW} \simeq \mathbf{R}i_* \mathbf{L}i^* \mathscr{E}_{MW}.$$

We define the rigid cohomology spectrum $\mathscr{E}_{rig}$ in $\operatorname{DM}_B(\operatorname{Spec}(k))$ by the formula

$$\mathscr{E}_{rig} = \mathbf{L}i^* \mathscr{E}_{MW}.$$

This is a ring spectrum associated to a K-linear mixed Weil cohomology: cohomology with coefficients in $\mathscr{E}_{rig}$ coincides with rigid cohomology in the sense of Berthelot, and the Künneth formula for rigid cohomology holds for smooth and projective k-schemes (as rigid cohomology coincides then with crystalline cohomology), from which we deduce the Künneth formula for smooth k-schemes of finite type; see [CD12, 3.2.10]. As before, we define

$$\operatorname{D}_{rig}(X) = \operatorname{D}_c(X, \mathscr{E}_{rig})$$

for any k-scheme of finite type X, and put

$$H^n_{rig}(X) = \operatorname{Hom}_{\operatorname{D}_{rig}(X)}(\mathscr{E}_{rig,X}, \mathscr{E}_{rig,X}[n]).$$

Here again, we have, by construction, *rigid realization functors*

$$R_{rig} : \operatorname{DM}_{B,c}(X) \longrightarrow \operatorname{D}_{rig}(X)$$

which preserve the six operations of Grothendieck (Theorem 4.4.25), as well as (higher) cycle class maps $H^q_B(X, \mathbf{Q}(p)) \longrightarrow H^q_{rig}(X)(p)$.

Throughout this part, $\mathscr{S}$ is assumed to be the category of noetherian schemes of finite dimension.

References

AGV73. M. Artin, A. Grothendieck, and J.-L. Verdier. *Théorie des topos et cohomologie étale des schémas*, volume 269, 270, 305 of *Lecture Notes in Mathematics*. Springer-Verlag, 1972–1973. Séminaire de Géométrie Algébrique du Bois–Marie 1963–64 (SGA 4).

ALP17. A. Ananyevskiy, M. Levine, and I. Panin. Witt sheaves and the η-inverted sphere spectrum. *J. Topol.*, 10(2):370–385, 2017.

And04. Y. André. *Une introduction aux motifs (motifs purs, motifs mixtes, périodes)*, volume 17 of *Panoramas et Synthèses*. Société Mathématique de France, Paris, 2004.

Aok19. K. Aoki. The weight complex functor is symmetric monoidal. arXiv:1904.01384, 2019.

Arn16. P. Arndt. *Abstract motivic homotopy theory*. PhD thesis, University of Osnabrück, Osnabrück, Germany, 2016.

Ayo07a. J. Ayoub. *Les six opérations de Grothendieck et le formalisme des cycles évanescents dans le monde motivique (I)*, volume 314 of *Astérisque*. Soc. Math. France, 2007.

Ayo07b. J. Ayoub. *Les six opérations de Grothendieck et le formalisme des cycles évanescents dans le monde motivique (II)*, volume 315 of *Astérisque*. Soc. Math. France, 2007.

Ayo10. J. Ayoub. Note sur les opérations de Grothendieck et la réalisation de Betti. *J. Inst. Math. Jussieu*, 9(2):225–263, 2010.

Ayo14. J. Ayoub. La réalisation étale et les opérations de Grothendieck. *Ann. Sci. École Norm. Sup.*, 47(1):1–145, 2014.

AZ12. J. Ayoub and S. Zucker. Relative Artin motives and the reductive Borel-Serre compactification of a locally symmetric variety. *Invent. Math.*, 188(2):277–427, 2012.

Bac18a. T. Bachmann. Motivic and real étale stable homotopy theory. *Compos. Math.*, 154(5):883–917, 2018.

Bac18b. T. Bachmann. Rigidity in etale motivic stable homotopy theory. arXiv:1810.08028, 2018.

Bal19. E. Balzin. Reedy model structures in families. arXiv:1803.00681v2, 2019.

Bar10. C. Barwick. On left and right model categories and left and right Bousfield localizations. *Homology, Homotopy Appl.*, 12(2):245–320, 2010.

BBD82. A. Beĭlinson, J. Bernstein, and P. Deligne. Faisceaux pervers. *Astérisque*, 100:5–171, 1982.

BD17. M. V. Bondarko and F. Déglise. Dimensional homotopy t-structures in motivic homotopy theory. *Adv. Math.*, 311:91–189, 2017.

Beĭ84. A. Beĭlinson. Higher regulators and values of L-functions. In *Current problems in mathematics, Vol. 24*, Itogi Nauki i Tekhniki, pages 181–238. Akad. Nauk SSSR, Vsesoyuz. Inst. Nauchn. i Tekhn. Inform., Moscow, 1984.

Beĭ87. A. Beĭlinson. Height pairing between algebraic cycles. In *Current trends in arithmetical algebraic geometry (Arcata, Calif., 1985)*, volume 67 of *Contemp. Math.*, pages 1–24. Amer. Math. Soc., 1987.

Beĭ12. A. Beĭlinson. Remarks on Grothendieck's standard conjectures. In *Regulators*, volume 571 of *Contemp. Math.*, pages 25–32. Amer. Math. Soc., Providence, RI, 2012.

Bek00. T. Beke. Sheafifiable homotopy model categories. *Math. Proc. Camb. Phil. Soc.*, 129:447–475, 2000.

BG73. K. S. Brown and S. M. Gersten. Algebraic K-theory as generalized sheaf cohomology. In *Higher K-theories I (Proc. Conf., Battelle Memorial Inst., Seattle, Wash., 1972)*, volume 341 of *Lecture Notes in Mathematics*, pages 266–292. Springer-Verlag, 1973.

BGI71. P. Berthelot, A. Grothendieck, and L. Illusie. *Théorie des intersections et théorème de Riemann–Roch*, volume 225 of *Lecture Notes in Mathematics*. Springer-Verlag, 1971. Séminaire de Géométrie Algébrique du Bois–Marie 1966–67 (SGA 6).

BI15. M. V. Bondarko and M. A. Ivanov. On Chow weight structures for cdh-motives with integral coefficients. *Algebra i Analiz*, 27(6):14–40, 2015. Reprinted in St. Petersburg Math. J. **27** (2016), no. 6, 869–888.

Bla03. B. Blander. Local projective model structures on simplicial presheaves. *K-Theory*, 24(3):283–301, 2003.

BMo3. C. Berger and I. Moerdijk. Axiomatic homotopy theory for operads. *Comment. Math. Helv.*, 78:805–831, 2003.
BMo9. C. Berger and I. Moerdijk. On the derived category of an algebra over an operad. *Georgian Math. J.*, 16(1):13–28, 2009.
Bon10. M. V. Bondarko. Weight structures vs. t-structures; weight filtrations, spectral sequences, and complexes (for motives and in general). *J. K-Theory*, 6(3):387–504, 2010.
Bon14. M. V. Bondarko. Weights for relative motives: relation with mixed complexes of sheaves. *Int. Math. Res. Not. IMRN*, (17):4715–4767, 2014.
Bon15. M. V. Bondarko. Mixed motivic sheaves (and weights for them) exist if 'ordinary' mixed motives do. *Compos. Math.*, 151(5):917–956, 2015.
Bou93. N. Bourbaki. *Éléments de mathématique. Algèbre commutative. Chapitres 8 et 9*. Masson, 193. Reprint of the 1983 original.
Bou98. N. Bourbaki. *Commutative algebra. Chapters 1–7*. Elements of Mathematics (Berlin). Springer-Verlag, Berlin, 1998. Translated from the French, Reprint of the 1989 English translation.
Bro74. K. S. Brown. Abstract homotopy theory and generalized sheaf cohomology. *Trans. Amer. Math. Soc.*, 186:419–458, 1974.
BRTV18. A. Blanc, M. Robalo, B. Toën, and G. Vezzosi. Motivic realizations of singularity categories and vanishing cycles. *J. Éc. polytech. Math.*, 5:651–747, 2018.
BVK16. L. Barbieri-Viale and B. Kahn. *On the derived category of 1-motives*, volume 381 of *Astérisque*. Soc. Math. France, 2016.
CD09. D.-C. Cisinski and F. Déglise. Local and stable homological algebra in Grothendieck abelian categories. *Homology, Homotopy and Applications*, 11(1):219–260, 2009.
CD12. D.-C. Cisinski and F. Déglise. Mixed Weil cohomologies. *Adv. Math.*, 230(1):55–130, 2012.
CD15. D.-C. Cisinski and F. Déglise. Integral mixed motives in equal characteristics. *Doc. Math.*, (Extra volume: Alexander S. Merkurjev's sixtieth birthday):145–194, 2015.
CD16. D.-C. Cisinski and F. Déglise. Étale motives. *Compos. Math.*, 152(3):556–666, 2016.
Cis03. D.-C. Cisinski. Images directes cohomologiques dans les catégories de modèles. *Annales Mathématiques Blaise Pascal*, 10:195–244, 2003.
Cis06. D.-C. Cisinski. *Les préfaisceaux comme modèles des types d'homotopie*, volume 308 of *Astérisque*. Soc. Math. France, 2006.
Cis08. D.-C. Cisinski. Propriétés universelles et extensions de Kan dérivées. *Theory and Applications of Categories*, 20(17):605–649, 2008.
Cis13. D.-C. Cisinski. Descente par éclatements en K-théorie invariante par homotopie. *Ann. of Math.*, 177, 2013.
Cis18. D.-C. Cisinski. Cohomological methods in intersection theory. Lecture notes from the LMS-CMI Research School 'Homotopy Theory and Arithmetic Geometry: Motivic and Diophantine Aspects', Imperial College London, 2018.
Cis19. D.-C. Cisinski. *Higher categories and homotopical algebra*, volume 180 of *Cambridge studies in advanced mathematics*. Cambridge University Press, 2019.
Con07. B. Conrad. Deligne's notes on Nagata compactifications. *J. Ramanujan Math. Soc.*, 22(3):205–257, 2007.
Cra95. S. E. Crans. Quillen closed model structures for sheaves. *J. Pure Appl. Algebra*, 101:35–57, 1995.
CT11. D.-C. Cisinski and G. Tabuada. Non connective K-theory via universal invariants. *Compos. Math.*, 147(4):1281–1320, 2011.
Dég07. F. Déglise. Finite correspondences and transfers over a regular base. In *Algebraic cycles and motives. Vol. 1*, volume 343 of *London Math. Soc. Lecture Note Ser.*, pages 138–205. Cambridge Univ. Press, Cambridge, 2007.
Dég08. F. Déglise. Around the Gysin triangle II. *Doc. Math.*, 13:613–675, 2008.
Del71. P. Deligne. Théorie de Hodge. II. *Inst. Hautes Études Sci. Publ. Math.*, (40):5–57, 1971.
Del74. P. Deligne. Théorie de Hodge. III. *Inst. Hautes Études Sci. Publ. Math.*, (44):5–77, 1974.

References

Del87. P. Deligne. Le déterminant de la cohomologie. In *Current trends in arithmetical algebraic geometry (Arcata, Calif., 1985)*, volume 67 of *Contemp. Math.*, pages 93–177. Amer. Math. Soc., 1987.

Del01. P. Deligne. Voevodsky lectures on cross functors. unpublished notes available at https://www.math.ias.edu/vladimir/node/94, Fall 2001.

DFJK19. F. Déglise, J. Fasel, F. Jin, and A. Khan. Borel isomorphism and absolute purity. arXiv:1902.02055, 2019.

dJ96. A. J. de Jong. Smoothness, semi-stability and alterations. *Publ. Math. IHES*, 83:51–93, 1996.

dJ97. A. J. de Jong. Families of curves and alterations. *Annales de l'institut Fourier*, 47(2):599–621, 1997.

DLØ+07. B. I. Dundas, M. Levine, P. A. Østvær, O. Röndigs, and V. Voevodsky. *Motivic homotopy theory*. Universitext. Springer-Verlag, Berlin, 2007. Lectures from the Summer School held in Nordfjordeid, August 2002.

Dre13. B. Drew. *Réalisations tannakiennes des motifs mixtes triangulés*. PhD thesis, Univ. Paris 13, 2013.

Dre18. B. Drew. Motivic hodge modules. arXiv:1801.10129, 2018.

Dug01. D. Dugger. Combinatorial model categories have presentations. *Adv. Math*, 164(1):177–201, 2001.

Dug06. D. Dugger. Spectral enrichment of model categories. *Homology, Homotopy, and Applications*, 8(1):1–30, 2006.

Eke90. T. Ekedahl. On the adic formalism. In *The Grothendieck Festschrift, Vol. II*, volume 87 of *Progr. Math.*, pages 197–218. Birkhäuser Boston, Boston, MA, 1990.

Ful98. W. Fulton. *Intersection theory*. Springer, second edition, 1998.

Gar19. G. Garkusha. Reconstructing rational stable motivic homotopy theory. arXiv:1705.01635v3, 2019.

GD60. A. Grothendieck and J. Dieudonné. Éléments de géométrie algébrique. I. Le langage des schémas. *Publ. Math. IHES*, 4, 1960.

GD61. A. Grothendieck and J. Dieudonné. Éléments de géométrie algébrique. II. Étude globale élémentaire de quelques classes de morphismes. *Publ. Math. IHES*, 8, 1961.

GD67. A. Grothendieck and J. Dieudonné. Éléments de géométrie algébrique. IV. Étude locale des schémas et des morphismes de schémas IV. *Publ. Math. IHES*, 20, 24, 28, 32, 1964-1967.

GG16. S. Gorchinskiy and V. Guletskiĭ. Symmetric powers in abstract homotopy categories. *Adv. Math.*, 292:707–754, 2016.

GG18. S. Gorchinskiy and V. Guletskiĭ. Positive model structures for abstract symmetric spectra. *Appl. Categ. Structures*, 26(1):29–46, 2018.

Gil81. H. Gillet. Riemann-Roch theorems for higher algebraic K-theory. *Adv. in Math.*, 40(3):203–289, 1981.

Gro57. A. Grothendieck. Sur quelques points d'algèbre homologique. *Tôhoku Math. J.*, 9(2):119–221, 1957.

Gro77. A. Grothendieck. *Cohomologie l-adique et Fonctions L*, volume 589 of *Lecture Notes in Mathematics*. Springer-Verlag, 1977. Séminaire de Géométrie Algébrique du Bois–Marie 1965–66 (SGA 5).

Gro03. A. Grothendieck. *Revêtements étales et groupe fondamental*, volume 3 of *Documents Mathématiques*. Soc. Math. France, 2003. Séminaire de Géométrie Algébrique du Bois–Marie 1960–61 (SGA 1). Édition recomposée et annotée du LNM 224, Springer, 1971.

GZ67. P. Gabriel and M. Zisman. *Calculus of fractions and homotopy theory*, volume 35 of *Ergebnisse der Mathematik*. Springer-Verlag, 1967.

Han95. M. Hanamura. Mixed motives and algebraic cycles I. *Math. Res. Lett.*, 2(6):811–821, 1995.

Han99. M. Hanamura. Mixed motives and algebraic cycles III. *Math. Res. Lett.*, 6(1):61–82, 1999.

[Han00] M. Hanamura. Homological and cohomological motives of algebraic varieties. *Invent. Math.*, 142(2):319–349, 2000.
[Han04] M. Hanamura. Mixed motives and algebraic cycles II. *Invent. Math.*, 158(1):105–179, 2004.
[Har66] R. Hartshorne. *Residues and duality*. Lecture notes of a seminar on the work of A. Grothendieck, given at Harvard 1963/64. With an appendix by P. Deligne. Lecture Notes in Mathematics, No. 20. Springer-Verlag, Berlin, 1966.
[Héb11] D. Hébert. Structure de poids à la Bondarko sur les motifs de Beilinson. *Compos. Math.*, 147(5):1447–1462, 2011.
[HKØ17] M. Hoyois, S. Kelly, and P. A. Østvær. The motivic Steenrod algebra in positive characteristic. *J. Eur. Math. Soc.*, 19(12):3813–3849, 2017.
[HMS17] A. Huber and S. Müller-Stach. *Periods and Nori motives*, volume 65 of *Ergebnisse der Mathematik und ihrer Grenzgebiete. 3. Folge. A Series of Modern Surveys in Mathematics [Results in Mathematics and Related Areas. 3rd Series. A Series of Modern Surveys in Mathematics]*. Springer, Cham, 2017. With contributions by Benjamin Friedrich and Jonas von Wangenheim.
[Hor13] J. Hornbostel. Preorientations of the derived motivic multiplicative group. *Algebr. Geom. Topol.*, 13(5):2667–2712, 2013.
[Hov99] M. Hovey. *Model Categories*, volume 63 of *Math. surveys and monographs*. Amer. Math. Soc., 1999.
[Hov01] M. Hovey. Spectra and symmetric spectra in general model categories. *J. Pure Appl. Algebra*, 165(1):63–127, 2001.
[Hoy17] M. Hoyois. The six operations in equivariant motivic homotopy theory. *Adv. Math.*, 305:197–279, 2017. Corrigendum in Adv. Math. 333 (2018).
[HS15] A. Holmstrom and J. Scholbach. Arakelov motivic cohomology I. *J. Algebraic Geom.*, 24(4):719–754, 2015.
[Hub00] Annette Huber. Realization of Voevodsky's motives. *J. Algebraic Geom.*, 9(4):755–799, 2000. Corrigendum in J. Algebraic Geom. 13 (2004).
[ILO14] Luc Illusie, Yves Laszlo, and Fabrice Orgogozo, editors. *Travaux de Gabber sur l'uniformisation locale et la cohomologie étale des schémas quasi-excellents*, volume 363–364 of *Astérisque*. Soc. Math. France, 2014. Séminaire à l'École Polytechnique 2006–2008. [Seminar of the Polytechnic School 2006–2008], With the collaboration of Frédéric Déglise, Alban Moreau, Vincent Pilloni, Michel Raynaud, Joël Riou, Benoît Stroh, Michael Temkin and Weizhe Zheng.
[Ivo07] F. Ivorra. Réalisation l-adique des motifs triangulés géométriques. I. *Doc. Math.*, 12:607–671, 2007.
[Jan90] U. Jannsen. *Mixed motives and algebraic K-theory*, volume 1400 of *Lecture Notes in Mathematics*. Springer-Verlag, Berlin, 1990. With appendices by S. Bloch and C. Schoen.
[Jan92] U. Jannsen. Motives, numerical equivalence, and semi-simplicity. *Invent. Math.*, 107(3):447–452, 1992.
[Jar00] J. F. Jardine. Motivic symmetric spectra. *Doc. Math.*, 5:445–553, 2000.
[Jar15] J. F. Jardine. *Local homotopy theory*. Springer Monographs in Mathematics. Springer, 2015.
[JKS94] Uwe Jannsen, Steven Kleiman, and Jean-Pierre Serre, editors. *Motives*, volume 55 of *Proceedings of Symposia in Pure Mathematics*. American Mathematical Society, Providence, RI, 1994.
[JY19] F. Jin and E. Yang. Künneth formulas for motives and additivity of traces. arXiv:1812.06441v2, 2019.
[Kah18] B. Kahn. *Fonctions zêta et L de variétés et de motifs*. Nano. Calvage et Mounet, Paris, 2018.
[Kel17] S. Kelly. *Voevodsky motives and ldh-descent*, volume 391 of *Astérisque*. Soc. Math. France, 2017.
[Kel18] S. Kelly. Un isomorphisme de Suslin. *Bull. Soc. Math. France*, 146(4):633–647, 2018.
[Kra01] H. Krause. On Neeman's well generated triangulated categories. *Doc. Math.*, 6:121–126, 2001.

References

Leh17. S. Pepin Lehalleur. Constructible 1-motives and exactness. arXiv:1712.01180, 2017.
Leh19. S. Pepin Lehalleur. Triangulated categories of relative 1-motives. arXiv:1512.00266v3, to appear in Advances in Mathematics, 2019.
Lev98. M. Levine. *Mixed motives*, volume 57 of *Math. surveys and monographs*. Amer. Math. Soc., 1998.
Lev01. M. Levine. Techniques of localization in the theory of algebraic cycles. *J. Alg. Geom.*, 10:299–363, 2001.
Lev06. M. Levine. Chow's moving lemma and the homotopy coniveau tower. *K-Theory*, 37(1-2):129–209, 2006.
Lev08. M. Levine. The homotopy coniveau tower. *J. Topol.*, 1(1):217–267, 2008.
Lev10. M. Levine. Slices and transfers. *Doc. Math.*, (Extra vol.: Andrei A. Suslin sixtieth birthday):393–443, 2010.
Lev13a. M. Levine. Convergence of Voevodsky's slice tower. *Doc. Math.*, 18:907–941, 2013.
Lev13b. M. Levine. Six lectures on motives. In *Autour des motifs—École d'été Franco-Asiatique de Géométrie Algébrique et de Théorie des Nombres/Asian-French Summer School on Algebraic Geometry and Number Theory. Vol. II*, volume 41 of *Panor. Synthèses*, pages 1–141. Soc. Math. France, Paris, 2013.
Lev18. M. Levine. Vladimir Voevodsky—an appreciation. *Bull. Amer. Math. Soc. (N.S.)*, 55(4):405–425, 2018.
Lip78. J. Lipman. Desingularization of two-dimensional schemes. *Annals of Math.*, 107:151–207, 1978.
Lur17. J. Lurie. Higher algebra. preprint, 2017.
LZ15. Y. Liu and W. Zheng. Gluing restricted nerves of ∞-categories. arXiv:1211.5294v4, 2015.
LZ16. Y. Liu and W. Zheng. Gluing pseudo functors via n-fold categories. arXiv:1211.1877v6, 2016.
LZ17. Y. Liu and W. Zheng. Enhanced six operations and base change theorem for higher Artin stacks. arXiv:1211.5948v3, 2017.
Mac98. S. MacLane. *Categories for the working mathematician*, volume 5 of *Graduate Texts in Mathematics*. Springer-Verlag, New York, second edition, 1998.
Mal05. G. Maltsiniotis. *La théorie de l'homotopie de Grothendieck*, volume 301 of *Astérisque*. Soc. Math. France, 2005.
Mat70. H. Matsumura. *Commutative algebra*. W. A. Benjamin, Inc., New York, 1970.
Mor03. F. Morel. An introduction to $\mathbf{A}^1$-homotopy theory. In *Contemporary Developments in Algebraic K-theory*, volume 15 of *ICTP Lecture notes*, pages 357–441. 2003.
Mor04a. F. Morel. On the motivic π_0 of the sphere spectrum. In *Axiomatic, enriched and motivic homotopy theory*, volume 131 of *NATO Sci. Ser. II Math. Phys. Chem.*, pages 219–260. Kluwer Acad. Publ., Dordrecht, 2004.
Mor04b. F. Morel. Sur les puissances de l'idéal fondamental de l'anneau de Witt. *Comment. Math. Helv.*, 79(4):689–703, 2004.
Mor05. F. Morel. The stable $\mathbf{A}^1$-connectivity theorems. *K-theory*, 35:1–68, 2005.
Mor06. F. Morel. Rational stable splitting of Grassmanians and the rational motivic sphere spectrum. statement of results, 2006.
Mor12. F. Morel. $\mathbf{A}^1$-*algebraic topology over a field*, volume 2052 of *Lecture Notes in Mathematics*. Springer, 2012.
MV99. F. Morel and V. Voevodsky. $\mathbf{A}^1$-homotopy theory of schemes. *Publ. Math. IHES*, 90:45–143, 1999.
MVW06. C. Mazza, V. Voevodsky, and C. Weibel. *Lecture notes on motivic cohomology*, volume 2 of *Clay Mathematics Monographs*. American Mathematical Society, Providence, RI, 2006.
Nee92. A. Neeman. The connection between the K-theory localization theorem of Thomason, Trobaugh and Yao and the smashing subcategories of Bousfield and Ravenel. *Ann. Sci. École Norm. Sup. (4)*, 25(5):547–566, 1992.
Nee96. A. Neeman. The Grothendieck duality theorem via Bousfield's techniques and Brown representability. *J. Amer. Math. Soc.*, 9(1):205–236, 1996.

Nee01. A. Neeman. *Triangulated categories*, volume 148 of *Annals of Mathematics Studies*. Princeton University Press, Princeton, NJ, 2001.
NSØ09. N. Nauman, M. Spitzweck, and P. A. Østvær. Motivic Landweber exactness. *Doc. Math.*, 14:551–593, 2009.
NV15. A.N. Nair and V. Vaish. Weightless cohomology of algebraic varieties. *J. Algebra*, 424:147–189, 2015.
Ols15. M. Olsson. Borel–Moore homology, Riemann–Roch transformations, and local terms. *Adv. Math.*, 273:56–123, 2015.
Ols16. M. Olsson. Motivic cohomology, localized Chern classes, and local terms. *Manuscripta Math.*, 149(1-2):1–43, 2016.
Par17a. D. Park. Construction of triangulated categories of motives using the localization property. arXiv:1708.00597, 2017.
Par17b. D. Park. Triangulated categories of motives over fs log schemes. arXiv:1707.09435, 2017.
Par19. D. Park. Equivariant cd-structures and descent theory. *J. Pure Appl. Algebra*, 2019. to appear.
PPR09. I. Panin, K. Pimenov, and O. Röndigs. On Voevodsky's algebraic K-theory spectrum. In *Algebraic Topology*, volume 4 of *Abel. Symp.*, pages 279–330. Springer, 2009.
PS18. D. Pavlov and J. Scholbach. Homotopy theory of symmetric powers. *Homology Homotopy Appl.*, 20(1):359–397, 2018.
Qui73. D. Quillen. Higher algebraic K-theory. In *Higher K-theories I (Proc. Conf., Battelle Memorial Inst., Seattle, Wash., 1972)*, volume 341 of *Lecture Notes in Mathematics*, pages 85–147. Springer-Verlag, 1973.
Rio10. J. Riou. Algebraic K-theory, $\mathbf{A}^1$-homotopy and Riemann–Roch theorems. *J. Topol.*, 3(2):229–264, 2010.
RØ08a. O. Röndigs and P. A. Østvær. Modules over motivic cohomology. *Adv. Math.*, 219(2):689–727, 2008.
RØ08b. O. Röndigs and P. A. Østvær. Rigidity in motivic homotopy theory. *Math. Ann.*, 341(3):651–675, 2008.
Rob15. M. Robalo. K-theory and the bridge from motives to noncommutative motives. *Adv. Math.*, 269:399–550, 2015.
Rön05. O. Röndigs. Functoriality in motivic homotopy theory. Preprint available at https://www.math.uni-bielefeld.de/~oroendig/functoriality.dvi, 2005.
RSØ10. O. Röndigs, M. Spitzweck, and P. A. Østvær. Motivic strict ring models for K-theory. *Proc. Amer. Math. Soc.*, 138(10):3509–3520, 2010.
Ryd10. D. Rydh. Submersions and effective descent of étale morphisms. *Bull. Soc. Math. France*, 138(2):181–230, 2010.
Sch15. J. Scholbach. Arakelov motivic cohomology II. *J. Algebraic Geom.*, 24(4):755–786, 2015.
Ser75. J.-P. Serre. *Algèbre locale , multiplicités*, volume 11 of *Lecture Notes in Mathematics*. Springer-Verlag, 1975. third edition.
Shi04. B. Shipley. A convenient model category for commutative ring spectra. *Contemp. Math.*, 346:473–483, 2004.
Spi99. M. Spivakovsky. A new proof of D. Popescu's theorem on smoothing of ring homomorphisms. *Journal of the Amer. Math. Soc.*, 12(2):381–444, 1999.
Spi01. M. Spitzweck. Operads, Algebras and Modules in Monoidal Model Categories and Motives. PhD thesis, Mathematisch Naturwissenschatlichen Fakultät, Rheinischen Friedrich Wilhelms Universität, Bonn, 2001.
Spi18. M. Spitzweck. A commutative $\mathbf{P}^1$-spectrum representing motivic cohomology over Dedekind domains. *Mém. Soc. Math. Fr. (N.S.)*, (157), 2018.
SS00. S. Schwede and B. Shipley. Algebras and modules in monoidal model categories. *Proc. London Math. Soc.*, 80:491–511, 2000.
SV96. A. Suslin and V. Voevodsky. Singular homology of abstract algebraic varieties. *Invent. Math.*, 123(1):61–94, 1996.

References

SV00a. A. Suslin and V. Voevodsky. *Bloch–Kato conjecture and motivic cohomology with finite coefficients*, volume 548 of *NATO Sciences Series, Series C: Mathematical and Physical Sciences*, pages 117–189. Kluwer, 2000.

SV00b. A. Suslin and V. Voevodsky. *Relative cycles and Chow sheaves*, volume 143 of *Annals of Mathematics Studies*, chapter 2, pages 10–86. Princeton University Press, 2000.

SVW18. W. Soergel, R. Virk, and M. Wendt. Equivariant motives and geometric representation theory. With an appendix by F. Hörmann and M. Wendt, arXiv:1809.05480, 2018.

SW18. W. Soergel and M. Wendt. Perverse motives and graded derived category $\mathcal{O}$. *J. Inst. Math. Jussieu*, 17(2):347–395, 2018.

Swa80. R. G. Swan. On seminormality. *J. Algebra*, 67(1):210–229, 1980.

Tot16. B. Totaro. The motive of a classifying space. *Geom. Topol.*, 20(4):2079–2133, 2016.

Tot18. B. Totaro. Adjoint functors on the derived category of motives. *J. Inst. Math. Jussieu*, 17(3):489–507, 2018.

TR19. J. Scholbach T. Richarz. The intersection motive of the moduli stack of shtukas. arXiv:1901.04919, 2019.

TT90. R. W. Thomason and Thomas Trobaugh. Higher algebraic K-theory of schemes and of derived categories. In *The Grothendieck Festschrift, Vol. III*, volume 88 of *Progr. Math.*, pages 247–435. Birkhäuser Boston, Boston, MA, 1990.

TV19. B. Toën and G. Vezzosi. Trace and Künneth formulas for singularity categories and applications. arXiv:1710.05902v2, 2019.

Vai16. V. Vaish. Motivic weightless complex and the relative Artin motive. *Adv. Math.*, 292:316–373, 2016.

Vai17. V. Vaish. Punctual gluing of t-structures and weight structures. arXiv:1705.00790, 2017.

Vai18. V. Vaish. On weight truncations in the motivic setting. *Adv. Math.*, 338:339–400, 2018.

Vez01. G. Vezzosi. Brown–Peterson spectra in stable $\mathbf{A}^1$-homotopy theory. *Rend. Sem. Mat. Univ. Padova*, 106:47–64, 2001.

Voe92. V. Voevodsky. A letter to Beilinson. K-theory preprint archive 0033, 1992.

Voe96. V. Voevodsky. Homology of schemes. *Selecta Math. (N.S.)*, 2(1):111–153, 1996.

Voe98. V. Voevodsky. $\mathbf{A}^1$-homotopy theory. In *Proceedings of the International Congress of Mathematicians, Vol. I (Berlin, 1998)*, pages 579–604 (electronic), 1998.

Voe02a. V. Voevodsky. Motivic cohomology groups are isomorphic to higher Chow groups in any characteristic. *Int. Math. Res. Not.*, 7:351–355, 2002.

Voe02b. V. Voevodsky. Open problems in the motivic stable homotopy theory. I. In *Motives, polylogarithms and Hodge theory, Part I (Irvine, CA, 1998)*, volume 3 of *Int. Press Lect. Ser.*, pages 3–34. Int. Press, Somerville, MA, 2002.

Voe10a. V. Voevodsky. Cancellation theorem. *Doc.Math. Extra volume: Andrei A. Suslin sixtieth birthday*, pages 671–685, 2010.

Voe10b. V. Voevodsky. Homotopy theory of simplicial sheaves in completely decomposable topologies. *J. Pure Appl. Algebra*, 214(8):1384–1398, 2010.

Voe10c. V. Voevodsky. Unstable motivic homotopy categories in Nisnevich and cdh-topologies. *J. Pure Appl. Algebra*, 214(8):1399–1406, 2010.

VSF00. V. Voevodsky, A. Suslin, and E. M. Friedlander. *Cycles, Transfers and Motivic homology theories*, volume 143 of *Annals of Mathematics Studies*. Princeton Univ. Press, 2000.

Wil17. J. Wildeshaus. Intermediate extension of Chow motives of Abelian type. *Adv. Math.*, 305:515–600, 2017.

Wil18. J. Wildeshaus. Weights and conservativity. *Algebr. Geom.*, 5(6):686–702, 2018.

Index

acyclic, $H_{\mathbb{B}}$-acyclic, 352
adequate, category of schemes, 32
adjunction
    of $\mathscr{P}$-fibred categories, see morphism of
    of premotivic categories, see morphism of
    Quillen adjunction, 88
admissible topology, see topology
admissible, class of morphisms, 3
algebra
    E_∞-algebra, 236
    $H_{\mathbb{B}}$-algebra, 355
alteration, 131
    Galois alteration, 131, 153
Auslander–Buchsbaum theorem, 320

base change
    $\mathscr{P}$-base change, 7
    proper base change, 9
    smooth base change, 9
bifibred category, 7
Bott isomorphism, 339
bounded (topology), 181
bounded generating family, 181
Brown representability theorem, 24, 25, 45, 52
bundle
    normal, 70, 71
    tangent, 71, 79
    virtual vector bundle, 61

cartesian morphism, see morphism
cd-structure, 35, 300
    lower, 35
    upper, 35
Chow's lemma, 31, 51
class
    Chern, 324, 346
    fundamental, 344, 359
coefficients, for Beilinson motives, 356, 363
cofibration, 173

termwise, 85
coherence, 5, 8, 11, 19
cohomology
    algebraic De Rham, 387
    analytic De Rham, 388
    Beilinson motivic, 354
    Betti, 379
    Chow group, see group
    effective motivic, 315
    higher Chow group, see group
    K-theory, see K-theory
    Landweber exact, 351
    mixed Weil, 381
    Monsky–Washnitzer, 390
    motivic, 315
    representable, 335
    rigid, 390
    stable, 381
commute, see functor
compact, 24, 183, 202, 312
compatible with (a topology) t, 172, 173, 179
compatible with transfers, see topology
compatible with twists, 19, 29
complex
    algebraic De Rham, 387
    Monsky–Washnitzer, 389
conservative, 48, 53, 54, 149, 150
constructibility, see constructible
constructible, see also τ-constructible 30
    $(\mathbf{Z} \times \tau)$-constructible, 218
    τ-constructible, 30, 133–162, 183, 202
    Beilinson motive, 360–363
    motive, 310, 312
    motivic complex, 310
$\mathbf{A}^1$-contractible, 194
cotransversality property, 9
cover, 98
    Galois cover, 115
    h-cover, 123, 130

pseudo-Galois cover, 115
qfh-cover, 121, 123
cycle
  Λ-cycle, 247, see also cycle248
  Λ-universal (morphism of), 262
  associated, 248
  Hilbert, 249
  pre-special (morphism of), 252
  pseudo-equidimensional, 276
  pullback, 256
    associativity, 265
    commutativity, 264
    of Hilbert cycles, 251
    projection formulas, 267
  pushforward, 248
  restriction, 249
  Samuel specialization, 274
  special (morphism of), 254
  specialization, 253
  standard form, 248
  trivial, 268

decomposition, Adams, 350, 356
deformation space, 70
derivator, Grothendieck, 95, 105, 107, 177
derived
  derived $\mathscr{P}$-premotivic category, 188
descent
  cdh-descent, 112, 113, 124
  cohomological h-descent, 383
  cohomological t-descent, 172, 177
  étale, 121, 123, 127, 358, 375
  Galois, 251
  h-descent, 124, 126, 127, 358, 375
  Nisnevich, 110, 111
  qfh-descent, 122, 124, 126, 314
  t-descent, 100, 177, 189, 213
dg-structure, 172, 187
diagram
  $\mathscr{S}$-diagram, 82, 176, 188
direct image with compact support, see functor, left exceptional

divisor
  Weil, 320
domain (of a Λ-cycle), 248
dual, strong, 68, 326
duality
  local duality, 160
duality, Grothendieck, 161, 362
dualizable, strongly, 68, 75
dualizing
  τ-dualizing, 155

embedding, Segre, 337
enlargement, of premotivic categories, see premotivic
equidimensional
  absolutely, 249
  flat morphism, 249
equivalence
  $\mathbf{A}^1$-equivalence, 193
  H_{B}-equivalence, 352
  of motivic categories, 376
  of triangulated monoidal categories, 354, 366
  strong $\mathbf{A}^1$-equivalence, 193
  termwise weak equivalence, 85
  weak equivalence of commutative monoids, 231
  weak equivalence of modules, 236
  weak equivalence of monoids, 230
  $\mathscr{W}$-equivalence, 186
equivalence, of categories, 34
exceptional functor, see functor
exchange
  isomorphism, 7, 13, 50, 226–229
  morphism, see exchange transformation
  transformation, 5, 8, 11, 12, 17, 18, 43

fibrant
  $\mathbf{A}^1$-fibrant, 193
fibration
  t-fibration, 173
  algebraic Hopf, 368

of commutative monoids, 231
of modules, 236
of monoids, 230
termwise, 85
$\mathscr{W}$-fibration, 186
fibred
    fibred category, 3
    monoidal pre-$\mathscr{P}$-fibred category, 10
    monoidal $\mathscr{P}$-fibred category, 12
        model (category), 26
        of finite correspondences, 289
    $\mathscr{P}$-fibred category, 7
        τ-generated, *see* generated
        abelian, 22
        abelian monoidal, 22
        canonical, 7
        canonical monoidal, 12
        complete, 7
        complete monoidal, 12
        finitely τ-presented, *see* finitely presented
        geometrically generated, *see* generated
        Grothendieck abelian, 22
        Grothendieck abelian monoidal, 22
        homotopy, 27
        homotopy monoidal, 27
        model, 26
        monoidal $\mathscr{P}$-fibred model category, 174
        triangulated, 23
        triangulated monoidal, 23
    pre-$\mathscr{P}$-fibred category, 4
filtration, γ-filtration, 350
finite correspondence, 280
    composition, 281
    finite S-correspondence, *see* finite correspondence
    graph functor, *see* functor
    tensor product, *see* tensor product
    transpose, *see* morphism
finitely presented

finitely τ-presented, 23, 183, 202, 217
    object of a category, 23
finiteness theorem, 145, 362
flasque, t-flasque complex, 172
formalism, Grothendieck 6 functors, 355
formalism, Grothendieck six functors, 78
functor
    commutes, 7
    evaluation, 83, 90, 205
    exceptional, 44, 79
    graph, 283
    infinite suspension, 29
    left exceptional, 38
    Quillen, 91
    t-exact endofunctor, 369

Galois group, *see* group
generated
    τ-generated, 15, 16, 20, 28, 54, 145
    compactly $(\mathbf{Z} \times \tau)$-generated, 217
    compactly τ-generated, 28
        triangulated $\mathscr{P}$-fibred, 24, 45, 55, 183, 202
    compactly generated, 24
        triangulated $\mathscr{P}$-fibred, 24, 361, 376, 378
    geometrically generated, 16
    well generated, 24
        triangulated $\mathscr{P}$-fibred, 24
global section, *see* section
group
    Chow, 316
    Galois group, 115
    H-group, 337
    higher Chow, 316
    Picard, 320
    relative Picard, 318

henselization, 149
homeomorphism, universal, 33, 274, 317
homotopic, $\mathbf{A}^1$-homotopic, 193

INDEX 401

homotopy
    colimit, 89
    limit, 89
    object of homotopy fixed points, 120
homotopy cartesian, 112, 120, 122, 126, 153, 191
    object over a diagram, 97
    square, 111
homotopy category, 4, 28, 70, 196
homotopy linear, 243
homotopy pullback, *see* homotopy cartesian
hypercover, 99, 189
    Čech *t*-hypercovers, 294

ind-constructible, 269
infinite suspension, *see also* functor29

K-theory
    homotopy invariant, 338
    Milnor, 371
    Milnor–Witt, 371
    Quillen, 338
    with support, 341

law, formal group, 337
linear
    Q-linear (stable model category), 106
    strongly **Q**-linear, 230
local, 172
    $\mathscr{W}$-local, 186
    $\mathbf{A}^1$-local, 193, 198
    H_{B}-local, 352
localization
    triangle, *see also* triangle, 47

map, trace, 329, 347
model structure
    t-descent, 173
    injective (diagrams), 87
    positive stable model structure, 236
    projective (diagrams), 86

$\mathscr{W}$-local, 187
module
    H_{B}-module, 353, 355
    strict H_{B}-module, 355
modules
    KGL-modules, 341
    over a homotopy cartesian commutative monoid, 240, 378
    over a monoid, 236
monoid, 230, 240
    cartesian, 240, 242
    cartesian commutative monoid, 203
    commutative monoid, 231
    homotopy cartesian, 241
monoid axiom, 230, 236, 238, 242
monoidal stable homotopy 2-functor, 78
morphism
    $\mathscr{T}$-pure, *see also* morphism, pure, 63
    cartesian —- of $\mathscr{S}$-diagrams, 92
    cocontinuous, 106
    continuous, 105
    degree, 284
    faithfully flat, 50
    finite Λ-universal, 316
    Gysin, 325
    of $\mathscr{P}$-fibred categories, 18
    of $\mathscr{P}$-fibred model categories, 26
    of $\mathscr{P}$-premotivic categories, 28, 29
    of Λ-cycles, 248
    of abelian $\mathscr{P}$-fibred categories, 22
    of abelian $\mathscr{P}$-premotivic categories, 28
    of abelian monoidal $\mathscr{P}$-fibred categories, 22
    of complete $\mathscr{P}$-fibred categories, 18
    of derivators, 105
    of monoidal $\mathscr{P}$-fibred model category, 26
    of $\mathscr{S}$-diagrams, 84

of triangulated $\mathscr{P}$-fibred categories, 23
of triangulated $\mathscr{P}$-premotivic categories, 28
of triangulated monoidal $\mathscr{P}$-fibred categories, 23
of triangulated premotivic categories, 81
pseudo-dominant, 248
pure, 65, 68, 74
pure (proper case), 63
Quillen — of $\mathscr{P}$-fibred model categories, 98
radical, 33, 50
separated, 38, 247
transpose, 281
universally $\mathscr{T}$-pure (proper case), 63
motive, 308
 Beilinson, 352, 372
 Chow (strong), 327
 constructible, *see* constructible
 effective h-motives, 193
 effective qfh-motives, 193
 generalized, 312
 geometric, 30, 311, 366
 geometric effective, 311
 h-motive, 214
 qfh-motive, 214
 Morel, 367–375
motivic complex, 308
 constructible, *see* constructible
 generalized, 312
 stable, *see also* motive308, 308
multiplicity
 geometric, 248
 Samuel (of a cycle), 273
 Samuel (of a module), 272
 Suslin–Voevodsky, 261

nilpotent, 337
Nisnevich
 distinguished square, *see* distinguished
 topology, *see* topology

orientable, 355
orientation, 74
 of a ring spectrum, 335
 of a triangulated premotivic category, 71, 74

perfect, 237, 241
perfect pairing, 68
Picard category, 61
point, 247
 fat point (of a cycle), 252
 generic (of a cycle), 247
 geometric, 247
 of a cycle, 252
pointed, smooth S-scheme, 57
prederivator, 104
premotive, 28
 Tate premotive, 62
premotivic
 case, 32
 category, 28
 category of h-motives, 215
 category of qfh-motives, 215
 enlargement of — category, 31, 222, 307, 313
 generalized — category, 28
 morphism, *see* morphism of premotivic categories
 $\mathscr{P}$-premotivic
  $\mathbf{A}^1$-derived category, 192
  abelian category, 28
  category, 28
  derived category, 175
  stable $\mathbf{A}^1$-derived category, 210
  triangulated category, 28
 stable $\mathbf{A}^1$-derived premotivic category, 214
presentation
 local presentation of a simplicial object, 99
presented, *see also* finitely presented, 23
presheaf
 Λ-presheaf, 170
 with transfers, 291
projection formula

INDEX

$\mathscr{P}$-projection formula, 11
projective system, of schemes, xx, 145, 184, 269, 292, 301
pseudo-Galois, see cover or distinguished
pullback
   of fundamental class, 359
purity
   absolute, 346, 359
   isomorphism (relative), 63, 71, 73, 325

quasi-excellent, 130
quotient
   Gabriel, 219

radicial, see morphism
realization functor
   (associated with a stable cohomology), 383
   Betti, 380
   de Rham, 388
   of constructible motives, 387
   rigid, 390
resolution of singularities, 130
   canonical —- up to quotient singularities, 132
   canonical dominant —- up to quotient singularities, 132
   wide —- up to quotient singularities, 132, 362
Riemann–Hilbert, 389
ring
   Grothendieck–Witt, 373

schematic closure, 247
scheme
   excellent, 130, 307, 314
   geometrically unibranch, 275, 280, 306, 307, 314, 366
   quasi-excellent, 130, 138, 358
   regular, 279, 280, 282, 320
   strictly local, 277
   unibranch, 275
section
   absolute derived global section, 109
   cartesian, 203
   geometric, 14, 15, 18, 53
   geometric derived global section, 104
sequence
   symmetric sequence, 205
sheaf
   étale sheaf with transfers, 292
   generalized sheaf with transfers, 302
   h-sheaf, 193, 214
   qfh-sheaf, 193, 214, 305
   sheaf with transfers, 292, 302
   t-sheaf of Λ-modules, 170
   t-sheaf with transfers, 292
sieve, 33, 35
singular
   Suslin singular complex, 200
specialization, 327
spectra, see spectrum
spectrum
   abelian Tate spectrum, 208
   absolute Tate spectrum, 207
   algebraic cobordism, 336
   Beilinson motivic cohomology spectrum, 350
   motivic cohomology ring spectrum, 322, 336
   rational, 338
   ring —-, 335
   ring —- (associated with a stable cohomology), 381
   strict ring —-, 335
   Tate spectrum, 209
   Tate Ω-spectrum, 211
   universal oriented ring —- with additive formal group law, 355
sphere
   simplicial, 335
square
   P-distinguished, 35
   cdh-distinguished, 112, 182

Nisnevich distinguished, 35, 110, 181
proper *cdh*-distinguished, 35
pseudo-Galois *qfh*-distinguished, 116
qfh-distinguished, 116, 126, 182
Tor-independent, 343, 359
stable homotopy category of schemes, 28
strict transform, 251, 256
strictification theorem, 236
strongly dualizable, *see* dualizable

Tate
motivic complex, 309
twist, *see* twist, 335
tensor product
of finite correspondences, 285
Thom
adjoint transformation, 57
class, 72
isomorphism, 71
premotive, 60
transformation, 57
tilting, 378, 384
topology
admissible, 170
cdh-topology, 35
compatible with transfers, 294, 300
h-topology, 115, 364
mildly compatible with transfers, 295, 298, 300
Nisnevich, 35
$\mathscr{P}$-admissible, 170
proper *cdh*, 35
qfh-topology, 115, 364
weakly compatible with transfers, 294
tractable, 97, 231
trait
of a cycle, 252
transfer, *see* presheaf *or* sheaf
transversal
$\mathscr{M}$-transversal square, 9

transversality property, 9
triangle
Gysin, 325
localization triangle, 47
Mayer–Vietoris triangle, 112
t-structure
heart, 371
non-degenerate, 369
t-structure, homotopy, 368
twist, 15, 28
commutes with τ-twists (*or* twists), 15, 17, 20
of a triangulated monoidal $\mathscr{P}$-fibred category, 23
Tate, 28, 29, 62, 212
τ-twisted, 15

underlying simplicial set
of a simplicial object, 99
universal, 190

weak equivalence, *see* equivalence

Notation

$\alpha \otimes_S^b S'$, 251
$\alpha \otimes_S^{\mathscr{L}} s$, 273
$\alpha \otimes_S^{tr} \alpha'$, 285
$\tilde{\alpha}$, 251
$\mathscr{A}^{\bar{z}}$, 205

$\beta \circ \alpha$, 281
$\beta_{R,k}$, 253
$\beta \otimes_\alpha \alpha'$, 257
$\langle Z \rangle_X$, 248

$c_S(X,Y)_\Lambda$, 280
$\underline{C}^*$, 200
$c_0(X/S,\Lambda)$, 279

$D_{\mathbf{A}^1}(\mathscr{A})$, 210
$D_{\mathbf{A}^1,gm}(\mathscr{A}_S)$, 217
$D_{\mathbf{A}^1,\Lambda}$, 214
$\underline{D}_{\mathbf{A}^1,\Lambda}$, 214
$D_{\mathbf{A}^1}(S,\Lambda)_+$, 368
$D_{Betti}(X)$, 379
$\deg_x(f)$, 284
$D_{\mathbf{A}^1}^{eff}(\mathscr{A})$, 192
$D_{\mathbf{A}^1,\Lambda}^{eff}$, 192
$DM_{B,c}(S)$, 361
$DM_B(S)$, 352
$DM_{gm}^{eff}(S,\Lambda)$, 311
$DM_{gm}(S,\Lambda)$, 311
$DM_{h,\Lambda}$, 215
DM_Λ, 308
$\underline{DM}_\Lambda$, 312
DM_Λ^{eff}, 308
$\underline{DM}_\Lambda^{eff}$, 312
$DM_{qfh,\Lambda}$, 215
$\underline{DM}_{h,\Lambda}^{eff}$, 193
$\underline{DM}_{qfh,\Lambda}^{eff}$, 193
$\bullet \longrightarrow$, 280
$D(X,\mathscr{E})$, 383

$e_q^\Lambda(M)$, 272

H_B, 353

$H_B^q(X,\mathbf{Q}(p))$, 354
$H_{\mathscr{M},eff}^{n,m}(S,\Lambda)$, 315
$H_\mathscr{M}^{n,m}(S,\Lambda)$, 315
$\mathrm{Hom}^\bullet(-,-)$, 169
$\mathscr{H}_\bullet(S)$, 4

KGL^β, 340
KGL', 340
$KGL_\mathbf{Q}$, 350
KGL_S, 338
$KGL_S^{(i)}$, 350

$\Lambda_S^t(X)$, 171
$\Lambda_S^{tr}(X)$, 291
$\underline{\Lambda}_S^{tr}(X)$, 292

$\mathscr{M}_{an}(X)$, 379
$\mathscr{M}_{Betti}^{eff}(X)$, 379
$\mathscr{M}_{Betti}(X)$, 379
MGL, 336
$m^{SV}(x;\beta \otimes_\alpha \alpha')$, 261
$M_S(X)$, 309
$\underline{M}_S(X)$, 313

$\mathscr{P}_{\Lambda,S}^{cor}$, 283
$\mathscr{P}_{cart}$, 94
$\mathrm{PSh}(\mathscr{P}/S,\Lambda)$, 170
$\mathrm{PSh}\left(\mathscr{P}_{\Lambda,S}^{cor}\right)$, 291

$R_{\mathbf{A}^1}$, 197

$SH(S)$, 28
$\mathrm{Sh}_t(\mathscr{P}/S,\Lambda)$, 170
$\mathrm{Sh}_t\left(\mathscr{P}_{\Lambda,S}^{cor}\right)$, 292
$\mathrm{Sh}^{tr}(-,\Lambda)$, 302
$\underline{\mathrm{Sh}}^{tr}(-,\Lambda)$, 302
$\mathrm{Sym}(A)$, 206

$\otimes^{\bar{z}}$, 206
Tot^π, 169
${}^t f$, 281
Tr_p^{KGL}, 347

$\underline{\Lambda}_S^t(X)$, 193

Index of properties of $\mathscr{P}$-fibred triangulated categories

Name	Symbol	Def.	related result	Remark
additive		2.1.1		
adjoint property	(Adj)	2.2.13	2.2.14	
adjoint property for f	(Adj$_f$)	2.2.13		f morphism of schemes
cotransversality property		1.1.17		defined for any $\mathscr{P}$-fibred category
homotopy property	(Htp)	2.1.3		
localization property	(Loc)	2.3.2	2.4.26 6.3.15	
localization property for i	(Loc$_i$)	§2.3.1		i closed immersion
motivic		2.4.45	2.4.50 14.2.11	for premotivic triangulated categories, means: (Htp), (Stab), (Loc), (Adj)
oriented		2.4.38	2.4.43	for triangulated premotivic categories satisfying (wLoc)
projection formula	(PF)	2.2.13		
projection formula for f	(PF$_f$)	2.2.13	2.4.26	f morphism of schemes
proper base change property	(BC)	2.2.13	2.4.26	
proper base change property for f	(BC$_f$)	2.2.13		f morphism of schemes
purity property	(Pur)	2.4.21	2.4.26	
separated	(Sep)	2.1.7	4.2.24 4.4.21 14.3.3	
semi-separated	(sSep)	2.1.7	3.3.33	
stability property	(Stab)	2.4.4		
support property	(Supp)	2.2.5	2.2.12 2.2.14 11.4.2	
τ-compatible		4.2.20	4.2.29	τ set of twists
τ-continuous		4.3.2	6.1.13 11.1.24 14.3.1	for homotopy $\mathscr{P}$-fibred categories, τ set of twists
τ-dualizable		4.4.13	4.4.21	τ set of twists
t-descent property		3.2.5		for homotopy $\mathscr{P}$-fibred categories, t topology
transversality property		1.1.17		for any $\mathscr{P}$-fibred category
t-separated	(t-sep)	2.1.5		t topology
weak localization property	(wLoc)	2.4.7	11.4.2	
weak purity property	(wPur)	2.4.21	2.4.26 2.4.43	